Studies in Choice and Welfare

For further volumes:
http://www.springer.com/series/6869

Jean-François Laslier • M. Remzi Sanver
Editors

Handbook on Approval Voting

Springer

Editors
Professor Jean-François Laslier
Laboratoire d'Econométrie
Ecole Polytechnique
91128 Palaiseau
France
jean-francois.laslier@polytechnique.edu

Professor M. Remzi Sanver
Istanbul Bilgi University
Department of Economics
Kurtuluş Deresi Cad. No. 47
34440 Istanbul
Turkey
sanver@bilgi.edu.tr

Studies in Choice and Welfare ISSN 1614-0311
ISBN 978-3-642-02838-0 e-ISBN 978-3-642-02839-7
DOI 10.1007/978-3-642-02839-7
Springer Heidelberg Dordrecht London New York

Library of Congress Control Number: 2010929049

Cover design: WMX Design GmbH, Heidelberg

Printed on acid-free paper

Springer is part of Springer Science+Business Media (www.springer.com)

To Ipek and Maryvonne: Approved love.

Preface

I am honored to write a preface to this remarkably broad and comprehensive volume on approval voting (AV). It has been almost 35 years since Peter C. Fishburn and I met in 1976 and began research on AV. Besides my 30-year collaboration with Fishburn, I have collaborated with several other scholars – including D. Marc Kilgour, Samuel Merrill, Jack H. Nagel, M. Remzi Sanver, and William S. Zwicker – on AV-related research. Over these years there has been a profusion of articles and books reporting on empirical and theoretical aspects of AV and their normative implications. This volume touches on all aspects of this research and will be a very helpful sourcebook to scholars who want to carry this research forward.

In Brams and Fishburn (1983/2007, p. 172), Fishburn and I were unabashed in our support of AV:

> Approval voting strikes at the heart of how political debate is resolved. It offers a new approach to the realization of democratic principles by redefining what constitutes a democratic choice. Indeed, the foundation on which representative government is built is periodic elections, and the central problem of elections today is how to translate voter preferences, with as little distortion as possible, into consensus choices in multicandidate races. We believe that approval voting is the best practical way for amalgamating these preferences, fairly and impartially, to produce a winner and thereby ameliorate the multicandate problem.

We added that "more than intellectual issues are at stake," pointing out that "there are some 500,000 elected officials serving in approximately 80,000 governments in the United States" (p. 171). Earlier I had brazenly predicted that AV "would be the election reform of the twentieth century" (Brams 1980, p. 105).

This was not to be, for reasons described in Brams and Fishburn (2005; reprinted in this volume); indeed, as we indicated, AV's success has been decidedly mixed. Consequently, I take this opportunity to move up the deadline for the widespread adoption of AV to the twenty-first century!

Of course, not everyone believes this should come to pass; AV, to say the least, remains controversial. In part, this is because AV is a radical reform – even if it does not require a constitutional amendment to implement in most democracies of the world - because the idea of judging each and every candidate as acceptable or not is fundamentally different from either

- Restricting a voter's approval to just one candidate, as under plurality voting; or
- Allowing voters to rank candidates – as under preference systems like the Hare system of single transferable vote (STV) or the Borda count – but not indicate where they would draw the line between those who are acceptable and those who are not

In my opinion, the advantages of AV over plurality voting, or plurality voting with a runoff, are compelling: AV is as simple as the former and less burdensome and costly than the latter, not to mention its appealing theoretical properties, such as its propensity to elect Condorcet winners (when they exist), its robustness against manipulation, and its monotonicity (STV fails this property). Less clear, however, is whether AV's merits extend to electing multiple winners to, say, a council or legislature.

In several chapters of Brams (2008), I analyze alternative methods of aggregating approval ballots – a subject that Kilgour, and Laffond and Lainé, also discuss in this volume – which would, among other things, facilitate the proportional representation (PR) of different factions in an electorate. While almost all parliamentary democracies seek to achieve PR, most limit the choice of voters to voting for parties, not candidates, and only one party at that. More research is clearly needed to assess the benefits of using AV ballots to elect representative committees.

Another direction that AV-related research has taken is to allow voters to rate candidates or other alternatives in terms of more than two grades. Range voting, which has been championed by Warren D. Smith (see http://rangevoting.org), lets voters grade candidates on a scale that might include as few as 3 gradations or as many as 100; the candidate with the highest overall rating, when all voters' ratings are summed up, is the winner. Under majority judgment voting, Balinski and Laraki (2010) suggest a 6-tier scale, but they emphasize that the ratings should not be numerical but verbal (e.g., from "excellent" to "poor"), provided that the voters share a common language that enables them to make similar judgments. Under their scheme, the winner is the candidate with the highest median ranking, not the highest overall (or average) ranking, as range voting prescribes.

While range voting and majority-judgment voting enable voters to make more nuanced judgments than does AV, they also have some less-than-desirable properties. Paradoxically, each voting system can elect a candidate preferred – based on the ratings – by only one voter when all the other voters favor a different candidate. Moreover, under range voting, voters may have a strategic incentive to dichotomize their ratings, giving their highest ratings to favored candidates and their lowest to nonfavored candidates, making it equivalent to AV. Under majority judgment voting, a voter can sometimes do better by not voting than by giving his or her favorite candidate the highest possible rating (the so-called no-show paradox). In sum, these more sophisticated variants of AV carry their own troublesome baggage (Brams 2009).

Besides these refinements of AV, an intellectual and practical challenge is to extend AV to new situations, such as voting on bills in a legislature, wherein there might be multiple alternatives to be voted upon (e.g., an original bill, amendments, and substitute amendments, which are allowed under different parliamentary rules).

Instead of voting on these alternatives serially, where the order of voting on these alternatives can critically affect the outcome, it would seem sensible to use AV to vote on these alternatives all at once.

As a case in point, there can be up to five alternatives on the floor in the US Congress and the United Nations. If a majority of members considered, say, three of five alternatives acceptable, one might declare this package to be the social choice - assuming that the different alternatives are consistent (i.e., one alternative in the package does not nullify another). I know of almost no research on this kind of AV application.

To conclude, I believe that empirical and theoretical research on AV, and the kinds of emendations and applications I have discussed, will continue apace and may even accelerate. But, as I have ruefully discovered, it is hard to predict when and where a new idea like AV will take hold and be implemented.

I have not lost hope and still feel that AV will be tried out in significant public elections. If so, we will learn quickly of any overlooked flaws. But the research over the past third of a century suggests, at least to me, that there are more likely to be some pleasant surprises, resulting in the election of consensus candidates who are better able to formulate and enact public policy. If so, then the contributors to this volume can – as academics whose contributions are not always taken seriously by policy makers – glow in the pride of making an intellectual contribution to an important public good.

Steven J. Brams

References

Balinski, M., & Laraki, R. (2010). *One-value, one-vote: measuring, electing, and ranking*. Cambridge, MA: MIT Press.

Brams, S. J. (1980). Approval voting in multicandidate elections. *Policy Studies Journal*, 9(1), 102–108.

Brams, S. J. (2008). *Mathematics and democracy: designing better voting and fair-division procedures*. Princeton, NJ: Princeton University Press.

Brams, S. J. (2009). *Comparison of three voting systems*. New York University, New York, NY, Preprint.

Brams, S. J., & Fishburn, P. C. (1983/2007). *Approval voting*. Boston/New York: Birkhaüser/Springer.

Brams, S. J., & Fishburn, P. C. (2005). Going from theory to practice: the mixed success of approval voting. *Social Choice and Welfare*, 25(2), 457–474.

Acknowledgements

The papers in this volume have been cross-reviewed by the contributors and by Johanna Goertz, Zeynep Gurguc and Mandar Oak. We thank all referees. We also thank Nick Baigent, Martina Bihn, Maryvonne Laslier, Ipek Ozkal-Sanver and Bennett Young for their contributions to the project. Remzi Sanver acknowledges the support of the Turkish Academy of Sciences Distinguished Young Scientist Award Program (TUBA-GEBIP) and the Scientific and Technological Research Council of Turkey (TUBITAK) through the project # 107K560.

Contents

Contributors

Carlos Alós–Ferrer Department of Economics, University of Konstanz, Box 150, D-78457 Konstanz, Germany, Carlos.Alos-Ferrer@uni-konstanz.de

Antoinette Baujard CREM, University of Caen Basse-Normandie, Caen, France, antoinette.baujard@unicaen.fr

Dorothea Baumeister Institut für Informatik, Heinrich-Heine-Universität Düsseldorf, 40225 Düsseldorf, Germany, baumeister@cs.uni-duesseldorf.de

Steven J. Brams Department of Politics, New York University, NY 10012, USA, steven.brams@nyu.edu

Arnaud Dellis Université Laval and CIRPEE 1025 Ave des Sciences Humaines, local 2174, Québec, QC, Canada G1V 0A6, arnaud.dellis@ecn.ulaval.ca

Mostapha Diss University of Caen Basse Normandie, CREM, UMR CNRS 6211, Caen, France, mostapha.diss@unicaen.fr

Gábor Erdélyi Institut für Informatik, Heinrich-Heine-Universität Düsseldorf, 40225 Düsseldorf, Germany, erdelyi@cs.uni-duesseldorf.de

Peter C. Fishburn AT&T Research, 180 Park Avenue, Florham Park, NJ 07932, USA

Charles Girard Université François Rabelais de Tours, Tours, France, Charles.Girard@univ-paris1.fr

Đura-Georg Granić Department of Economics, University of Konstanz, Box 150, 78457 Konstanz, Germany, dura-georg.granic@uni-konstanz.de

Edith Hemaspaandra Department of Computer Science, Rochester Institute of Technology, Rochester, NY 14623, USA, eh@cs.rit.edu

Lane A. Hemaspaandra Department of Computer Science, University of Rochester, Rochester, NY 14627, USA, lane@cs.rochester.edu

Herrade Igersheim CNRS and BETA, University of Strasbourg, Strasbourg, France, igersheim@unistra.fr

Biung-Ghi Ju Department of Economics, Korea University, Anam-dong 5-1, Seong-buk-gu, Seoul 136-701, Korea, bgju@korea.ac.kr

D. Marc Kilgour Wilfrid Laurier University, 75 University Avenue West, Waterloo, ON, Canada N2L 3C5, mkilgour@wlu.ca

Gilbert Laffond Conservatoire National des Arts et Métiers, 292 rue Saint-Martin, 75141 Paris, France, gilbert.laffond@cnam.fr

Jean Lainé Department of Economics, Istanbul Bilgi University, Dolapdere Campus, Kurtulus Deresi Cad No 47, Istanbul 34440, Turkey, jean@bilgi.edu.tr

Jean-François Laslier Laboratoire d' Econométrie, École Polytechnique, 91128 Palaiseau, France, jean-francois.laslier@polytechnique.edu

Aki Lehtinen Department of social and moral philosophy, P.O. Box 24, University of Helsinki, 00014 Helsinki, Finland, aki.lehtinen@helsinki.fi

François Maniquet CORE, Voie du Roman Pays, 34, 1348 Louvain-la-Neuve, Belgique, francois.maniquet@uclouvain.be

Vincent Merlin CNRS and University of Caen Basse Normandie, CREM, UMR CNRS 6211, Caen, France, vincent.merlin@unicaen.fr

Matías Núñez CNRS, THEMA, Université Cergy-Pontoise, 95011 Cergy-Pontoise, France, matias.nunez@polytechnique.edu

Jörg Rothe Institut für Informatik, Heinrich-Heine-Universität Düsseldorf, 40225 Düsseldorf, Germany, rothe@cs.uni-duesseldorf.de

M. Remzi Sanver Department of Economics, Istanbul Bilgi University, Dolapdere Campus, Kurtulus Deresi, Cad No 47, Istanbul 34440, Turkey, sanver@bilgi.edu.tr

Fabrice Valognes University of Caen Basse Normandie, CREM, UMR CNRS 6211, Caen, France, fabrice.valognes@unicaen.fr

Yongsheng Xu Department of Economics, Andrew Young School of Policy Studies, Georgia State University, Atlanta, GA 30303, USA, yxu3@gsu.edu

Chapter 1
Introduction to the Handbook on Approval Voting

Jean-François Laslier and M. Remzi Sanver

> …the elementary part of each science, which all men can access, becoming more and more expanded, will in a more complete manner contain all that each man can be required to know to be able to manage his life and exert his intelligence in total independence. Condorcet (1793)

Since the publication, in 1983, of Steven Brams and Peter Fishburn's seminal work *Approval Voting*, a variety of theoretical and empirical studies have enhanced our understanding of the various aspects of this voting system: its axiomatic properties have been analyzed; its ballot structure has been examined; its strategic aspects have been scrutinized; the electoral competition structures it induces have been explored; and the patterns of voter behavior entailed by it have been observed both in the laboratory and the field. This research has also engendered various academic controversies, some of which will be mentioned in this introduction. In brief, the merest glance at the literature since 1983 reveals a remarkable accumulation of results, obtained through efforts in a remarkably diverse range of fields: social choice theory, game theory, computer science, political science and experimental economics. This book, then, presents a collection of essays intended to summarize the current state of knowledge on this system of voting.

Under Approval Voting, each voter says either "yes" or "no" *for each candidate*. The voter can thus approve as many candidates as she wishes, and, for single-winner elections, the candidate elected is the one who is approved by the largest number of voters. Initial arguments in favor of this voting system were based on considerations both from the voter's point of view and with respect to its consequences for political parties (see for instance Brams and Fishburn 1978; Cox 1985).

1. Because the voter is given the opportunity to provide more information about her opinion than with a single-name ballot, adoption of Approval Voting might increase voter turnout in general elections. Given the generally accepted view that the quality of a democracy is linked to the number of voters participating

J.-F. Laslier (✉)
Laboratoire d' Econométrie, École Polytechnique, 91128 Palaiseau, France
e-mail: jean-francois.laslier@polytechnique.edu

J.-F. Laslier and M.R. Sanver (eds.), *Handbook on Approval Voting*, Studies in Choice and Welfare, DOI 10.1007/978-3-642-02839-7_1,

and their level of satisfaction with the electoral process, this suggests that Approval Voting can contribute to strengthening democracy.

2. By eliminating the wasted-vote effect, Approval Voting might broaden the span of candidates running for office, thereby contributing to the richness of the political debate. This point is related to the standard observation that the one-round Plurality system makes third parties nonviable, a critical point in U.S. politics.
3. By eliminating the squeezing effect, Approval Voting would encourage the election of consensual candidates. The squeezing effect is typically observed in multiparty elections with a runoff. The runoff tends to prevent extremist candidates from winning, but a centrist candidate who would win any pairwise runoff (the "Condorcet winner") is also often "squeezed" between the left-wing and the right-wing candidates and so eliminated in the first round. This point is critical in countries using two-round Plurality.

The validity of such claims would normally be demonstrated by appeal to empirical and historical studies. Approval Voting and related systems have, indeed, occasionally been used in the past. The complicated rules for the election of the Dogi of Venice from 1268 to 1789, analyzed by Lines (1986), included the use of Approval Voting. Aleskerov (2005) depicts the use of Approval Voting in the eighteenth century, during the reign of Catherine the Great, for local elections. There, instead of ballot papers, there was a double urn for each candidate, made up of two compartments, "yes" and "no." The voter was given a ball for each candidate, and was required to place one ball in the "yes" or in the "no" compartment of each candidate's urn, with his hand covered by a cloth. The same system – with urns, balls and cloth – was used in Greece from the 1864 Constitution to the 1923 elections, before the country turned to proportional representation (Pantelis 2007; Voloudakis 1977). The 1800 U.S. presidential election, in which a version of Approval Voting was used, is discussed by Nagel (2007) and Brams (2008). For nineteenth century England, Cox (1987) analyzes the effect of the possibility of casting two votes. Closer to present day, the election procedure of the secretary-general of the United Nations embodies a variant of Approval Voting. Unfortunately, our information regarding instances of the use of Approval Voting in history remains based almost entirely on anecdote: anecdotes that are, indeed, not numerous, not always very well documented, and which do not give rise to reserves of empirical knowledge comparable to our knowledge of one-round and two-round Plurality. Currently, Approval Voting is not used in any large election.

Due to the lack of historical evidence, and despite the interesting data provided by the adoption of Approval Voting in some academic societies, the subject has essentially remained a matter for theoreticians. There is, indeed, a curious contrast between the complexity of the academic and intellectual debate devoted to Approval Voting, and the simplicity of the voting system itself. Except for the Plurality rule, in which voters are asked to vote for a single candidate, Approval Voting is certainly simpler than all the systems which ask voters to rank the candidates, to grade them, or to name a limited number of them. Nevertheless, among all the voting rules discussed in the literature, Approval Voting would be ranked quite high – indeed, arguably at the top – regarding the amount of academic attention it has received.

We do not aim to explain this contrast, but, as is discussed in this volume, we note that the meaning of Approval Voting as a voting rule affords of an unusually wide variety of interpretations, due to the fact that its implementation in ballots blurs the distinction between Approval Voting and the traditional Arrovian model of social choice theory (Arrow 1952).

This blurriness is at the root of a scholarly controversy that arose in the 1980s. As well as its obvious significance for our subject, this controversy is also interesting for anyone curious about the advantages and disadvantages of using mathematical formalism in the social sciences. The two sides to the controversy can safely be called pro-AV and anti-AV. The pro-AV scholars are Steven Brams, a political scientist trained in formal methods, and Peter Fishburn and Sam Merrill, both mathematicians. The anti-AV scholars are Richard Niemi, also a political scientist trained in formal methods, and Donald Saari, a mathematician.

Soon after Brams and Fishburn's seminal publications (Brams and Fishburn 1978, 1983; Brams 1980), *The American Political Science Review* published a paper by Niemi (1984) entitled "The problem of strategic behavior under Approval Voting." The main thrust of the paper is a critique of the definition of "sincerity" used by Brams and Fishburn. Niemi argues that this definition says nothing about whether a voter "approves" of a candidate or not, but only about whether the voter is willing to vote for that candidate.[1] He then goes on to study the possible strategic behavior patterns that may be adopted by voters endowed with *preferences* in the usual sense (which are complete rankings of candidates), together with *opinions* about each candidate taking the form "approved" or "not approved." Preferences and approval opinions are related in the sense that a voter who approves some candidate x also approves any candidate she prefers to x. This requirement is very similar to the definition of sincerity, except that Niemi now distinguishes "approving x" from "voting for x under Approval Voting." This framework is explored in Chap. 20 of this volume. To adopt, for a moment, the vocabulary of Chap. 20, "approving x" refers to *intrinsic approbation* and "voting for x under Approval Voting" refers to the vote cast by the voter. Given this distinction, there is no reason why strategic behavior based on individual preferences should bear any relation to intrinsic approbations, which leads Niemi (1984, p. 958) to conclude regarding Approval Voting that "in the general case it is neither honest, strategy proof, nor wise." This negative conclusion is not specific to Approval Voting. Since intrinsic approbations are largely disconnected from preferences, any behavior based on preferences (such as strategic voting) may produce results largely unrelated to intrinsic approbations. The reason why Niemi raised this point in connection to Approval Voting, rather than in full

[1] Niemi (1984, p. 952) quotes the following definition of Sincere Approval Voting from Brams (1982, p. 10): "A voter votes sincerely 'if and only if whenever he votes for some candidate, he votes for all candidates preferred to that candidate'" and writes "Note that this definition includes nothing about approval as such; it does not require voting only for 'approved' alternatives." See also Brams and Fishburn (1985) and Niemi (1985).

generality, seems to be related to Niemi's conception of the link between the nature of voters' behavior and the form of the ballots.[2]

The controversy that arose from Niemi's criticism turns on the following question: Is it a good thing or a bad thing that the machinery of Approval Voting leaves it open for the voter to approve any number of candidates? This question led to a further step in the development of this controversy when, in 1988, *Public Choice* published a paper by Saari and Van Newenhizen entitled "The problem of indeterminacy in approval, multiple and truncated voting systems". The same issue of this journal also published comments by Brams, Fishburn and Merrill on this paper, a response by Saari and Van Newenhizen to these comments, and a further rejoinder by Brams, Fishburn and Merrill. Saari and Van Newenhizen (1988a, p. 101) start from the idea that for voting rules such as Approval Voting "there are several ways to tally each voter's preferences." The paper then deploys geometrical and mathematical arguments based on Saari's previous publications in order to derive basically the same "indeterminacy" conclusion as did Niemi, namely that one cannot infer much about the outcome of an Approval Voting election under the sole hypothesis that voters use sincere approval strategies in the usual sense.[3] Brams et al. (1988a,b) respond to this argument by claiming that indeterminacy in Saari and Van Newenhizen's sense is the consequence of the existence of multiple sincere strategies, which is a good thing because it reflects the freedom the voter has to express further information she might find relevant.

A closer examination of this controversy suggests the conclusion that the matter is not susceptible to answer purely through the application of formalism and mathematics, as was attempted in the 1980s. It is noticeable that this discussion manifests limited contact with the usual themes of research within Political Science; for instance, it makes no connection with the then-contemporary work of Gary Cox on Approval Voting (Cox 1984, 1985; Weber 1995). Moreover, the discussion is unconstrained by reference to facts and observation; the relationship with reality being exclusively through the prism of the scholar's intuition. This is quite surprising, given that what is at stake is precisely the conception of the rule from the voter's point of view. Fortunately, we now have pieces of evidence that were not available twenty years ago. Several academic societies have used this voting system and made the corresponding data available, and various experiments have been conducted to investigate voters' understanding and use of the rule. It is now quite clear that voters consider the possibility of giving an opinion about all candidates as a good thing, and not as some embarrassing flaw in the system.

---

[2] In the abstract to the same paper he writes "the existence of multiple sincere strategies almost begs the voter to behave strategically."

[3] The meaning of "indeterminacy" is very clearly defined by Saari and Van Newenhizen (1988b, p. 135): "The omnipresent danger of indeterminacy: An Approval Voting election can be indeterminate. The essential idea is this. Suppose that we know each voter's ranking of the candidates. Armed with this information, we can compute the unique election outcome. We cannot do this with Approval Voting."

This conclusion arises from the few pieces of empirical knowledge we have about Approval Voting, some of which are presented in this book. But it should come as no surprise to anyone familiar with the variety of voting rules which are actually used in practice and considered satisfactory. Many voting systems currently in use require much more from the voter than a single candidate's name – or maybe one should say rather that they *allow* the voter to provide more than a single name. Small committees often reach a decision after several rounds of voting[4] but this is hardly possible for political elections and, in such cases, a one-round, or at most two-round, system is most common. A closer look at the systems actually in use throughout the world (Lijphart 1994; Farrell 2001) reveals that a variety of election procedures are actually in use for political elections. Most countries in the world have direct presidential elections, and the most frequent direct rule is majority rule with a runoff (Blais et al. 1997). To citizens (and scholars) familiar with single-name, first-past-the-post elections, these rules may appear complicated. In fact, the diversity of voting systems is reflected in the various form taken by the ballot papers. The simplest and most common form is a small sheet of paper with a single candidate name. More complex ballot papers are used when several elections are held simultaneously, which may give rise to huge sheets of paper, as occurs sometimes in the U.S. In New Zealand and Australia, some elections are held under Transferable Vote systems, and voters submit, as their vote, a ranking of candidates. This requires complex ballots, and voters actually have the option to write simply that they adhere to the ranking suggested by one party (Farrell and McAllister 2006).

In Germany, some elections require relatively complex ballots. See for instance the ballot used in Heidelberg in June 2009 in Fig. 1.1.[5] For the election of the 40 members of the municipal council, each voter has 40 votes. In June 2009, there were 10 lists of 40 candidates. The voter can either give her or his votes to one list only, in which case she or he returns only the ballot for this list. She or he can also give votes to candidates of different lists. The voter has 40 votes but can give up to three votes to a candidate; this is done by writing the number 2 or 3 in front of the candidate's name. The ballot is invalid if these conditions are not met (more than 40 votes in total or more than three votes for a candidate).[6] The existence of such voting systems shows that the "simplicity" argument in favor of single-name voting is weak.

On the contrary, the evidence shows that the public is eager to understand the consequences of using different voting rules, and is ready to switch rules – provided, of course, that it is done for sound reasons. The role of academic research

---

[4] A famous instance of this phenomenon is the election of the Pope by the cardinals, where the number of rounds is not limited. In 1271 the process was endless and the inhabitants of Vitorbo, the city in the papal states where the election was taking place, had to lock the cardinals in, and stop bringing them food until they reached a decision. Three days later, Gregory X was elected.

[5] We thank Martina Bihn for providing this ballot.

[6] The leaflet provided to the voters specifies this rule. Notice that it does not mention the fact that the 40 elected candidates are those who receive the largest number of votes. The logic of adding points or votes is obvious.

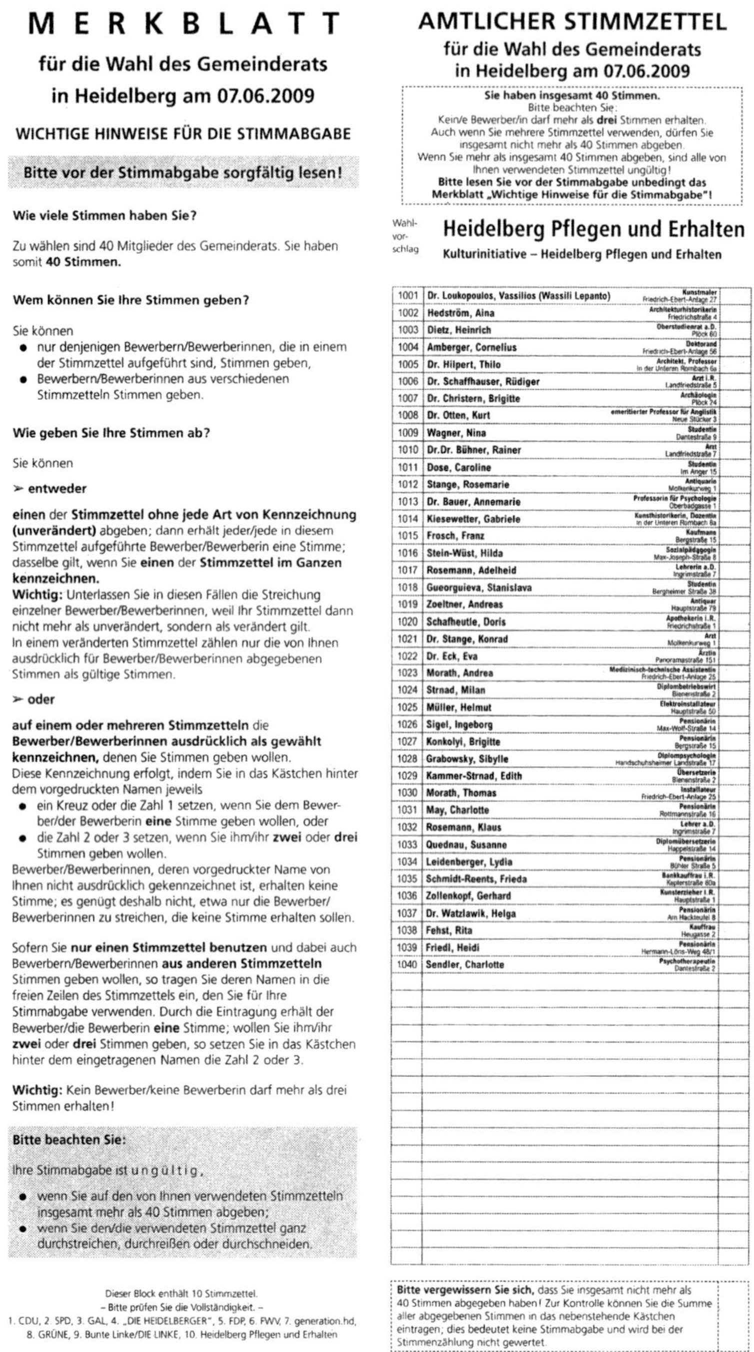

MERKBLATT

für die Wahl des Gemeinderats in Heidelberg am 07.06.2009

WICHTIGE HINWEISE FÜR DIE STIMMABGABE

Bitte vor der Stimmabgabe sorgfältig lesen!

Wie viele Stimmen haben Sie?

Zu wählen sind 40 Mitglieder des Gemeinderats. Sie haben somit **40 Stimmen.**

Wem können Sie Ihre Stimmen geben?

Sie können

- nur denjenigen Bewerbern/Bewerberinnen, die in einem der Stimmzettel aufgeführt sind, Stimmen geben,
- Bewerbern/Bewerberinnen aus verschiedenen Stimmzetteln Stimmen geben.

Wie geben Sie Ihre Stimmen ab?

Sie können

➢ **entweder**

einen der **Stimmzettel ohne jede Art von Kennzeichnung (unverändert)** abgeben; dann erhält jeder/jede in diesem Stimmzettel aufgeführte Bewerber/Bewerberin eine Stimme; dasselbe gilt, wenn Sie **einen** der **Stimmzettel im Ganzen kennzeichnen.**
Wichtig: Unterlassen Sie in diesen Fällen die Streichung einzelner Bewerber/Bewerberinnen, weil Ihr Stimmzettel dann nicht mehr als unverändert, sondern als verändert gilt.
In einem veränderten Stimmzettel zählen nur die von Ihnen ausdrücklich für Bewerber/Bewerberinnen abgegebenen Stimmen als gültige Stimmen.

➢ **oder**

auf einem oder mehreren Stimmzetteln die **Bewerber/Bewerberinnen ausdrücklich als gewählt kennzeichnen,** denen Sie Stimmen geben wollen.
Diese Kennzeichnung erfolgt, indem Sie in das Kästchen hinter dem vorgedruckten Namen jeweils

- ein Kreuz oder die Zahl 1 setzen, wenn Sie dem Bewerber/der Bewerberin **eine** Stimme geben wollen, oder
- die Zahl 2 oder 3 setzen, wenn Sie ihm/ihr **zwei** oder **drei** Stimmen geben wollen.

Bewerber/Bewerberinnen, deren vorgedruckter Name von Ihnen nicht ausdrücklich gekennzeichnet ist, erhalten keine Stimme; es genügt deshalb nicht, etwa nur die Bewerber/Bewerberinnen zu streichen, die keine Stimme erhalten sollen.

Sofern Sie **nur einen Stimmzettel benutzen** und dabei auch Bewerbern/Bewerberinnen **aus anderen Stimmzetteln** Stimmen geben wollen, so tragen Sie deren Namen in die freien Zeilen des Stimmzettels ein, den Sie für Ihre Stimmabgabe verwenden. Durch die Eintragung erhält der Bewerber/die Bewerberin **eine** Stimme; wollen Sie ihm/ihr **zwei** oder **drei** Stimmen geben, so setzen Sie in das Kästchen hinter dem eingetragenen Namen die Zahl 2 oder 3.

Wichtig: Kein Bewerber/keine Bewerberin darf mehr als drei Stimmen erhalten!

Bitte beachten Sie:

Ihre Stimmabgabe ist u n g ü l t i g,

- wenn Sie auf den von Ihnen verwendeten Stimmzetteln insgesamt mehr als 40 Stimmen abgeben;
- wenn Sie den/die verwendeten Stimmzettel ganz durchstreichen, durchreißen oder durchschneiden.

Dieser Block enthält 10 Stimmzettel.
– Bitte prüfen Sie die Vollständigkeit. –
1. CDU, 2. SPD, 3. GAL, 4. „DIE HEIDELBERGER", 5. FDP, 6. FWV, 7. generation.hd, 8. GRÜNE, 9. Bunte Linke/DIE LINKE, 10. Heidelberg Pflegen und Erhalten

AMTLICHER STIMMZETTEL

für die Wahl des Gemeinderats in Heidelberg am 07.06.2009

Sie haben insgesamt 40 Stimmen.
Bitte beachten Sie:
Kein/e Bewerber/in darf mehr als **drei** Stimmen erhalten.
Auch wenn Sie mehrere Stimmzettel verwenden, dürfen Sie insgesamt nicht mehr als 40 Stimmen abgeben.
Wenn Sie mehr als insgesamt 40 Stimmen abgeben, sind alle von Ihnen verwendeten Stimmzettel ungültig!
Bitte lesen Sie vor der Stimmabgabe unbedingt das Merkblatt „Wichtige Hinweise für die Stimmabgabe"!

Wahl-vor-schlag

Heidelberg Pflegen und Erhalten

Kulturinitiative – Heidelberg Pflegen und Erhalten

1001	**Dr. Loukopoulos, Vassilios (Wassili Lepanto)**	**Kunstmaler** Friedrich-Ebert-Anlage 27	
1002	**Hedström, Aina**	**Architekturhistorikerin** Friedrichstraße 4	
1003	**Dietz, Heinrich**	**Oberstudienrat a.D.** Plöck 60	
1004	**Amberger, Cornelius**	**Doktorand** Friedrich-Ebert-Anlage 56	
1005	**Dr. Hilpert, Thilo**	**Architekt, Professor** In der Unteren Rombach 6a	
1006	**Dr. Schaffhauser, Rüdiger**	**Arzt i.R.** Landfriedstraße 5	
1007	**Dr. Christern, Brigitte**	**Archäologin** Plöck 24	
1008	**Dr. Otten, Kurt**	**emeritierter Professor für Anglistik** Neue Stücker 3	
1009	**Wagner, Nina**	**Studentin** Dantestraße 9	
1010	**Dr.Dr. Bühner, Rainer**	**Arzt** Landfriedstraße 7	
1011	**Dose, Caroline**	**Studentin** Im Anger 15	
1012	**Stange, Rosemarie**	**Antiquarin** Molkenkurweg 1	
1013	**Dr. Bauer, Annemarie**	**Professorin für Psychologie** Oberbadgasse 1	
1014	**Kiesewetter, Gabriele**	**Kunsthistorikerin, Dozentin** In der Unteren Rombach 8a	
1015	**Frosch, Franz**	**Kaufmann** Bergstraße 15	
1016	**Stein-Wüst, Hilda**	**Sozialpädagogin** Max-Joseph-Straße 8	
1017	**Rosemann, Adelheid**	**Lehrerin a.D.** Ingrimstraße 7	
1018	**Gueorguieva, Stanislava**	**Studentin** Bergheimer Straße 38	
1019	**Zoeltner, Andreas**	**Antiquar** Hauptstraße 79	
1020	**Schafheutle, Doris**	**Apothekerin i.R.** Friedrichstraße 1	
1021	**Dr. Stange, Konrad**	**Arzt** Molkenkurweg 1	
1022	**Dr. Eck, Eva**	**Ärztin** Panoramastraße 151	
1023	**Morath, Andrea**	**Medizinisch-technische Assistentin** Friedrich-Ebert-Anlage 25	
1024	**Strnad, Milan**	**Diplombetriebswirt** Bienenstraße 2	
1025	**Müller, Helmut**	**Elektroinstallateur** Hauptstraße 50	
1026	**Sigel, Ingeborg**	**Pensionärin** Max-Wolf-Straße 14	
1027	**Konkolyi, Brigitte**	**Pensionärin** Bergstraße 15	
1028	**Grabowsky, Sibylle**	**Diplompsychologin** Handschuhsheimer Landstraße 17	
1029	**Kammer-Strnad, Edith**	**Übersetzerin** Bienenstraße 2	
1030	**Morath, Thomas**	**Installateur** Friedrich-Ebert-Anlage 25	
1031	**May, Charlotte**	**Pensionärin** Rottmannstraße 16	
1032	**Rosemann, Klaus**	**Lehrer a.D.** Ingrimstraße 7	
1033	**Quednau, Susanne**	**Diplomübersetzerin** Happelstraße 14	
1034	**Leidenberger, Lydia**	**Pensionärin** Bühler Straße 5	
1035	**Schmidt-Reents, Frieda**	**Bankkauffrau i.R.** Keplerstraße 80a	
1036	**Zollenkopf, Gerhard**	**Kunsterzieher i.R.** Hauptstraße 1	
1037	**Dr. Watzlawik, Helga**	**Pensionärin** Am Hackteufel 8	
1038	**Fehst, Rita**	**Kauffrau** Heugasse 2	
1039	**Friedl, Heidi**	**Pensionärin** Hermann-Löns-Weg 48/1	
1040	**Sendler, Charlotte**	**Psychotherapeutin** Dantestraße 2	

Bitte vergewissern Sie sich, dass Sie insgesamt nicht mehr als 40 Stimmen abgegeben haben! Zur Kontrolle können Sie die Summe aller abgegebenen Stimmen in das nebenstehende Kästchen eintragen; dies bedeutet keine Stimmabgabe und wird bei der Stimmenzählung nicht gewertet.

Fig. 1.1 A ballot used in Germany. A voter can split his or her 40 votes between different such lists, and give up to three votes to a candidate

in this case is to understand what may be the consequences of choosing one voting rule over another. This requires a type of research which is not only descriptive, but which also attempts to imagine the world as it could be under the new circumstances. Accordingly, skills from different research fields are required. Voting rules have peculiar contents which are not always captured by formal models. There are peculiarities vis-à-vis the rule's meaning to the voter and to the society in which it is employed: different voting rules cause political systems to adjust differently, and, together with other factors, shape thereby the actual political system as well as the way citizens think about the act of voting (Myerson 1995). Ultimately, this may influence the citizen's conception of democracy itself.

Content of the Book

To better understand the diversity of voting rules, it is worth going beyond formal models and studying the history of their application in various societies. The first part of the volume (*Part I, History of Approval Voting*) is a modest attempt in this direction. It comprises two chapters. Chapter 2 is a short reminder, by Charles Girard, of how elections for the council of the elders were conducted in ancient Sparta. There, the voting rule was not exactly by approbation but by *shouting*. Candidates came in one by one, in random order, and voters would shout more or less loudly in favor of them. The candidate with the greatest and loudest acclamation was elected. As Girard notes, this non-anonymous rule is a single-round plurinominal election in which the voter can support as many candidates as he wishes, hence exhibiting the basic features of Approval Voting. Chapter 3, by Steven Brams and Peter Fishburn, was originally published in the journal *Social Choice and Welfare* in 2006. The authors describe the recent history of Approval Voting at work, in particular its use, since the eighties, in academic and engineering societies. This paper is also a reflection on reformism, exploring the paths leading from science to application when the problem at stake is institutional design.

The second part of the volume (*Part II, Axiomatic Theory*) is devoted to axiomatic approaches which conceive of Approval Voting as a method for aggregating judgments provided by the voters about candidates. Chapter 4, written by Biung-Ghi Ju, is a reasonably complete overview of the literature on the aggregation of dichotomous and trichotomous preferences. In Chap. 5, Yongsheng Xu focuses on the dichotomous case – which is the basis for Approval Voting – and surveys the various axiomatic characterizations of Approval Voting.

The third part of the volume (*Part III, Committees*) diverges from the general direction of the book. Approval Voting, as generally studied, is essentially designed to elect precisely one candidate among the existing ones or to choose precisely one option among a given set of feasible options. By contrast, approval balloting – where a ballot is a list of any number of approved candidates – can be used as a basis to elect a committee made up of several candidates. Part III addresses the problem of electing a committee through approval balloting. In Chap. 6, Marc Kilgour collects and classifies the plethora of procedures than can be used for counting approval ballots. In Chap. 7, Gilbert Laffond and Jean Lainé study the properties that one can expect from these procedures. Their analysis is related to the structure of the voters' preferences and it also applies to the case of multiple, simultaneous or sequential referenda.

The three chapters presented in the fourth part of the volume (*Part IV, Strategic Voting*) return to the case of electing a single candidate, and deal with voters' strategic behavior. In Chap. 8, Jean-François Laslier and Remzi Sanver survey the results on Approval Voting obtained within the framework of classical noncooperative game theory. Here, restricting the set of ballots to those which are undominated strategies and sincere does not help in predicting the outcome of an election under Approval Voting. Neither does using Nash equilibrium and its usual refinements. On the other hand, Strong Nash equilibrium, if it exists, predicts the election of the Condorcet winner, which means that, under Approval Voting, voters who can coordinate will not miss the chance to select a Condorcet winner. The poor predictive power of Nash equilibrium in most voting games with many voters has led scholars to propose alternative models, which are presented in Chap. 9 by Matías Núñez. The works of Myerson, Weber, Laslier and Núñez himself help to understand why voters who vote strategically in popular elections under Approval Voting would often vote sincerely and elect the Condorcet winner, if one exists; and why this is not always true. This approach allows for predictive models of voter behavior which can be tested in the laboratory (see Part VI of this book) and serve as building blocks in models of electoral competition (see Part VII and the further studies of voting under incomplete information by Bouton and Castanheira 2009 or Goertz and Maniquet 2009). Chapter 10 also studies strategic issues for voters, but from another perspective. Dorothea Baumeister, Gábor Erdélyi, Edith Hemaspaandra, Lane Hemaspaandra and Jörg Rothe, all of whom work in the new field of *Computational Social Choice*, apply the tools of complexity theory to study how hard it is to modify the outcome of Approval Voting elections through manipulation, control, or bribery.

As general theory rarely can pretend to be able to predict the outcomes of Approval Voting elections, scholars aiming for more precise conclusions have resorted to specific assumptions about voters' preferences and ballot choices. These assumptions are usually of a probabilistic nature (Regenwetter and Tsetlin 2004; Regenwetter et al. 2006). The fifth part of the volume (*Part V, Probabilistic Exercises*) is devoted to this line of research. An often-used hypothesis is that voters' preferences are independently and uniformly distributed. In Chap. 11, Mostapha Diss, Vincent Merlin and Fabrice Valognes follow this tradition, à la Gehrlein (2006), for studying the three-candidate case, assuming further that voters choose at random how many candidates they approve. They compute analytically the Condorcet efficiency (i.e., the probability that the Condorcet winner is chosen given that one exists) of several voting rules. Analytical computations are only possible for specific assumptions on preferences and behavior. The following two chapters go beyond those cases by using computer simulations. In Chap. 12, Aki Lehtinen, again with three candidates and utilities being randomly and independently chosen, contrasts what happens when all voters behave "sincerely" with the case in which some of them behave "strategically". Unlike the rest of the book, this chapter defines sincere behavior as approving the most preferred candidate and approving the second best if and only if his utility is larger than the average utility, and strategic behavior as voting for any candidate whose utility for the voter is larger than a specified threshold. In Chap. 13, Jean-François Laslier simulates different sorts of random electorates: *Rousseauist* cultures model situations of common interest, *impartial* cultures are often used in the social choice literature, and *distributive* cultures and *spatial* cultures are often used in the political science literature. Several voting rules are compared in these different situations, assuming various voting behaviors. The study confirms that while voting rules that improve over the Plurality rule do exist, voter behavior is of primary importance in assessing the quality of a voting rule. It is also remarked that comparisons may be quite different from one culture to another: the best rule to resolve a conflict of interest need not be the best rule to aggregate information. In situations of moderate conflict, Approval Voting usually elects the Condorcet winner, which represents a moderate candidate, and is relatively robust to strategizing.

The sixth part of the volume (*Part VI, Experiments*) deals with empirical observations, and hence brings us closer to reality. In Chap. 14, Jean-François Laslier surveys laboratory experiments in which subjects vote and the experimenter (more or less) controls their preferences through monetary incentives. This is usually called the Experimental Economics methodology. These experiments all show that, at least in the laboratory, Approval Voting makes it easier, compared to other voting rules, to reach consensual voting outcomes. They also make clear that individual approval decisions are well described by strategic theories, a point that highlights the importance of information such as previous elections or preelection polls. The next two chapters are devoted to the description of original experiments that were performed on the occasion of real elections in France and in Germany. Chapter 15, by Antoinette Baujard and Herrade Igersheim, deals with the French presidential elections in 2002 and 2007, while Chap. 16, written by Carlos Alos-Ferrer and Đura-Georg Granić, deals with a German state election in 2008. These experiments, which are here called "field experiments", took place on election day itself, and voters were asked to vote using Approval Voting (or some other rule) as if this was the official system in operation. As votes are anonymous and any voter in the real election could participate in the field experiment (excepting constraints imposed by the location of the experimental polling stations), neither the set of candidates nor the sample could be controlled. Several lessons can nevertheless be drawn: It is clear that the idea of experimenting on voting rules, as well as the idea of Approval Voting, are welcomed by voters. Moreover, extrapolation of the results to the whole country indicates that, in France, the use of Approval Voting would have changed the relative position of several candidates and might even have changed the identity of the elected candidate in favor of a more moderate one. In Germany, there would have been four (rather than two) main parties of comparable size, and small parties would have obtained parliamentary representation.

All the preceding supposes that the set of candidates and voters' preferences over the candidates are independent of the voting system. This neglects an important issue: in practice, as noted by political scientists (Duverger 1951), the set of existing candidates or political parties, and their policy positions, may depend on the formalities of the electoral system and the voting rule in operation. A recurrent theme of the political science literature since Black (1958) is that first-past-the-post essentially kills any third party and drives the only two serious parties to propose political platforms close to the center of the political space. It is therefore important to try to figure out what may be the outcome of electoral competition under Approval Voting. This is the subject of *Part VII* (*Electoral Competition*), which is, obviously enough, purely theoretical. The chapters in this part draw on what has been learned about voters' behavior from theory and experiments, and concentrate on the behavior of candidates and parties. In Chap. 17, Jean-François Laslier and François Maniquet use the classical Downsian model of competition among office-motivated candidates (Downs 1957). They show that if a Condorcet winner policy exists, then there exists an electoral competition equilibrium supporting this policy. Moreover, if the set of policies is one-dimensional and voters have single-peaked preferences, then it is the only electoral competition equilibrium. In Chap. 18, Arnaud Dellis studies an alternative model of electoral competition, in which each voter decides herself whether to be a candidate or not, and has to weigh the cost of being a candidate with the expected payoff from winning the election. In this so-called "citizen-candidate model", where the candidates have policy preferences, Dellis shows that Approval Voting induces the choice of a moderate policy, provided that two conditions are met: the first condition is that a candidate enters the race as soon as she anticipates that she has some chance of winning the election, and the second condition is a reasonable assumption about the voters' strategic behavior. These two papers on electoral competition provide arguments in favor of the claim that Approval Voting, compared to Plurality Voting, is immune to the squeezing and wasted-vote effects.

The two last chapters of this book are gathered in *Part VIII* (*Meaning for Individual and Society*). In Chap. 19, Jean-François Laslier recalls that the data obtained from an election held under Approval Voting is much richer than the mere list of the scores of candidates. Using Approval Voting, a society can obtain a richer description of itself than can be obtained with a Plurality rule or Plurality with a runoff. The chapter argues that such data can and should be published as the "result" of an AV election. In Chap. 20, Remzi Sanver considers Approval Voting within the context of a widening of the standard ordinal non-comparable framework of Arrovian social choice. In the normative perspective, the informational basis of Approval Voting embodies one element of inter-individual comparability: the alternatives being intrinsically deemed "good" or "bad" by the individuals, and the notions of "good" or "bad" are common to all individuals and can therefore used at the level of the society. This allows us to revisit Approval Voting as well as to define and study other interesting voting systems which combine preference and approval-type inputs.

Throughout these 20 chapters, the book raises many different questions about Approval Voting and proposes answers to most of them. For obvious reasons, the societal questions can, for the moment, receive only theoretical answers; but these answers are informed by the theoretical and empirical knowledge we have now amassed about Approval Voting from the voter's point of view.

References

Aleskerov, F. (2005). The history of social choice in Russia and the Soviet Union. *Social Choice and Welfare, 25*, 419–431.

Arrow, K. J. (1952). *Social choices and individual values*. New York: Wiley.

Black, D. (1958). *The theory of committees and elections*. Cambridge, MA: Cambridge University Press.

Blais, A., Massicotte, L., & Dobrzynska, A. (1997). Direct presidential elections: A world summary. *Electoral Studies, 16*, 441–455.

Bouton, L., & Castanheira, M. (2009). *One person, many votes: Divided majority and information aggregation* (ECARES working paper).

Brams, S. J. (1980). Approval voting in multicandidate elections. *Policy Studies Journal, 9*, 102–108.

Brams, S. J. (1982). Strategic information and voting behavior. *Society, 19*, 4–11.

Brams, S. J. (2008). *Mathematics and democracy*. Princeton, NJ: Princeton University Press.

Brams, S. J., & Fishburn, P. C. (1978). Approval voting. *American Political Science Review, 72*, 831–847.

Brams, S. J., & Fishburn, P. C. (1983). *Approval voting*. Boston: Birkhäuser.

Brams, S. J., & Fishburn, P. C. (1985). Comments on the problem of strategic behavior under approval voting. *American Political Science Review, 79*, 816–818.

Brams, S. J., Fishburn, P. C., & Merrill, S., III (1988a). The responsiveness of approval voting. *Public Choice, 59*, 121–131.

Brams, S. J., Fishburn, P. C., & Merrill, S., III (1988b). Rejoinder to Saari and Van Newenhizen. *Public Choice, 59*, 149.

Condorcet (1793). *Esquisse d'un tableau historique des progrès de l'esprit humain*. Paris: Flamarion. (Reprinted in 1978).

Cox, G. W. (1984). Strategic electoral choice in multi-member districts: Approval voting in practice? *American Journal of Political Science, 28*, 722–738.

Cox, G. W. (1985). Electoral equilibrium under approval voting. *American Journal of Political Science, 29*, 112–118.

Cox, G. W. (1987). *The efficient secret: The cabinet and development of political parties in Victorian England*. New York: Cambridge University Press.

Downs, A. (1957). *An economic theory of democracy*. New York: Harper and Row.

Duverger, M. (1951). *Les partis politiques*. Paris: Armand Colin.

Farrell, D. M. (2001). *Electoral systems: A comparative introduction*. New York: Palgrave.

Farrell, D., & McAllister, I. (2006). *The Australian electoral system: Origins, variations, and consequences*. Sydney: University of New South Wales Press.

Gehrlein, W. V. (2006). *Condorcet's paradox*. Heidelberg: Springer Verlag.

Goertz, J., & Maniquet, F. (2009). *On the informational efficiency of simple scoring rules* (CORE discussion paper 2009/26).

Lijphart, A. (1994). *Electoral systems and party systems*. Oxford: Oxford University Press.

Lines, M. (1986). Approval voting and strategy analysis: A Venitian example. *Theory and Decision, 20*, 155–172.

Myerson, R. (1995). Analysis of democratic institutions: Structure, conduct, and performance. *The Journal of Economic Perspectives, 9*, 77–89.

Nagel, J. (2007). The Burr dilemma in approval voting. *Journal of Politics, 69*, 43–58.

Niemi, R. G. (1984). The problem of strategic behavior under approval voting. *American Political Science Review, 78*, 952–958.

Niemi, R. G. (1985). Reply to Brams and Fishburn. *American Political Science Review, 79*, 818–819.

Pantelis, A. (2007). Approval voting in Greece. *Oral communication at the CAS seminar "Vote par note et vote par approbation"*, Paris, June 28, 2007.

Regenwetter, M., & Tsetlin, I. (2004). Approval voting and positional voting methods: Inference, relationship, examples. *Social Choice and Welfare, 22*, 539–566.

Regenwetter, M., Grofman, B., Marley, A. A. J., & Tsetlin, I. M. (2006). *Behavioral social choice. Probabilistic models, statistical inference and applications*. Cambridge, MA: Cambridge University Press.

Saari, D. G., & Van Newenhizen, J. (1988a). The problem of indeterminacy in approval, multiple, and truncated voting systems. *Public Choice, 59*, 101–120.

Saari, D. G., & Van Newenhizen, J. (1988b). Is approval voting an 'unmitigated evil'? A response to Brams, Fishburn, and Merrill. *Public Choice, 59*, 133–147.

Voloudakis, E. (1977). *Recherche sur le suffrage politique en Grèce (1910–1976)* (Thesis, University of Paris II).

Weber, R. J. (1995). Approval voting. *The Journal of Economic Perspectives, 9*, 39–49.

Part I
History of Approval Voting

Chapter 2
Acclamation Voting in Sparta: An Early Use of Approval Voting

Charles Girard

An early form of approval voting was arguably used in Ancient Greece, as is described in Plutarch's account of the elections to the *Gerousia*, Sparta's Council of Elders.

In his *Life of Illustrious Men*, Plutarch credits the legendary lawgiver Lycurgus with having fixed the rules for electing the Spartan Council's members, the *Gerontes*. The office of Elder was seen as a reward for virtue and as a high honor, since the *Gerontes* were elected for life and the *Gerousia* was a powerful institution – it prepared legislation for approval by the Assembly and acted as a high court in serious cases such as those of homicide (Staveley 1972 p. 76). Plutarch recounts that, after having filled the Council with men chosen among his followers, Lycurgus ordered that the future vacancies "be supplied out of the best and most deserving men past sixty years old":

> We need not wonder if there was much striving for it; for what more glorious competition could there be amongst men, than one in which it was not contested who was swiftest among the swift or strongest of the strong, but who of many wise and good was wisest and best, and fittest to be entrusted for ever after, as the reward of his merits, with the supreme authority of the commonwealth, and with power over the lives, franchises, and highest interests of all his countrymen? (Plutarch 1876 p. 40)

Although it is likely that only aristocrats were entitled to put themselves forward as candidates, the rules organizing the selection of the 28 *Gerontes* were designed to favor merit, and not only birth or wealth. And while elections to the *Gerousia* involved some form of lottery, it was not exactly an election by lot, as was often the case in Greek democracies. Sparta resorted to a very specific form of election, in which voting was conducted by shouting:

> The manner of their election was as follows: The people being called together, some selected persons were locked up in a room near the place of election, so contrived that they could neither see nor be seen, but could only hear the noise of the assembly without; for they decided this, as most other affairs of moment, by the shouts of the people. This done, the

C. Girard
Université François Rabelais de Tours, Tours, France
e-mail: Charles.Girard@univ-paris1.fr

J.-F. Laslier and M.R. Sanver (eds.), *Handbook on Approval Voting*, Studies in Choice and Welfare, DOI 10.1007/978-3-642-02839-7_2,

> competitors were not brought in and presented all together, but one after another by lot, and passed in order through the assembly without speaking a word. Those who were locked up had writing-tables with them, in which they recorded and marked each shout by its loudness, without knowing in favour of which candidate each of them was made, but merely that they came first, second, third, and so forth. He who was found to have the most and loudest acclamations was declared senator duly elected. (Plutarch 1876 p. 40)

Elections to the *Gerousia* thus involved "an early form of applaudometer" (Elster 1989 p. 85), in which whoever was judged to have received the loudest acclaim was elected. Aristotle notoriously denounced this procedure as "childish" (Aristotle 1984), probably because he considered it a blatantly inappropriate way to select the most competent among the citizens – he was also critical of the fact that individuals had to put themselves forward as candidates to the *Gerousia*, a rule more suitable for ambitious men than for virtuous ones.

Aristotle is not alone in judging the Spartan Shout harshly. Historians have deemed the procedure "primitive" (Staveley 1972 p. 74), interpreting it as a sign that Spartans had "no notion of 'one man one vote'" (Cartledge 2001 p. 51). Contemporary political theorists even invoke the "Spartan Shout" as a negative model, contrasting it with the positive example of Athenian democracy. In Fishkin's view "missing in the Spartan method was the entire social context of careful debate and deliberative argument fostered by the Athenian institutions.... Yet if we ask which model of ancient democracy we have come closer to realizing in our modern quest for direct democracy, we must concede that there are ways in which the Spartan model is closer than the Athenian to contemporary practices" (Fishkin 1997 p. 24). Elections to the *Gerousia* are even seen as the precursors of today's media "applaudometers," in which citizens influence politicians in proportion to their loudness. "The sting of an offensive sound bite arouses a populace that is only sound-bitten. The ire of talk-show democracy has given us a mass electronic version of the Shout" (Fishkin 1997 p. 25).

Critics of the Shout thus question the ability of the Spartan Assembly to select the wisest and the best among the candidates on at least two counts: Because it required citizens to shout as the candidates were presented in random order, it precluded careful consideration and deliberation; and because it was based on the auditory evaluation of a collective shout, it did not give an equal voice to each citizen but favored the loudest and the most motivated among the Assembly's members.

However, far from being inexplicably primitive, the Spartan way of proceeding seems to have been designed to respond to specific concerns regarding the impartiality of the election. The use of the lottery to decide the candidates' order of appearance was most likely meant to ensure the spontaneity of the procedure. "The order in which the candidates appeared was the key to the whole proceedings. Despite the opportunities which an allotment behind closed doors might have presented for collusion, it is most unlikely that lots were drawn in the presence of the Assembly itself, if only because foreknowledge of the order in which the candidates would present themselves would inevitably have detracted from the element of spontaneity in the shouting which was so necessary to the credibility of the vote" (Staveley 1972 p. 74). Such precautions may not have been enough to prevent strategies of manipulation, especially by Sparta's kings and their families (Birgalias 2007 p. 347), but spontaneity was not the Shout's only upside.

Although Plutarch's succinct account does not allow us to draw very precise conclusions, it is clear that the Assembly's members were free to acclaim several candidates. To this extent, the shout appears as an early form of approval voting, in which each elector either approves or disapproves each candidate, the winner being the one that is approved by the greatest number of electors. In the absence of ballots, the Spartan version of approval voting appears clearly imperfect: individual voices may differ in loudness, and one can imagine the Assembly's members adjusting the intensity of their cheering to the intensity of their support for each of them. Furthermore, it involved a very approximate method of aggregation: clumping voices rather than counting ballots. Nonetheless, the Shout roughly satisfies the formal requisites of approval voting: it was a multi-candidate contest consisting in a single round of voting that allowed each participant to support as many candidates as desired, leading to an – admittedly crude – summation of the support received by each candidate. Looking to fill the *Gerousia* with men recognized by all as virtuous, Sparta's lawgivers might have been aware of the advantages offered by such a form of election, which favors the candidate with the greatest overall support.

If we are to trust Plutarch's account of the way they celebrated the winner, the Spartans themselves apparently believed that acclamation voting was a satisfying form of election. "Upon [his election] he had a garland set upon his head, and went in procession to all the temples to give thanks to the gods; a great number of young men followed him with applauses, and women, also, singing verses in his honour, and extolling the virtue and happiness of his life" (Plutarch 1876 p. 40).

References

Aristotle (1984). Politics. In J. Barnes (Ed.), B. Jowett (Trans.), *The complete works of Aristotle*. Princeton: Princeton University Press.

Birgalias, N. (2007). La Gérousia et les gérontes de Sparte. *Ktema*, 32, 341–349.

Cartledge, P. (2001). *Spartan reflections*. Berkeley: University of California Press.

Elster, J. (1989). *Solomonic judgements: Studies in the limitations of rationality*. Cambridge: Cambridge University Press.

Fishkin, J. (1997). *The voice of the people: Public opinion and democracy*. New Haven: Yale University Press.

Plutarch (1876). Lycurgus. In A. H. Clough (Ed.), J. Dryden (Trans.), *Lives of illustrious men*. London: Sampson Law, Marston, Searle, & Rivington.

Staveley, E. S. (1972). *Greek and Roman voting and elections*. London: Thames and Hudson.

Chapter 3
Going from Theory to Practice: The Mixed Success of Approval Voting*

Steven J. Brams and Peter C. Fishburn

3.1 Background

Approval voting (AV) is a voting procedure in which voters can vote for, or approve of, as many candidates as they like in multicandidate elections (i.e., those with more than two candidates). Each candidate approved of receives one vote, and the candidate with the most votes wins.

Beginning in 1987, several scientific and engineering societies adopted AV, including the

- Mathematical Association of America (MAA), with about 32,000 members;
- American Mathematical Society (AMS), with about 30,000 members;
- Institute for Operations Research and Management Sciences (INFORMS), with about 12,000 members;
- American Statistical Association (ASA), with about 15,000 members;
- Institute of Electrical and Electronics Engineers (IEEE), with about 377,000 members.

Smaller societies that use AV include the Society for Judgment and Decision Making, the Social Choice and Welfare Society, the International Joint Conference on Artificial Intelligence, and the European Association for Logic, Language and Information.

Additionally, the Econometric Society has used AV (with certain emendations) to elect fellows since 1980 (Gordon 1981); likewise, since 1981 the selection of members of the National Academy of Sciences (1981) at the final stage of balloting has been based on AV. Coupled with many colleges and universities that now use AV – from the departmental level to the school-wide level – it is no exaggeration to say that several hundred thousand individuals have had direct experience with AV.

---

*Reprinted with permission from Brams and Fishburn (2005); see also Brams (2004), which includes more recent studies on AV and other voting systems that use an AV ballot.

S.J. Brams (✉)
Department of Politics, New York University, New York, NY 10012, USA
e-mail: steven.brams@nyu.edu

J.-F. Laslier and M.R. Sanver (eds.), *Handbook on Approval Voting*, Studies in Choice and Welfare, DOI 10.1007/978-3-642-02839-7_3,

Probably the best-known official elected by AV today is the secretary-general of the United Nations (Brams and Fishburn 1983). AV has also been used in internal elections by the political parties in some states, such as Pennsylvania, where a presidential straw poll using AV was conducted by the Democratic State Committee in 1983 (Nagel 1984).

Bills to implement AV have been introduced in several state legislatures (see Section 3.2). In 1987, a bill to enact AV in certain statewide elections passed the Senate but not the House in North Dakota. In 1990, Oregon used AV in a statewide advisory referendum on school financing, which presented voters with five different options and allowed them to vote for as many as they wished (Wright 1990).

In the late 1980s, AV was used in some competitive elections in countries in Eastern Europe and the Soviet Union, where it was effectively "disapproval voting," because voters were permitted to cross off names on ballots but not to vote for candidates (Shabad 1987; Keller 1987, 1988; White 1989; Federal Election Commission 1989). But this procedure is logically equivalent to AV: candidates not crossed off are, in effect, approved of, although psychologically there is almost surely a difference between approving and disapproving of candidates.

With this information as background, we trace in Section 3.2 our early involvement, and that of several associates, with AV. After outlining the arguments we and others have made for AV, we discuss in Section 3.3 how AV came to be adopted by the different societies.

In Section 3.4, we report on empirical analyses of ballot data of some professional societies that adopted AV; they help to answer the question of when AV can make a difference in the outcome of an election. In Section 3.5, we investigate the extent to which AV elects "lowest common denominators." In Section 3.6, we discuss whether voting is "ideological" under AV.

The confrontation between theory and practice offers some interesting lessons on "selling" new ideas. The rhetoric of AV supporters has been opposed not only by those supporting extant systems like plurality voting (PV) – including incumbents elected under PV – but also by those with competing ideas, particularly proponents of other voting systems like the Borda count and the Hare system of single transferable vote.

We conclude that academics probably are not the best sales people for two reasons: (1) they lack the skills and resources, including time, to market their ideas, even when they are practicable; and (2) they squabble among themselves. Because few if any ideas in the social sciences are certifiably "right" under all circumstances, squabbles may well be grounded in serious intellectual differences. Sometimes, however, they are not.

3.2 Early History and Rhetoric

In 1976, one of us (Brams) was attracted by the concept of "negative voting" (NV), proposed in a brief essay by Boehm (1976) that was passed on to me by the late Oskar Morgenstern. Under NV, voters can either vote for one candidate or against one candidate, but they cannot do both. Independently, Robert J. Weber had begun working on AV (he was apparently the first to coin the term "approval voting").

When Brams and Weber met in the summer of 1976 at a workshop at Cornell University under the direction of William F. Lucas, it quickly became apparent that NV and AV are equivalent when there are three candidates. Under both systems, a voter can vote for just one candidate. Under NV, a voter who votes against one candidate has the same effect as a voter who votes for the other two candidates under AV. And voting for all three candidates under AV has the same effect as abstaining under both systems.

When there are four candidates, however, AV enables a voter better to express his or her preferences. While voting against one candidate under NV has the same effect as voting for the other three candidates under AV, there is no equivalent under NV for voting for two of the four candidates. More generally, everything that a voter can do under NV he or she can do under AV, but not vice versa, so AV affords voters more opportunity to express themselves.

Brams and Weber wrote up their results separately, as did three other analysts who worked independently on AV in the 1970s (discussed in Brams and Fishburn 1983; see also Weber 1995). But the *idea* of AV did not spring forth full-blown only about 25 years ago; its provenance is much earlier. Indeed, AV was actually *used*, beginning in the 13th century, in Venice (Lines 1986) and in papal elections (Colomer and McLean 1998); it was also used in elections in 19th-century England (Cox 1987), among other places.

In the summer of 1977, after we met at a conference on Hilton Head Island, SC, under the direction of James S. Coleman, we began a long collaboration, which resulted in one book (Brams and Fishburn 1983) and many articles on AV and other voting procedures (Brams and Fishburn 2002).

Our first article (Brams and Fishburn 1978) was a formal analysis of the properties of AV that included, as an illustration, its application to the 1968 U.S. presidential election, in which there were three significant candidates (Richard M. Nixon, Hubert H. Humphrey, and George Wallace). Our analysis of this election was based on empirical research of Brams's former Yale student, D. Roderick Kiewiet (1979), who showed that Nixon's popular-vote and electoral-vote victory in 1968 would have been much more substantial under AV than it was under PV.[1]

Even at this early stage AV generated academic controversy (Tullock 1979; Brams and Fishburn 1979), which we will say more about later. Nevertheless, we became convinced that AV is a simple and practicable election reform that could ameliorate, if not solve, serious problems in multicandidate elections.

Brams began a "campaign" in 1979 to get it adopted in public elections, beginning with New Hampshire's first-in-the-nation presidential primaries in February 1980, which had multiple candidates running in both the Democratic and Republican primaries. Although his efforts received both national coverage (e.g., in the

[1] For other retrospective studies of elections, including the 1992 presidential election involving Bill Clinton, George Bush, and Ross Perot, see the citations in Brams and Fishburn (2002).

New York Times and *Los Angeles Times*) and in several New Hampshire newspapers (e.g., the *Manchester Union-Leader* and *Concord Monitor*), he was not successful in getting an AV bill out of committee, despite being a native of New Hampshire ("prodigal son returns"), testifying before Senate and House committees in New Hampshire's General Court (legislature), and meeting with the governor. Later testimony Brams gave before legislative committees in other states (e.g., New York and Vermont) was similarly unavailing in effecting reform.

Arguments we and others have made for AV proved more persuasive in convincing professional societies to adopt AV. Our rhetoric has remained relatively constant over the years and can be summarized by the following six propositions:

1. AV gives voters more flexible options. They can do everything they can under PV – vote for a single favorite – but if they have no strong preference for one candidate, they can express this fact by voting for all candidates they find acceptable. In addition, if a voter's most-preferred candidate has little chance of winning, then that voter can vote for both a first choice *and* a more viable candidate without worrying about wasting his or her vote on the less popular candidate.
2. AV helps elect the strongest candidate. Under PV, the candidate supported by the largest minority often wins, or at least makes the runoff if there is one. Under AV, by contrast, the candidate with the greatest overall support will generally win. In particular, *Condorcet candidates*, who can defeat every other candidate in separate pairwise contests, almost always win under AV, whereas under PV they often lose because they split the vote with one or more other centrist candidates.
3. AV will reduce negative campaigning. AV induces candidates to try to mirror the views of a majority of voters, not just cater to minorities whose votes could give them a slight edge in a crowded plurality contest. AV is therefore likely to cut down on negative campaigning, because candidates will have an incentive to broaden their appeals by reaching out for approval to voters who might have a different first choice. Lambasting such a choice, rather than being more expansive, risks alienating this candidate's supporters, thereby losing their approval.
4. AV will increase voter turnout. By being better able to express their preferences, voters are more likely to vote in the first place. Voters who think they might be wasting their votes, or who cannot decide which of several candidates best represents their views, will not have to despair about making a choice.[2] By not being forced to make a single – perhaps arbitrary – choice, they will feel that the election system allows them to be more honest, which will make voting more meaningful and encourage greater participation in elections.

[2] Perhaps the best recent example of voters who faced this dilemma were supporters of Ralph Nader in the 2000 U.S. presidential election. Although Nader received less than 3% of the popular vote in this election, polls show that if his supporters could have voted for a second choice, Al Gore would have been the choice of most. Thereby Gore would have won Florida and its electoral votes, making him rather than George W. Bush the winner.

5. AV will give minority candidates their proper due. Minority candidates will not suffer under AV: their supporters will not be torn away simply because there is another candidate who, though less appealing to them, is generally considered a stronger contender. Because AV allows these supporters to vote for *both* candidates, they will not be tempted to desert the one who is weak in the polls, as under PV. Hence, minority candidates will receive their true level of support under AV, even if they cannot win. This will make election returns a better reflection of the overall acceptability of candidates, relatively undistorted by strategic voting, which is important information often denied to voters today.
6. AV is eminently practicable. Unlike more complicated ranking systems, which suffer from a variety of theoretical as well as practical defects, AV is simple for voters to understand and use. Although more votes must be tallied under AV than under PV, AV can readily be implemented on existing voting machines. Because AV does not violate any state constitutions in the United States (or, for that matter, the constitutions of most countries in the world), it requires only an ordinary statute to enact.

Voting systems that involve ranking candidates may appear, at first blush, to be more appealing than AV. One, the Borda count or Borda voting (BV), awards points to candidates according to their ranking. Another, the Hare system of single transferable vote (STV; also called the "alternative vote" or "instant runoff"), progressively eliminates candidates with the fewest first-choice votes and transfers their votes to second choices – and lower choices if necessary – until one candidate emerges with a majority.

Proponents of AV argue that these systems have serious drawbacks. BV fosters "insincere voting" – when, for example, a voter moves a second choice down to last place to minimize that candidate's threat to his or her top choice – and is also vulnerable to "irrelevant candidates," who cannot win but can affect the outcome. STV may eliminate a centrist candidate early on and thereby elect one less acceptable to a majority. In addition, STV suffers from "nonmonotonicity," in which voters, by raising the ranking of a candidate, may actually cause that candidate to lose – just the opposite of what one would want to happen.

PV is also vulnerable to insincere voting, whereby a voter may switch to a second choice if his or her first choice appears to be a long shot, as indicated, for example, by polls. While AV encourages sincere voting – voting for all candidates above the lowest-ranked candidate one considers acceptable – it does not eliminate strategic calculations altogether. Because approval of a less-preferred candidate can hurt a more-preferred candidate, the voter still faces the decision under AV of where to draw the line between acceptable and nonacceptable candidates.

The pros and cons of AV vs. other voting systems have been debated over the last twenty years in numerous publications.[3] But this is not the subject of this paper,

---

[3] For a sampling of this debate, see Arrington and Brenner (1984) and Brams and Fishburn (1984); Niemi (1984, 1985) and Brams and Fishburn (1985); Saari and Van Newenhizen (1988), Brams et al. (1988); Brams and Fishburn (2001) and Saari (2001a); and Brams and Herschbach (2001a,

except insofar as the rhetoric has influenced the history of adoptions (and nonadoptions) of AV.[4] We next discuss the adoption decisions of the first societies to use AV in the late 1980s.

3.3 The Adoption Decisions in the Societies[5]

Elections are not a burning issue in most scientific societies, with participation rates often considerably below 50% of the membership and sometimes closer to about 10%. For the candidates, on the other hand, who are often luminaries in their disciplines, outcomes are usually more consequential and sometimes represent, especially if the office is president, recognition of professional achievements over one's career.

It is not surprising, then, that candidates are willing to make subdued versions of what, in political life, would be called campaign statements. In the more rarefied atmosphere of an academic or professional society, these statements, which usually accompany a mailed ballot, tend more to emphasize broad goals than specific programs, although candidates often pledge to undertake new initiatives. Most candidates, while listing their past offices and qualifications for the new office, generally do not seek to disparage the opposition.

Genteel as most of these campaigns are, candidates do, nonetheless, try to garner support by highlighting their qualifications, and proposing new approaches or ideas, that differentiate them from their opponents. When AV was first proposed as a reform in the four societies that adopted AV in the late 1980s, no candidates or factions, with one major exception, identified AV as a threat either to their candidacies or points of view.

Of course, after AV's use, there are winners and losers, and some losers, undoubtedly, see themselves as victims of this reform. In one society (The Institute of Management Sciences, or TIMS, before it merged with the Operations Research Society of America, or ORSA, to become INFORMS), this logic worked in reverse:

2001b) and Richie et al. (2001). Recent popular accounts of the controversy over voting systems by science writers include MacKenzie (2000), Guterman (2002), Klarreich (2002), and Begley (2003).

[4] Donald G. Saari has been a proponent of BV, most recently in Saari (2001b), but we know of no recent adoptions of BV, though it and a variant have been used in two small Pacific Island countries, beginning about 30 years ago (Reilly 2002). Proponents of instant runoff voting (IRV), based on STV, recently succeeded in getting it enacted in elections in San Francisco; they formed an organization, the Center for Voting and Democracy (CV&D), which now has a staff of about ten people that includes the authors of Richie et al. (2001) and Hill (2002). As noted in Brams and Herschbach (2001a), IRV supporters have done little serious analysis to back up their claims, although other studies of STV (e.g., Dummett 1984) have been more probing. On the other hand, CV&D does have human and monetary resources that few academics can claim.

[5] This and the next two sections are based on Brams and Fishburn (1992a) as well as earlier and later studies that we cite.

the winner under PV, before AV was adopted, would almost certainly have lost under AV – and this became an argument made for the adoption of AV!

We hasten to add that this argument against PV was not a personal argument directed against the PV winner. Rather, the argument was that another candidate commanded broader support and thereby "deserved" to win.

Next we briefly recount the adoption decisions of the first four societies to use AV:

1. Mathematical Association of America (MAA). In 1985, the president of the MAA, Lynn Arthur Steen, who was familiar with work on AV, asked the Board of Governors of the MAA to consider adoption of AV in its biennial elections for president-elect and other national offices. After "heated but not acrimonious" debate (Steen 1985), AV was approved by the Board in 1985, passed by the membership in 1986, and used for the first time in the 1987 MAA elections.

   Steen earlier had written an article in *Scientific American* (Gardner 1980) on the mathematics of elections, in which he discussed AV. Before the MAA's consideration of AV, he asked Brams to look into the use of STV by the American Mathematical Society (AMS), the major research society of mathematicians.[6] Brams (1982) demonstrated via two counterexamples that the "Instructions to Voters" accompanying the 1981 ballot used by the AMS to elect a nominating committee contained an erroneous statement about a property of STV, which led to an exchange with Chandler Davis (1982), who had been a proponent of STV when it was adopted by the AMS several years earlier. The erroneous statement was deleted from future instructions, but AV was not adopted by the AMS until 1992.[7]

   Both Steen's knowledge and his position as president of the MAA made him a crucial player in the MAA's adoption of AV. So, also, was Steen's successor as president of the MAA, Leonard Gillman, who was a strong advocate of AV and played an active role in its eventual implementation in the 1987 elections of the Association. For example, he wrote a description of AV for mathematicians, which included results of his own analysis (Gillman 1987).
2. The Institute of Management Sciences (TIMS), which is now part of INFORMS. The use of AV by TIMS in 1988 was preceded by an experiment in which members were sent a nonbinding AV ballot, along with the regular PV ballot, in the 1985 elections. Although the AV ballot did not count, 85% of the members who voted in these elections returned the AV ballot. This permitted Fishburn and Little (1988) to compare the results of voting under the two different systems.

[6] The MAA is the more teaching-oriented of the two major American mathematical societies at the college-university level.

[7] It was adopted in part because counting votes by hand under STV proved to be too onerous, and computerizing the counting was not feasible at the time. Even so, AV was adopted only for those offices of the AMS that did not require an amendment to the bylaws, which would have required considerable effort to enact; voting for other offices is still by PV (Daverman 2002, and Fossum 2002). Patently, pragmatic considerations played a key role in the AMS's choices.

On the basis of their empirical analysis, which will be discussed later, Fishburn and Little (1988) concluded that AV did a better job of electing Condorcet candidates than did PV. Not only was the experiment "remarkably successful" (Little and Fishburn 1986), but the results also convinced TIMS Council to adopt AV in 1987, leading to its later adoption by INFORMS when it formed in 1995. In fact, an argument for conducting the experiment in the first place was that management scientists should "practice what we preach" (Jarvis 1984): before deciding on its usage, TIMS should collect the information necessary to make an informed judgment about the applicability of the theoretical analysis of AV to its own elections.

Both the consideration and adoption of AV by TIMS were certainly helped by the fact that the president of TIMS in 1984–1985, John D. C. Little, was interested in AV and collaborated with Fishburn on the experiment and its analysis. Before undertaking the experiment, inquiries were made of the candidates to ask their permission to participate in it. Because of its research potential, all agreed, prefiguring AV's eventual adoption.

3. American Statistical Association (ASA). The former chair of the ASA's Committee on Elections, Richard F. Potthoff, had read about AV and brought it to the attention of his committee. This committee recommended its adoption first in "internal" ASA elections; the ASA Board of Directors approved this recommendation.

   After AV's successful use in 1986 in three elections for Council governors, the election of two editors to serve on the Board, and the election of a Board member to serve on the Executive Committee, the Committee on Elections recommended that AV be used in Association-wide elections, which was approved by the Board (Amendment to ASA By-Laws 1987) and ratified as an amendment in 1987. Unlike the other societies, the ASA has had no Association-wide multicandidate elections since the adoption of AV, though some internal elections and single-winner section elections have had more than two candidates.

4. Institute of Electrical and Electronics Engineers (IEEE). The adoption of AV by the IEEE has a politically charged history (Brams and Nagel 1991). Beginning in 1984, AV was considered, along with other voting systems, for possible use in multicandidate elections. But not until the 1986 elections – when a petition candidate, Irwin Feerst, ran against two candidates for president-elect who were nominated by the Board of Directors – did the issue of election reform take center stage. The reason is that Feerst, with 35% of the vote, defeated one of the two Board-nominated candidates and came within 242 votes (of 52,405 cast) of defeating the other candidate. This result starkly illustrated to the Board how vulnerable their nominees, who together might win a substantial majority in an election, are to a minority candidate if these nominees should split the majority vote more or less evenly.

   In 1987 the Board reverted to nominating only one candidate for president-elect, breaking a tradition of nominating two candidates that it had begun in 1982. Feerst was instrumental in bringing the question of how many nominees the Board must nominate to a vote of the entire membership in the 1987 election, in which he did not run and there were no other petition candidates. By a 57-percent majority, members supported a constitutional amendment requiring that the Board nominate at least two candidates, but this fell short of the 2/3's majority needed to amend the IEEE's constitution.

Nevertheless, it was clear that there was strong member support for making IEEE elections more competitive, which renewed interest in AV should the Board return to nominating two candidates and have petition candidates run as well. In 1987, Brams was invited by the then president of the IEEE, Henry L. Bachman, to attend an Executive Council meeting to discuss AV.

Unable to do so, he suggested that Jack H. Nagel of the University of Pennsylvania, who had done extensive research on AV, take his place. Nagel did; he also attended a later meeting of the full Board of Directors, which adopted AV in November 1987. (AV had previously been used in internal IEEE elections, sometimes in modified form.) With its adoption, the Board voted to nominate at least two candidates for each office.

When the IEEE's adoption of AV was announced at a December 1987 IEEE press conference in New York City that Brams and Nagel attended, Feerst objected strenuously to its use, arguing that it was a deliberate move to undermine his candidacy and the interests of "working engineers," whom he claimed to represent. When Feerst ran in 1988 for president-elect under AV, he came in fourth in a field of four candidates.

To recapitulate, the paths to adoption of AV in the different societies have been diverse. Only in the MAA did full-scale use of AV begin before it was first tried out in an experiment (TIMS) or in internal elections (ASA and IEEE).

The presidents of the MAA, TIMS, and the IEEE played active roles in AV's adoption in their societies, and each received assistance from an advocate of AV. In the ASA, on the other hand, it was writings on AV that sparked initial interest, which turned into adoption without much controversy.

Controversy was the hallmark of the IEEE deliberations. While the IEEE's adoption of AV was in part a response to a perceived threat to its established leadership, it is important to realize that the IEEE did not view it as its only alternative.

In fact, several other election systems had been considered before AV was selected. For example, a runoff election between the two top contenders, if neither received a majority in the initial balloting under PV, was also seriously considered, but it was viewed as too costly to have a second round of voting and also would have required a constitutional change. Ultimately, a majority of Board members concluded that AV better fit the needs of the organization than any other voting system, and that is why it was adopted.[8]

[8] By no means do we suggest that AV is a panacea in all elections, especially those involving multiple winners; for such elections, see the AV-related reforms in Brams (1990), Fishburn and Brams (1991), Brams and Fishburn (1992b), and Potthoff and Brams (1998).

This quick overview does not do justice to the serious debates that occurred over the merits of AV, particularly in the MAA and the IEEE. Indeed, although there has been dissent over AV's use in some societies (Kiely 1991), no society that adopted AV ever rescinded its decision, with one notable exception (the IEEE).[9] Looking at what has AV wrought in them may offer some explanation of why it has been generally, but not universally, accepted.

3.4 Does Approval Voting Make a Difference?

Clearly, a new voting procedure makes a difference if it leads to the selection of a different winner. The best evidence we have that AV would have elected a different winner is from the 1985 TIMS experiment, in which ballot data for both the PV official elections and the AV nonbinding elections were compared (Fishburn and Little 1988).

In one of the three 1985 elections, the official PV and actual AV ballot totals are shown in Table 3.1 for candidates A, B, and C. Also shown are the AV totals extrapolated from the 85-percent sample of members who returned their AV non-binding ballots, which is a very high figure. The extrapolation is a straightforward one: approval votes are added to the actual AV totals for each candidate based on the propensity of the sample respondents who voted for one particular candidate on the PV ballot to vote for each of the other candidates on the AV ballot. This extrapolation is justified by the finding that there are no major differences in voting patterns on the official PV ballot between AV respondents and nonrespondents.

Observe that candidate C wins the official PV election by a bare eight votes (0.4%), but B would have won under AV by a substantial 170 votes (6.1%). By itself, the fact that C wins more plurality votes and B wins more approval votes does not single out one candidate as the manifestly preferred choice. But on the experimental ballot, voters were asked one piece of additional information: to rank the candidates from best to worst by marking next to their names 1) for their first choice, 2) for their second choice, and so on.

These data can be used to reconstruct who would defeat whom in hypothetical pairwise contests, which is not evident from the PV totals. For example, the fact that C edges out B in presumed first choices, based on the PV totals, does not mean that C would hold his or her lead when the preferences of the 166 A voters are taken into account. In fact, the experimental ballots of these 166 voters show that

[9] According to the IEEE Executive Director, Daniel J. Senese, AV was abandoned in 2002 because "few of our members were using it and it was felt that it was no longer needed." Brams responded in an e-mail exchange (June 2, 2002) that since "candidates now can get on the ballot with 'relative ease' [according to former IEEE president Henry L. Bachman in the same e-mail exchange] ... the problem of multiple candidates [in the late 1980s] might actually be exacerbated ... and come back to haunt you [IEEE] some day."

Table 3.1 PV and AV vote totals in 1985 TIMS election

Candidates	Official PV	Actual AV	Extrapolated AV
A	166	417	486
B	827	1,038	1,224
C	835	908	1,054
Total	1,828	2,363	2,764
No. of voters	1,828	1,567	1,828

1. 70 provided rankings in the order ABC;
2. 66 provided rankings in the order ACB;
3. 3 provided no rankings but approved both A and B;
4. 27 made no distinction between B and C by rankings or approval.

In the B-vs.-C comparison, it is reasonable to credit (1) and (3) to B (73 votes), (2) to C (66 votes), and (4) to neither candidate. When added to the PV totals, these credits give C (901 votes) exactly one more vote than B (900 votes). However, assuming the 27 voters in (4) split their votes between B and C in the pattern of the 139 voters (70 + 66 + 3) who ranked A first and also expressed a preference between B and C, B would pick up an additional vote (rounded to the nearest vote), resulting in a 914–914 tie.

This extrapolation indicates that there is not a single Condorcet candidate.[10] While surprising, the lack of a single Condorcet candidate should not obscure the fact that 170 more voters approved of B rather than C in the extrapolated AV returns, albeit C won the PV contest by eight votes.

The reason for this discrepancy between the AV and PV results is that whereas C has slightly more *stalwart* supporters (i.e., those who vote only for one candidate) than B, supporters of the third candidate, A, more approve of B than C (36% to 23%). Furthermore, because more of C's supporters approve of B than B's do of C, B would have won handily under AV.

Is this desirable? In the absence of a Condorcet candidate, Fishburn and Little (1988, pp. 559–560) concluded that approval voting picks a clear winner on the basis of second choices. These show that B has a broader acceptance in the electorate than C. Therefore, the approval process, by eliciting more information from the voters, leads to the election of the candidate with the widest support.

[10] It is worth noting that the usual reason for the nonexistence of a Condorcet candidate is because of a Condorcet paradox, whereby majorities cycle. In this election, however, it is a projected tie that precludes one candidate from defeating the others in pairwise contests. That there is no cycle, and that A in fact would lose to both B and C, is shown by ranking data in Fishburn and Little (1988).

Although it is theoretically possible in close elections that the Condorcet candidate will not be the most approved candidate, it has almost never occurred.[11] But the legitimacy of the AV winner may be questioned on other grounds.

3.5 Does Approval Voting Elect the Lowest Common Denominator?

One fear that has been expressed about the use of AV is that while it may help elect candidates more broadly representative than PV, these candidates could turn out to be rather bland and uninspiring. They may win simply because they offend the fewest voters, not because they excite the passions of many.

It is difficult to say whether, in principle, a compromise candidate is a better or worse social choice than a more extreme candidate who is the darling of some voters but the bane of others. In practice, fortunately, this dichotomous choice seems rarely to arise, as the data from the AV elections of the four societies demonstrate. Specifically, the winners under AV were candidates who were generally popular among *all* voters, however many candidates they voted for in the different elections. Thus, a divergence between forceful minority candidates, approved of by few, and "wishy-washy" majority candidates, approved of by many, is probably an infrequent event.

There are, however, examples of elections in which the winner was not strong among all classes of voters. Consider the 1987 MAA election shown in Table 3.2 (Brams et al. 1988), wherein the votes received by the five candidates in this election are broken down by the votes each of the candidates received from voters' casting exactly one vote (1-voters), voters' casting exactly two votes (2-voters), and so on. Excluded from these totals are nine voters who voted for all the candidates, whose undifferentiated support obviously has no effect on the outcome.

[11] The 1999 election for president of the Social Choice and Welfare Society, which was decided by two approval votes among 76 cast, is the only exception we know of: the second-place AV candidate in this election would have defeated the AV winner by four votes in a head-to-head contest, based on the hypothetical use of BV, for which voters ranked candidates. Brams and Fishburn (2001) deem this "nail-biting" election essentially a toss-up, whereas Saari (2001a) argues that most positional methods would have chosen the Condorcet candidate (including BC, wherein the Condorcet winner would have defeated the AV winner 60–59); see Laslier (2003a) for more details on voting patterns in this election. Regenwetter and Grofman (1998), using a random-utility model to reconstruct voter preferences in several elections – including some discussed here – show that AV, BV, and Condorcet winners generally coincide. Laslier (2003b) and Laslier and Van der Straeten (2003) analyze data from a field experiment with AV in the 2002 French presidential election, which involved over 5,000 voters in two French towns, and conclude that AV was easily understood, readily accepted, and provided a more complete picture of the "political space." Earlier theoretical analyses as well as computer simulations (Brams and Fishburn 1983; Lijphart and Grofman 1984; Nurmi 1987; Merrill 1988) demonstrate that AV almost always elects a Condorcet winner if there is one. If there is not one, as in the 1985 TIMS election experiment, then proponents of AV argue that AV provides a compelling way to break either a cycle or a tie.

Table 3.2 AV vote totals in 1987 MAA election

Candidates	1-Voters	2-Voters	3-Voters	4-Voters	Total
A	848	276	122	21	1,267
B	618	275	127	32	1,052
C	652	264	134	34	1,084
D	660	273	118	31	1,082
E	303	132	87	30	552
Total	3,081	1,220	588	148	5,037
No. of voters	3,081	610	196	37	3,924

In this election, 3,081 of the 3,924 voters (79%) were 1-voters, while the remaining 843 voters cast 1,956 votes, or an average of 2.3 votes each. Thus, the multiple voters cast 39% of the votes, though they constituted only 21% of the electorate.

Did the multiple voters make a difference? It would appear not, because the winner (A) received 28% more votes from 1-voters than the 1-voters' runner-up (D) did, just edged out B among 2-voters, but lost to several candidates among 3-voters and among 4-voters. A's victory, then, is largely attributable to the substantial margin received from 1-voters, not from the presumably more lukewarm support received from multiple voters.

Define a candidate who wins among all classes of voters – those who cast few votes (narrow voters) and those who cast many votes (wide voters) – as *AV-dominant*. In the MAA election, we assume narrow voters are those who cast one or two votes, and wide voters are those who cast three or four votes.

It turns out that A is not AV-dominant, because he or she wins among narrow but not among wide voters. Does this vitiate A's winning status? In winning so decisively among 1-voters, whose preference intensities would seem to be greatest, it would be hard to argue that A is any kind of lowest common denominator. It should be noted, however, that some of the 37 voters who voted for four of the five candidates probably also had intense preferences – but against the one candidate they chose to leave off their approved lists.

In 12 of the 16 multicandidate AV elections analyzed in the four societies, the winners were AV-dominant. In the four elections in which there was not an AV-dominant winner, the pattern is similar to that in the 1987 MAA election shown in Table 3.2: the winner won by virtue of receiving greater support among narrow voters than among wide voters. These AV-nondominant winners, therefore, do not fit the mold of lowest common denominators – the choice of many wide voters but few narrow voters – but rather the opposite, which reinforces, not undermines, their legitimacy as winners.

The fact that the winners in three-quarters of the elections were AV-dominant is perhaps not surprising, because one would expect such candidates would do better than losers across different types of voters. A little reflection, however, shows that this need not be the case. Paradoxically, a candidate may lose among every possible class of voters – that is, be *AV-dominated* – and still be the AV winner. For example, A might be the victor over C among narrow voters, and B might be the victor over

C among wide voters. But C could emerge as the AV winner if A did badly among wide voters, B did badly among narrow voters, but C was a close second among both types.

No winners in the 16 elections were AV-dominated. As already noted, even the support of the four AV-nondominant winners appeared to be more intense and heartfelt (i.e., from narrow voters) than that of the losers, so AV does not appear to elect lowest common denominators.

3.6 Is Voting Ideological?

Consider again the 1987 MAA election. As can be calculated from Table 3.2, 2-voters gave the candidates 22–26% of all their votes, 3-voters 10–16%, and 4-voters 2–5%. Venn diagrams (not shown here) indicate the shared support among the 10 subsets of two candidates, 10 subsets of three candidates, five subsets of four candidates, and one of all five candidates. Examination of the *sources* of this support, as shown in the Venn diagrams, does not reveal any particular pairs, triples, or quadruples that received unusually great support, indicating that there was not obvious coalitional voting.

On the contrary, multiple votes are spread about as one would expect according to the null hypothesis that votes are distributed in proportion to the candidates' totals. In the case of A, for example, there were 82 shared votes with just B, 91 with just C, 80 with just D, and 23 with just E, which is roughly in accord with the candidates' overall totals. Indeed, every one of the 32 subsets in this election – including the 2.6% who abstained – got at least three votes.

The story is very different for the 1988 IEEE election shown in Table 3.3 (Brams and Nagel 1991), wherein the approval vote totals are shown for all 16 subsets of the four candidates in this race. Consider first the 3-voters, and note that nearly everyone in this category voted for ABD – 5,605 voters, to be precise. By contrast, only 148, 143, and 89 voters, respectively, supported the other 3-subsets of ABC, ACD, and BCD that contain C.

Evidently, the numerous supporters of ABD voted against C by voting for everybody except C. This essentially negative kind of voting against C can also be seen in

Table 3.3 Numbers of voters who voted for 16 different subsets in 1988 IEEE election and AV totals

Subsets					
None = 1,100					
A = 10,738	B = 6,561	C = 7,626	D = 8,521		
AB = 3,578	AC = 659	AD = 6,679	BC = 1,425	BD = 1,824	CD = 608
ABC = 148	ABD = 5,605	ACD = 143	BCD = 89		
All = 523					
Totals					
A = 28,073	B = 19,753	C = 11,221	D = 23,992		

voting for the six 2-subsets. The three 2-subsets that do not include C (AB, AD, and BD) had an average of 4,027 voters each, whereas the three that included C (AC, BC, and CD) had an average of only 897 voters each.

In addition to the predominant clustering of support around A, B, and D, there are some subtle differences in the sharing of support. For each pair of candidates, Brams and Nagel (1991) computed an index of shared support by taking the ratio of ballots approving both candidates by 2-voters and 3-voters to total ballots, excluding abstentions and votes for all four candidates. By this measure, A and D have the most affinity, with 22.9% shared support. They are followed by A and B, with 17.2%; and then by B and D, with 13.9%. Although A, B, and D share much less support with C, B at 3.1% shares slightly more with C than do A (1.8%) and D (1.5%).

From these results, one might infer an underlying dimension on which D and C occupy opposite extremes, whereas A and B are located at intermediate positions. A is somewhat closer than B to D, but both B and A are much closer to D than to C, as shown in the following hypothetical continuum:

This representation corresponds to certain facts about the candidates. D and A were both Board nominees, whereas C was a vociferous critic of IEEE officers, Board, and staff. B, though like C a petition candidate, was in other ways close to the IEEE establishment, having previously served on the Board. As for the slight distinction between D and A, judging from the candidates' biographies and statements it may reflect D's emphasis on technical research, which perhaps made him seem most distant from C, who sought to champion the working engineer.

Of the 54,204 ballots analyzed in this election, only 3,323 (6.1%) are "inconsistent" with the assumption that voters' preferences are based on the foregoing DABC ordering of candidates. *Inconsistent ballots* include approval of two nonadjacent candidates without including the adjacent candidate(s) between them, notably DC (608), AC (659), DAC (143), and DBC (89). Accounting for more than half the inconsistencies is the relatively minor inconsistency – in terms of perceived differences – represented by the pattern DB (1,824). Of the multiple voters, 17,435 (84.0%) cast ballots consistent with the hypothetical ordering.

Thus, candidates with obvious affinities tended disproportionately to share approval from multiple voters. In this sense voting was ideological: it reflected a pattern consistent with an underlying ordering of the candidates. Only in this election, however, was such a pattern found; far more typically, voting in the societies is nonideological, which is consistent with the null hypothesis alluded to earlier. But if AV is used in public elections, their more political character could well lead to the kind of ideological cleavages observed in the IEEE election.

It is important to note, however, that nonideological voting may mirror regularities not evident in the AV data themselves. As a case in point, the winner in the 1987 MAA election (Table 3.2) was a woman, and this pattern was repeated in the next MAA election in 1989. We have not analyzed data from the latter election, but the 1987 winner's victory, as shown earlier, cannot be impeached on grounds that she won mostly because of lukewarm support from wide voters. Nonetheless, as the only women in each of the two races, it may be the case that they were helped by their uniqueness: by some they were perceived as the single best choice; by others they were seen as broadly acceptable.

3.7 Summary and Conclusions

AV has proved to be a practical and viable election reform in the four scientific and engineering societies that used it for the first time in 1987 and 1988. While AV supporters played a role in its adoption in three of the four societies (TIMS, MAA, and IEEE), none of its proponents was even aware of its consideration in the fourth society (ASA) until its adoption was imminent.

In all these societies, AV's adoption rested principally on the arguments – summarized earlier – that it is preferable to PV in multicandidate races. In the IEEE, a petition candidate's near-win with vocal but only minority support certainly gave urgency to these arguments, accelerating AV's adoption after the Board's attempt to limit the number of Board-nominated candidates to one person met with the membership's disapprobation. Only in the case of the AMS's 1992 adoption of AV did practical considerations give it an edge over STV, and then only in some elections that were relatively easy to change.

The empirical analyses of election returns from the different societies indicate that AV may make a difference. So far it seems not to have elected candidates who can be characterized as lowest common denominators but instead candidates who either enjoyed support among all classes of voters, or who did particularly well among narrow voters whose support is presumed to be more intense. Although voting seems generally nonideological in most society elections, a clear ordering of positions was identified in the IEEE election, and voting tended to be only for adjacent candidates in this ordering.

Condorcet candidates almost always win under AV, with the only known exception being the 1999 Social Choice and Welfare election, which was a near-tie under both AV (the official procedure) and BV (the hypothetical procedure). If there is no single Condorcet candidate, as was illustrated in the 1985 TIMS election experiment, then AV provides a way of determining which candidate receives the most support from all voters, not just those who rank this person first.

Not all societies that have been approached about adopting AV, including three that Brams belongs to – the American Political Science Association (APSA), the International Studies Association (ISA), and the Public Choice Society (PCS) – have been amenable to election reform, much less the adoption of AV. Significantly, these societies are dominated, or heavily populated by, academic political scientists; none holds competitive elections unless a petition candidate challenges the official slate (this has never happened in the ISA or PCS; in the APSA, the last challenges occurred more than 25 years ago).

Among the lessons we draw from our experience is that the adoption of AV, and probably any election reform, requires key support from within an organization. We never received this kind of support from politicians or political parties in our attempts to get AV adopted in public elections. By contrast, the society adoptions would not have occurred without influential members of each society favoring reform, sometimes for practical or political reasons. Of course, they also needed to make their cases with arguments based on democratic principles; we like to believe that both the rhetoric of AV supporters as well as their analyses helped in this regard.

Acknowledgements We thank Richard F. Potthoff for valuable comments. Brams gratefully acknowledges the support of the C.V. Starr Center for Applied Economics at New York University.

References

Amendment to ASA By-Laws (1987) AMSTAT News 135:1

Arrington TS, Brenner S (1984) Another look at approval voting; Arrington and Brenner to Brams and Fishburn. Polity 17(1):118–134, 144

Begley S (2003) Why we sometimes get tofu for president when we want beef. Wall Street J B1

Boehm GAW (1976) One fervent vote against wintergreen. Preprint

Brams SJ (1982) The AMS nomination procedure is vulnerable to truncation of preferences; Rejoinder [to Chandler Davis]. Not Am Math Soc 29(2):136–138

Brams SJ (1988) MAA elections produce decisive winners. Focus: the newsletter of the mathematical association of America 8(3):1–2

Brams SJ (1990) Constrained approval voting: a voting system to elect a governing board. Interfaces 20(5):67–80

Brams SJ (2008) Mathematics and democracy: designing better voting and fair-division procedures. Princeton University Press, Princeton, NJ

Brams SJ, Fishburn PC (1978) Approval voting. Am Polit Sci Rev 72(3):831–847

Brams SJ, Fishburn PC (1979) Reply [to Gordon Tullock]. Am Polit Sci Rev 73(2):552

Brams SJ, Fishburn PC (1983) Approval voting. Birkhäuser Boston, Cambridge, MA

Brams SJ, Fishburn PC (1984) A careful look at another look at approval voting. Polity 17(1): 135–143

Brams SJ, Fishburn PC (1985) Comment on the problem of strategic voting under approval voting; Rejoinder to Niemi. Am Polit Sci Rev 79(3):816–819

Brams SJ, Fishburn PC (1988) Does approval voting elect the lowest common denominator? PS Polit Sci Polit 21(2):277–284

Brams SJ, Fishburn PC (1992a) Approval voting in scientific and engineering societies. Group Decis Negot 1:41–55

Brams SJ, Fishburn PC (1992b) Coalition voting. Math Comput Model (Formal Theory of Politics II: Mathematical Modelling in Political Science) 16:15–26

Brams SJ, Fishburn PC (2001) A nail-biting election. Soc Choice Welf 18(3):409–414

Brams SJ, Fishburn PC (2002) Voting procedures. In: Arrow K, Sen A, Suzumura K (eds) Handbook of social choice and welfare. Elsevier Science, Amsterdam, pp 175–236

Brams SJ, Fishburn PC (2005) Going from theory to practice: the mixed success of approval voting. Soc Choice Welf 25(2):457–474

Brams SJ, Herschbach DR (2001a) Response to Richie, Bouricius, and Macklin. Science 294: 305–306

Brams SJ, Herschbach DR (2001b) The science of elections. Science 292:1449

Brams SJ, Nagel JH (1991) Approval voting in practice. Public Choice 71(1–2):1–17

Brams SJ, Fishburn PC, Merrill S III (1988) The responsiveness of approval voting: comments on Saari and Van Newenhizen; Rejoinder to Saari and Van Newenhizen. Public Choice 59: 121–131, 149
Colomer JM, McLean I (1998) Electing popes: approval balloting and qualified-majority rule. J Interdiscip Hist 29(1):1–22
Cox GW (1987) The Cabinet and the development of political parties in Victorian England. Cambridge University Press, New York
Daverman RJ (2002) Private communication
Davis C (1982) Comment Not Am Math Soc 29(2):138
Dummett M (1984) Voting procedures. Oxford University Press, Oxford, UK
Federal Election Commission (1989) Report on the visit by the Federal Election Commission to the Soviet Union, June 1989. Federal Election Commission, Washington, DC
Fishburn PC, Brams SJ (1991) Yes–no voting. Soc Choice Welf 10:35–50
Fishburn PC, Little JDC (1988) An experiment in approval voting. Manage Sci 34(5):555–568
Fossum RM (2002) Private communication
Gardner M (written by Lynn Arthur Steen) (1980) Mathematical games (From counting votes to making votes count: the mathematics of elections). Sci. Am. 243(4):16ff
Gillman L (1987) Approval voting and the coming MAA elections. Focus: the newsletter of the Mathematical Association of America 7, no. 2 (March-April) 2, 5
Gordon JP (1981) Report of the secretary. Econometrica 48(1):229–233
Guterman L (2002) When votes don't add up: mathematical theory reveals problems in election procedures. Chron High Educ A18–A19
Hill S (2002) Fixing elections: the failure of America's winner take all politics. Routledge, New York
Jarvis JJ (1984) Council—actions and issues: approval voting. OR/MS Today 11(4):16
Keller B (1987) In Southern Russia, a glimpse of democracy. N Y Times 1, 4
Keller B (1988) Moscow says changes in voting usher in many new local leaders. N Y Times A1, A7
Kiely T (1991) A choice, not an echo? Technol Rev 94(6):19–20
Kiewiet RD (1979) Approval voting: the case of the 1968 election. Polity 12(1):528–537
Klarreich E (2002) Election selection: are we using the worst voting procedure? Sci News 162 (18):280–282
Laslier J-F (2003a) Analysing a preference and approval profile. Soc Choice Welf 20(2):229–242
Laslier J-F (2003b) Spatial approval voting. Working paper 2003–001, Laboratoire d'Econométrie, Ecole Polytechnique, Paris
Laslier J-F, Van der Straeten K (2003) Approval voting: an experiment during the French 2002 presidential election. Laboratoire d'Econométrie, Ecole Polytechnique, Paris
Lijphart A, Grofman B (eds) (1984) Choosing an electoral system: issues and alternatives. Praeger, New York
Lines M (1986) Approval voting and strategy analysis: a venetian example. Theory Decis 20:155–172
Little J, Fishburn P (1986) TIMS tests voting method. OR/MS Today 13(5):14–15
MacKenzie D (2000) May the best man lose. Discover 21(11):84–91
Merrill S III (1988) Making multicandidate elections more democratic. Princeton University Press, Princeton, NJ
Nagel J (1984) A debut for approval voting. PS Polit Sci Polit 17(1):62–65
National Academy of Sciences (1981) Constitution and Bylaws
Niemi RG (1984) The problem of strategic voting under approval voting. Am Polit Sci Rev 78(4):952–958
Niemi RG (1985) Reply to Brams and Fishurn. Am Polit Sci Rev 79(3):818–819
Nurmi H (1987) Comparing voting systems. D. Reidel, Dordrecht, Holland
Potthoff RF, Brams SJ (1998) Proportional representation: broadening the options. J Theor Polit 10(2):147–178

Regenwetter M, Grofman B (1998) Approval voting, Borda winners, and Condorcet winners: evidence from seven elections. Manage Sci 44(4):520–533

Reilly B (2002) Social choice in the south seas: electoral innovation and the Borda Count in the Pacific Island countries. Int Polit Sci Rev 23(4):355–372

Richie R, Bouricius T, Macklin P (2001) Candidate number 1: instant runoff voting. Science 294:303–304

Saari DG (2001a) Analyzing a nail-biting election. Soc Choice Welf 18(3):415–430

Saari DG (2001b) Chaotic elections! a mathematician looks at voting. American Mathematical Society, Providence, RI

Saari DG, Van Newenhizen Jill (1988) The problem of indeterminacy in approval, multiple, and truncated voting systems; Is approval voting an 'unmitigated evil'?: a response to Brams, Fishburn, and Merrill. Public Choice 59:101–120, 133–147

Shabad T (1987) Soviets to begin multi-candidate election experiment in June. N Y Times A6

Steen LA (1985) Private communication

Tullock G (1979) Comment on Brams and Fishburn and Balinski and Young. Am Polit Sci Rev 73(2):552–553

Weber RJ (1995) Approval voting. J Econ Perspect 9(1):39–49

White S (1989) Reforming the electoral system [USSR]. J Communist Stud 4:1–17

Wright J (1990) School funding reform options unpopular. Register-Guard (Eugene, OR):1C, 3C

Part II
Axiomatic Theory

Chapter 4
Collective Choice for Simple Preferences

Biung-Ghi Ju

4.1 Introduction

Individual preferences often take simple structures in some restricted environments. The so-called universal domain assumption in the three impossibility results by Arrow (1951), Sen (1970a,b), and Gibbard (1973) and Satterthwaite (1975) have been scrutinized and (partially) abandoned in numerous later studies, which do not intend to identify well-behaved social welfare functions that "would be universal in the sense that it would be applicable to any community" (Arrow 1951, p. 24). Important breakthroughs have been made in this line of research: Gaertner (2002) provides a comprehensive survey of the literature on domain restrictions.

Of our central interest in this survey are simple preferences with few indifference classes such as the so-called dichotomous or trichotomous preferences as studied by Inada (1964, 1969, 1970)[1]. Later investigations on collective choice with dichotomous preferences have been closely connected to studies of the normative and strategic advantages of majority and approval voting systems and of their axiomatic foundation: see Brams and Fishburn (2002) for an extensive survey of this literature as well as Brams and Fishburn (1978) and Fishburn (1978a,b, 1979) among others. This survey connects old and recent theoretical developments in this literature with a single but comprehensive perspective.

The survey starts with a brief overview of the classical impossibility results. Section 4.2 discusses some possibility results on several domains of dichotomous preferences. Section 4.3 discusses axiomatic foundations for majority and approval voting systems. We investigate the logical relationship among the existing axiomatic characterizations. In the process, we discover ways of strengthening existing results and we offer new characterization results. Readers are referred to Xu (2010) in this

[1]Dichotomous preferences are also considered by Bogomolnaia and Moulin (2004) and Bogomolnaia et al. (2005) in their investigation of well-behaved randomization mechanisms.

B.-G. Ju
Department of Economics, Korea University, Anam-dong 5-1, Seong-buk-gu, Seoul 136-701, Korea
e-mail: bgju@korea.ac.kr

J.-F. Laslier and M.R. Sanver (eds.), *Handbook on Approval Voting*, Studies in Choice and Welfare, DOI 10.1007/978-3-642-02839-7_4,

volume for a compact overview of the literature on axiomatic characterizations of majority voting.[2] Section 4.4 discusses strategic voting and the robustness of voting systems. Some results associated with the Condorcet principle and realizability of Condorcet winners in strategic voting environments are included. Section 4.5 discusses some recent developments in unconstrained multi-issue problems with separable preferences. The section deals with strategy-proof voting schemes and shows the conflict between Pareto efficiency and strategy-proofness on the entire domain of separable preferences and on restricted domains of "dichotomous" or "trichotomous" preferences. Section 4.6 discusses dichotomous opinion aggregation problems that have drawn some attention recently among scholars interested in group identification.

4.1.1 Preliminaries

Let $\mathbb{X}$ be the set of all alternatives. There are infinitely many "potential" agents, identified with the natural numbers in $\mathbb{N}$. Let $\mathcal{X}$ and $\mathcal{N}$ be the set of finite subsets of $\mathbb{X}$ and of $\mathbb{N}$ respectively. Each agent $i \in \mathbb{N}$ has a preference ordering R_i that is a complete, reflexive, and transitive binary relation over $\mathbb{X}$. Let $\mathcal{R}$ be the set of all preference orderings over $\mathbb{X}$. We sometimes consider binary relations that are not necessarily transitive. Let $\underline{\mathcal{R}}$ be the set of all complete and reflexive binary relations. For each $N \in \mathcal{N}$ and each $X \in \mathcal{X}$, let $\mathcal{R}^N$ be the set of profiles of preference orderings of agents in N and let $\mathcal{U}_{N,X} \equiv \mathcal{R}^N \times \{X\}$. Let $\mathcal{U}_N \equiv \bigcup_{X\in\mathcal{X}} \mathcal{U}_{N,X}$, $\mathcal{U}_X \equiv \bigcup_{N\in\mathcal{N}} \mathcal{U}_{N,X}$, and $\mathcal{U} \equiv \bigcup_{N\in\mathcal{N},X\in\mathcal{X}} \mathcal{U}_{N,X}$. Subsets of $\mathcal{U}_{N,X}$, $\mathcal{U}_N$, $\mathcal{U}_X$, and $\mathcal{U}$ are denoted respectively by $\mathcal{D}_{N,X}$, $\mathcal{D}_N$, $\mathcal{D}_X$, and $\mathcal{D}$. Elements of $\mathcal{R}^N$ are denoted by R_N, R'_N, R''_N, etc., and also by R, R', R'', etc., when N is clear from the context. Elements of $\mathcal{R}$ are denoted by R_0, R'_0, R''_0, etc., and also by R_i, R'_i, R''_i, etc., when they belong to agent i.

A *social decision function* on $\mathcal{D}_{N,X}$ is a function $f : \mathcal{D}_{N,X} \to \underline{\mathcal{R}}$ associating with each profile $(R, X) \in \mathcal{D}_{N,X}$ a social preference relation $f(R, X) \in \underline{\mathcal{R}}$. A *social welfare function* on $\mathcal{D}_{N,X}$ is a function $f : \mathcal{D}_{N,X} \to \mathcal{R}$ associating with each profile $(R, X) \in \mathcal{D}_{N,X}$ a social preference *ordering* $f(R, X) \in \mathcal{R}$. We often denote a social preference relation by $\succeq$, its strict counterpart by $\succ$, and its indifference by $\sim$, in order to distinguish them from individual preference relations.

Let $P(X)$ be the set all subsets of X and $\bar{P}(X)$ the set of all non-empty subsets of X. A social preference relation $\succeq$ generates a choice rule $C(\cdot;\succeq)\colon P(X) \to P(X)$ as follows: for all $Y \subseteq X$,

$$C(Y;\succeq) \equiv \{x \in Y : \text{ for all } y \in Y,\ x \succeq y\}. \tag{4.1}$$

By finiteness of X, if $\succeq$ is transitive, the choice rule is non-empty valued at each non-empty $Y \subseteq X$. For non-empty valuedness, each of the following weaker conditions

[2] Thomson (2001) offers an extensive survey and discussion on the axiomatic method in Social Choice Theory and Game Theory.

is also sufficient. Preference relation $\succsim$ is *quasi-transitive* if its strict counterpart $\succ$ is transitive, that is, for all $x, y, z \in X$, $x \succ y$ and $y \succ z$ imply $x \succ z$. It is *acyclic* if there is no sequence of finite alternatives, $x_1, \ldots, x_T \in X$ such that $x_1 \succ x_2$, $x_2 \succ x_3, \ldots, x_{T-1} \succ x_T, x_T \succ x_1$. Clearly, transitivity implies quasi-transitivity, which implies acyclicity. Quasi-transitivity is sufficient but not necessary for the non-emptiness of the choice rule in (4.1) (so is transitivity). Acyclicity is necessary and sufficient for the non-emptiness of the choice rule (Sen 1970a,b, Lemma 1*1).

A *collective choice quasi-rule* on $\mathcal{D}_{N,X}$ is a function $c\colon \mathcal{D}_{N,X} \times \bar{P}(X) \to P(X)$ associating with each profile $(R, X, Y) \subseteq \mathcal{D}_{N,X} \times \bar{P}(X)$ a subset of Y, that is, $c(R, X, Y) \subseteq Y$. A *collective choice rule* on $\mathcal{D}_{N,X}$ is a non-empty valued collective choice quasi-rule, namely a function $c\colon \mathcal{D}_{N,X} \times \bar{P}(X) \to \bar{P}(X)$ associating with each profile $(R, X, Y) \subseteq \mathcal{D}_{N,X} \times \bar{P}(X)$ a nonempty subset of Y, that is, $\emptyset \neq c(R, X, Y) \subseteq Y$. We sometimes use notation $c_{R,X}(Y) \equiv c(R, X, Y)$.

Each choice rule $C : \bar{P}(X) \to \bar{P}(X)$ generates a binary relation $R(C)$ as follows:

$$xR(C)y \text{ if and only if } x \in C(\{x, y\}).$$

Call $R(C)$ the *base relation* of $C(\cdot)$ (as in Herzberger 1973). Choice rule C is *normal* if $C(\cdot) = C(\cdot; R(C))$. Unless specified otherwise, we consider collective choice rules generating normal choice rules. Necessary and sufficient conditions for a choice rule to be normal are summarized in Sen (1977, pp. 64–65, Propositions 8 and 9).[3]

4.1.2 Classical Impossibility Results

Consider social decision functions or collective choice rules over $\mathcal{D}_{N,X}$. Here are some basic axioms for social decision functions considered in Arrow (1951), Sen (1970a,b), Gibbard (1973) and Satterthwaite (1975). In defining the axioms, we will only state the properties needed for a social decision function f. That is, instead of stating "f is said to satisfy Axiom A if it satisfies property A," we simply state property A.

Unrestricted Domain: $\mathcal{D}_{N,X} = \mathcal{U}_{N,X}$.

Transitive Social Preferences, (briefly, **Transitivity):** For all $R \in \mathcal{D}_{N,X}$, the social preference relation at R, $f(R, X)$, is transitive.

Replacing transitivity with quasi-transitivity or acyclicity, we define the axioms of *quasi-transitive social preferences* (briefly, *quasi-transitivity*) and *acyclic social preferences* (briefly, *acyclicity*), respectively.

Weak Pareto: For all $x, y \in X$, if everyone strictly prefers x to y at R, then x is strictly preferred to y under the social preference relation $f(R, X)$.

Non-dictatorship: There is no person $i \in N$ - such a person would be a dictator - such that for all $R \in \mathcal{D}_{N,X}$ and all $x, y \in X$, if i strictly prefers x to y, then x is strictly preferred to y under the social preference relation $f(R, X)$.

[3] One necessary and sufficient condition in citetSen77 (properties $\alpha 2$ and $\gamma 2$) is the following: for all $Y \in \bar{P}(X)$, $x \in C(Y)$ if and only if for all $y \in Y$, $x \in C(\{x, y\})$.

Note that if a collective choice rule $c(\cdot)$ generates a dictatorial base relation $R(c)$, then by normality, the dictator's preferred choices constitute $c(R, X, Y)$ for all $Y \in \bar{P}(X)$.

Independence of Irrelevant Alternatives: For all $R, R' \in \mathcal{D}_{N,X}$ and all $Y \subseteq X$, if individual preferences of both profiles R and R' are identical over Y, then the two social preference relations at the two profiles generate the same choice over Y, that is, $C(Y; f(R, X)) = C(Y; f(R', X))$.[4]

Replacing $f(R, X)$ in the above axioms with the base relation $R(c)$ generated from a collective choice rule $c(\cdot)$, we define the corresponding axioms for collective choice rules. The same names are used for these axioms.

4.1.2.1 Arrow's Theorem

Arrow (1951) investigates the existence of social decision functions satisfying the five basic axioms in the previous section. When there are at least three alternatives, such a function does not exists.

Theorem 4.1.1 (Arrow's Impossibility Theorem). *If there are at least three alternatives, then no social decision function (or collective choice rule) satisfies unrestricted domain, transitive social preferences, weak Pareto, non-dictatorship, and independence of irrelevant alternatives.*[5]

When the axiom of transitive social preferences is weakened to quasi-transitivity, there does exist a social decision function satisfying the other four axioms. For example, "Pareto dominance" gives a quasi-transitive, but not necessarily transitive, social preference relation. Later works in this direction (Gibbard 1969; Guha 1972; Mas-Colell and Sonnenschein 1972) deliver a characterization of "oligarchic" social decision functions where a group, namely oligarchy, is decisive and each member of the group has veto power. Replacing quasi-transitivity with acyclicity leads to a larger family of social decision functions that may not be oligarchic but close to oligarchy, in the sense that all decisive groups share some core members as shown by Brown (1975) and Banks (1995).

Further progress has been made in the line of research that focuses on restricted preferences in some specialized environments. Gaertner (2002) provides a comprehensive survey of the literature on restricted domains. Sections 4.2–4.6 give an overview of results pertaining to dichotomous preferences. Section 4.2 provides some possibility results on dichotomous domains. We list several definitions of dichotomous domains that admit some social decision functions satisfying Arrow's axioms except for the axiom of unrestricted domain. Moreover, as we will see in Sect. 4.3, majority decision stands out among other decision functions as the unique one satisfying Arrow's axioms and other standard axioms in the literature.

---

[4] See (4.1) for the definition of the choice rule generated by a social preference relation.

[5] Note that normality assumption for collective choice rules allows us to state this result for both social decision function and collective choice rule at once.

4.1.2.2 Gibbard–Satterthwaite Theorem

Here we consider a domain $\mathcal{D}_{N,X} \subseteq \mathcal{U}_{N,X}$ such that for some $\bar{\mathcal{R}} \subseteq \mathcal{R}$, $\mathcal{D}_{N,X} = \bar{\mathcal{R}}^N \times \{X\}$. Preferences are primitive variables for collective choice or social decision but they are often unobservable. Agents or voters seek their own private interests and may vote untruthfully whenever advantageous. Collective choice procedures may not work properly unless they have a certain embedded property in themselves preventing untruthful voting. An important line of research has been devoted to the search for truthful collective choice procedures. The seminal work of Gibbard (1973) and Satterthwaite (1975) show that when there are at least three alternatives, there is no truthful procedure that also satisfies unrestricted domain, non-dictatorship, and the full-range condition.

A collective choice rule $c: \mathcal{D}_{N,X} \times \bar{P}(X) \to \bar{P}(X)$ is *resolute* if it always picks a single alternative, that is, for all $(R, X) \in \mathcal{D}_{N,X}$ and all $Y \in \bar{P}(X)$, $c(R, X, Y)$ is a singleton. For truthful procedures, Gibbard (1973) and Satterthwaite (1975) require that for all possible reported preferences of others, each agent $i \in N$ always prefers the outcome that results from the truthful announcement of his preferences to any outcome that he could obtain by lying.

Strategy-Proofness: For all $R \in \mathcal{D}_{N,X}$, all $Y \in \bar{P}(X)$, all $i \in N$, and all $R_i' \in \bar{\mathcal{R}}$, $c((R_i, R_{-i}), X, Y)\ R_i\ c((R_i', R_{-i}), X, Y)$.

An extension of strategy-proofness for set-valued rules is discussed in Sect. 4.4.

Theorem 4.1.2 (Gibbard–Satterthwaite Theorem). *If there are at least three alternatives, no resolute collective choice rule satisfies unrestricted domain, non-dictatorship, strategy-proofness and the full-range condition.*

Important positive results are derived in later works pertaining to specialized environments that accommodate some natural restrictions on preferences. We will survey the results pertaining to dichotomous domains in Sects. 4.2 and 4.4 and the domain of separable preferences in Sect. 4.5. Moulin (1980) characterizes a large family of strategy-proof rules on the domain of single-peaked preferences over public alternatives that are ordered on a line. Any such rule chooses a "generalized Condorcet winner." In the case of private good rationing model with single peaked preferences, the family of strategy-proof rules is much more restricted as shown by Sprumont (1991).

4.1.2.3 Sen's Paretian Liberal Paradox

Sen (1970a,b) investigates the existence of a social decision function that satisfies weak Pareto and a minimal form of liberalism, as well as the condition of acyclic social preferences and unrestricted domain. Again the result is negative.

His minimal notion of liberalism requires that there should be at least two agents who are decisive when making social comparison of a pair of alternatives. Formally, we say that agent i is *decisive for x and y with $x \neq y$*, if for all $R \in \mathcal{D}_{N,X}$, $xP_i y$ implies $x \succ_{f(R,X)} y$.

Minimal Liberalism: There are at least two agents who are decisive for a pair of alternatives.

His main result, known as the Paretian liberal paradox, is the following.

Theorem 4.1.3 (Sen's Paradox). *No social decision function (or collective choice rule) satisfies unrestricted domain, acyclic social preferences, weak Pareto, and minimal liberalism.*

Gibbard (1974) pushes this negative result to the most extreme form by showing that Sen's liberalism, properly extended in the model of collective decision with personal components, cannot be well-defined. He provides a simple preference profile for which any choice of an alternative necessarily violates at least one liberal right: this is known as Gibbard's paradox.

We will ask whether Sen's paradox holds on the dichotomous preferences domain for the problems of unconstrained choice of multiple issues in Sects. 4.5 and 4.6.

4.2 Possibility Results on Some Dichotomous Domains

Consider $\mathcal{D}_{N,X} \subseteq \mathcal{U}_{N,X}$ such that for some $\bar{\mathcal{R}} \subseteq \mathcal{R}$, $\mathcal{D}_{N,X} = \bar{\mathcal{R}}^N \times \{X\}$. Throughout this section, we consider several examples of "dichotomous" domains. On these domains, there do exist some social decision functions satisfying Arrow's axioms (in Theorem 4.1.1) except for unrestricted domain. This is shown by some existing results that we overview here. We also offer some characterizations imposing Arrow's axiom of independence of irrelevant alternatives together with other standard axioms.

For all $R \in \mathcal{D}_{N,X}$ and all $x, y \in X$, let $N_{x,y}(R) \equiv \{i \in N : x\ P_i\ y\}$ be the set of agents who prefer x to y (or vote for x against y) and $n_{x,y}(R) \equiv |\{i \in N : x\ P_i\ y\}|$ the number of agents who prefer x to y (or the number of votes x wins against y). Independence of irrelevant alternatives can be restated as follows:

Independence of Irrelevant Alternatives: For all $R, R' \in \mathcal{D}_{N,X}$ and all $x, y \in X$, if $N_{x,y}(R) = N_{x,y}(R')$ and $N_{y,x}(R') = N_{y,x}(R)$, then $x \succsim_{f(R)} y$ implies $x \succsim_{f(R')} y$.

The next axiom is stronger and is crucial for strategy-proofness.

Monotonicity: For all $R, R' \in \mathcal{D}_{N,X}$ and all $x, y \in X$, if $N_{x,y}(R) \subseteq N_{x,y}(R')$ and $N_{y,x}(R') \subseteq N_{y,x}(R)$, then $x \succsim_{f(R)} y$ implies $x \succsim_{f(R')} y$.

Applying monotonicity when $N_{x,y}(R) = N_{x,y}(R')$ and $N_{y,x}(R') = N_{y,x}(R)$ yields independence of irrelevant alternatives.

The next axiom, considered by MaY (1952), plays a key role in his and other axiomatic characterizations of majority decision.

Positive Response: For all $R, R' \in \mathcal{D}_{N,X}$ and all $x, y \in X$, if $N_{x,y}(R) \subseteq N_{x,y}(R')$, $N_{y,x}(R') \subseteq N_{y,x}(R)$, and at least one of the two inclusions is strict, then $x \succsim_{f(R)} y$ implies $x \succ_{f(R')} y$.

Note that when there are only two alternatives, say x and y, positive response implies monotonicity because then, $N_{x,y}(R) = N_{x,y}(R')$ and $N_{y,x}(R') = N_{y,x}(R)$, which is the case relevant not to positive response but to monotonicity, imply $R = R'$ and so application of monotonicity in this case is trivial. However, with more than two alternatives, this implication no longer holds, and there is no logical relation between the two axioms.

The next two axioms require symmetric treatment of individuals and of alternatives, respectively.

Anonymity: For all permutations on N, $\pi : N \to N$, and all $R \in \mathcal{D}_{N,X}$, $f(R) = f(R_\pi)$, where $R_\pi \in \mathcal{D}_{N,X}$ is such that for all $i \in N$, $R_{\pi i} = R_{\pi(i)}$.

Neutrality: For all $R \in \mathcal{D}_{N,X}$ and all $x, y, x', y' \in X$, if $R' \in \mathcal{D}_{N,X}$ is the preferences profile obtained after relabeling x and y in profile R with x' and y' respectively, then $x \succeq_{f(R)} y$ if and only if $x' \succeq_{f(R')} y'$.

The best known decision function satisfying the above axioms is *majority decision function* $f_{MAJ}(\cdot)$, which maps each $R \in \mathcal{D}_{N,X}$ into a social preference relation $\succeq_{f_{MAJ}(R)}$ defined as follows: for all $x, y \in X$,

$$x \succeq_{f_{MAJ}(R)} y \text{ if and only if } n_{x,y}(R) \geq n_{y,x}(, R).$$

In fact, there is a large family of monotonic decision functions, of which the special example is majority decision. In order to define this family, we need the following notation and concepts. Let $\mathfrak{d}^* \equiv \{(L_1, L_2) : L_1, L_2 \in P(N), L_1 \cap L_2 = \emptyset\}$ be the set of all pairs of disjoint subsets of N. A *decisive structure for a pair* $x, y \in X$, $\mathfrak{d}_{x,y}$ is a non-empty subset of $\mathfrak{d}^*$such that for all $(L_1, L_2), (L'_1, L'_2) \in \mathfrak{d}^*$,

$$\text{if } (L_1, L_2) \in \mathfrak{d}_{x,y},\ L_1 \subseteq L'_1, \text{ and } L'_2 \subseteq L_2, \text{ then } (L'_1, L'_2) \in \mathfrak{d}_{x,y}. \tag{4.2}$$

Call this property $\mathfrak{d}$-*monotonicity*. A *decisive structure* $\mathfrak{d} \equiv (\mathfrak{d}_{x,y})_{x,y \in X}$ is a profile of decisive structures for pairs of alternatives such that for all $x, y \in X$ and all $(L_1, L_2) \in \mathfrak{d}^*$,

$$\text{if } (L_1, L_2) \notin \mathfrak{d}_{x,y}, \text{ then } (L_2, L_1) \in \mathfrak{d}_{y,x}. \tag{4.3}$$

Call this property $\mathfrak{d}$-*completeness*. A decisive structure $\mathfrak{d}$ represents the social decision function $f^{\mathfrak{d}}$ defined as follows: for all $R \in \mathcal{D}_{N,X}$ and all $x, y \in X$, $x \succeq_{f^{\mathfrak{d}}(R)} y$ if and only if $(N_{x,y}(R), N_{y,x}(R)) \in \mathfrak{d}_{x,y}$.

Note that by (4.2), $f^{\mathfrak{d}}$ is monotonic and that by (4.3), the social preference relations chosen by $f^{\mathfrak{d}}$ are complete. It is easy to show that neutrality of $f^{\mathfrak{d}}$ requires $\mathfrak{d}_{x,y} = \mathfrak{d}_{x',y'}$ for all $x, y, x', y' \in X$. Conversely, any monotonic social decision function f generates a decisive structure $\mathfrak{d}^f$ and is represented by it. To show this, define $\mathfrak{d}^f_{x,y}$ as follows: for all $(L_1, L_2) \in \mathfrak{d}^*$, $(L_1, L_2) \in \mathfrak{d}^f_{x,y}$ if and only if for some $R \in \mathcal{D}_{N,X}$, $x \succeq_{f(R)} y$, $N_{x,y}(R) \subseteq L_1$, and $L_2 \subseteq N_{y,x}(R)$. To prove (4.3), suppose $(L_1, L_2) \in \mathfrak{d}^* \backslash \mathfrak{d}^f_{x,y}$. Consider $R \in \mathcal{D}_{N,X}$ such that $N_{x,y}(R) = L_1$ and

$N_{y,x}(R) = L_2$.[6] Then not $x \succeq_{f(R)} y$ and by completeness of $f(R)$, $y \succ_{f(R)} x$. This shows $(L_2, L_1) \in \mathfrak{d}^f_{y,x}$. Monotonicity of f directly implies (4.2). Therefore we obtain:

Proposition 4.2.1. *A social decision function satisfies monotonicity if and only if it is represented by a decisive structure.*

For anonymous and neutral decision functions, decisive structures representing them take a simple form. Let $\mathfrak{n}^* \equiv \{(n_1, n_2) : n_1, n_2 \in \{0, 1, \ldots, n\}$, and $n_1 + n_2 \leq n\}$. A *decisive index structure for a pair* x, y, $\mathfrak{n}_{x,y}$ is a non-empty subset of $\mathfrak{n}^*$ such that for all $(n_1, n_2), (n'_1, n'_2) \in \mathfrak{n}^*$,

$$\text{if } (n_1, n_2) \in \mathfrak{n}_{x,y},\ n_1 \leq n'_1, \text{ and } n'_2 \leq n_2, \text{ then } (n'_1, n'_2) \in \mathfrak{n}_{x,y}. \tag{4.4}$$

Call this $\mathfrak{n}$-*monotonicity*. A *decisive index structure* $\mathfrak{n} \equiv (\mathfrak{n}_{x,y})_{x,y \in X}$ is a profile of decisive index structures for pairs of alternatives such that for all $x, y \in X$ and all $(n_1, n_2) \in \mathfrak{n}^*$,

$$\text{if } (n_1, n_2) \notin \mathfrak{n}_{x,y}, \text{ then } (n_2, n_1) \in \mathfrak{n}_{y,x}. \tag{4.5}$$

Call this $\mathfrak{n}$-*completeness* Note that for neutral social decision functions represented by a decisive index structure $\mathfrak{n}$, neutrality and $\mathfrak{n}$-completeness imply the following: for all $k \in \{0, 1, \ldots, [n/2]\}$ and all $x, y, x', y' \in X$ with $x \neq y$ and $x' \neq y'$,

$$\mathfrak{n}_{x,y} = \mathfrak{n}_{x',y'} \text{ and } (k, k) \in \mathfrak{n}_{x,y}, \tag{4.6}$$

where $[n/2]$ is the greatest integer that is less than or equal to $n/2$. Call this $\mathfrak{n}$-*neutrality*. This property and $\mathfrak{n}$-monotonicity together imply that for all $k \in \{0, 1, \ldots, [(n-1)/2]\}$ and all $x, y, x', y' \in X$ with $x \neq y$ and $x' \neq y'$, $(k+1, k) \in \mathfrak{n}_{x,y} = \mathfrak{n}_{x',y'}$. Combining this with (4.6), we get: for all $(n_1, n_2) \in \mathfrak{n}^*$ and all $x, y, x', y' \in X$ with $x \neq y$ and $x' \neq y'$,

$$\mathfrak{n}_{x,y} = \mathfrak{n}_{x',y'}, \text{ and if } n_1 \geq n_2, \text{ then } (n_1, n_2) \in \mathfrak{n}_{x,y}. \tag{4.7}$$

Therefore we obtain:

Proposition 4.2.2. *A social decision function satisfies monotonicity and anonymity if and only if it is represented by a decisive index structure. Adding neutrality, we characterize the subfamily of social decision functions represented by an* $\mathfrak{n}$*-neutral decisive index structure. Moreover, these* $\mathfrak{n}$*-neutral index structures satisfy* (4.7).

When preferences are linear (no indifference), Propositions 4.2.1 and 4.2.2 give characterizations of what are known as "monotonic simple games." Since we will mostly focus on dichotomous domains where indifference is prevalent, decisive structures are more relevant to our later discussion.

[6] Existence of such R is the basic richness assumption for $\mathcal{D}_{N,X}$ that we need in order to obtain Proposition 4.2.1.

4.2.1 Two Alternatives

The simplest example of dichotomous domains is of course when there are only two alternatives, say a, b (that is, $X = \{a, b\}$). Then any social decision function satisfies *transitive social preferences* trivially. There are numerous social decision functions satisfying all other axioms in Arrow's theorem. For example, the social decision functions represented by a monotonic and non-dictatorial decisive structure satisfy all of Arrow's axioms. There are also numerous strategy-proof and non-dictatorial collective choice functions. An important property for strategy-proofness in this binary choice framework is monotonicity. Since there are only two alternatives, Sen's minimal liberalism is hardly satisfied unless the set of admissible preferences is extremely restricted.

Majority decision function stands out among other well-behaved social decision functions, as shown by MaY (1952). The key axiom in his axiomatic characterization of majority decision is positive response.

Now, to find out the implication of positive response, consider a function f represented by decisive structure $\mathfrak{d}$. Let R, R' be the two profiles in the premise of the axiom of positive response. Assume $x \succeq_{f(R)} y$, that is, $(N_{x,y}(R), N_{y,x}(R)) \in \mathfrak{d}_{x,y}$. Positive response, then requires $x \succ_{f(R')} y$, which implies $(N_{y,x}(R'), N_{x,y}(R')) \notin \mathfrak{d}_{y,x}$. Thus positive response implies the following extra condition on decisive structures: for all $x, y \in X$ and all $(L_1, L_2), (L'_1, L'_2) \in \mathfrak{d}^*$ with $x \neq y$ and $(L_1, L_2) \neq (L'_1, L'_2)$,

$$\text{if } (L_1, L_2) \in \mathfrak{d}_{x,y},\ L_1 \subseteq L'_1, \text{ and } L'_2 \subseteq L_2, \text{ then } (L'_2, L'_1) \notin \mathfrak{d}_{y,x}. \tag{4.8}$$

For decisive index structures, this condition can be written as: for all $x, y \in X$ and all $(n_1, n_2), (n'_1, n'_2) \in \mathfrak{n}^*$ with $x \neq y$ and $(n_1, n_2) \neq (n'_1, n'_2)$,

$$\text{if } (n_1, n_2) \in \mathfrak{n}_{x,y},\ n_1 \leq n'_1, \text{ and } n'_2 \leq n_2, \text{ then } (n'_2, n'_1) \notin \mathfrak{n}_{y,x}. \tag{4.9}$$

For a neutral social decision function represented by a decisive index structure $\mathfrak{n}$, if there is $(n_1, n_2) \in \mathfrak{n}^*$ such that $n_1 < n_2$ and $(n_1, n_2) \in \mathfrak{n}_{x,y}$, then by (4.9), $([\frac{n_1+n_2}{2}] + 1, [\frac{n_1+n_2}{2}]) \notin \mathfrak{n}_{y,x}$, which contradicts to (4.7). Therefore, neutrality and positive response together imply the following: for all $x, y \in X$ and all $(n_1, n_2) \in \mathfrak{n}^*$,

$$\text{if } n_1 < n_2, \text{ then } (n_1, n_2) \notin \mathfrak{n}_{x,y}. \tag{4.10}$$

Combining (4.7) and (4.10), we obtain:

Theorem 4.2.1 (MaY 1952). *When there are two alternatives, a social decision function on $\mathcal{U}_{N,X}$ satisfies anonymity, neutrality, and positive response if and only if it is majority decision function.*

An extended version of this result with more than two alternatives is provided in Theorem 4.3.1. Aşan and Sanver (2002) replaces positive response with the combination of "path independence" and Pareto (if no voter prefers b to a and some voter prefers a to b, then a should be socially preferred to b). In the same framework, Sanver (2009) imposes weak Pareto, anonymity, neutrality, and monotonicity, together with some additional axioms, and characterizes variants of majority decision function.

4.2.2 Two Fixed Indifference Classes

In this section, we assume that there are two types of alternatives and all alternatives of a type are indifferent. This assumption is formulated by the following domain property.

Definition 4.2.1. A domain $\mathcal{D}_{N,X}^{2fic} \subseteq \mathcal{U}_{N,X}$ has the property of *two-fixed-indifference-class* if alternatives are partitioned into two fixed classes and for all $R \in \mathcal{D}_{N,X}^{2fic}$ and all $i \in N$, the two classes constitute the two indifference sets of R_i. Let $a, b \in X$ be two representative alternatives and the two fixed classes are denoted by X_a and X_b.

On such a domain, majority decision function do satisfy transitivity (as is implied by Theorem 4.2.2). Hence there does exist a social decision function satisfying all of Arrow's axioms. A characterization of a family of transitive social decision functions is provided in the next proposition.

Let $f : \mathcal{D}_{N,X}^{2fic} \to \underline{\mathcal{R}}$ be a social decision function satisfying monotonicity and transitivity. Let $\eth$ be a decisive structure representing f. Suppose that for some $x, y \in X_a$ and some $R \in \mathcal{D}_{N,X}^{2fic}$, $x \succ_{f(R)} y$. Then by monotonicity, the strict social ranking holds at all other preference profiles, that is, for all $R' \in \mathcal{D}_{N,X}^{2fic}$, $x \succ_{f(R')} y$. This is because $N_{x,y}(R) = N_{x,y}(R') = N_{y,x}(R) = N_{y,x}(R') = \emptyset$. Then the ranking between x and y can be decided by $\eth_{x,y}$ and $\eth_{y,x}$ such that $(\emptyset, \emptyset) \in \eth_{x,y}$ and $(\emptyset, \emptyset) \notin \eth_{y,x}$. Similarly, if for some $R \in \mathcal{D}_{N,X}^{2fic}$, $x \sim_{f(R)} y$, then this social indifference holds at all other preference profiles and $(\emptyset, \emptyset) \in \eth_{x,y}$ and $(\emptyset, \emptyset) \in \eth_{y,x}$. Therefore, there is a fixed social preference relation over alternatives in X_a and over alternatives in X_b, which holds at all $R \in \mathcal{D}_{N,X}^{2fic}$. Since social decision function f satisfies transitivity, we may order elements in the two sets X_a and X_b in the same order of their fixed social rankings; that is, elements of X_a are $a_1 \succeq a_2 \succeq \cdots \succeq a_q$ and elements of X_b are $b_1 \succeq b_2 \succeq \cdots \succeq b_r$. Strict ranking among $a_1, \ldots, a_q$ or among $b_1, \ldots, b_r$ is excluded when we require the following mild axiom:

Indifference Unanimity: For all R and all $x, y \in X$, if for all $i \in N$, $x \, I_i \, y$, then $x \sim_{f(R)} y$.

The next result characterizes a family of functions satisfying transitivity, monotonicity, and indifference unanimity.

Proposition 4.2.3. *Consider a domain with the property of two-fixed-indifference-class. Denote two representative alternatives in the two fixed classes by a and b and* *the two fixed classes by* X_a *and* X_b. *A social decision function satisfies transitivity, monotonicity, and indifference unanimity if and only if it is represented by a decisive structure* $\eth \equiv (\eth_{x,y})_{x,y \in X}$ *such that for all* $x, x' \in X_a$ *and all* $y, y' \in X_b$, $(\emptyset, \emptyset) \in \eth_{x,x'} = \eth_{y,y'}$ *and for all* $x \in X_a$ *and all* $y \in X_b$, $\eth_{x,y} = \eth_{a,b}$ *and* $\eth_{y,x} = \eth_{b,a}$.

Proof. By indifference unanimity, for all $R \in \mathcal{D}^{2fic}_{N,X}$, alternatives in X_a are all socially indifferent and similarly for X_b. Define $\eth_{x,y}$ as follows: for all $(L_1, L_2) \in \eth^*$, $(L_1, L_2) \in \eth_{x,y}$ if and only if for some $R \in \mathcal{D}^{2fic}_{N,X}$, $x \succeq_{f(R)} y$, $N_{x,y}(R) \subseteq L_1$, and $L_2 \subseteq N_{y,x}(R)$. By indifference unanimity, for all $x, x' \in X_a$ and all $y, y' \in X_b$, $(\emptyset, \emptyset) \in \eth_{x,x'} = \eth_{y,y'}$. Now let $x \in X_a$ and $y \in X_b$. For all $R \in \mathcal{D}^{2fic}_{N,X}$, since $x \sim_{f(R)} a$ and $y \sim_{f(R)} b$, then by transitivity, $x \succeq_{f(R)} y$it if and only if $a \succeq_{f(R)} b$. This and the construction of $\eth$ imply $\eth_{x,y} = \eth_{a,b}$. Similarly, $\eth_{y,x} = \eth_{b,a}$. □

4.2.3 Two Indifference Classes

We now consider domains where individual preferences can have at most two indifference classes.

Definition 4.2.2. A domain $\mathcal{D}^{2ic}_{N,X} \subseteq \mathcal{U}_{N,X}$ has the property of *two-indifference-class* if for all $R \in \mathcal{D}^{2ic}_{N,X}$, all triples $x, y, z \in X$ and all $i \in N$, R_i partitions $\{x, y, z\}$ into at most two indifference classes.

Clearly any domain with the property of two-fixed-indifference-class has this property, but not vice versa. On such domains, majority decision function always generates a transitive social preference relation.

Theorem 4.2.2 (Inada 1964). *On any domain with the property of two-indifference-class, majority decision function satisfies transitivity.*

Proof. Let $x, y, z \in X$ be three distinct alternatives. If $R \in \mathcal{D}^{2ic}_{N,X}$, then for all $i \in N$, R_i is one of the following seven "dichotomous" preference orderings: (1) $xI_i yI_i z$, (2) $xI_i yP_i z$, (3) $xP_i yI_i z$, (4) $xI_i zP_i y$, (5) $yP_i xI_i z$, (6) $yI_i zP_i x$, (7) $zP_i xI_i y$. Let $n_1, \ldots, n_7$ be the numbers of agents of each type. Note that $n_{x,y}(R) = n_3 + n_4$, $n_{y,x}(R) = n_5 + n_6$, $n_{y,z}(R) = n_2 + n_5$, $n_{z,y}(R) = n_4 + n_7$, $n_{x,z}(R) = n_2 + n_3$, and $n_{z,x} = n_6 + n_7$. To show transitivity of social preference relation, suppose $x \succeq_{f_{MAJ}(R)} y$ and $y \succeq_{f_{MAJ}(R)} z$. Then

$$n_3 + n_4 \geq n_5 + n_6 \text{ and } n_2 + n_5 \geq n_4 + n_7. \tag{4.11}$$

Combining the two inequalities, we obtain $n_2 + n_3 + n_4 + n_5 \geq n_4 + n_5 + n_6 + n_7$, that is,

$$n_2 + n_3 \geq n_6 + n_7. \tag{4.12}$$

This implies $n_{x,z}(R) \geq n_{z,x}(R)$. Therefore, $x \succeq_{f_{MAJ}(R)} z$. □

In fact, majority decision is the only transitive social decision function satisfying monotonicity, anonymity, and neutrality, except for *degenerate indifference function* that is the constant social decision function taking the complete indifference as its value (alternatives are all indifferent).

Theorem 4.2.3 (Ju 2009b). *Consider a domain with the property of two-indifference-class.*[7] *A social decision function satisfies monotonicity, anonymity, neutrality and transitivity if and only if it is either majority decision function or degenerate indifference function.*

Proof. By Theorem 4.2.2, majority decision function is transitive. It also satisfies the other axioms by Proposition 4.2.2. In order to prove the converse, let f be a social decision function on $\mathcal{D}^{2ic}_{N,X}$ satisfying monotonicity, anonymity, neutrality and transitivity. By Proposition 4.2.2, f is represented by an $\mathfrak{n}$-neutral index structure $\mathfrak{n} \equiv (\mathfrak{n}_{x,y})_{x,y \in X}$ and the index structure satisfies (4.7). Let $\mathfrak{n}_0 \equiv \mathfrak{n}_{x,y}$ for all $x, y \in X$ with $x \neq y$. Throughout the proof, we follow the same classification of dichotomous preferences over $\{x, y, z\}$ as in the proof of Theorem 4.2.2. For all $k = 1, \ldots, 7$, let n_k the number of persons with the dichotomous preferences of type k. Recall $n_{x,y}(R) = n_3 + n_4$, $n_{y,x}(R) = n_5 + n_6$, $n_{y,z}(R) = n_2 + n_5$, $n_{z,y}(R) = n_4 + n_7$, $n_{x,z}(R) = n_2 + n_3$, and $n_{z,x}(R) = n_6 + n_7$.

Step 1: If $(p, q) \in \mathfrak{n}_0$ *and* $p, q \geq 1$, *then* $(p - 1, q - 1) \in \mathfrak{n}_0$.

Let $(p, q) \in \mathfrak{n}_0$ be such that $p, q \geq 1$. Consider a profile R consisting of $p - 1$ agents of type 3, $q - 1$ agents of type 6, 1 agent of type 4 and type 5, and $n - (p + q)$ agents of type 1 (thus there is no type 1 agent if $p + q = n$). That is, at R, $n_1 = n - (p + q)$, $n_2 = 0$, $n_3 = p - 1$, $n_4 = 1$, $n_5 = 1$, $n_6 = q - 1$, and $n_7 = 0$. Then $n_{x,y}(R) = p$, $n_{y,x}(R) = q$, $n_{y,z}(R) = 1$, $n_{z,y}(R) = 1$, $n_{x,z}(R) = p - 1$, and $n_{z,x}(R) = q - 1$. Since $(p, q) \in \mathfrak{n}_0$, $x \succeq_{f(R)} y$. By (4.7), $(1, 1) \in \mathfrak{n}_0$ and so $y \succeq_{f(R)} z$. Then by transitivity, $x \succeq_{f(R)} z$, which means $(p - 1, q - 1) \in \mathfrak{n}_0$.

Step 2: $\max\{q - p : (p, q) \in \mathfrak{n}_0\} = n$ *or* 0.

Let $(p^*, q^*) \in \mathfrak{n}_0$ be such that

$$q^* - p^* = \max\{q - p : (p, q) \in \mathfrak{n}_0\}. \tag{4.13}$$

Suppose $q^* - p^* \neq 0$. Then applying Step 1 repeatedly p^*-times, we show $(0, q^* - p^*) \in \mathfrak{n}_0$. Then since $q^* - p^* \geq 1$, by $\mathfrak{n}$-monotonicity, $(0, 1) \in \mathfrak{n}_0$.

Suppose by contradiction $q^* - p^* \neq n$. Then evidently $q^* \leq n - 1$. Thus there is a profile R consisting of $q^* - p^*$ agents of type 6, 1 agent of type 7, and the rest of $n - (q^* - p^* + 1)$ agents of type 1 (note that $q^* - p^* + 1 \leq q^* + 1 \leq n$ and so the number of agents of type 1 is a non-negative integer and the total number of agents is n). Then at R, $n_1 = n - (q^* - p^* + 1)$, $n_2 = n_3 = n_4 = n_5 = 0$, $n_6 = q^* - p^*$, and $n_7 = 1$. Thus $n_{x,y}(R) = 0$, $n_{y,x}(R) = q^* - p^*$, $n_{y,z}(R) = 0$,

[7] A stronger property, adding a domain richness to the property of two-indifference-class, is needed to prove this result. See Ju (2009b) for details.

$n_{z,y}(R) = 1, n_{x,z}(R) = 0$, and $n_{z,x}(R) = q^* - p^* + 1$. Since $(0, q^* - p^*), (0, 1) \in \mathfrak{n}_0$, then $x \succsim_{f(R)} y$ and $y \succsim_{f(R)} z$. By transitivity, $x \succsim_{f(R)} z$, which implies $(0, q^* - p^* + 1) \in \mathfrak{n}_0$, contradicting (4.13).

Step 3: f is either majority decision function or degenerate indifference function.

When $q^* - p^* = 0$, this and (4.7) imply that f is majority decision function. When $q^* - p^* = n$, $(p^*, q^*) = (0, n)$. Thus by $\mathfrak{n}$-monotonicity, $\mathfrak{n}_0 = \mathfrak{n}^*$. Hence for all $R \in \mathcal{D}^{2ic}_{N,X}$ and all $x, y \in X$, $(n_{x,y}(R), n_{y,x}(R)), (n_{y,x}(R), n_{x,y}(R)) \in \mathfrak{n}_0$; so $x \sim_{f(R)} y$. Therefore, f is degenerate indifference function. □

Remark 4.2.1. Maskin (1995) proved that on the domain of *linear* preference profiles with an *odd* number of voters, majority decision function is "most transitive" among social decision functions satisfying monotonicity, anonymity, and neutrality (in fact, he considers independence of irrelevant alternatives and weak Pareto instead of monotonicity). A similar result without the odd-number-assumption is obtained by Campbell and Kelly (2000). These results rely on some domain richness properties that our dichotomous domain does not have; e.g., Campbell and Kelly's characterization relies on the availability of single-peaked preferences in the domain. In addition, dichotomous preferences do not have linearity assumed in the above two papers. Moreover, our result is with transitivity on the "entire domain under consideration" and for both odd or even numbers of voters.

Other social decision functions satisfying monotonicity, anonymity, and neutrality violate transitivity. However, all these functions satisfy acyclicity.

Theorem 4.2.4 (Ju 2009b). *On any domain with the property of two-indifference-class, all social decision functions with monotonicity, anonymity, and neutrality satisfy acyclicity.*

Proof. Let f be a social decision function on $\mathcal{D}^{2ic}_{N,X}$ satisfying the three axioms. By Proposition 4.2.2, f is represented by an $\mathfrak{n}$-neutral index structure $\mathfrak{n} \equiv (\mathfrak{n}_{x,y})_{x,y \in X}$ which satisfies (4.7). For all $x, y \in X$ with $x \neq y$, let $\mathfrak{n}_0 \equiv \mathfrak{n}_{x,y}$.

Step 1: For all $x, y \in X$ and all $R \in \mathcal{D}^{2ic}_{N,X}$, if $x \succ_{f(R)} y$, then $n_{x,y}(R) > n_{y,x}(R)$.

This follows directly from (4.7).

Step 2: For all $x, y, z \in X$ and all $R \in \mathcal{D}^{2ic}_{N,X}$, if $n_{x,y}(R) > n_{y,x}(R)$ and $n_{y,z}(R) > n_{z,y}(R)$, then $n_{x,z}(R) > n_{z,x}(R)$.

The proof of this step uses a similar argument as in the proof of Theorem 4.2.2. Let $n_{x,y}(R) > n_{y,x}(R)$ and $n_{y,z}(R) > n_{z,y}(R)$. Then the two inequalities in (4.11) hold with strict inequality and from them, (4.12) is obtained as a strict inequality, which means $n_{x,z}(R) > n_{z,x}(R)$.

Step 3: If $R \in \mathcal{D}^{2ic}_{N,X}$ and a sequence of finite alternatives, $x_1, \ldots, x_T \in X$ are such that $x_1 \succ_{f(R)} x_2, x_2 \succ_{f(R)} x_3, \ldots, x_{T-1} \succ_{f(R)} x_T$, then $x_1 \succsim_{f(R)} x_T$; thus $x_T \succ_{f(R)} x_1$ does not hold.

If $x_1 \succ_{f(R)} x_2$ and $x_2 \succ_{f(R)} x_3$, then by Step 1, $n_{x_1,x_2}(R) > n_{x_2,x_1}(R)$ and $n_{x_2,x_3}(R) > n_{x_3,x_2}(R)$, which imply by Step 2, $n_{x_1,x_3}(R) > n_{x_3,x_1}(R)$. Applying this argument iteratively, we obtain, $n_{x_1,x_T}(R) > n_{x_T,x_1}(R)$, which implies, by (4.7), $x_1 \succsim_{f(R)} x_T$. □

4.2.4 Two Fixed Classes Separated by Strict Preferences

Now we consider a domain where alternatives are separated into two fixed subsets and any alternative in one is always preferred to any alternative in the other. Formally:

Definition 4.2.3. A domain $\mathcal{D}^{2fcs}_{N,X} \subseteq \mathcal{U}_{N,X}$ has the property of *two-fixed-class-separation* if for all distinct triples $x, y, z \in X$, there is a nonempty proper subset $A \subsetneq \{x, y, z\}$ such that for all $R \in \mathcal{D}^{2fcs}_{N,X}$ and all $i \in N$, either [for all $a \in A$ and all $b \in \{x, y, z\}\backslash A$, $aP_i b$] or [for all $a \in A$ and all $b \in \{x, y, z\}\backslash A$, $bP_i a$].[8]

Theorem 4.2.5 (Inada 1964). *On any domain with the property of two-fixed-class-separation, if the number of agents is odd, majority decision function satisfies transitivity.*

Proof. Let $x, y, z \in X$ be three distinct alternatives. Without loss of generality, assume that the two fixed classes are $A \equiv \{x\}$ and $B \equiv \{y, z\}$. Let $R \in \mathcal{D}^{2fcs}_{N,X}$. To prove transitivity, we need to consider the following six cases: (1) $x \succsim_{f_{MAJ}(R)} y$ and $y \succsim_{f_{MAJ}(R)} z$, (2) $x \succsim_{f_{MAJ}(R)} z$ and $z \succsim_{f_{MAJ}(R)} y$, (3) $y \succsim_{f_{MAJ}(R)} z$ and $z \succsim_{f_{MAJ}(R)} x$, (4) $z \succsim_{f_{MAJ}(R)} y$ and $y \succsim_{f_{MAJ}(R)} x$, (5) $y \succsim_{f_{MAJ}(R)} x$ and $x \succsim_{f_{MAJ}(R)} z$, (6) $z \succsim_{f_{MAJ}(R)} x$ and $x \succsim_{f_{MAJ}(R)} y$. Arguments for (1) and (2) are similar and also the arguments for (3) and (4) and for (5) and (6) are similar. Thus we only consider (1), (3), and (5) below.

Note that by the property of two-fixed-class-separation, $N_{x,y}(R) = N_{x,z}(R)$ and $N_{y,x}(R) = N_{z,x}(R)$. Thus by independence of irrelevant alternatives and neutrality of f_{MAJ},

$$x \succsim_{f_{MAJ}(R)} y \iff x \succsim_{f_{MAJ}(R)} z. \tag{4.14}$$

Case 1: $x \succsim_{f_{MAJ}(R)} y$ and $y \succsim_{f_{MAJ}(R)} z$.

By (4.14), $x \succsim_{f_{MAJ}(R)} y$ implies $x \succsim_{f_{MAJ}(R)} z$.

Case 2: $y \succsim_{f_{MAJ}(R)} z$ and $z \succsim_{f_{MAJ}(R)} x$.

By (4.14), $z \succsim_{f_{MAJ}(R)} x$ implies $y \succsim_{f_{MAJ}(R)} x$.

Case 3: $y \succsim_{f_{MAJ}(R)} x$ and $x \succsim_{f_{MAJ}(R)} z$.

By (4.14), $y \succsim_{f_{MAJ}(R)} x$ implies $z \succsim_{f_{MAJ}(R)} x$. Hence $z \sim_{f_{MAJ}(R)} x$, which implies $n_{z,x}(R) = n_{x,z}(R)$. By the property of two-fixed-class-separation, $n_{z,x}(R) + n_{x,z}(R) = n$. Therefore n is an even number, contradicting the initial assumption. Therefore, Case 3 does not occur on the domain. □

With a stronger condition on the domain, we can show that except for degenerate indifference function, majority decision function is the only transitive social decision function satisfying the three standard axioms.

[8] Sakai and Shimoji (2006) study "dichotomous domains" that are close to domains with two-fixed-class-separation. Assuming that the domain of individual preferences can be either dichotomous or universal, they find some domain conditions for the existence of Arrovian social welfare function.

Theorem 4.2.6 (Ju 2009b). *Consider a domain with the property of two-fixed-class-separation.*[9] *Assume that all preferences in this domain are linear and that there is an odd number of agents. Then a social decision function satisfies monotonicity, anonymity, neutrality, and transitivity if and only if it is either majority decision function or degenerate indifference function.*

Proof. Let f be a social decision function on $\mathcal{D}_{N,X}$ satisfying the four axioms. By Proposition 4.2.2, f is represented by a decisive structure $\mathfrak{n} \equiv (\mathfrak{n}_{x,y})_{x,y \in X}$ satisfying (4.7). Let $\mathfrak{n}_0 \equiv \mathfrak{n}_{x,y}$ for all distinct $x, y \in X$. By (4.7) and the assumption that all preference orderings on domain $\mathcal{D}_{N,X}$ are linear, in order to show that f is majority decision function, we only have to show that for all $(n_1, n_2) \in \mathfrak{n}^*$ with $n_1 + n_2 = n$, if $n_1 < n_2$, then $(n_1, n_2) \notin \mathfrak{n}_0$. Suppose that f is not majority function and so for some $(n_1, n_2) \in \mathfrak{n}^*$, $n_1 + n_2 = n$, $n_1 < n_2$, and $(n_1, n_2) \in \mathfrak{n}_0$. Let $x, y, z \in X$ be three distinct alternatives. Without loss of generality, assume that the two fixed classes are $A \equiv \{x\}$ and $B \equiv \{y, z\}$. Let $R \in \mathcal{D}_{N,X}$ be such that $n_{x,y}(R) = n_2$, $n_{y,x}(R) = n_1$, and for all $i \in N$, yP_iz. Then by the property of two-fixed-class-separation, $n_{x,z}(R) = n_2$ and $n_{z,x}(R) = n_1$. Since $(n_1, n_2) \in \mathfrak{n}_0$, $y \succsim_{f(R)} x$ and $z \succsim_{f(R)} x$. By (4.7), the reverse relations also hold and therefore $y \sim_{f(R)} x$ and $z \sim_{f(R)} x$. Finally by transitivity, $y \sim_{f(R)} z$. Since every agent prefers y to z at R by construction, this implies that $(0, n) \in \mathfrak{n}_0$, which means that f is degenerate indifference function. □

When there are even number of agents, the result does not hold, as shown by the following example due to Inada (1964). There are four agents with xP_iyP_iz and four agents with yI_izP_ix. Then majority decision gives $x \sim_{f_{MAJ}(R)} y$, $y \succ_{f_{MAJ}(R)} z$, and $x \sim_{f_{MAJ}(R)} z$, violating transitivity. However, note that this social preference relation is quasi-transitive. In fact, for quasi-transitivity, we do not need the odd number assumption. Moreover, any social decision function satisfying monotonicity and neutrality is quasi-transitive.

Theorem 4.2.7 (Ju 2009b). *On any domain with the property of two-fixed-class-separation, all social decision functions with monotonicity and neutrality satisfy quasi-transitivity.*

Proof. The proof is similar to the proof of the above theorem with the replacement of weak majority preference relation with the strict one. Note that the arguments used for Cases 1–2 in the above proof do not depend on the fact that the social decision function is majority decision function. The same arguments go through for any social decision function as long as it is represented by a decisive structure and is neutral. Case 3 will not occur now because $y \succ_{f(R)} x$ implies $z \succ_{f(R)} x$, which contradicts $x \succ_{f(R)} z$. □

Note that monotonicity in Theorem 4.2.7 can be weakened to independence of irrelevant alternatives.

[9] A stronger property, adding a domain richness to two-fixed-class-separation, is needed to prove this result. See Ju (2009b) for details.

4.3 Axiomatic Foundations for Majority Decision and Approval Voting

Throughout this section, assume that the set of alternatives X is fixed. Assume further that as in Sect. 4.2.3, preferences can have at most two indifference classes. These preferences are called *dichotomous preferences.* Each dichotomous preference is characterized by the set of best, or preferred alternatives. Thus we use $B_0 \in \bar{P}(X)$ to denote the dichotomous preference of which the set of preferred alternatives is B_0 and use $\mathfrak{D} \equiv \bar{P}(X)$ to denote the set of dichotomous preferences. In what follows, we fix the feasibility set to be equal to X and focus on characteristics of collective choice rules on the set of admissible preference profiles. Thus given a domain of dichotomous preferences $\mathcal{D} \subseteq \mathfrak{D}$, a *collective choice rule in this section* is a non-empty valued correspondence $c\colon \bigcup_{N \in \mathcal{N}} \mathcal{D}^N \to \bar{P}(X)$. Similarly a *collective choice quasi-rule* is a correspondence $c\colon \bigcup_{N \in \mathcal{N}} \mathcal{D}^N \to P(X)$ that may take the empty set as its value.

Rule $c(\cdot)$ is *anonymous* if the identities of persons are inessential, that is, for all $N, N' \in \mathcal{N}$ with $|N| = |N'|$ and all one-to-one functions $\lambda\colon N \to N'$, $c((B_i)_{i \in N}) = c((B_{\lambda(i)})_{i \in N})$. A profile of dichotomous preferences may be reduced to a function $\pi\colon \mathcal{D} \to \{0, 1, 2, \ldots\}$ mapping each dichotomous preference in the domain to the number of agents who have this preference. Let $\Pi(\mathcal{D})$ be the set of all such functions. With a slight abuse, we refer to elements in $\Pi(\mathcal{D})$ preference profiles. We often denote an anonymous rule (or quasi-rule) $c\colon \Pi(\mathcal{D}) \to P(X)$ as a function on $\Pi(\mathcal{D})$ instead of its original domain $\bigcup_{N \in \mathcal{N}} \mathcal{D}^N$.

A *voting system* is a pair of a set of valid ballots $\mathfrak{B} \subseteq P(X)$ and a non-empty valued correspondence $\phi\colon \bigcup_{N \in \mathcal{N}} \mathfrak{B}^N \to \bar{P}(X)$ on the set of all possible ballot profiles. We call $\phi(\cdot)$ a *ballot aggregator.* Voting system $(\mathfrak{B}, \phi)$ is *anonymous* if for all $N, N' \in \mathcal{N}$ with $|N| = |N'|$ and all one-to-one functions $\lambda\colon N \to N'$, $\phi((B_i)_{i \in N}) = \phi((B_{\lambda(i)})_{i \in N})$. For an anonymous voting system, the identities of voters are inessential. Reducing this information, a *ballot response profile* $\pi : \mathfrak{B} \to \{0, 1, 2, \ldots\}$ maps each valid ballot into the number of voters casting this ballot. Let $\Pi(\mathfrak{B})$ be the set of all ballot response profiles. For an anonymous voting system $(\mathfrak{B}, \phi)$, for all pairs $N, N' \in \mathcal{N}$, if $(B_i)_{i \in N}$ and $(B'_i)_{i \in N'}$ generate the same ballot response profile, then $\phi((B_i)_{i \in N}) = \phi((B'_i)_{i \in N'})$. Therefore we may define a ballot aggregator ϕ as a function φ on the set of ballot response profiles $\Pi(\mathfrak{B})$. Conversely, any such function $\varphi\colon \Pi(\mathfrak{B}) \to \bar{P}(X)$ defines an anonymous ballot aggregator. We call $\varphi\colon \Pi(\mathfrak{B}) \to \bar{P}(X)$ a *voting rule*. When voters have dichotomous preferences and reveal their true preferences using ballot response profiles in $\Pi(\mathfrak{B})$, a voting system $(\mathfrak{B}, \varphi)$ gives the collective choice rule identical to the voting rule φ.

Throughout Sects. 4.3 and 4.4, we assume that ballot space $\mathfrak{B}$ satisfies the *basic richness*, consisting of the following two properties: for all distinct pairs $x, y \in X$ and all permutations $\lambda\colon X \to X$,

$$\text{There is } B_0 \in \mathfrak{B} \text{ such that } x \in B_0 \text{ and } y \notin B_0. \tag{4.15}$$

$$\text{For all } B_0 \in \mathfrak{B},\ \lambda(B_0) \in \mathfrak{B}. \tag{4.16}$$

Given a profile $\pi \in \Pi(\mathcal{D})$, for all $x \in X$, let $n(x,\pi) \equiv \sum_{B_0 \in \mathcal{D}: x \in B_0} \pi(B_0)$ be the number of votes x wins at π. *Majority rule* on $\mathcal{D}$, $c_{MAJ}\colon \Pi(\mathcal{D}) \to \bar{P}(X)$, maps each profile $\pi \in \Pi(\mathcal{D})$ into $c_{MAJ}(\pi) \equiv \{x \in X : \text{for all } y \in X, n(x,\pi) \geq n(y,\pi)\}$. In the case of voting systems, majority rule is denoted by ϕ_{MAJ} or φ_{MAJ}. Note that for dichotomous preferences, there always exists a Condorcet winner since majority decision function is transitive (Theorem 4.2.2). Thus the Condorcet rule $CW(\cdot)$ mapping each preference profile into the set of Condorcet winners is well-defined, and it coincides with majority rule. In general, any transitive social decision function $f : \mathcal{D}^N \to \mathcal{R}$ on the restricted domain of dichotomous preferences $\mathcal{D} \subseteq P(X)$ generates a collective choice rule as in (4.1). Since we fix the set of alternatives X in our definition of collective choice rules, not all social decision functions can be generated by collective choice rules. A collective choice rule can be considered as generating a social decision function of which the social preferences are dichotomous.

In the following two subsections, we overview some important axiomatic characterizations for majority rule and approval voting. A more focused overview of the literature considering the ballot space $\mathfrak{B} = \bar{P}(X)$ and approval voting is provided in Xu (2010) in this volume. Most of the characterizations we overview are accompanied by some conditions on ballot space $\mathfrak{B}$ that are sufficient for the characterization. Thus, we will clarify to what ballot spaces (or voting procedures) each characterization of majority rule applies, which was not all clear in the literature. We will find that some of the results apply to a very wide variety of ballot spaces (voting procedures) and others apply only to the ballot space for approval voting.

Throughout this section, our discussion is focused on voting systems. However, most results on voting systems also apply to collective choice rules after the straightforward extension of axioms and conditions we state for voting systems. When there is no need of distinguishing ballot space $\mathfrak{B}$ and the same domain of dichotomous preferences, we use $\mathfrak{B}$ to denote both the ballot space and the preference domain.

4.3.1 Characterizations of Majority Voting Systems

4.3.1.1 Basic Axioms in the Fixed Population Model

In this section, we define basic axioms for voting systems in a fixed population framework. Let $N \equiv \{1, 2, \ldots, n\}$ be the set of voters.

The first axiom says that alternatives should be treated equally. In other words, changing their labels should not make any essential change in the voting outcome.

Neutrality: For all $B \in \mathfrak{B}^N$ and all permutations $\lambda\colon X \to X$, $\lambda(\phi(B)) = \phi(\lambda(B))$.

The next axiom introduced by Baigent and Xu (1991) has the flavor of anonymity. It embodies the condition that each vote for an alternative by a voter has the same weight independently of what other alternatives are in his ballot.

Independence of Vote Exchange: For all $B \in \mathfrak{B}^N$ and all $i, j \in N$, if $x \in B_i \backslash B_j$ and $y \in B_j \backslash B_i$, then letting $B_i' \equiv [B_i \backslash \{x\}] \cup \{y\}$ and $B_j' \equiv [B_j \backslash \{y\}] \cup \{x\}$,[10] $\phi(B_i', B_j', B_{-\{i,j\}}) = \phi(B)$.[11]

When only singleton ballots are available, this axiom coincides with anonymity. The next axiom pertains to even more drastic vote reallocations than vote exchange.

Independence of Vote Reallocation: For all $B, B' \in \mathfrak{B}^N$, if for all $x \in X$, $n(x, B) = n(x, B')$, then $\phi(B) = \phi(B')$.

Clearly, independence of vote reallocation implies independence of vote exchange.

The next axiom says that when two alternatives win the same number of votes, they should be treated equally.

Equal Treatment of Equal Votes: For all $B \in \mathfrak{B}^N$ and all $x, y \in X$, if $n(x, B) = n(y, B)$, then $x \in \phi(B)$ if and only if $y \in \phi(B)$.[12]

This axiom is an implication of neutrality and independence of vote exchange as shown by the next lemma. Baigent and Xu (1991) obtain this implication in a richer setting with choice aggregation procedures.

Lemma 4.3.1. *Neutrality and independence of vote exchange together imply equal treatment of equal votes.*

Proof. Let $B \in \mathfrak{B}^N$ and $x, y \in X$ be such that $n(x, B) = n(y, B)$. Since $n(x, B) = n(y, B)$, then $N(x, B) \backslash N(y, B)$ and $N(y, B) \backslash N(x, B)$ have the same cardinality. Thus it is possible to exchange one x-vote and one y-vote between agents in the former set and agents in the latter set one by one. Let $B' \in \mathfrak{B}^N$ be the profile obtained after these vote exchanges. Applying the reverse iterative vote exchanges at B', we return to B.

It is clear that B' can also be obtained after the transposition of x and y at B, that is, letting $\tau\colon X \to X$ be such that $\tau(x) = y$, $\tau(y) = x$, and $\tau(z) = z$ for all $z \in X \backslash \{x, y\}$, we have $B' = \tau B \equiv (\tau(B_i))_{i \in N}$. Clearly, $\tau B' = B$.

By neutrality, $x \in \phi(B)$ if and only if $\tau(x) = y \in \phi(\tau B) = \phi(B')$. By independence of vote exchange, $y \in \phi(B')$ if and only if $y \in \phi(B)$. Therefore, $x \in \phi(B)$ if and only if $y \in \phi(B)$. □

Baigent and Xu (1991) reformulate May's (1952) positive response for social decision function in the current framework as follows.

Positive Response to Vote Addition: For all $B \in \mathfrak{B}^N$ and all $i \in N$, if $x \notin B_i$ and $B_i' \equiv B_i \cup \{x\} \in \mathfrak{B}$, then $x \in \phi(B)$ implies $\phi(B_i', B_{-i}) = \{x\}$.

Note that this axiom has bite when the ballot space $\mathfrak{B}$ is closed under the addition of an alternative (vote) to any ballot. For example, if $\mathfrak{B} \equiv \{\{x\} : x \in X\}$, any ballot

---

[10] The two ballots B_i', B_j' are admissible in $\mathfrak{B}$ because of assumption (4.16).

[11] Xu (2010) in this volume and Baigent and Xu (1991) call this axiom "independence of symmetric substitution."

[12] The same axiom is called as "equal treatment" in Xu (2010) in this volume.

aggregator satisfies this axiom trivially. The next axiom is an alternative formulation that has a wider applicability.

Positive Response* to Vote Addition: For all $B \in \mathfrak{B}^N$, all $i \in N$, and all $x, y \in X$, if $x \notin B_i$, $x \in B_i' \in \mathfrak{B}$, and $B_i \cap \{y\} = B_i' \cap \{y\}$, then $x \in \phi(B)$ implies $\phi(B_i', B_{-i}) \cap \{x, y\} = \{x\}$.

The next axiom says that any additional vote for another alternative does not do any good for alternative x.

Negative Response to Competing Vote Addition: For all $B \in \mathfrak{B}^N$, all $i \in N$, and all $x, y \in X$, if $y \notin B_i$ and $B_i \cup \{y\} \in \mathfrak{B}$, then $[x \notin \phi(B)$ or $y \in \phi(B)]$ implies $x \notin \phi(B_i \cup \{y\}, B_{-i})$ (i.e., $x \in \phi(B_i \cup \{y\}, B_{-i})$ implies $x \in \phi(B)$ and $y \notin \phi(B)$).

Equivalently, for all $B \in \mathfrak{B}^N$, all $i \in N$, and all $x, y \in X$, if $y \in B_i$ and $B_i \backslash \{y\} \in \mathfrak{B}$, then $x \notin \phi(B_i \backslash \{y\}, B_{-i})$ or $y \in \phi(B_i \backslash \{y\}, B_{-i})$ implies $x \notin \phi(B)$ (i.e., $x \in \phi(B)$ implies $x \in \phi(B_i \backslash \{y\}, B_{-i})$ and $y \notin \phi(B_i \backslash \{y\}, B_{-i})$). Like positive response, this axiom has bite when the ballot space is closed under the addition of an alternative. Here is an alternative formulation with wider applicability.

Negative Response* to Competing Vote Addition: For all $B \in \mathfrak{B}^N$, all $i \in N$, all $B_i' \in \mathfrak{B}$, and all $x, y \in X$, if $y \notin B_i$, $y \in B_i'$, and $B_i \cap \{x\} = B_i' \cap \{x\}$, then $[x \notin \phi(B)$ or $y \in \phi(B)]$ implies $x \notin \phi(B_i', B_{-i})$ (equivalently, $x \in \phi(B_i', B_{-i})$ implies $x \in \phi(B)$ and $y \notin \phi(B)$).

4.3.1.2 Characterization Results: Voting Systems

We first show that May's Theorem (Theorem 4.2.1) for the binary choice framework can be extended in the current framework in a fairly straightforward manner. This result is based on Propositions 4.2.1 and 4.2.2. Since there can be more than two alternatives, we need independence of irrelevant alternatives in addition to May's three axioms.

Theorem 4.3.1. *A social decision function on $\mathfrak{B}^N$ satisfies independence of irrelevant alternatives, anonymity, neutrality, and positive response if and only if it is majority decision function on $\mathfrak{B}^N$. Moreover, majority decision function on $\mathfrak{B}^N$ satisfies transitivity and generates majority voting system $(\mathfrak{B}, \phi_{MAJ})$ as its choice rule.*

Proof. By Theorem 4.2.2, majority decision function satisfies transitivity on dichotomous domain $\mathfrak{B}$ as well as the other three axioms. To prove the converse, let f be a social decision function on $\mathfrak{B}^N$ satisfying the four stated axioms. Independence of irrelevant alternatives and positive response together imply monotonicity. Due to the richness of ballot space $\mathfrak{B}$ stated in (4.15) and (4.16), Proposition 4.2.2

holds, and f can be represented by a decisive structure. By anonymity, neutrality, and Proposition 4.2.2, f can be represented by a decisive index structure satisfying (4.7). Following the same argument as is given before Theorem 4.2.1, we show (4.10). □

As a corollary, we obtain:

Corollary 4.3.1. *A voting system* $(\mathfrak{B}, \phi)$ *is generated by a social decision function on* $\mathfrak{B}^N$ *satisfying independence of irrelevant alternatives, anonymity, neutrality, and positive response if and only if it is a majority voting system, that is,* $\phi = \phi_{MAJ}$. *Thus when* $\mathfrak{B} = \bar{P}(X)$, *it is approval voting system.*

In the framework of collective aggregation procedures, Baigent and Xu (1991) obtain a similar axiomatic characterization of *approval* voting imposing positive response to vote addition. In the current framework, their result can be stated as follows:

Theorem 4.3.2 (Baigent and Xu 1991). *Assume that ballot space* $\mathfrak{B}$ *is closed under the addition of a single vote, that is, for all* $B_0 \in \mathfrak{B}$ *and all* $x \in X$, $B_0 \cup \{x\} \in \mathfrak{B}$.[13] *Then the following are equivalent*:

(i) *Voting system* $(\mathfrak{B}, \phi)$ *satisfies neutrality, independence of vote exchange, and positive response to vote addition.*[14]
(ii) *Voting system* $(\mathfrak{B}, \phi)$ *satisfies equal treatment of equal votes and positive response to vote addition.*
(iii) *Voting system* $(\mathfrak{B}, \phi)$ *is a majority voting system,* $\phi = \phi_{MAJ}$.

Proof. Lemma 4.3.1 shows that (i) implies (ii). It is easy to show (iii) implies (i). We only prove (ii) implies (iii) below. Let $\mathfrak{B}$ be the ballot space with the stated property.

Let ϕ be the ballot aggregator in part (ii). Let $B \in \mathfrak{B}^N$. We need to show that $x \in \phi(B)$ if and only if for all $y \in X$, $n(x, B) \geq n(y, B)$. By equal treatment of equal votes, we only have to show the "only if" part. Suppose to the contrary that $x \in \phi(B)$ and for some $y \in X, n(y, B) > n(x, B)$. Then $N(y, B) \backslash N(x, B) \neq \emptyset$ and there are at least $[n(y, B) - n(x, B)]$ agents in this set. Change ballots of these agents from B_i to $B'_i \equiv B_i \cup \{x\}$. For all other i's, let $B'_i \equiv B_i$. Thus by construction, $n(x, B') = n(y, B')$. By positive response to vote addition, $\phi(B') = \{x\}$. On the other hand, since $n(x, B') = n(y, B')$, then by equal treatment of equal votes, $y \in \phi(B') = \{x\}$, which is a contradiction. □

Unlike Theorem 4.3.1, this result uses the assumption that the ballot space is closed under vote addition.

---

[13] Thus we need to allow $X \in \mathfrak{B}$. The assumption is needed to prove that (ii) implies (iii). It is not needed for other implications.

[14] Universal domain axiom is added in Baigent and Xu (1991).

Remark 4.3.1. Instead of the assumption on $\mathfrak{B}$ in the above theorem, require that for all distinct pairs $x, y \in X$, there is $B_0 \in \mathfrak{B}$ such that $x, y \in B_0$. Then the same result holds if positive response in parts (i) and (ii) is replaced with positive response*. To prove this, we only replace B_i' in the proof with $B_i' \in \mathfrak{B}$ such that $x, y \in B_i'$ and replace $\phi(B')$ in the proof with $\phi(B') \cap \{x, y\}$. The rest of the proof is the same.

The equivalence between (i) and (iii) is also stated in Theorem 5 of Xu (2010) in this volume, focusing on $\mathfrak{B} = P(X)\backslash\{\emptyset\}$. In fact, as stated in Theorem 4.3.2, the equivalence holds for a broader set of ballot spaces. The equivalence between (ii) and (iii) is somewhat close to Theorem 4 of Xu (2010) in this volume, focusing on $\mathfrak{B} = P(X)\backslash\{\emptyset\}$.

An alternative characterization with negative response to competing vote addition is obtained with a different assumption on $\mathfrak{B}$.

Theorem 4.3.3. *Assume that ballot space $\mathfrak{B}$ is closed under the deletion of a single vote, that is, for all $B_0 \in \mathfrak{B}$ and all $x \in B_0$, $B_0\backslash\{x\} \in \mathfrak{B}$.*[15] *Then the following are equivalent:*

(i) *Voting system $(\mathfrak{B}, \phi)$ satisfies neutrality, independence of vote exchange, and negative response to competing vote addition.*
(ii) *Voting system $(\mathfrak{B}, \phi)$ satisfies equal treatment of equal votes and negative response to competing vote addition.*
(iii) *Voting system $(\mathfrak{B}, \phi)$ is a majority voting system, $\phi = \phi_{MAJ}$.*

Proof. By Lemma 4.3.1, (i) implies (ii). We only prove that (ii) implies (iii). Let $\mathfrak{B}$ be given as stated above.

Let ϕ be the ballot aggregator in part (ii). Let $B \in \mathfrak{B}^N$. We need to show that $x \in \phi(B)$ if and only if for all $y \in X$, $n(x, B) \geq n(y, B)$. By equal treatment of equal votes, we only have to show the "only if part." Suppose to the contrary that $x \in \phi(B)$ and for some $y \in X$, $n(y, B) > n(x, B)$. Then $N(y, B)\backslash N(x, B) \neq \emptyset$ and there are at least $[n(y, B) - n(x, B)]$ agents in this set. Change ballots of these agents from B_i to $B_i' \equiv B_i\backslash\{y\}$ (this is possible by the assumption on $\mathfrak{B}$). For all other i's, let $B_i' \equiv B_i$. Thus by construction, $n(x, B') = n(y, B')$. Applying negative response to competing vote addition repeatedly, we show $x \in \phi(B')$ and $y \notin \phi(B')$, contradicting equal treatment of equal votes for $n(x, B') = n(y, B')$. □

Remark 4.3.2. Assume instead that for all distinct pairs $x, y \in X$, there is $B_0 \in \mathfrak{B}$ such that $B_0 \cap \{x, y\} = \emptyset$. Then the same result holds if negative response in parts (i) and (ii) is replaced with negative response*. To prove this, we only replace B_i' in the proof with $B_i' \in \mathfrak{B}$ such that $B_i' \cap \{x, y\} = \emptyset$. The rest of the proof is the same.

4.3.1.3 Extension in the Variable Population Framework

We now consider voting systems on a variable population domain. All the axioms defined in the fixed population framework can be extended to that framework by

[15] Thus we need to allow $\emptyset \in \mathfrak{B}$.

simply adding the quantifier "for all $N \in \mathcal{N}$." All results in the previous section can be extended in the variable population framework. In particular, Theorem 4.3.2 can be so extended. Moreover, the result applies to more variety of ballot spaces by adding the extra, but mild condition that the empty ballot is allowed ($\emptyset \in \mathfrak{B}$), and the following natural axiom pertaining to the effect of the empty ballot. It says that adding an empty (abstention) vote does not affect the voting outcome.

Null Consistency: For all $N \in \mathcal{N}$ and all $B \in \mathfrak{B}^N$, if $i \notin N$ and $B_i = \emptyset$, then $\phi(B) = \phi(B, B_i)$.

The next two results extend Theorem 4.3.2 in the variable population framework.

Theorem 4.3.4. *Assume that* $\emptyset \in \mathfrak{B}$ *and for all* $x \in X$, $\{x\} \in \mathfrak{B}$. *Then on the domain* $\bigcup_{N \in \mathcal{N}} \mathfrak{B}^N$, *the following are equivalent:*

(i) *Voting system* $(\mathfrak{B}, \phi)$ *satisfies null consistency, neutrality, independence of vote exchange, and positive response to vote addition.*
(ii) *Voting system* $(\mathfrak{B}, \phi)$ *satisfies null consistency, equal treatment of equal votes, and positive response to vote addition.*
(iii) *Voting system* $(\mathfrak{B}, \phi)$ *is a majority voting system,* $\phi = \phi_{MAJ}$.

Proof. The proof is similar to the proof of Theorem 4.3.2. To prove that (ii) implies (iii), suppose to the contrary that $x \in \phi(B)$ and for some $y \in X$, $n(y, B) > n(x, B)$. Let N' be a set of $[n(y, B) - n(x, B)]$ agents such that $N' \cap N = \emptyset$. Let $B^0 \equiv (\emptyset, \ldots, \emptyset) \in \mathfrak{B}^{N'}$ and $B' \in \mathfrak{B}^{N'}$ be such that for each $i \in N'$, $B'_i = \{x\}$. By construction, $n(x, (B, B')) = n(y, (B, B'))$ and by null consistency, $\phi(B, B^0) = \phi(B)$ and so $x \in \phi(B, B^0)$. Applying positive response to vote addition repeatedly at (B, B^0), we get $\phi(B, B') = \{x\}$. On the other hand, by equal treatment of equal votes, $y \in \phi(B, B') = \{x\}$, which is a contradiction. □

Replacing positive response with positive response* to vote addition, we obtain a similar result. Unlike in Theorem 4.3.4, we do not need any assumption on the ballot space except for the availability of the empty ballot.

Theorem 4.3.5. *Assume that* $\emptyset \in \mathfrak{B}$. *On the domain* $\bigcup_{N \in \mathcal{N}} \mathfrak{B}^N$, *the following are equivalent:*

(i) *Voting system* $(\mathfrak{B}, \phi)$ *satisfies null consistency, neutrality, independence of vote exchange, and positive response* to vote addition.*
(ii) *Voting system* $(\mathfrak{B}, \phi)$ *satisfies null consistency, equal treatment of equal votes, and positive response* to vote addition.*
(iii) *Voting system* $(\mathfrak{B}, \phi)$ *is a majority voting system,* $\phi = \phi_{MAJ}$.

Proof. To prove this, we only have to replace $B'_i = \{x\}$ in the proof of Theorem 4.3.4 with $B'_i \in \mathfrak{B}$ such that $x \in B'_i$ and $y \notin B'_i$ (such B'_i exists by (4.15)) and replace $\phi(B, B')$ with $\phi(B, B') \cap \{x, y\}$. The rest of the proof is the same. □

4.3.2 *Characterizations of Majority Voting in the Variable Population Framework*

In this section, we consider anonymous voting systems in the variable population framework. Recall that such a voting system can be described by a pair of a ballot space $\mathfrak{B}$ and a voting rule (an anonymous ballot aggregator) $\varphi\colon \Pi(\mathfrak{B}) \to \bar{P}(X)$. We will use the following concept and notation.

A *null response profile* is a profile π where no alternative is supported by anyone, that is, for all $A \in \bar{P}(X)$, $\pi(A) = 0$. The *empty response profile* $\pi^{\emptyset}$ is the null response profile with no vote, that is, for all $A \in P(X)$, $\pi^{\emptyset}(A) = 0$. For all $A \in P(X)$, let π_A be such that $\pi_A(A) = 1$ and for all other ballots $B \in P(X)\backslash\{A\}$, $\pi_A(B) = 0$. For all $\pi \in \Pi(\mathfrak{B})$ and all $x \in X$, let $n(x,\pi) \equiv \sum_{A:x\in A} \pi(A)$ and $n(\pi) \equiv \sum_{x\in X} n(x,\pi)$.

4.3.2.1 Basic Axioms of Voting Rules

Let $m \equiv |X|$ be the number of alternatives. The following axioms have been considered by numerous authors in the literature on approval voting.

First, if there is only one voter, that voter's ballot should be fully respected.

Faithfulness: For all $A \in \mathfrak{B}\backslash\{\emptyset\}$, $\varphi(\pi_A) = A$.

Neutrality can be defined in the same way in the current framework as in earlier sections. A much weaker axiom requires that decisions at a null response profile should be neutral.

Null-Neutrality: For all null response profiles $\pi \in \Pi(\mathfrak{B})$, $\varphi(\pi) = X$.

The next axiom plays a key role in some characterizations of approval voting to be presented later. It pertains to a merger of two groups of voters. If a rule has a common recommendation for the two groups before the merger, the common recommendation should be the recommendation after the merger.

Consistency: For all $\pi, \pi' \in \Pi(\mathfrak{B})$, if $\varphi(\pi) \cap \varphi(\pi') \neq \emptyset$, then $\varphi(\pi + \pi') = \varphi(\pi) \cap \varphi(\pi')$.[16]

The next one is a weaker version of consistency considered by Sertel (1988).

Weak Consistency: For all $\pi \in \Pi(\mathfrak{B})$ and all $A \in \mathfrak{B}$, if $\varphi(\pi) \cap \varphi(\pi_A) \neq \emptyset$, then $\varphi(\pi + \pi_A) = \varphi(\pi) \cap \varphi(\pi_A)$.

The next axiom says that when there are two voters casting disjoint ballots, a voting rule should recommend the union of the two ballots.

Disjoint Equality: For all $A, B \in \mathfrak{B}\backslash\{\emptyset\}$, if $A \cap B = \emptyset$, then $\varphi(\pi_A + \pi_B) = A \cup B$.

The next axiom proposed by Sertel (1988) captures a similar idea but in a much stronger form.

[16] This axiom and other axioms of consistency were studied also by Ching (1996) and Yeh (2006) for characterizations of plurality voting rule on the standard domain of preferences.

Sertel Disjoint Equality: For all $\pi \in \Pi(\mathfrak{B})$ and all $A \in \mathfrak{B}$ if $\varphi(\pi) \cap \varphi(\pi_A) = \emptyset$, then $x \in \varphi(\pi + \pi_A)$ if and only if $x \in \varphi(\pi)$ or $[x \in \varphi(\pi_A)$ and $\max_{y \in \varphi(\pi)} n(y, \pi) = 0]$ or $[x \in \varphi(\pi_A)$ and $n(x, \pi) = \max_{y \in \varphi(\pi)} n(y, \pi) - 1 \geq 0]$.

The next axiom pertains to a special case of ballot responses where all alternatives receive the same number of votes. It requires in this case that a voting rule should treat all alternatives equally by recommending all of them.

Cancellation: For all $\pi \in \Pi(\mathfrak{B})$, if all alternatives receive the same number of votes at π, that is, for all $x, y \in X$, $n(x, \pi) = n(y, \pi)$, then $\varphi(\pi) = X$.

Cancellation implies that the choice at any null response profile should be X as in approval voting.

The next axiom requires that a voting rule should make the same decision when two voters merge their ballots and cast the merged ballot as a single voter.

Independence of Pairwise Vote Merge: For all $\pi \in \Pi(\mathfrak{B})$ and all $A, B \in \mathfrak{B}$, if $A \cap B = \emptyset$ and $A \cup B \in \mathfrak{B}$, then $\varphi(\pi + \pi_A + \pi_B) = \varphi(\pi + \pi_{A \cup B})$.

Vote merge is a type of vote reallocation. The next independence axiom pertains to more drastic vote reallocations.

Independence of Vote Reallocation: For all $\pi, \pi' \in \Pi(\mathfrak{B})$, if for all $x \in X$, $n(x, \pi) = n(x, \pi')$, then $\varphi(\pi) = \varphi(\pi')$.

Note that independence of pairwise vote merge together with faithfulness and consistency imply cancellation.[17]

4.3.2.2 Scoring Rules

Majority rule is an example in the large family of voting rules based on scoring methods. Characterization of this family is quite useful for our later discussion of majority or approval voting.

A score function $s\colon \{1, \ldots, m\} \to \mathbb{R}$ maps each natural number of a ballot size into a real number (the score of the ballot). For all $\pi \in \Pi(\mathfrak{B})$ and all $x \in X$, let

$$p(x, \pi; s) \equiv \sum_{B \in \mathfrak{B}: x \in B} s(|B|)\pi(B) = \sum_{k=1}^{m} \sum_{\substack{B \in \mathfrak{B}: x \in B, \\ |B|=k}} s(k)\pi(B)$$

be the total points x wins at π under score function s. A voting rule φ is a *scoring rule* if there is a score function $s\colon \{1, \ldots, m\} \to \mathbb{R}$ such that for all $\pi \in \Pi(\mathfrak{B})$,

[17] To show this let $\pi \in \Pi(\mathfrak{B})$ be such that for all $x, y \in X$, $n(x, \pi) = n(y, \pi)$. Note that by independence of pairwise vote merge (when $B = \emptyset$), we may assume that $\pi(\emptyset) = 0$. Let $n \equiv n(x, \pi)$ for all $x \in X$. Applying this axiom again repeatedly, we obtain $\varphi(\pi) = \varphi(\sum_{x \in X} n\pi_{\{x\}}) = \varphi(n \sum_{x \in X} \pi_{\{x\}}) = \varphi(n\pi_X)$. By faithfulness and consistency, $\varphi(n\pi_X) = X$.

$$\varphi(\pi) \equiv \{x \in X : p(x, \pi; s) \geq p(y, \pi; s) \text{ for all } y \in X\}.$$

Note that when φ is represented by score function s, φ is also represented by $a \times s$ for all $a > 0$. Majority rule is the scoring rule represented by a positive and constant score function s such as $s(k) = 1$ for all $k = 1, \ldots, m$. When $\mathfrak{B} = P(X)$ or $P(X)\backslash\{\emptyset\}$ or $P(X)\backslash\{\emptyset, X\}$, majority rule on $\Pi(\mathfrak{B})$ is called as approval voting rule.[18]

Given a finite sequence of score functions $s^1, \ldots, s^T$, for all $x, y \in X$ and all $\pi \in \Pi$, $(p(y, \pi; s^t))_{t=1}^T$ *lexicographically dominates* $(p(x, \pi; s^t))_{t=1}^T$ if there is $t_0 \in \{1, \ldots, T\}$ such that $p(y, \pi; s^{t_0}) > p(x, \pi; s^{t_0})$ and for all $t = 1, \ldots, t_0 - 1$, $p(y, \pi; s^t) \geq p(x, \pi; s^t)$. A voting rule φ is a *lexicographic scoring rule* if there are $T \geq 1$ score functions $s^1, \ldots, s^T$ such that for all $\pi \in \Pi(\mathfrak{B})$, $x \in \varphi(\pi)$ if and only if there is no $y \in X$ such that $(p(y, \pi; s^t))_{t=1}^T$ lexicographically dominates $(p(x, \pi; s^t))_{t=1}^T$.

Young (1975) characterizes (lexicographic) scoring rules in the framework of ranked voting procedures, where voters can express their preferences in their ballots. The key axioms in his result are neutrality and consistency. In the current "non-ranked" voting procedures, the next two results are counterparts of Young's characterization.

Theorem 4.3.6 (Fishburn 1979). *Given a ballot space $\mathfrak{B} \subseteq P(X)\backslash\{\emptyset, X\}$, a voting rule satisfies neutrality and consistency if and only if it is a lexicographic scoring rule. Moreover, the number of score functions representing the rule is at most the number of possible sizes of ballots, namely, $|\{|B| : B \in \mathfrak{B}\}|$.*

Due to the nature of lexicographic comparison, "overwhelming majority" may not be enough to influence the voting outcome under lexicographic scoring rules. In order to avoid this unnatural feature, we impose the next axiom.[19]

Continuity: For all $\pi, \pi' \in \Pi(\mathfrak{B})$ and all $x \in X$, if $x \notin \varphi(\pi)$, then there is an integer $K > 0$ such that for all $k \geq K$, $x \notin \varphi(k\pi + \pi')$.

It is clear that scoring rules satisfy continuity since increasing k, the difference between the score of x and the score of another winning alternative at π gets arbitrarily larger. No other lexicographic scoring rules can satisfy continuity and we obtain:

Theorem 4.3.7 (Fishburn 1979). *Given a ballot space $\mathfrak{B} \subseteq P(X)\backslash\{\emptyset, X\}$, a voting rule satisfies neutrality, consistency, and continuity if and only if it is a scoring rule.*

Suppose that a lexicographic scoring rule is represented by score functions $s^1, \ldots,$ s^T, and for some $k \in \{|B| : B \in \mathfrak{B}\}$ and $t \in \{1, \ldots, T\}$, $s^1(k) = \cdots = s^{t-1}(k) = 0$

---

[18] Admissibility of $\emptyset$ or X in the ballot space does not make any essential difference in the choices made by majority rule.

[19] Myerson (1995) calls it "overwhelming majority."

and $s^t(k) < 0$. Then for any $B \in \mathfrak{B}$ with $|B| = k$ and any $b \in B$, $p(b, \pi_B; s^1) = \cdots = p(b, \pi_B; s^{t-1}) = 0$ and $p(b, \pi_B; s^t) = s^t(k) < 0$. Then b is not chosen by this lexicographic scoring rule, violating faithfulness. Thus faithfulness implies that the first non-zero component in $(s^1(k), \ldots, s^T(k))$ is positive. Consequently, a scoring rule satisfies faithfulness if and only if it is represented by a positive score function.

Corollary 4.3.2. *Assume that* $\mathfrak{B} \subseteq P(X)\backslash\{\emptyset, X\}$. *Then*:

(i) *A voting rule satisfies neutrality, consistency, and faithfulness if and only if it is a lexicographic scoring rule represented by a finite sequence of score functions* $s^1, \ldots, s^T$ *such that for all* $k \in \{1, \ldots, m\}$, *there is* $t \in \{1, \ldots, T\}$ *such that* $s^1(k) = \cdots = s^{t-1}(k) = 0 < s^t(k)$.
(ii) *A voting rule satisfies neutrality, consistency, continuity, and faithfulness if and only if it is a scoring rule represented by a positive score function.*

4.3.2.3 Characterizations of Majority Voting

If a positive score function s gives different score points for different ballot sizes, then there is $\pi \in \Pi$ such that all alternatives win the same number of votes but the alternatives winning a ballot with a higher score point have the greatest total score point. These alternatives are chosen and other alternatives are not chosen. For example, when $s(1) > s(2)$, let π be such that $\pi(\{a, b\}) = 1$, for all $x \in X\backslash\{a, b\}$, $\pi(\{x\}) = 1$, and for all other ballots Y, $\pi(Y) = 0$. Then the scoring rule will choose $X\backslash\{a, b\}$, which is a violation of cancellation. Thus, in order to satisfy cancellation, score function s must be constant. Therefore, the scoring rule represented by s is majority rule. The next result is similar to Young's characterization of the Borda rule for linear preferences.

Theorem 4.3.8 (Fishburn 1979). *Given a ballot space* $\mathfrak{B} \subseteq P(X)\backslash\{\emptyset, X\}$, *a voting rule satisfies neutrality, consistency, faithfulness, and cancellation if and only if it is majority rule.*

Note that this result holds for any arbitrary ballot space satisfying the richness conditions (4.15) and (4.16). For example, the ballot space consisting of only singleton ballots is rich. When the ballot space has no restriction on ballot sizes, the theorem yields a characterization of approval voting. The proof of Theorem 4.3.8 is relatively long. A much simpler proof is provided by Alos-Ferrer (2006) for unrestricted ballot space $\mathfrak{B} = \bar{P}(X)$. Moreover, he shows that neutrality in Fishburn's result can be dropped. The next theorem is based on the main results in Alos-Ferrer (2006).

Theorem 4.3.9. *Assume* $\mathfrak{B} \equiv \bar{P}(X)$. *Consider a voting rule* φ *on* $\Pi(\mathfrak{B})\backslash\{\pi^{\emptyset}\}$. *The following are equivalent*:

(i) *Voting rule* φ *satisfies faithfulness, consistency, and cancellation.*
(ii) *Voting rule* φ *satisfies faithfulness, consistency, and independence of pairwise vote merge.*

(iii) *Voting rule* φ *satisfies faithfulness, consistency, and independence of vote reallocation.*

(iv) *Voting rule* φ *is a majority rule,* $\varphi = \varphi_{MAJ}$.

Proof. It is easy to show (iv) $\Rightarrow$(i). In what follows, show (i) $\Rightarrow$ (ii) $\Rightarrow$ (iii) $\Rightarrow$ (iv).

Step 1: (i) $\Rightarrow$ (ii)

We only have to show that the three axioms in (i) imply independence of pairwise vote merge. Let $A, B \in P(X)$ be such that $A \cap B = \emptyset$. By cancellation,

$$\varphi(\pi_{A\cup B} + \pi_{X\setminus(A\cup B)}) = X = \varphi(\pi_A + \pi_B + \pi_{X\setminus(A\cup B)}). \tag{4.17}$$

Hence,

$$\begin{aligned} \varphi(\pi + \pi_A + \pi_B) &= \varphi(\pi + \pi_A + \pi_B) \cap \varphi(\pi_{A\cup B} + \pi_{X\setminus(A\cup B)}); \\ \varphi(\pi + \pi_{A\cup B}) &= \varphi(\pi + \pi_{A\cup B}) \cap \varphi(\pi_A + \pi_B + \pi_{X\setminus(A\cup B)}). \end{aligned} \tag{4.18}$$

Then by consistency,

$$\begin{aligned} &\varphi(\pi + \pi_A + \pi_B) \cap \varphi(\pi_{A\cup B} + \pi_{X\setminus(A\cup B)}) \\ &\quad = \varphi(\pi + \pi_A + \pi_B + \pi_{A\cup B} + \pi_{X\setminus(A\cup B)}); \\ &\varphi(\pi + \pi_{A\cup B}) \cap \varphi(\pi_A + \pi_B + \pi_{X\setminus(A\cup B)}) \\ &\quad = \varphi(\pi + \pi_{A\cup B} + \pi_A + \pi_B + \pi_{X\setminus(A\cup B)}). \end{aligned} \tag{4.19}$$

Finally, since $\pi+\pi_A+\pi_B+\pi_{A\cup B}+\pi_{X\setminus(A\cup B)} = \pi+\pi_{A\cup B}+\pi_A+\pi_B+\pi_{X\setminus(A\cup B)}$, then (4.18) and (4.19) give $\varphi(\pi + \pi_A + \pi_B) = \varphi(\pi + \pi_{A\cup B})$.

Step 2: (ii) $\Rightarrow$ (iii)

We only have to show that the three axioms in (ii) imply independence of vote reallocation. Let $\pi \in \Pi$. By independence of pairwise vote merge, we may assume that $\pi(\emptyset) = 0$. Iterative application of independence of pairwise vote merge gives $\varphi(\pi) = \varphi(\sum_{A\in\bar{P}(X)} \pi(A) \sum_{x\in A} \pi_{\{x\}})$. Since $\sum_{A\in\bar{P}(X)} \pi(A) \sum_{x\in A} \pi_{\{x\}} = \sum_{x\in X} n(x,\pi)\pi_{\{x\}}$,

$$\varphi(\pi) = \varphi(\sum_{x\in X} n(x,\pi)\pi_{\{x\}}).$$

Thus $\varphi(\pi)$ depends only on $n(x,\pi)$. Therefore, when π and π' satisfy $n(x,\pi) = n(x,\pi')$ for all $x \in X$, $\varphi(\pi) = \varphi(\pi')$.

Step 3: (iii) $\Rightarrow$ (iv)

Let $\pi \in \Pi(\mathfrak{B})$ and $K \equiv max_{x\in X} n(x,\pi)$. Since $\emptyset \notin \mathfrak{B}$ and $\pi^{\emptyset}$ is assumed to be out of the domain, $K > 0$.[20] For each $k \in \{1,\ldots,K\}$, let $X_k \equiv \{x \in X :$

[20] If $\pi^{\emptyset}$ is in the domain, neither independence of pairwise vote merge nor independence of vote reallocation implies $\varphi(\pi^{\emptyset}) = \varphi_{MAJ}(\pi^{\emptyset}) = X$, while cancellation does. Thus the equivalence

$n(x, \pi) = k\}$. Then $X_0, X_1, \ldots, X_K$ partition X. Now construct a non-decreasing sequence of subsets as follows:

$$\begin{array}{ll} Y_K & \equiv X_K \\ Y_{K-1} & \equiv X_K \cup X_{K-1} \\ Y_{K-2} & \equiv X_K \cup X_{K-1} \cup X_{K-2} \\ \vdots & \vdots \quad \vdots \\ Y_1 & \equiv X_K \cup X_{K-1} \cup X_{K-2} \cup \cdots \cup X_1 \end{array}$$

Note that $Y_K \cap Y_{K-1} = X_K, Y_K \cap Y_{K-1} \cap Y_{K-2} = X_K, \cdots, Y_K \cap Y_{K-1} \cap Y_{K-2} \cap \cdots \cap Y_1 = X_K$. By faithfulness,

$$\varphi(\pi_{Y_K}) = Y_K, \varphi(\pi_{Y_{K-1}}) = Y_{K-1}, \ldots, \varphi(\pi_{Y_1}) = Y_1.$$

Applying consistency,

$$\begin{aligned} \varphi(\pi_{Y_K} + \pi_{Y_{K-1}} + \cdots + \pi_{Y_1}) &= \varphi(\pi_{Y_K}) \cap \varphi(\pi_{Y_{K-1}}) \cap \cdots \cap \varphi(\pi_{Y_1}) \\ &= Y_K \cap Y_{K-1} \cap \cdots \cap Y_1 \\ &= X_K. \end{aligned}$$

Finally, since π and $\pi_{Y_K} + \pi_{Y_{K-1}} + \cdots + \pi_{Y_1}$ give the same number of votes for each alternative, by independence of vote reallocation, $\varphi(\pi) = X_K = \varphi_{MAJ}(\pi)$. □

The proof relies heavily on the richness of the ballot space $\mathfrak{B} \equiv \bar{P}(X)$. In particular, the ballot space is closed under union.[21] Therefore the result cannot be applied to restricted ballot spaces such as the space of singleton ballots. The equivalence between (i) and (iv) is also stated in Theorem 1 in Xu (2010) of this volume.

The next characterization of approval voting rule uses disjoint equality. Unlike Theorem 4.3.8, the result applies only to the ballot space $P(X)\backslash\{\emptyset, X\}$.

Theorem 4.3.10 (Fishburn 1978a, 1979). *Assume that* $\mathfrak{B} = P(X)\backslash\{\emptyset, X\}$ and consider voting rules over $\Pi(\mathfrak{B})\backslash\{\pi^{\emptyset}\}$.

(i) *Assume* $|X| = 2$. *Then a voting rule satisfies neutrality, consistency, and faithfulness if and only if it is majority rule.*
(ii) *Assume* $|X| \geq 3$. *Then a voting rule satisfies neutrality, consistency, and disjoint equality if and only if it is majority rule.*

cannot be established. If $\pi^{\emptyset}$ is in the domain, the result may be changed by replacing cancellation with a slightly weaker version by requiring $\pi \neq \pi^{\emptyset}$ in the definition of the axiom and weakening (iv) by allowing for any arbitrary choice at $\pi^{\emptyset}$.

[21] Alos-Ferrer (2006) assumes $X \notin \mathfrak{B}$. But then Y_1 in the above proof may not be an admissible ballot (the ballot space is not closed under union) and the proof does not go through. This is why we assume $X \in \mathfrak{B}$.

Sertel (1988) replaces disjoint equality with a stronger axiom, Sertel disjoint equality, and characterizes the rule that coincides with majority rule except when the empty set is the only ballot response. In this case, the rule selects the empty set (taking the empty value is allowed in his definition of voting rules). Although Sertel disjoint equality is less primitive and harder to motivate than disjoint equality, his proof is remarkably simpler than the proof of Theorem 4.3.10. Here we present his result in a different way in order to give a more clear comparison. Unlike Theorem 4.3.10, Sertel's characterization holds with an arbitrary ballot space $\mathfrak{B}$ and with null-neutrality, which is much weaker than neutrality.

Theorem 4.3.11. *Given any ballot space $\mathfrak{B}$, a voting rule φ satisfies null-neutrality, faithfulness, weak consistency, and Sertel disjoint equality if and only if $\varphi = \varphi_{MAJ}$.*

Remark 4.3.3. Sertel's faithfulness says that when there is only one ballot response A that is possibly the empty set, voting rule must choose A (recall that in our definition, faithfulness pertains to non-empty A). Clearly φ_{MAJ} does not satisfy this axiom. Sertel (1988) shows that his approval voting rule (identical to the standard approval voting rule except at null response profiles) is the only quasi-rule satisfying his faithfulness together with neutrality, weak consistency, and Sertel disjoint equality. In fact, dropping the requirement of non-empty valuedness in the definition of voting rule (thus among quasi-rules) and replacing null-neutrality in Theorem 4.3.11 either with neutrality or with "$\varphi(\pi) = \emptyset$ or X at all null response profiles π," we obtain a joint characterization of the two rules, Sertel's approval voting rule and the standard approval voting rule. The proof is essentially the same.

Proof. Let $\mathfrak{B}$ be a ballot space and φ a rule on $\Pi(\mathfrak{B})$ satisfying the four axioms. In what follows, for all $k \in \mathbb{N}$, we prove the claim that for all $\pi \in \Pi(\mathfrak{B})$ with $n(\pi) \leq k$, $\varphi(\pi) = \varphi_{MAJ}(\pi)$. The proof is by induction on k. The claim with $k = 1$ follows directly from null-neutrality and faithfulness. Let $k \geq 2$. Suppose by induction that for all $\pi \in \Pi(\mathfrak{B})$ with $n(\pi) \leq k$, $\varphi(\pi) = \varphi_{MAJ}(\pi)$. Let $\pi \in \Pi(\mathfrak{B})$ be such that $n(\pi) = k + 1$. We prove that $\varphi(\pi) = \varphi_{MAJ}(\pi)$. Note that there are $\pi' \in \Pi(\mathfrak{B})$ and $A \in \mathfrak{B}$ such that $n(\pi') = k$ and $\pi = \pi' + \pi_A$. Then by the induction hypothesis, $\varphi(\pi') = \varphi_{MAJ}(\pi')$ and $\varphi(\pi_A) = \varphi_{MAJ}(\pi_A)$.

Case 1: $\varphi(\pi') \cap \varphi(\pi_A) \neq \emptyset$. Then $\varphi_{MAJ}(\pi') \cap \varphi_{MAJ}(\pi_A) \neq \emptyset$. Since both φ and φ_{MAJ} satisfy weak consistency, $\varphi(\pi' + \pi_A) = \varphi(\pi') \cap \varphi(\pi_A)$ and $\varphi_{MAJ}(\pi' + \pi_A) = \varphi_{MAJ}(\pi') \cap \varphi_{MAJ}(\pi_A)$. Since $\varphi(\pi') = \varphi_{MAJ}(\pi')$, $\varphi(\pi_A) = \varphi_{MAJ}(\pi_A)$, and $\pi = \pi' + \pi_A$, then $\varphi(\pi) = \varphi_{MAJ}(\pi)$.

Case 2: $\varphi(\pi') \cap \varphi(\pi_A) = \emptyset$. Then $\varphi_{MAJ}(\pi') \cap \varphi_{MAJ}(\pi_A) = \emptyset$. By Sertel disjoint equality of φ, $x \in \varphi(\pi' + \pi_A)$ if and only if (i) $x \in \varphi(\pi')$ or (ii) $x \in \varphi(\pi_A)$ and $max_{y \in \varphi(\pi')} n(y, \pi') = 0$ or (iii) $x \in \varphi(\pi_A)$ and $n(x, \pi') = max_{y \in \varphi(\pi')} n(y, \pi') - 1 \geq 0$. Note that since $\varphi(\pi') = \varphi_{MAJ}(\pi')$ and $\varphi(\pi_A) = \varphi_{MAJ}(\pi_A)$, then (i), (ii), and (iii) are equivalent respectively to (i′) $x \in \varphi_{MAJ}(\pi')$, (ii′) $x \in \varphi_{MAJ}(\pi_A)$ and $max_{y \in \varphi_{MAJ}(\pi')} n(y, \pi') = 0$, and (iii′) $x \in \varphi_{MAJ}(\pi_A)$ and $n(x, \pi') = max_{y \in \varphi_{MAJ}(\pi')} n(y, \pi') - 1 \geq 0$. Therefore since both φ and φ_{MAJ} satisfy Sertel disjoint equality, $x \in \varphi(\pi' + \pi_A)$ if and only if $x \in \varphi_{MAJ}(\pi' + \pi_A)$. Since $\pi = \pi' + \pi_A$, we obtain $\varphi(\pi) = \varphi_{MAJ}(\pi)$. □

4.4 Strategic Voting and Condorcet Principle

In this section, we overview important supports for approval voting from the point of view of robustness to strategic voting as well as satisfying the Condorcet principle.

Consider a ballot space $\mathfrak{B}$. For all profiles of dichotomous preferences $\pi \in \Pi(\mathfrak{B})$, the *Condorcet set* $CW(\pi) \equiv \{x : n(x,\pi) \geq n(y,\pi),\ \text{for all } y \in X\}$ is the set of Condorcet winners at π. By Theorem 4.2.2, the Condorcet set is non-empty and it coincides with the choice made by majority voting rule, $\varphi_{MAJ}(\pi) = CW(\pi)$. The Condorcet principle requires that a voting rule should select the Condorcet set.

Condorcet: For all $\pi \in \Pi(\mathfrak{B})$, $\varphi(\pi) = CW(\pi)$.

Evidently, a voting rule on $\mathfrak{B}$ satisfies Condorcet if and only if it is the majority rule on $\mathfrak{B}$. A weaker requirement is that a voting rule should select some Condorcet winners.

Weak Condorcet: For all $\pi \in \Pi(\mathfrak{B})$, $\varphi(\pi) \cap CW(\pi) \neq \emptyset$.

A voting rule φ on $\mathfrak{B}$ is *minimally selective* if for some $\pi \in \Pi(\mathfrak{B})$, $\varphi(\pi) \neq X$. Clearly, any non-constant voting rule is minimally selective. Fishburn (1979) obtains the following characterization of majority voting rule.

Theorem 4.4.1 (Fishburn 1979). *A voting rule on $\mathfrak{B}$ satisfies neutrality, consistency, continuity, minimal selectiveness, and weak Condorcet if and only if it is majority voting rule.*

In the strategic voting environment, Condorcet, not to speak of weak Condorcet, does not guarantee a Condorcet winner to be a final voting outcome. To investigate strategic voting behavior under a voting rule that sometimes produces tied outcomes, understanding how voters evaluate subsets of alternatives is needed. For a dichotomous preference relation $B \in \mathfrak{D}$ of voter i there are five natural assumptions about its extension over subsets of alternatives. Denote the extended preference relation of B by R_i^B. The five assumptions are as follows: for all $x, y \in X$,

P1. $\{x\} P_i^B \{y\}$ if and only if $x \in B$ and $y \notin B$;
P2. $\{x\} P_i^B \{x, y\}$ and $\{x, y\} P_i^B \{y\}$ if $x \in B$ and $y \notin B$;
P3. $A R_i^B A'$ if $A \subseteq B$ or $A' \subseteq X \backslash B$ or $[A \backslash A' \subseteq B$ and $A' \backslash A \subseteq X \backslash B]$,
P4. $A \cup \{a\} I_i^B A \cup \{a'\}$ if $a, a' \notin A \cup B$ or $a, a' \in B \backslash A$,
P5. $A P_i^B X \backslash B$ if $A \cap B \neq \emptyset$; $B P_i^B A$ if $A \cap [X \backslash B] \neq \emptyset$, where P_i^B is the strict counterpart of R_i^B.

The first three assumptions, P1–P3, are considered by Fishburn (1979). Two additional assumptions, P4–P5, are needed to extend his result on the unrestricted ballot space to general ballot spaces.

4.4.1 Strategic Voting Under Anonymous Voting Systems

Given a ballot space $\mathfrak{B}$ and an agent with dichotomous preference $B \in \mathfrak{D}$, a ballot response $A \in \mathfrak{B}$ is *dominated* by ballot response $A' \in \mathfrak{B}$ if for all

$\pi \in \Pi(\mathfrak{B}) \cup \{\pi^{\emptyset}\}$, $\varphi(\pi + \pi_{A'}) R^B \varphi(\pi + \pi_A)$ with strict relation for at least one $\pi \in \Pi(\mathfrak{B}) \cup \{\pi^{\emptyset}\}$. Let $\mathfrak{B}^{ud}(B, \varphi)$ be the set of all undominated ballots in $\mathfrak{B}$. Since $\mathfrak{B}$ is a finite set, there exists at least one undominated ballot and $\mathfrak{B}^{ud}(B, \varphi) \neq \emptyset$. For each dichotomous preference profile $\pi \in \Pi(\mathfrak{D})$, let $\Pi^{ud}(\mathfrak{B})(\pi, \varphi)$ be the set of all ballot response profiles consisting of undominated ballots of all agents. A voting system $(\mathfrak{B}, \varphi)$ and the correspondence of undominated ballot response profiles $\Pi^{ud}(\mathfrak{B})(\cdot, \varphi)$ generate a collective choice rule for dichotomous preferences in $\mathcal{D} \subseteq \mathfrak{D}$, $c\colon \Pi(\mathcal{D}) \to P(X)\backslash\{\emptyset\}$ defined as follows: for all $\pi \in \Pi(\mathcal{D})$, $c(\pi) \equiv \bigcup\{\varphi(\hat{\pi}) : \hat{\pi} \in \Pi^{ud}(\mathfrak{B})(\pi, \varphi)\}$. It is natural to assume that each voter will not cast a dominated ballot and that the outcomes from strategic voting will be within the set of outcomes from undominated ballot profiles, that is, $c(\pi)$. Strategic voting is not an issue for agents who have complete indifference over all outcomes, namely agents with *unconcerned* dichotomous preference X because any two ballots will be indifferent independently of others' ballots. In what follows, we will focus on *concerned* agents who have dichotomous preferences with a preferred set $B \neq X$.

The set of undominated outcomes $c(\pi)$ may be quite different from the set of outcomes from truthful voting, $\varphi(\pi)$ and so the voting system $(\mathfrak{B}, \varphi)$ may lead to too different an outcome from the truthful outcome. Particular attention has been paid to voting systems that do not have this problem. A voting rule φ (or an anonymous collective choice rule c) on domain $\Pi(\mathcal{D})$ is *realizable in undominated strategies* by voting system $(\mathfrak{B}, \varphi)$ if for all profiles of concerned preferences $\pi \in \Pi(\mathcal{D}\backslash\{X\})$, $\hat{\varphi}(\hat{\pi}) \subseteq \varphi(\pi)$ for all undominated ballot response profiles $\hat{\pi} \in \Pi(\mathfrak{B})$ *at* π. Voting rule $\varphi(\cdot)$ on $\Pi(\mathcal{D})$ is *strategy-proof* if for all profiles of concerned preferences $\pi \in \Pi(\mathcal{D}\backslash\{X\})$, it is realizable in undominated strategies by voting system $(\mathfrak{B}, \varphi)$ and there is a unique undominated ballot response profile at π. We say that voting system $(\mathfrak{B}, \varphi)$ is *strategy-proof on* $\mathcal{D}$ if it always has a unique undominated profile at all $\pi \in \Pi(\mathcal{D}\backslash\{X\})$. Formally:

Strategy-Proofness on $\mathcal{D}$: For all $\pi \in \Pi(\mathcal{D}\backslash\{X\})$, $\Pi^{ud}(\mathfrak{B})(\pi, \varphi) = \{\pi'\}$ and $\varphi(\pi') \subseteq \varphi(\pi)$.

The next lemma shows that if a neutral and faithful voting system is strategy-proof on $\mathcal{D}$, then there should be no constraint on expressing one's concerned preferences in $\mathcal{D}$. That is,

No Ballot Constraint on $\mathcal{D}$: $\mathcal{D}\backslash\{X\} \subseteq \mathfrak{B}$.

Now we are ready to state the lemma.

Lemma 4.4.1. *If a voting system $(\mathfrak{B}, \varphi)$ satisfies neutrality, faithfulness, and strategy-proofness on $\mathcal{D}$, then it has no ballot constraint and for all $B \in \mathcal{D}\backslash\{X\}$, B is the unique undominated strategy for dichotomous preference B.*

Proof. Let B' be the undominated strategy for a concerned preference B and $B' \notin \{B, X\backslash B, \emptyset, X\}$. Then there exist $c, d \in X$ such that (i) $c \in B' \cap [X\backslash B]$ and $d \in [X\backslash B]\backslash B'$ or (ii) $c \in B' \cap B$ and $d \in B\backslash B'$. Consider the first case (i) (similar argument applies to case (ii)). Let $\lambda : X \to X$ be such that $\lambda(c) = d$, $\lambda(d) = c$, and for all other $x \in X\backslash\{c, d\}$, $\lambda(x) = x$. Then since B' is the only undominated strategy for B, for some π, $\varphi(\pi_{B'}+\pi)P^B\varphi(\pi_{[B'\backslash\{c\}]\cup\{d\}}+\pi)$. Note that $\pi_{B'}\lambda = \pi_{[B'\backslash\{c\}]\cup\{d\}}$ and $\pi_{B'} = \pi_{[B'\backslash\{c\}]\cup\{d\}}\lambda$. Hence $\varphi((\pi_{B'}+\pi)\lambda) = \varphi(\pi_{[B'\backslash\{c\}]\cup\{d\}}+\pi\lambda)$ and $\varphi((\pi_{[B'\backslash\{c\}]\cup\{d\}}+\pi)\lambda) = \varphi(\pi_{B'}+\pi\lambda)$. By neutrality, $\varphi(\pi_{[B'\backslash\{c\}]\cup\{d\}}+\pi\lambda) = \lambda(\varphi(\pi_{B'}+\pi))$ and $\varphi(\pi_{B'}+\pi\lambda) = \lambda(\varphi(\pi_{[B'\backslash\{c\}]\cup\{d\}}+\pi))$. By P4, $\lambda(\varphi(\pi_{B'}+\pi))I^B\varphi(\pi_{B'}+\pi)$ and $\lambda(\varphi(\pi_{[B'\backslash\{c\}]\cup\{d\}}+\pi))I^B\varphi(\pi_{[B'\backslash\{c\}]\cup\{d\}}+\pi)$. Hence $\varphi(\pi_{[B'\backslash\{c\}]\cup\{d\}}+\pi\lambda)P^B\varphi(\pi_{B'}+\pi\lambda)$, which shows that $[B'\backslash\{c\}]\cup\{d\}$ is not dominated by B', contradicting that B' is the unique undominated strategy for B.

If $X\backslash B$ is the unique undominated strategy for B, $\varphi(\pi_{X\backslash B}+\pi^{\emptyset})R^B\varphi(\pi_A+\pi^{\emptyset})$ for any $A \in \mathfrak{B}$ with $A\cap B \neq \emptyset$. If φ is faithful, then $X\backslash BR^BA$, contradicting P5.

Therefore, if a neutral and faithful voting system $(\mathfrak{B},\varphi)$ is strategy-proof on $\mathcal{D}$, then for all concerned preference $B \in \mathcal{D}\backslash\{X\}$, B should be the unique undominated strategy; so $B \in \mathfrak{B}$. Hence there should be no constraint in expressing one's concerned preferences in $\mathcal{D}$. □

Brams and Fishburn (1978, Theorems 2 and 6) offer a necessary and sufficient conditions for undominated ballots under majority voting systems. For dichotomous preferences, their condition roughly says that undominated ballots for each dichotomous preference B are the ballots that best approximate B either from above or from below in the ballot space $\mathfrak{B}$. Formally:

Lemma 4.4.2 (Brams and Fishburn 1978). *Given a majority voting system* $(\mathfrak{B},\varphi_{MAJ})$, *for each dichotomous preference* $B \in P(X)\backslash\{\emptyset,X\}$, *a ballot* $\hat{B} \in \mathfrak{B}$ *is undominated if and only if* (i) $\hat{B} \subseteq B$ *and there is no* $A \in \mathfrak{B}\backslash\{\hat{B}\}$ *such that* $\hat{B} \subsetneq A \subseteq B$ *or* (ii) $B \subseteq \hat{B}$ *and there is no* $A \in \mathfrak{B}\backslash\{\hat{B}\}$ *such that* $B \subseteq A \subsetneq \hat{B}$.

Thus if dichotomous preference B is in ballot space $\mathfrak{B}$, then B is the only undominated strategy, $\mathfrak{B}^{ud}(B,\varphi_{MAJ}) = \{B\}$. Similarly, if $\pi \in \Pi(\mathfrak{B})$, $\Pi^{ud}(\mathfrak{B})(\pi,\varphi_{MAJ}) = \{\pi\}$. Thus if $\mathcal{D} \subseteq \mathfrak{B}$, majority voting system $(\mathfrak{B}, \varphi_{MAJ})$ is strategy-proof on $\mathcal{D}$. Conversely, if $\mathcal{D} \nsubseteq \mathfrak{B}$, then by Lemma 4.4.2, majority voting system $(\mathfrak{B},\varphi_{MAJ})$ has more than one undominated ballot response profiles at a profile π consisting of some B in $\mathcal{D}\backslash\mathfrak{B}$. Therefore, we obtain:

Theorem 4.4.2. *Majority voting system* $(\mathfrak{B},\varphi_{MAJ})$ *is strategy-proof on a subdomain of dichotomous preferences* $\mathcal{D} \subseteq \mathfrak{D}$ *if and only if there is no ballot constraint, i.e.,* $\mathcal{D} \subseteq \mathfrak{B}$. *Thus approval voting is the only strategy-proof majority voting system on the entire domain of dichotomous preferences,* $\mathfrak{D}$.

Note that Condorcet winners at π coincide with the alternatives selected by majority voting rule at π. Thus when π is in the space of ballot response profiles, by Lemma 4.4.2, π is the only undominated ballot response profile and thus any undominated ballot response profile at π gives the set of Condorcet winners. However, if a voter has a dichotomous preference that is not in the ballot space, then this equivalence between the set of Condorcet winners and the set of alternatives obtained by an undominated strategy profile in the majority voting system fails. Moreover, the failure can be so drastic that some undominated ballot response profile does not give any Condorcet winner. A stronger version of this claim is established for voting systems satisfying neutrality and the following basic axiom.

Strong Pareto: For all $\pi \in \Pi(\mathfrak{B})$ and all $x, y \in X$, if for all $A \in \mathfrak{B}$ with $\pi(A) > 0$, $x \in A$ or $x, y \in X \setminus A$, and there is $A \in \mathfrak{B}$ with $\pi(A) > 0$ such that $x \in A$ and $y \in X \setminus A$, then $y \notin \varphi(\pi)$.[22]

Lemma 4.4.3 (Fishburn 1979, Theorem 9). *Consider a voting system* $(\mathfrak{B}, \varphi)$ *satisfying neutrality and strong Pareto. Assume that for a dichotomous preference,* $B \in P(X)\setminus\{\emptyset, X\}$, *there is an undominated ballot* A *different from* B *(i.e.,* $A \in \mathfrak{B}^{ud}(\varphi, B)$ *and* $A \neq B$*), then there is a profile of dichotomous preferences* $\pi \in \Pi$ *such that for some undominated ballot response profile* $\hat{\pi} \in \Pi^{ud}(\mathfrak{B})(\pi, \varphi)$, $\varphi(\hat{\pi}) \cap CW(\pi) = \emptyset$.

Proof. Consider a voting system $(\mathfrak{B}, \varphi)$ satisfying neutrality and strong Pareto, and a dichotomous preference $B \in P(X)\setminus\{\emptyset, X\}$. Suppose $A \in \mathfrak{B}^{ud}(\varphi, B)$ and $A \neq B$.

Case 1: There is $b \in B\setminus A$.

Let π be such that for all $B' \in P(X)$ with $b \in B'$ and $|B'| = |B|$, $\pi(B') = 1$ and for all other $C \in P(X)$, $\pi(C) = 0$. Since $A \in \mathfrak{B}^{ud}(\varphi, B)$, then for all $B' \in P(X)\setminus\{\emptyset, X\}$ with $b \in B'$ and $|B'| = |B|$, by neutrality, there is $A(B') \in \mathfrak{B}$ such that $A(B') \in \mathfrak{B}^{ud}(f, B')$ and $b \in B'\setminus A(B')$. Let $\hat{\pi}$ be such that for all B' with $b \in B'$ and $|B'| = |B|$, $\hat{\pi}(A(B')) = 1$ and for all other $C \in \mathfrak{B}$, $\hat{\pi}(C) = 0$. Then by construction of π, $CW(\pi) = \{b\}$. Also by construction, $\hat{\pi} \in \Pi^{ud}(\mathfrak{B})(\varphi, \pi)$ and by strong Pareto, $b \notin \varphi(\hat{\pi})$. Therefore, $\varphi(\hat{\pi}) \cap CW(\pi) = \emptyset$.

Case 2: There is $a \in A\setminus B$.

Let π be such that for all $A' \in \mathfrak{B}$ with $a \in A'$ and $|A'| = |A|$, $\pi(A') = 1$ and for all other ballots $B' \in \mathfrak{B}$, $\pi(B') = 0$. Since $A \in \mathfrak{B}^{ud}(\varphi, B)$, then by neutrality, for all $A' \in \mathfrak{B}$ with $a \in A'$ and $|A'| = |A|$, there is $B(A') \in P(X)\setminus\{\emptyset, X\}$ such that $A' \in \mathfrak{B}^{ud}(\varphi, B(A'))$ and $a \in A'\setminus B(A')$. Let $\hat{\pi}$ be such that for all A' with $\pi(A') > 0$, $\hat{\pi}(B(A')) = 1$ and for all other ballots $C \in P(X)$, $\hat{\pi}(C) = 0$. Then by strong Pareto, $\varphi(\pi) = \{a\}$. Also by construction of $\hat{\pi}$, $\pi \in \Pi^{ud}(\mathfrak{B})(\varphi, \hat{\pi})$ and $a \notin CW(\hat{\pi})$. Therefore, $\varphi(\pi) \cap CW(\hat{\pi}) = \emptyset$. □

We now return to the Condorcet principle in the strategic voting environment.

Condorcet realizability on $\mathcal{D}$: For all $\pi \in \Pi(\mathcal{D})$ and all $\hat{\pi} \in \Pi^{ud}(\mathfrak{B})(\pi, \varphi)$, $\varphi(\hat{\pi}) = CW(\pi)$.

The next axiom is weaker and corresponds to weak Condorcet.

Weak Condorcet realizability on $\mathcal{D}$: For all $\pi \in \Pi(\mathcal{D})$ and all $\hat{\pi} \in \Pi^{ud}(\mathfrak{B})(\pi, \varphi)$, $\varphi(\hat{\pi}) \cap CW(\pi) \neq \emptyset$.

The next lemma is important for establishing the next characterization of majority voting.

[22] If for all $A \in \mathfrak{B}$ with $\pi(A) > 0$, $y \notin A$, then using any $x \in A$ for some $A \in \mathfrak{B}$ with $\pi(A) > 0$, we can show that the premise of strong Pareto is met. Thus in this case $y \notin \varphi(\pi)$.

Lemma 4.4.4. *If voting system* $(\mathfrak{B}, \varphi)$ *satisfies neutrality, strong Pareto, and weak Condorcet realizability on* $\mathcal{D}$*, then the system has no ballot constraint on* $\mathcal{D}$ *(that is,* $\mathcal{D} \subseteq \mathfrak{B}$*), strategy-proofness on* $\mathcal{D}$ *and weak Condorcet. Replacing weak Condorcet realizability with Condorcet realizability, we obtain* $\varphi = \varphi_{MAJ}$*. Thus when* $\mathcal{D} = P(X)\backslash\{\emptyset, X\}$*, it is approval voting.*

Proof. For any voting system $(\mathfrak{B}, \varphi)$ satisfying neutrality and strong Pareto, if there is an undominated strategy A that differs from the voter's dichotomous preference $B \in \mathcal{D}$, by Lemma 4.4.3, voting system $(\mathfrak{B}, \varphi)$ violates weak Condorcet realizability. Note that there can be such an undominated strategy $A \neq B$ if $B \in \mathcal{D}\backslash\mathfrak{B}$ or the voting system is not strategy-proof. Hence, neutrality, strong Pareto, and weak Condorcet realizability together imply both no ballot constraint, $\mathcal{D} \subseteq \mathfrak{B}$, and strategy-proofness. Moreover, the unique undominated strategy for $B \in \mathcal{D}$ is B itself. Therefore, at all dichotomous preference profiles $\pi \in \Pi(\mathcal{D})$, $\Pi^{ud}(\mathfrak{B})(\pi, \varphi) = \{\pi\}$ and weak Condorcet realizability implies $\varphi(\pi) \cap CW(\pi) \neq \emptyset$, that is, the voting system satisfies weak Condorcet. If $(\mathfrak{B}, \varphi)$ satisfies Condorcet realizability, then the last conclusion is strengthened to $\varphi(\pi) = CW(\pi)$, that is the voting system satisfies Condorcet; it is majority voting. □

We now obtain the following characterization of majority voting based on Condorcet realizability and strategy-proofness.

Theorem 4.4.3. *Consider a subdomain of dichotomous preferences* $\mathcal{D} \subseteq \mathfrak{D}$ *and a ballot space* $\mathfrak{B} \subseteq \mathcal{D}$*. The following are equivalent.*

(i) *Voting system* $(\mathfrak{B}, \varphi)$ *satisfies neutrality, strong Pareto, and Condorcet realizability on* $\mathcal{D}$*.*
(ii) *Voting system* $(\mathfrak{B}, \varphi)$ *satisfies neutrality, consistency, continuity, minimal selectiveness, and weak Condorcet realizability on* $\mathcal{D}$*.*
(iii) *Voting system* $(\mathfrak{B}, \varphi)$ *satisfies neutrality, consistency, faithfulness, and strategy-proofness on* $\mathcal{D}$*.*
(iv) *Voting system* $(\mathfrak{B}, \varphi)$ *is majority voting without ballot constraint on* $\mathcal{D}$*.*

Proof. The proof of the equivalence between (i) and (iv) is established using Theorem 4.4.2 and Lemma 4.4.4. The equivalence between (ii) and (iv) is obtained from Lemma 4.4.4 and Theorem 4.4.1. Finally, the next lemma states that (iii) implies (iv), and the converse follows from Theorem 4.4.2. □

The next lemma is an extension of a result in Fishburn (1979, Theorem 10, pp. 216–217), which is for the ballot space $\bar{P}(X)\backslash\{X\}$. Our result is for any arbitrary ballot space satisfying the richness conditions, (4.15) and (4.16).

Lemma 4.4.5. *Consider a subdomain of dichotomous preferences* $\mathcal{D} \subseteq \mathfrak{D}$ *and a ballot space* $\mathfrak{B} \subseteq \mathcal{D}$*. If voting system* $(\mathfrak{B}, \varphi)$ *satisfies neutrality, consistency, faithfulness, and strategy-proofness on* $\mathcal{D}$*, then it is majority voting without ballot constraint on* $\mathcal{D}$*.*

Proof. By Theorem 4.3.6, there are scoring functions $s^1, \ldots, s^T$ that represent φ as the lexicographic scoring rule. The case where all scoring functions are zero functions can be treated easily.[23] Excluding this case, without loss of generality, assume that no s^t is uniformly zero.

By Lemma 4.4.1, the voting system has no ballot constraint ($\mathcal{D}\backslash\{X\} = \mathfrak{B}\backslash\{X\}$) and for all $B \in \mathcal{D}\backslash\{X\}$, B is the unique undominated strategy for dichotomous preference B.

Suppose that for some $B \in \mathcal{D}\backslash\{\emptyset, X\}$, $s^1(|B|) < 0$. Then $\varphi(\pi_B) = X\backslash B$ (since $B \neq \emptyset, X$ and so $X\backslash B \neq \emptyset, X$), contradicting faithfulness.

Suppose that for some $A, B \in \mathcal{D}\backslash\{\emptyset, X\}$, $0 \leq s^1(|A|) < s^1(|B|)$. Without loss of generality, assume there exist $a \in A\backslash B$ and $b \in B\backslash A$. Consider dichotomous preference A. Construct a ballot response profile π such that for all $C \in \mathfrak{B}$, if $\pi(C) > 0$, then $|C| = |B|$, and $n(b,\pi) = n(a,\pi) + 1$ and $n(x,\pi) \leq n(a,\pi) - 1$ for all $x \in X\backslash\{a,b\}$.[24] Existence of such a profile is guaranteed by (4.15) and (4.16) because $\mathcal{D} \subseteq \mathfrak{B}$. Since $s^1(|A|) < s^1(|B|)$, $\varphi(\pi + \pi_A) = \{b\}$. On the other hand, $\varphi(\pi + \pi_{[B\backslash\{b\}]\cup\{a\}}) = \{a,b\}$, which is preferred to $\{b\}$ by the agent with preferences A. Therefore, A does not dominate $[B\backslash\{b\}] \cup \{a\}$, which implies that A is not the only undominated ballot, contradicting Lemma 4.4.1.

The above argument shows that $s^1(\cdot)$ is a constant and positive valued function over $\{|B| : B \in \mathcal{D}\backslash\{\emptyset, X\}\}$. The same argument can be used to show that the remaining score functions $s^2(\cdot), \ldots, s^T(\cdot)$ are constant functions over $\{|B| : B \in \mathcal{D}\backslash\{\emptyset, X\}\}$, which is sufficient to conclude that $\varphi = \varphi_{MAJ}$. □

It follows from Theorem 4.4.3 and Lemma 4.4.5 that:

Corollary 4.4.1. *Consider the domain of all dichotomous preferences, $\mathfrak{D}$. Then the following are equivalent.*

(i) *Voting system $(\mathfrak{B}, \varphi)$ satisfies neutrality, strong Pareto, and Condorcet realizability on $\mathfrak{D}$.*
(ii) *Voting system $(\mathfrak{B}, \varphi)$ satisfies neutrality, consistency, continuity, minimal selectiveness, and weak Condorcet realizability on $\mathfrak{D}$.*
(iii) *Voting system $(\mathfrak{B}, \varphi)$ satisfies neutrality, consistency, faithfulness, and strategy-proofness on $\mathfrak{D}$.*
(iv) *Voting system $(\mathfrak{B}, \varphi)$ is approval voting.*

4.5 Unconstrained Multi-issue Problems and Voting by Committees

In this section, we consider a collective decision model where there are multiple issues and for each issue, a binary decision needs to be made. This model is studied

---

[23] If all scoring functions are zero functions, then φ will always choose X, in which case no ballot is dominated and all ballots are undominated.

[24] Consider a ballot response profile π such that $\pi(B) = 1$, $\pi([B\backslash\{x\}] \cup \{a\}) = 2$ for each $x \in B\backslash\{b\}$, and $\pi(B') = 0$ for all other ballots B'. Then $n(b,\pi) = 2(|B|-1)+1 = n(a,\pi)+1$, and for each $x \in B\backslash\{b\}$, $n(x,\pi) = 2(|B|-2)+1 = n(a,\pi)-1$.

by Barberà et al. (1991, 1997), etc., and an extended model by Barberà et al. (1993), Le Breton and Sen (1999), etc.

Let $M \equiv \{1, \ldots, m\}$ be the set of issues. A collective decision is a vector of 1 or -1, that is,

$$x \equiv (x_1, \ldots, x_m) \in \{-1, 1\}^M,$$

where 1 in the kth component means accepting the kth issue and -1 means rejecting it. Thus $S_x \equiv \{k \in M : x_k = 1\}$ is the set of accepted issues at x. Let $X \equiv \{-1, 1\}^M$ be the set of all possible decisions. There is no constraint on the set of accepted issues in this model and any number or none of the issues can be accepted.[25]

On the unrestricted domain of preferences, the three impossibility results, Theorems 4.1.1–4.1.3, apply. Moreover, an even more disturbing paradox, known as Gibbard's paradox, holds (Gibbard 1974): the mere assignment of Sen's liberal rights to each person cannot be made coherently under any collective choice function. Sen (1983, p. 14) points out that Gibbard's paradox does not hold on the restricted domain of preferences for which each issue affects a person's welfare separately from other issues, the so-called separable preferences. Moreover, on the restricted domain of separable and linear preferences, Gibbard–Satterthwaite theorem does not hold and there do exist non-dictatorial and well-behaved strategy-proof rules (Barberà et al. 1991).

Formally, a preference R_0 is *separable* if for all $x, x' \in X$ and all $k \in M$, $(x_k, x_{-k}) R_0 (-x_k, x_{-k})$ if and only if $(x_k, x'_{-k}) R_0 (-x_k, x'_{-k})$. An issue $k \in M$ is a *good* (resp. a *bad* or a *null*) if for all $x \in X$ with $x_k = 1$, $(1, x_{-k}) P_0 (-1, x_{-k})$ (resp. $(-1, x_{-k}) P_0 (1, x_{-k})$ or $(1, x_{-k}) I_0 (-1, x_{-k})$). Let $G(R_0)$ be the set of goods for R_0 and $B(R_0)$ the set of bads. Let $\mathcal{S}$ be the set of separable preferences and $\mathcal{S}_L$ the set of *linear* separable preferences. Given a domain of separable preference profiles, $\mathcal{D} \subseteq \mathcal{S}^N$, a *collective choice function* $c \colon \mathcal{D} \to X$ associates with each preference profile a single collective decision.

Collective choice functions that can be practiced through a simple non-ranked voting procedure have been of central interest in the literature. A *voting scheme* is a collective choice function that only uses information about which issues are good or bad and so can be applied through a voting procedure under which voters express, in their ballots, which issues are goods and which are bads. That is, a voting scheme is a collective choice function satisfying:

Votes-Only: For all $R, R' \in \mathcal{S}^N$, if for all $i \in N$, $G(R_i) = G(R'_i)$ and $B(R_i) = B(R'_i)$, then $c(R) = c(R')$.

On the domain of linear separable preferences $\mathcal{S}_L^N$, this property is known as the "tops-only" property because $G(R_i)$ is the top alternative for $R_i \in \mathcal{S}_L$. When there are nulls, adding some or all nulls to $G(R_i)$ makes no difference from $G(R_i)$

[25] Barberà et al. (2005) consider a similar model with some constraints on the number of issues to be accepted.

and makes another top alternative for R_i. Thus there are multiple top alternatives consisting of all goods, some or all nulls, and no bads.

The following two additional axioms pin down an important family of voting schemes. The first axiom says that if for each agent, there are more goods and less bads, then more issues should be accepted.

Issues Monotonicity: For all $R, R' \in \mathcal{S}^N$, if for all $i \in N$, $G(R_i) \subseteq G(R'_i)$ and $B(R_i) \supseteq B(R'_i)$, then $c(R) \leq c(R')$.

The next axiom says that decisions on each issue should be made independently from the other issues, relying on who is in favor of the issue and who is against it. For all $R \in \mathcal{S}^N$ and all $k \in M$, let $N_k^G(R)$ be the set of agents for whom issue k is a good, and $N_k^B(R)$ the set of agents for whom issue k is a bad.

Issues Independence: For all $R, R' \in \mathcal{S}^N$ and all $k \in M$, if $N_k^G(R) = N_k^G(R')$ and $N_k^B(R) = N_k^B(R')$, then $c_k(R) = c_k(R')$.

Note that each of the above two axioms implies votes-only. The family of collective choice functions satisfying the two axioms can be represented by an issue-wise decisive structure, similar to the decisive structures in Sect. 4.2, defined as follows. For all $k \in M$, a *decisive structure for issue* k is a nonempty subset of $\mathfrak{d}^*$, $\mathfrak{d}_k \subseteq \mathfrak{d}^*$ satisfying: for all $(L_1, L_2), (L'_1, L'_2) \in \mathfrak{d}^*$,

$$\text{if } (L_1, L_2) \in \mathfrak{d}_k,\ L_1 \subseteq L'_1 \text{ and } L_2 \supseteq L'_2, \text{ then } (L'_1, L'_2) \in \mathfrak{d}_k.$$

Call this property $\mathfrak{d}$*-monotonicity*, as in Sect. 4.2. An *issue-wise decisive structure* is a list of decisive structures for all issues, $\mathfrak{d} = (\mathfrak{d}_k)_{k \in M}$. An issue-wise decisive structure $\mathfrak{d} = (\mathfrak{d}_k)_{k \in M}$ represents the function $c(\cdot)$ defined as follows: for all $R \in \mathcal{S}^N$ and all $k \in M$, $c_k(R) = 1$ if and only if $(N_k^G(R), N_k^B(R)) \in \mathfrak{d}_k$.

Proposition 4.5.1 (Ju 2003). *A collective choice function on* $\mathcal{S}^N$ *(or on* $\mathcal{S}_L^N$*) satisfies issues monotonicity and issues independence if and only if it is represented by an issue-wise decisive structure.*

The proof is similar to the proof of Proposition 4.2.1, and it holds on numerous subdomains of $\mathcal{S}^N$ (Ju 2003). On the domain of linear separable preferences $\mathcal{S}_L^N$, voting schemes represented by an issue-wise decisive structure consisting of *proper* subsets $\mathfrak{d}_k$ of $\mathfrak{d}^*$ for all $k \in M$ are called *schemes of voting-by-committees* (Barberà et al. 1991). Note that because for all $k \in M$, $\mathfrak{d}_k$ is a proper subset of $\mathfrak{d}^*$, schemes of voting-by-committees have full-range. *Issue-wise majority voting scheme* is the voting scheme represented by $\mathfrak{d} \equiv (\mathfrak{d}_k)_{k \in M}$ such that for all $k \in M$, $(L_1, L_2) \in \mathfrak{d}_k$ if and only if $|L_1| > |L_2|$. An axiomatization of issue-wise majority voting scheme can be established with the combination of issues monotonicity, issues independence, anonymity, neutrality, and a duality-type axiom, on the domain of separable linear preferences with an odd number of agents.

4.5.1 Strategy-Proofness and Separable Preferences

Barberà et al. (1991) show that on the domain of linear separable preferences $\mathcal{S}_L^N$, all strategy-proof collective choice functions with full-range satisfy the votes-only property, and so they are voting schemes. Based on this result, they also show that strategy-proofness and the full-range condition together imply issues independence as well as issues monotonicity. This leads to the following characterization of voting-by-committees.

Theorem 4.5.1 (Barberà et al. 1991). *A collective choice function on the domain of linear separable preferences satisfies the full-range condition and strategy-proofness if and only if it is a scheme of voting-by-committees.*

Barberà et al. (1993) and Le Breton and Sen (1999) generalize this result in the extended model of multi-issue problems where more than two alternatives are available on each issue. In particular, Le Breton and Sen (1999) identify a general domain condition under which their characterization holds. The key argument is to prove an extended version of issues independence, called "decomposability," of strategy-proof collective choice functions. All these works assume linearity of preferences, which plays a crucial role.

When preferences are not linear, as shown by Le Breton and Sen (1995), issues independence of a strategy-proof collective choice function is not guaranteed, which makes it hard to obtain a result like Theorem 4.5.1 on the domain of separable "weak" orderings $\mathcal{S}^N$. In fact, we need an additional axiom to characterize voting schemes represented by an issue-wise decisive structure.

Null-Independence: For all $i \in N$, all $k \in M$, all $R_i, R_i' \in \mathcal{S}$, and all $R_{-i} \in \mathcal{S}^{N\setminus\{i\}}$, if k is a null issue for both R_i and R_i', then $c_k(R_i, R_{-i}) = 1$ if and only if $c_k(R_i', R_{-i}) = 1$.

Among voting schemes, the combination of strategy-proofness and null-independence is equivalent to the combination of issues monotonicity and issues independence (Ju 2003, Proposition 4, p. 485). Thus it follows from Proposition 4.5.1 that:

Theorem 4.5.2 (Ju 2003). *A voting scheme on the domain of separable preferences satisfies strategy-proofness and null-independence if and only if it is represented by an issue-wise decisive structure.*

After identifying a domain $\mathcal{D}$ where well-behaved strategy-proof functions exist, it is important to understand whether this existence result may be extended to a larger domain. In fact, as shown by Barberà et al. (1991, Theorem 3), the domain of separable linear preferences is the unique maximal "rich" domain (of linear preferences) where well-behaved strategy-proof functions exist. Dropping the linearity assumption, yet focusing on voting schemes, Ju (2003, Theorem 3) shows that the domain of separable preferences (weak orderings) is the unique maximal "rich" domain where well-behaved strategy-proof *voting schemes* exist. Maximal domain results are also established in the extended model of multi-issue problems by Serizawa (1995) and Le Breton and Sen (1999).

4.5.2 Strategy-Proofness Versus Efficiency and Domain Restrictions

Although strategy-proof collective choice functions on the domain of separable preferences are numerous, only dictatorial ones are efficient.

Theorem 4.5.3 (Barberà et al. 1991). *When there are at least three issues, a collective choice function on the domain of linear separable preferences is strategy-proof and efficient if and only if it is dictatorial.*

Le Breton and Sen (1999) obtain this result in their extended model of multi-issue problems. Shimomura (1996) weakens efficiency by requiring it on the subdomain where agents' preferences bear some degrees of resemblance and pins down a small family of schemes of voting by committees, which includes non-dictatorial voting schemes. When the set of alternatives is variable, Ju (2005b) requires efficiency only for problems with at most two alternatives. He shows that only those voting schemes that are quite close to the issue-wise majority function satisfy this restricted notion of efficiency as well as strategy-proofness, anonymity, and two additional axioms pertaining to agenda variation.

Ju (2005a) further restricts the domain of separable preferences to domains of "dichotomous" preferences. An *additive* preference is a separable preference represented by a utility vector $u \equiv (u_1, \ldots, u_m) \in \mathbb{R}^m$ as follows: for all $x, x' \in \{-1, 1\}^M$, xR_0x' if and only if $u{\cdot}x \geq u{\cdot}x'$. An additive preference is *trichotomous* if all goods are indifferent and all bads are indifferent. A trichotomous additive preference is *dichotomous* if all issues are either goods or bads. Although we use the same term as in the earlier sections, dichotomous preferences here are not dichotomous in the sense we use in Sect. 4.3. Dichotomous preferences in this section have at most two indifference sets of issues but may have more than two indifference sets in the alternative space $\{-1, 1\}^M$. Considering some examples of restricted domains consisting of dichotomous or trichotomous additive preferences, Ju (2005a) proves that only those voting schemes that are very close to issue-wise majority voting scheme satisfy efficiency as well as issues independence, anonymity, and neutrality (neutrality is needed only in the case of dichotomous preferences). Whether this result or a similar result holds for other simple domains such as the domain of dichotomous separable preferences that are not necessarily additive is open for future research.

4.6 Simple Opinion Aggregation and Decision by Powers and Consent

In this section, we consider the problem of aggregating dichotomous or trichotomous opinions, introduced by Wilson (1975) and further studied by Rubinstein and Fishburn (1986), Aleskerov et al. (2007), and Ju (2005a, 2008, 2010). The model is similar to the unconstrained multi-issue problems except that here, we deal with opinions rather than preferences. For separable linear preferences, issues are either goods or bads. Thus opinions can be interpreted as restricted preference revelations.

A special example of opinion aggregation is the problem of group identification. A finite number of potential members have to decide who among themselves belong to a certain collective through aggregating their dichotomous or trichotomous opinions. This problem is introduced by Kasher and Rubinstein (1997) and further studied by Samet and Schmeidler (2003), Sung and Dimitrov (2003), Dimitrov et al. (2007), Çengelci and Sanver (2010), Miller (2008), and Ju (2008, 2009a, 2010).

We continue using the same notation as in Sect. 4.5. Each person $i \in N$ has his *opinion* on issues in M, represented by an $1 \times m$ row vector V_i consisting of 1, 0, or -1. A *problem* is an $n \times m$ opinion matrix V consisting of n row vectors $V_1, \ldots, V_n$. Let $\mathcal{V}_{\text{Tri}}$ be the set of all opinion matrices, called, the *trichotomous* (opinion) *domain*. An *alternative* is a vector of 1 and -1, $x \equiv (x_1, \ldots, x_m) \in \{-1, 1\}^M$, where 1 (resp. -1) in the kth component means accepting the kth issue (resp. rejecting the kth issue). For all $V \in \mathcal{V}_{\text{Tri}}$ and all $k \in M$, V^k denotes the kth column vector of V. Let

$$||V_+^k|| \equiv \sum_{i \in N : V_{ik} = 1} V_{ik}, \quad ||V_-^k|| \equiv \sum_{i \in N : V_{ik} = -1} -V_{ik}, \quad \text{and} \; ||V_{+,-}^k|| \equiv ||V_+^k|| + ||V_-^k||.$$

For example, in the *group identification* model, $M = N$ and an alternative describes who belongs to the collective and who does not.

Let $\mathcal{V}_{\text{Di}}$ be the subset of $\mathcal{V}_{\text{Tri}}$, consisting of the opinion matrices whose entries are either 1 or -1, called the *dichotomous* (opinion) *domain*. Let $\mathcal{D}$ be either one of the two domains. Samet and Schmeidler (2003) consider the dichotomous domain of the group identification model.[26] A *collective choice function* on $\mathcal{D}$, $c : \mathcal{D} \to \{-1, 1\}^M$, associates with each problem in the domain a single alternative. A collective choice function satisfies *non-degeneracy* if for each $i \in N$, there exist $V, V' \in \mathcal{D}$ such that $c_i(V) = 1$ and $c_i(V') = -1$.

Section 4.5 provides the definition of a collective choice function represented by an issue-wise decisive structure. The same definition applies here, treating all issues k with $V_{ik} = 1$ as goods for person i and all issues l with $V_{il} = -1$ as bads, and all other issues as nulls. In the same way, we can extend the definitions of issues monotonicity and issues independence, which together characterize the family of collective choice functions represented by an issue-wise decisive structure (Proposition 4.5.1).

A subfamily of these collective choice functions plays an important role in the literature on opinion aggregation. In particular, an *issue-wise dictatorial function* $c(\cdot)$ is represented by an issue-wise decisive structure conferring on a person the full decision power over an issue: that is, for each $k \in M$, there is $i \in N$ such that

---

[26] Dichotomous opinions in Samet and Schmeidler (2003) are described by vectors of 1 and 0, where 0 has the same meaning as -1 in our model.

for all $V \in \mathcal{D}$ with $V_{ik} \in \{-1, 1\}$, $c_k(V) = V_{ik}$. In group identification problems with dichotomous opinions, the issue-wise dictatorial function where person i has the decision power on his own qualification is called as the *liberal* function. A milder notion of decision powers is discussed in the next subsection.

4.6.1 Social Decision by Powers and Consent for Dichotomous or Trichotomous Opinion Aggregation

Here we define a milder notion of decision powers. We first focus on dichotomous opinions. After this, we give the general definition.

Given a collective choice function c defined on the dichotomous domain $\mathcal{V}_{\mathrm{Di}}$, person $i \in N$ has the "power to influence the social decision on the kth issue", briefly, the *power on the k-th issue* if the decision on the kth issue is made following person i's opinion whenever person i's opinion obtains sufficient consent from society: formally, there exist $q_+, q_- \in \{1, \ldots, n+1\}$ such that for all $V \in \mathcal{V}_{\mathrm{Di}}$,

$$\begin{aligned} &\text{(i) when } V_{ik} = 1,\ c_k(V) = 1 \Leftrightarrow ||V_+^k|| \geq q_+ ; \\ &\text{(ii) when } V_{ik} = -1,\ c_k(V) = -1 \Leftrightarrow ||V_-^k|| \geq q_- . \end{aligned} \tag{4.20}$$

The two numbers q_+ and q_- are called *consent-quotas*. The greater q_+ or q_- is, the higher degree of social consent is required for the exercise of the power. There are three extreme cases. When $q_+ = q_- = 1$, i's opinion determines social decision independently of social consent. Thus the power is *decisive*. When $q_+ = n+1$ and $q_- = n+1$, the power is *anti-decisive* because i's opinion is reflected reversely in the social decision. When $q_+ + q_- = n+1$, the two parts in (4.20) coincide and all persons can have the same powers as person i (changing i with any j in (4.20) makes no difference). In this case, all persons have the equal power on the same issue; so such a power is said to be non-exclusive (formal definition will be provided later).[27]

The total number of positive or negative votes, denoted by ν, always equals n on the dichotomous domain. However, on the trichotomous domain, it is variable. We allow consent-quotas to vary relative to the total number of votes. Given a collective choice function c defined on $\mathcal{V}_{\mathrm{Tri}}$, a person $i \in N$ has the *power on the k-th issue* if there exist three functions $q_+ : N \cup \{0\} \to N \cup \{0, n+1\}$, $q_0 : N \cup \{0\} \to N \cup \{0, n+1\}$, and $q_- : N \cup \{0\} \to N \cup \{0, n+1\}$ such that for all $\nu \in N \cup \{0\}$, and all $V \in \mathcal{V}_{\mathrm{Tri}}$ with $||V_{+,-}^k|| = \nu$,

$$\begin{aligned} &\text{(i) when } V_{ik} = 1,\ c_k(V) = 1 \Leftrightarrow ||V_+^k|| \geq q_+(\nu); \\ &\text{(ii) when } V_{ik} = 0,\ c_k(V) = 1 \Leftrightarrow ||V_+^k|| \geq q_0(\nu); \\ &\text{(iii) when } V_{ik} = -1,\ c_k(V) = -1 \Leftrightarrow ||V_-^k|| \geq q_-(\nu). \end{aligned} \tag{4.21}$$

[27] Consent quotas are closely related with the power index by Shapley and Shubik (1954) as discussed in Ju (2010).

We call the list of the three functions $q(\cdot) \equiv (q_+(\cdot), q_0(\cdot), q_-(\cdot))$ the *consent-quotas function*. The power is *decisive* if for all $\nu \in N$, both $q_+(\nu)$ and $q_-(\nu)$ take the value of 1. The power is *anti-decisive* if for all ν, both $q_+(\nu)$ and $q_-(\nu)$ take the value of $\nu + 1$. To avoid unnecessary complication, we assume that for all $\nu \in N$,

$$q_+(\nu), q_-(\nu) \in \{1, \dots, \nu + 1\},\ q_0(\nu) \in \{0, 1, \dots, \nu + 1\},\ \text{and } q_0(0) \in \{0, 1\},$$

and

$$q_+(0) = q_+(1),\ q_-(0) = q_-(1),\ \text{and } q_0(n) = q_0(n - 1).$$

Let $\mathcal{Q}$ be the family of consent-quota functions satisfying these assumptions.

Definition 4.6.1 (System of Powers). A *system of powers* representing a collective choice function c on $\mathcal{V}_{\text{Tri}}$ is a function $W: M \to N \times \mathcal{Q}$ mapping each issue $k \in M$ into a pair of the person, $W_1(k)$, who has the power on the kth issue, and the consent-quotas function, $W_2(k) = (q_+(\cdot), q_0(\cdot), q_-(\cdot))$, associated with the power. That is, when $W_1(k) = i$, for all $\nu \in \{0, 1, \dots, n\}$ and all $V \in \mathcal{V}_{\text{Tri}}$ with $||V^k_{+,-}|| = \nu$, the social decision on the kth issue is made as described in (4.21).

The power on the kth issue is (fully) *exclusive* if there is a person i who has the power on the kth issue and no one else does. It is (fully) *non-exclusive* if all persons have the "equal" power on the kth issue associated with a single consent-quotas function (or, on the dichotomous domain, a list of consent-quotas). The power on an issue is either exclusive or non-exclusive (Remark 1 of Ju (2010)). Thus either only one person has the power or all persons have equal power. Two systems of powers W and W' are *equivalent*, denoted by $W \sim W'$, if for all k with $W_1(k) \neq W_1'(k)$, the power on the kth issue is *non-exclusive* (so, $W_2(k) = W_2'(k)$); otherwise, $W_2(k) = W_2'(k)$. If a collective choice function is represented by a system of powers, the system of powers is unique up to this equivalence relation (Ju 2010, Proposition 2). The following two extreme systems are notable. Under a *non-exclusive system of powers*, everyone has non-exclusive power on every issue. Under a *monocentric system of powers*, one person has exclusive power on every issue.

A necessary and sufficient condition for *issues monotonicity* is composed of the following two properties of consent-quotas functions (Ju 2010, Proposition 3). A consent-quotas function $q(\cdot) \equiv (q_+(\cdot), q_0(\cdot), q_-(\cdot))$ has the *component ladder property* if for all $\nu \in \{1, \dots, n\}$, the following three inequalities hold:

$$\begin{aligned} &\text{(i) } q_+(\nu - 1) \leq q_+(\nu) \leq q_+(\nu - 1) + 1; \\ &\text{(ii) } q_0(\nu - 1) \leq q_0(\nu) \leq q_0(\nu - 1) + 1; \\ &\text{(iii) } q_-(\nu - 1) \leq q_-(\nu) \leq q_-(\nu - 1) + 1. \end{aligned} \tag{4.22}$$

When this property fails, the decision may not respond monotonically after other persons' opinions become more favorable. The function has the *intercomponent ladder property* if for all $\nu \in \{1, \dots, n\}$,

$$q_+(\nu) \leq q_0(\nu - 1) + 1 \leq \nu - q_-(\nu) + 2. \tag{4.23}$$

When this property fails, the issue, initially accepted, may be rejected after the person who has the power on the issue becomes more favorable. For example, any anti-decisive power which has $q_{+}(\nu) = q_{-}(\nu) = \nu + 1$ for all ν violates intercomponent ladder property. On the dichotomous domain, component ladder property has no bite and intercomponent ladder property reduces to $q_{+} + q_{-} \leq n + 2$. The *ladder property* refers to the conjugation of the two ladder properties.

In the Arrovian framework, Sen (1970a,b, 1976, 1983) and many of his critics formulate individual rights based on (1) the existence of the so-called recognized personal spheres and (2) 'how the recognition of the personal spheres of different individuals should be reflected in the choices made by the society' (Gaertner et al. 1992, p. 162). Here, to formulate such recognized personal spheres, we use a function mapping each issue into a person, $\lambda: M \rightarrow N$, called a *linkage*. The next axiom requires the existence of recognized personal spheres. However, it does not impose any specific condition regarding what form the recognition should take, except for a minimal "symmetry" condition, which says that collective choice functions should treat person i and i's issues (constituting i's personal sphere) symmetrically to any other person j and j's issues. Technically, when names of person i and all i's issues are switched simultaneously to names of person j and all j's issues, social decision should also be switched accordingly. Given a linkage $\lambda \in \Lambda$, for all $i \in N$, let us call elements in $\lambda^{-1}(i)$ person i's issues. Let $\pi: N \rightarrow N$ and $\delta: M \rightarrow M$ are permutations on N and on M such that for all $i \in N$, δ maps the set of person i's issues *onto* the set of person $\pi(i)$'s issues. Let ${}_{\pi}^{\delta}P$ be the matrix such that for all $i \in N$ and all $k \in M$, ${}_{\pi}^{\delta}P_{ik} \equiv P_{\pi(i)\delta(k)}$. Then person i and his issue k play the same role in ${}_{\pi}^{\delta}P$ as person $\pi(i)$ and his issue $\delta(k)$ do in P.

Symmetric Linkage There is a linkage $\lambda: M \rightarrow N$ such that for all permutations $\pi: N \rightarrow N$ and all permutations $\delta: M \rightarrow M$, if for all $i \in N$, δ maps the set of i's issues $\lambda^{-1}(i)$ *onto* the set of $\pi(i)$'s issues $\lambda^{-1}(\pi(i))$, then for all $k \in M$, $f_k\left({}_{\pi}^{\delta}P\right) = f_{\delta(k)}(P)$.

Symmetry holds in the model of group identification if the collective choice function satisfies symmetric linkage and the linkage is the identity function, which means that the qualification of i is recognized as i's personal sphere, as is natural in this model.

A condition on systems of powers that is necessary and sufficient for symmetric linkage is *horizontal equality*: for all pair of persons i and $j \in N$ with the same number of issues under W_1, that is, $|W_1^{-1}(i)| = |W_1^{-1}(j)|$, their powers are associated with the same consent-quotas function, that is, for all $k \in W_1^{-1}(i)$ and all $l \in W_1^{-1}(j)$, $W_2(k) = W_2(l)$ (Ju 2010, Proposition 4).[28] When $i = j$, this

[28] A linkage creates primitive differences among persons and among issues in this setting; except for this, all other aspects of the model give equal standing to all persons (they share the same set of potential opinion vectors) and to all issues. A linkage differentiates persons vertically depending on the number of issues one is associated with. Horizontal equality allows us to incorporate this vertical differentiation in systems of powers not harming equality too much among persons.

property says that person i's powers on two different issues are associated with the same consent-quotas function.

Adding symmetric linkage to issues monotonicity and issues independence provides a characterization of collective choice functions represented by a system of powers.

Theorem 4.6.1 (Ju 2010). *Let $\mathcal{D} \in \{\mathcal{V}_{Di}, \mathcal{V}_{Tri}\}$. A collective choice function on $\mathcal{D}$ satisfies issues monotonicity, issues independence, and symmetric linkage if and only if it is represented by a system of powers satisfying the ladder property and horizontal equality. Moreover, the system is unique up to the equivalence relation $\sim$.*

4.6.2 Group Identification

We now consider group identification problems, where $M = N$. Several recent studies on group identification introduced by Kasher and Rubinstein (1997) formulate principles of liberalism in this specific model and establish axiomatic characterizations of "liberal" collective choice function.

4.6.2.1 Liberalism and Axiomatic Characterizations

A system of powers W on the domain of dichotomous problems $\mathcal{V}_{Di}$ is *liberal* if $W_1(i) = i$ for all $i \in N$ and all powers are decisive. The *liberal* collective choice function on $\mathcal{V}_{Di}$ is represented by the liberal system of powers.

The next axiom incorporates the minimal sense of liberalism by requiring only that if someone qualifies (disqualifies) herself, then not everyone should be disqualified (qualified), in other words, there should be someone, possibly the same person, who is qualified (disqualified).

Semi-Liberal Principle: For all $V \in \mathcal{V}_{Di}$, if for some $i \in N$, $V_{ii} = 1$, then for some $j \in N$, $c_j(V) = 1$; if for some $i \in N$, $V_{ii} = -1$, then for some $j \in N$, $c_j(V) = -1$.

Sung and Dimitrov (2003) establish the following characterization of the liberal collective choice function.

Theorem 4.6.2 (Sung and Dimitrov 2003). *Assume $M = N$. A collective choice function on $\mathcal{V}_{Di}$ satisfies independence, symmetry, and semi-liberal principle if and only if it is the liberal function.*

This is a strengthening of the characterization by Kasher and Rubinstein (1997) where they impose monotonicity and unanimity as well as the three axioms above. Sung and Dimitrov (2003) show that these two additional axioms are redundant and that the three remaining axioms are logically independent.

Samet and Schmeidler (2003) propose the following two interesting axioms.[29] The first axiom says, in their words, that non-Hobbits' opinions about Hobbits do not matter in determining who are Hobbits.

Exclusive Self-Determination: If $V, V' \in \mathcal{D}$ are such that for all $i, j \in N$, $V_{ij} \neq V'_{ij}$ only if $c_i(V) = -1$ and $c_j(V) = 1$, then $c(V) = c(V')$.

The next axiom says that the two groups, of Hobbits and of qualifiers of Hobbits, should coincide.

Affirmative Self-Determination: For all $V \in \mathcal{D}$, $c(V) = c(V^t)$, where V^t is the transpose of V.

Imposing either one of the two self-determination axioms together with other axioms defined earlier, we have the following characterization of liberal choice function:

Theorem 4.6.3 (Samet and Schmeidler 2003). *Assume $M = N$. A collective choice function on $\mathcal{V}_{Di}$ satisfies monotonicity, independence, non-degeneracy and exclusive self-determination (or affirmative self-determination) if and only if it is the liberal function.*

Ju (2010) extends this result on the domain of trichotomous opinions $\mathcal{V}_{Tri}$. Çengelci and Sanver (2010) introduces the axiom of positive weak equal treatment property requiring that all persons should be qualified when everyone qualifies himself. Based on this axiom or a variant, they establish a characterization of the liberal choice function. Ju (2009a) weakens monotonicity and non-degeneracy in Theorem 4.6.3 and obtains an alternative characterization result.

For all $x, x' \in \{-1, 1\}^N$, let $x \wedge x' \equiv (\min\{x_i, x'_i\})_{i \in N}$ and $x \vee x' \equiv (\max\{x_i, x'_i\})_{i \in N}$. Similarly, for all $V, V' \in \mathcal{V}_{Di}$, let $V \wedge V' \equiv (\min\{V_{ij}, V'_{ij}\})_{i \in N, j \in N}$ and $V \vee V' \equiv (\max\{V_{ij}, V'_{ij}\})_{i \in N, j \in N}$. Miller (2008) considers an extended framework where a collective choice function is used to identify more than one groups. The key axiom in Miller (2008) pertains to the two methods of identifying a collective consisting of persons with feature a and feature b. One method is to identify the collective with feature a and the collective with feature b separately and take the intersection of the two groups. The other method is to identify the collective with feature a and feature b at once. The next axiom requires that both methods should yield the same group.

Meet Separability: For all $V, V' \in \mathcal{D}$, $c(V) \wedge c(V') = c(V \wedge V')$.

The same requirement for identifying a collective consisting of persons with feature a *or* feature b is as follows.

Join Separability: For all $V, V' \in \mathcal{D}$, $c(V) \vee c(V') = c(V \vee V')$.

Miller (2008) shows that the liberal function is the only collective choice function satisfying the two separability axioms as well as non-degeneracy and anonymity.

[29] See Samet and Schmeidler (2003, pp. 222–224), for detailed discussion and motivation for the two axioms.

Theorem 4.6.4 (Miller 2008). *Assume $M = N$. A collective choice function on $\mathcal{V}_{Di}$ satisfies meet separability, join separability, non-degeneracy, and symmetry if and only if it is the liberal function.*

Miller (2008, Theorem 2.5) shows that collective choice functions satisfying the first three axioms depend only on a single vote and call them one-vote rules. From this result, Theorem 4.6.4 directly follows because only the liberal function among these one-vote rules can satisfy symmetry. Note that independence is not needed in this characterization result and in fact, it is implied by the four axioms.

4.6.2.2 Consent-Based Choice Functions

Samet and Schmeidler (2003) introduce a spectrum of choice functions connecting issue-wise majority function and the liberal function as two extreme functions of the family. A *consent-based choice function* on the domain of dichotomous opinions $\mathcal{V}_{Di}$ is represented by a system of powers W such that for all $i \in N$, $W_1(i) = i$ and $q_+ + q_- \leq n + 2$, where $(q_+, q_-) = W_2(k)$ for all $k \in N$.

Theorem 4.6.5 (Samet and Schmeidler 2003). *Assume $M = N$. A collective choice function on $\mathcal{V}_{Di}$ satisfies monotonicity, independence, and symmetry if and only if it is a consent-based choice function.*

A collective choice function c on $\mathcal{V}_{Di}$ satisfies *self-duality* if for all $V \in \mathcal{V}_{Di}$, $c(-V) = -c(V)$. Adding self-duality to the three axioms above, Samet and Schmeidler (2003, Theorem 2) characterize the subfamily of consent-based choice functions of which the consent quotas functions satisfy the following property: for all $i, j \in N$, $W_2(i) = W_2(j) = (q_+, q_-)$ and $q_+ = q_-$.

Note that self-dual consent-based choice functions have the same consent quotas for all persons. Allowing for different consent quotas across persons, a slightly larger family can be defined. This family is characterized by Çengelci and Sanver (2010, Theorem 4.1) with the set of four axioms, monotonicity, independence, self-duality and a weaker version of anonymity axiom. It should be noted that in this characterization, they do not impose symmetry, which plays a crucial role in Samet and Schmeidler (2003).

When $q_+ + q_- = n + 1$, parts (i) and (ii) of (4.20) are identical to the single condition that for all $V \in \mathcal{V}_{Di}$ and all $i \in N$, $c_i(V) = 1$ if and only if $||V^i_+|| \geq q_+$. Thus social decisions are made anonymously. Conversely, anonymous consent-based choice functions have consent quotas with $q_+ + q_- = n + 1$.

When there is an odd number of persons, the two conditions $q_+ = q_-$ and $q_+ + q_- = n + 1$ are satisfied only by the issue-wise majority function. Therefore, the issue-wise majority function is characterized by adding anonymity to the four axioms of monotonicity, independence, symmetry, and self-duality (Samet and Schmeidler 2003, p. 225). Replacing anonymity with neutrality, gives an alternative characterization of the issue-wise majority function.

4.6.3 Simple Preferences and the Paradox of Paretian Liberal

Compatibility of *Pareto efficiency* and existence of the so-called libertarian rights (decisive powers) is widely studied by a number of authors after Sen (1970a,b). We investigate Sen's paradox of Paretian liberal (Sen 1970a,b) in the current opinion aggregation framework by considering separable preference relations. We formulate Sen's liberal rights as a person's decisive power on a certain issue (Gibbard 1974). Note that each separable preference R_0 is associated with an opinion vector V_0, each positive (resp. negative or zero) component of V_0 representing the corresponding issue as a good (resp. a bad or a null). Obviously, there are a number of separable preference relations corresponding to a single opinion vector.

Sen's paradox holds on the separable preferences domain.[30] Sen's (1970a,b) *minimal liberalism* postulates that there should be at least two persons who have decisive powers. Assume that persons 1 and 2 are given the decisive powers on the first issue and the second issue respectively. Consider the following preference relations of the two persons. For person 1, the first issue is a bad and the second issue is a good. But person 1 cares so much about the second issue (person 2's issue) that he prefers the positive decision on this issue to the negative decision no matter what decisions are made on the other issues. For person 2, the second issue is a bad and the first issue is a good. But person 2 cares so much about the first issue (person 1's issue) that he prefers the positive decision on this issue to the negative decision no matter what decisions are made on the other issues. Then by the decisive powers of the two persons, decisions on the first and the second issues are both negative. But the two persons will be better off at any decision with positive components for both issues. This confirms that minimal liberalism and Pareto efficiency are incompatible on the separable preferences domain.

Preference relations in the above example are "meddlesome" (Blau 1975). One may hope that without such relations, the paradox of Paretian liberal will not occur. Unfortunately, the paradox holds even in a substantially restricted environment where only trichotomous or dichotomous preference relations are admissible. Consider a *trichotomous preference relation* R_0 that is a separable preference relation represented by a function $U_0\colon \{-1,1\}^M \to \mathbb{R}$ such that for each $x \in \{-1,1\}^M$, $U_0(x) = \sum_{k\in M: x_k=1} V_{0k}$, where $V_0 \in \{-1,0,1\}^M$ is the opinion vector corresponding to R_0.[31] Let $\mathcal{A}^*_{Tri}$ be the family of all such trichotomous preference relations. Let $\mathcal{A}^*_{Di}$ be the subfamily of dichotomous preferences in $\mathcal{A}^*_{Tri}$.

Proposition 4.6.1 (Ju 2008). *When there are at least three persons, no Pareto efficient collective choice function on* $\mathcal{A}^*_{Di}$ *or* $\mathcal{A}^*_{Tri}$ *satisfies minimal liberalism.*

Proof. Suppose that persons 1 and 2 have the decisive powers respectively on issue 1 and issue 2. Consider the profile of dichotomous preference relations

---

[30] This was originally proven by Gibbard (1974, Theorem 2).

[31] Equivalently, $U_0(x) = |\{k \in M : x_k = 1 \text{ and } P_{0k} = 1\}| - |\{k \in M : x_k = 1 \text{ and } P_{0k} = -1\}|$.

$(R_i)_{i \in N}$ given by the following opinion vectors: $V_1 \equiv (1, -1, -1, \ldots, -1)$, $V_2 \equiv (-1, 1, -1, \ldots, -1)$, and for all $i \in N \backslash \{1, 2\}$, $V_i \equiv (-1, \ldots, -1)$. Then by the decisive powers of persons 1 and 2, $c_1(R) = c_2(R) = 1$. If $c(\cdot)$ is *Pareto efficient*, for all $k \in M \backslash \{1, 2\}$, $c_k(R) = -1$. Thus $c(R) = (1, 1, -1, \ldots, -1)$. Note that this alternative is indifferent to $x \equiv (-1, \ldots, -1)$ for both person 1 and person 2 and x is preferred to $c(R)$ by all others. This contradicts *Pareto efficiency*. □

Note that unlike the previous paradox on the separable preferences domain, we need the assumption $n \geq 3$. The case with two persons ruled out by this assumption is very limited. In fact, the paradox does not apply in the two-person case (decisiveness is quite close to majority principle since one person's opinion accounts for 50%).

Collective choice functions that are represented by a system of powers do not satisfy minimal liberalism if no power is decisive. However they capture a somewhat weak sense of liberalism because they allow limited powers to individuals. Ju (2008) shows among these collective choice functions, there do exist Pareto efficient ones on $\mathcal{A}^*_{Tri}$. Issue-wise majority function is an example and all other Pareto efficient functions are very close to the issue-wise majority function. The only difference is when the number of voters in favor of an issue is the same as the number of voters against the issue, in which case the person who has the power on the issue dictates the social decision. Thus the exercise of a person's power is most limited. To be compatible with Pareto efficiency, exclusive powers that can be assigned to individuals should be limited so extremely that the resulting collective choices are very close to the issue-wise majority function where no individual has an exclusive power.

Acknowledgments I am grateful to William Thomson for helpful comments. I also thank Seong-Jae Oh and Seongkyu Park for their careful reading, discussion and comments.

References

Aleskerov, F., Yakuba, V., & Yuzbashev, D. (2007). A 'threshold aggregation' of three-graded rankings. *Mathematical Social Science, 53*, 106–110.

Alos-Ferrer, C. (2006). A simple characterization of approval voting. *Social Choice and Welfare, 27*, 621–625.

Arrow, K. J. (1951). *Social choice and individual values*. Wiley, New York.

Aşan, G., & Sanver, M. R. (2002). Another characterization of the majority rule. *Economic Letters, 75*, 409–413.

Banks, J. S. (1995). Acyclic social choice from finite sets. *Social Choice and Welfare, 12*, 293–310.

Baigent, N., & Xu, Y. (1991). Independent necessary and sufficient conditions for approval voting. *Mathematical Social Science, 21*, 21–29.

Barberà, S., Masso, J., & Neme, A. (1997). Voting under Constraints. *J Econ Theory, 76*, 298–321.

Barberà, S., Masso, J., & Neme, A. (2005). Voting by committees under constraints. *Journal of Economic Theory, 122*, 185–205.

Barberà, S., Sonnenschein, H., & Zhou, L. (1991). Voting by committees. *Econometrica, 59*, 595–609.

Barberà, S., Gul, F., & Stacchetti, E. (1993). Generalized median voter schemes and committees. *Journal of Economic Theory, 76*, 298–321.

Blau, J. H. (1975). Liberal values and independence. *Review of Economic Studies, 42,* 395–402.

Bogomolnaia, A., & Moulin, H. (2004). Random matching under dichotomous preferences. *Econometrica, 72,* 257–279.

Bogomolnaia, A., Moulin, H., & Stong, R. (2005). Collective choice under dichotomous preferences. *Journal of Economic Theory, 122,* 165–184.

Brams, S. J., & Fishburn, P. C. (1978). Approval voting. *American Political Science Review, 72,* 831–847.

Brams, S. J., & Fishburn, P. C. (2002). Voting procedures. In K. J. Arrow, A. K. Sen., & K. Suzumura (Eds.), *Handbook of social choice of welfare* (pp. 173–236). Amsterdam: Elsevier Science B.V.

Brown, D. J. (1975). Aggregation of preferences. *Quarterly Journal of Economics, 89,* 456–469.

Campbell, D. E., & Kelly, J. S. (2000). A simple characterization of majority rule. *Economic Theory, 15,* 689–700.

Çengelci, M. A., & Sanver, R. (2010). Simple collective identity functions. *Theory and Decision, 68,* 417–443.

Ching, S. (1996). A simple characterization of plurality rule. *Journal of Economic Theory, 71,* 298–302.

Dimitrov, D., Sung, S. C., & Xu, Y. (2007). Procedural group identification. *Mathematical Social Sciences, 54,* 137–146.

Fishburn, P. C. (1978a). Axioms for approval voting: Direct proof. *Journal of Economic Theory, 19,* 180–185; Corrigendum 45 (1988), 212.

Fishburn, P. C. (1978b). A strategic analysis of nonranked voting systems. *SIAM Journal on Applied Mathematics, 35,* 488–495.

Fishburn, P. C. (1979). Symmetric and consistent aggregation with dichotomous voting. In J.-J. Laffont (Ed.), *Aggregation and revelation of preferences* (pp. 201–218). Amsterdam: North-Holland.

Gaertner, W. (2002). Domain restrictions. In K. J. Arrow, A. K. Sen, & K. Suzumura (Eds.), *Handbook of social choice and welfare* (Vol. I, pp. 131–170). Amsterdam: Elsevier Science B.V.

Gaertner, W., Pattanaik, P. K., & Suzumura, K. (1992). Individual rights revisited. *Economica, 59*(234), 161–177.

Gibbard, A. (1969). *Social choice and the Arrow condition.* Mimeograph, Harvard University.

Gibbard, A. (1973). Manipulation of voting schemes: A general result. *Econometrica, 41,* 587–602.

Gibbard, A. (1974). A Pareto-consistent libertarian claim. *Journal of Economic Theory, 7,* 388–410.

Guha, A. S. (1972). Neutrality, monotonicity, and the right of veto. *Econometrica, 40,* 821–826.

Herzberger, H. G. (1973). Ordinal preference and rational choice. *Econometrica, 41,* 187–237.

Inada, K.-I. (1964). A note on the simple majority decision rule. *Econometrica, 32,* 525–531.

Inada, K.-I. (1969). The simple majority decision rule. *Econometrica, 37,* 490–506.

Inada, K.-I. (1970). Majority rule and rationality. *Journal of Economic Theory, 2,* 27–40.

Ju, B.-G. (2003). A characterization strategy-proof voting rules for separable weak orderings. *Social Choice and Welfare, 21*(3), 469–499.

Ju, B.-G. (2005a). An efficiency characterization of plurality social choice on simple preference domains. *Economic Theory, 26*(1), 115–128.

Ju, B.-G. (2005b). A characterization of plurality-like rules based on non-manipulability, restricted efficiency, and anonymity. *International Journal of Game Theory, 33,* 335–354.

Ju, B.-G. (2008). *Individual powers and social consent: Sen's paradox reconsidered* (working paper). Korea University.

Ju, B.-G. (2009a). *On the characterization of liberalism by Samet and Schmeidler* (working paper). Korea University.

Ju, B.-G. (2009b). *Collectively rational voting rules for simple preferences* (working paper). Korea University.

Ju, B.-G. (2010). Individual powers and social consent: An axiomatic approach. *Social Choice and Welfare, 34,* 571–596.

Kasher, A., & Rubinstein, A. (1997). On the question 'Who is a j', a social choice approach. *Logique et Analyse, 160*, 385–395.

Le Breton, M., & Sen, A. (1995). Strategyproofness and decomposability: weak ordering. Discussion Papers in Economics No. 95–04, Indian Statistical Institute, Delhi Centre.

Le Breton, M., & Sen, A. (1999). Separable preferences, strategyproofness, and decomposability. *Econometrica 67*(3): 605–628.

Mas-Colell, A., & Sonnenschein, H. (1972). General possibility theorems for group decisions. *Review of Economic Studies, 39*, 185–192.

Maskin, E. S. (1995). Majority rule, social welfare functions, and game forms. In: Basu, K., Pattanaik, P. K., & Suzumura, K. (Eds.), *Choice, welfare, and development*. Oxford: The Clarendon Press.

May, K. O. (1952). A set of independent necessary and sufficient conditions for simple majority decision. *Econometrica, 20*, 680–684.

Miller, A. D. (2008). Group identification. *Games and Economic Behavior, 63*, 188–202.

Moulin, H. (1980). On strategy-proofness and single peakedness. *Public Choice, 35*, 437–455.

Myerson, R. B. (1995). Axiomatic derivation of scoring rules without the ordering assumption. *Social Choice and Welfare, 12*, 59–74.

Rubinstein, A., & Fishburn, P. C. (1986). Algebraic aggregation theory. *Journal of Economic Theory, 38*, 63–77.

Sakai, T., & Shimoji, M. (2006). Dichotomous preferences and the possibility of Arrovian social choice. *Social Choice and Welfare, 26*, 435–445.

Samet, D., & Schmeidler, D. (2003). Between liberalism and democracy. *Journal of Economic Theory, 110*(2), 213–233.

Sanver, M. R. (2009). Characterizations of majoritarianism: A unified approach. *Social Choice and Welfare, 33*, 159–171.

Satterthwaite, M. A. (1975). Strategy-proofness and Arrow's condition: Existence and correspondence theorems for voting procedures and social welfare functions. *Journal of Economic Theory, 10*, 187–217.

Sen, A. K. (1970a). *Collective choice and social welfare*. San Francisco: Holden-Day.

Sen, A. K. (1970b). The impossibility of a Paretian Liberal. *Journal of Political Economy, 78*, 152–157.

Sen, A. K. (1976). Liberty, unanimity and rights. *Economica, 43*, 217–245.

Sen, A. K. (1977). Social Choice Theory: A Re-Examination. *Econometrica, 45*, 53–89.

Sen, A. K. (1983). Liberty and social choice. *Journal of Philosophy, 80*(1), 5–28.

Serizawa, S. (1995). Power of voters and domain of preferences where voting by committees is strategy-proof. *J Econ Theory, 67*, 599–608.

Sertel, M. R. (1988). Characterizing approval voting. *Journal of Economic Theory, 45*, 207–211.

Shapley, L. S., & Shubik, M. (1954). A method for evaluating the distribution of power in a committee system. *American Political Science Review, 48*, 787–792.

Shimomura, K-I. (1996). Partially efficient voting by committees, *Social Choice and Welfare, 13*, 327–342.

Sprumont, Y. (1991). The division problem with single-peaked preferences: A characterization of the uniform allocation rule. *Econometrica, 59*, 509–519.

Sung, S. C., & Dimitrov, D. (2003). *On the axiomatic characterization of "Who is a J?"* (working paper). Tilburg University.

Thomson, W. (2001). On the axiomatic method and its recent applications to game theory and resource allocation. *Social Choice and Welfare, 18*, 327–386.

Wilson, R. (1975). On the theory of aggregation. *Journal of Economic Theory, 10*, 89–99.

Xu, Y. (2010). Axiomatizations of approval voting. In J. F. Laslier & M. R. Sanver (Eds.), *Handbook on approval voting*. Heildelberg: Springer-Verlag

Yeh, C.-H. (2006). Reduction-consistency in collective choice problems. *Journal of Mathematical Economics, 42*, 637–652.

Young, H. P. (1975). Social choice scoring functions. *SIAM Journal of Applied Mathematics, 28*, 824–838.

Chapter 5
Axiomatizations of Approval Voting

Yongsheng Xu

5.1 Introduction

There has been a number of axiomatic studies of approval voting since its introduction by Brams and Fishburn (1978). The axiomatic characterizations of approval voting have given researchers a better understanding of the structure of approval voting, and have made the pros and cons of approval voting much sharper. In this article, we present a survey of various axiomatic characterizations of approval voting that are there in the literature.

Approval voting discussed in this article is referred to the way that ballots are aggregated, and is therefore a particular example of a broad class of methods that aggregate individual ballots. An individual ballot can be regarded as consisting of those candidates or options that are *acceptable* to a particular voter. Interpreted in this way, a voter can be viewed as expressing a *dichotomous preference* over all relevant candidates or options by dividing them into acceptable and unacceptable ones. The problem of aggregating ballots is therefore naturally linked with the classical social choice problem of aggregating individual preferences under the domain restriction of dichotomous preferences (interested readers should consult Chap. 20). In this article, however, we shall focus on approval voting as a method of aggregating ballots.

The existing axiomatic characterizations of approval voting based on ballot aggregation can be categorized into three groups. In the first place, Fishburn (1978a,b) develops a framework with *variable electorate* to analyze approval voting. The essence of the framework is that it assumes electorates consist of all non-empty and finite collections of voters, and voters cast their ballots for their approved candidates. Within this framework, Fishburn (1978a) characterizes approval voting by using three axioms: neutrality, consistency and disjoint equality (see Sect. 5.3 for formal definitions). Sertel (1988) uses similar axioms to the ones used in Fishburn

Y. Xu
Department of Economics, Andrew Young School of Policy Studies, Georgia State University, Atlanta, GA 30303, USA
e-mail: yxu3@gsu.edu

J.-F. Laslier and M.R. Sanver (eds.), *Handbook on Approval Voting*, Studies in Choice and Welfare, DOI 10.1007/978-3-642-02839-7_5,

(1978a) to characterize a slightly different version of approval voting: in Fishburn (1978a), approval voting picks up every candidate as a winning candidate when none is approved by any voter; while in Sertel (1988), when every voter approves no candidate, approval voting picks none as well. Within the same framework, Fishburn (1978b) presents another axiomatization of approval voting by using four axioms: neutrality, consistency, faithfulness and cancellation (see Sect. 5.3 for formal definitions). As shown by Alós-Ferrer (2006), neutrality in this characterization is redundant: approval voting is characterized by consistency, faithfulness and cancellation. Within this framework, a slightly different characterization of approval voting is presented: approval voting is characterized by consistency, faithfulness and disjoint equality. The second group of axiomatizations of approval voting is developed in Baigent and Xu (1991) where they work with a *fixed electorate*. The set of voters is assumed to be given and fixed. In this framework, they characterize approval voting by neutrality, independence of symmetric substitutions and positive responsiveness (see Sect. 5.4 for formal definitions). Based on the work by Baigent and Xu (1991), we present a variant of their characterization by using neutrality, equal treatment (a stronger version of independence of symmetric substitutions; see Sect. 5.4 for a formal definition) and monotonicity. The third group of axiomatization of approval voting has been developed in Xu (2008) for variable electorates that are drawn from a fixed set of voters. He works in a single profile of voters' ballots and axiomatically characterizes approval voting by four axioms: faithfulness, weak consistency, disjoint inclusion and dual consistency.

It may be noted that approval voting has been studied from various other perspectives in the literature. For example, in the framework of aggregating dichotomous preferences, Brams and Fishburn (1978) examine whether approval voting is strategy-proof (defined on the domain of dichotomous preferences), and more recently, Vorsatz (2007) shows that approval voting can be characterized by strategy-proofness together with some other well-known axioms characterizing approval voting. In our approach, ballots are aggregated into a set of winning candidates. One could also consider the possibility of aggregating ballots into *probability distributions* over candidates. This is the approach adopted by Bogomolnaia et al. (2005) where, among other things, they present an axiomatic characterization of approval voting in the framework of aggregating dichotomous preferences: approval voting is characterized by anonymity, neutrality, strategy-proofness, and efficiency (under the domain restriction of dichotomous preferences). Lastly, it may be noted that approval voting is a member of *scoring rules* in the classical social choice approach of aggregating individual preferences and has been studied along this line as well. For a summary of the results along this approach, see Chebotarev and Shamis (1998).

The structure of the remaining article is as follows. In Sect. 5.2, we present the basic notion and definitions. Section 5.3 is devoted to axiomatizations of approval voting with variable electorate developed in Fishburn (1978a,b). Section 5.4 focuses on axiomatizations with fixed electorate considered in Baigent and Xu (1991), and Sect. 5.5 presents a characterization result based on variable electorate developed in Xu (2008).

5.2 Notation and Definitions

Let X be a finite set of alternatives. It is assumed that X contains at least two alternatives. Alternatives in X in our context may be interpreted as *candidates* in an election. The set of all non-empty subsets of X will be denoted by $\mathcal{K}$. An element A of $\mathcal{K}$ is called a *ballot.*

An *electorate* is understood as a collection of *voters* with at least one but a finite number of them. Each voter in an electorate is assumed to cast one and only one ballot. It is assumed that a ballot is non-empty so that abstention is represented by the full ballot X in the present context.

Let $\mathbb{N}$ be the set of all non-negative integers. A *ballot response profile* is a function $\pi : \mathcal{K} \to \mathbb{N}$. For any $A \in \mathcal{K}$, $\pi(A)$ is the number of voters who cast ballot A. The number of voters who participated in the election is $\sum_{A \in \mathcal{K}} \pi(A)$. The set of all ballot response profiles is denoted by Π. For any $\pi, \pi' \in \Pi$, $\pi + \pi'$ is to denote the following profile: for all $A \in \mathcal{K}$, $(\pi + \pi')(A) = \pi(A) + \pi'(A)$.

Example 5.2.1. An example would be helpful to illustrate the above concepts. Let $X = \{x, y, z\}$. Then $\mathcal{K} = \{\{x, y, z\}, \{x, y\}, \{x, z\}, \{y, z\}, \{x\}, \{y\}, \{z\}\}$. Faced with the set X of the three candidates and when a voter approves both x and y, the voter casts the ballot $\{x, y\}$. Consider the following ballot response profile: $\pi(\{x, y, z\}) = 0, \pi(\{x, y\}) = 3, \pi(\{x, z\}) = 5, \pi(\{y, z\}) = 2, \pi(\{x\}) = 4,$ $\pi(\{y\}) = 2, \pi(\{z\}) = 0$. In this example, there are no voters casting the ballot $\{x, y, z\}$, 3 voters casting the ballot $\{x, y\}$, 5 voters casting the ballot $\{x, z\}$, 2 casting $\{y, z\}$, 4 casting $\{x\}$, 2 casting $\{y\}$, and 0 casting $\{z\}$. The number of participating voters is: $0 + 3 + 5 + 2 + 4 + 2 + 0 = 16$.

Let $D(\Pi)$ be a non-empty subset of Π. A *ballot aggregation function* is a function $f : D(\Pi) \to \mathcal{K}$ such that, for every ballot response profile $\pi \in D(\Pi)$, $f(\pi) \in \mathcal{K}$ is the (*winning*) ballot. Candidates in $f(\pi)$ are often called *winning candidates* under π. Note that a ballot response profile π does not retain voters' identities. It is therefore clear that a ballot aggregation function f is necessarily *anonymous* by definition: the names of voters have no impact on the aggregation result.

For any $x \in X$ and $\pi \in D(\Pi)$, let $n(x, \pi) = \sum_{A \in \mathcal{K},\, x \in A} \pi(A)$. Thus, $n(x, \pi)$ is the number of voters who approve of x when a ballot response profile is π. f is said to be *approval voting* if, for all $\pi \in \Pi$, $f(\pi) = \{x \in X : n(x, \pi) \geq n(y, \pi)$ for all $y \in X\}$.

Example 5.2.2. In Example 5.2.1, for the given X and ballot response profile π, we have $n(x, \pi) = 3 + 5 + 4 = 12, n(y, \pi) = 3 + 2 + 2 = 7$, and $n(z, \pi) = 5 + 2 = 7$. Then, $\{x\}$ is the outcome of approval voting.

5.3 Variable Electorate

In this section, we present a summary of axiomatizations of approval voting with a *variable electorate* for a given, fixed set of candidates X based on Fishburn (1978a,b). It is further assumed that electorates consist of all non-empty and finite collections of voters who cast their ballots for their approved candidates. The domain of a ballot aggregation function is taken to be the set of all ballot response profiles; that is, $D(\Pi) = \Pi$. We start by defining several axioms to be imposed on a ballot aggregation function f.

A ballot aggregation function satisfies:

- *Faithfulness* if and only if, for all $A \in \mathcal{K}$ and all $\pi \in \Pi$, if $\pi(A) = 1$ and $\pi(B) = 0$ for all $B \in (\mathcal{K} - \{A\})$, then $f(\pi) = A$.
- *Consistency* if and only if, for all $\pi, \pi' \in \Pi$, if $f(\pi) \cap f(\pi') \neq \emptyset$ then $f(\pi + \pi') = f(\pi) \cap f(\pi')$.
- *Cancellation* if and only if, for all $\pi \in \Pi$, if $[n(x, \pi) = n(y, \pi)$ for all $x, y \in X]$ then $f(\pi) = X$.
- *Disjoint equality* if and only if, for all $A, B \in \mathcal{K}$ with $A \cap B = \emptyset$, and all $\pi \in \Pi$, if $\pi(A) = \pi(B) = 1$, and $\pi(C) = 0$ for all $C \in (\mathcal{K} - \{A, B\})$, then $f(\pi) = A \cup B$.
- *Neutrality* if and only if, for all $\pi, \pi' \in \Pi$, if π' is such that $\pi'(A) = \pi(\sigma(A))$ for all $A \in \mathcal{K}$, where σ is a permutation on X, then $f(\pi') = \sigma(f(\pi))$.

The above axioms are introduced in Fishburn (1978a,b). Faithfulness requires that, if an electorate has just one voter who casts a ballot A, then every candidate in A must be a winning candidate. Consistency says that if there are overlapping winning candidates under ballot response profiles π and π', then the winning candidates for a ballot response profile that joins the two original ballot response profiles are exactly those who overlap. Cancellation stipulates that if all the candidates get the same number of votes from participating voters, then every candidate is a winning candidate. Disjoint equality requires that, if an electorate has two voters who cast two disjoint ballots A and B, then every candidate approved by either of the two voters is a winning candidate. Neutrality basically says that the names of candidates should not play any role in determining winning candidates.

Theorem 5.3.1 (Fishburn 1978b, Alós-Ferrer 2006). *A ballot aggregation function f is approval voting if and only if f satisfies Faithfulness, Consistency and Cancellation.*

Proof. It can be checked that if f is approval voting, then f satisfies Faithfulness, Consistency and Cancellation. We next show that, if f satisfies Faithfulness, Consistency and Cancellation, then f must be approval voting.

Let f be a ballot aggregation function satisfying Faithfulness, Consistency and Cancellation. We first show that, for any $\pi_1, \pi_2, \pi_3 \in \Pi$ and any $A, B \in \mathcal{K}$ with $A \cap B = \emptyset$, if $[\pi_2(A) = 1$ and $\pi_2(C) = 0$ for all $C \in (\mathcal{K} - \{A\})]$ and $[\pi_3(B) = 1$ and $\pi_3(C) = 0$ for all $C \in (\mathcal{K} - \{B\})]$, then

$$f(\pi_1 + \pi_2 + \pi_3) = f(\pi_1 + (\pi_2 + \pi_3)) \tag{5.1}$$

Consider $\pi \in \Pi$ such that $\pi(X-(A \cup B))=1$ and $\pi(C)=0$ for all $C \in \mathcal{K}-\{X-(A \cup B)\}$. If $A \cup B = X$, then, by Faithfulness, $f((\pi_2+\pi_3))=X$. Otherwise, by Cancellation, $f((\pi_2+\pi_3)+\pi)=X$. By Consistency and noting that $f(\pi_1+\pi_2+\pi_3) \cap f((\pi_2+\pi_3)+\pi) \neq \emptyset$, it follows that

$$f(\pi_1+\pi_2+\pi_3) = f(\pi_1+\pi_2+\pi_3+(\pi_2+\pi_3)+\pi) \tag{5.2}$$

Similarly, by Cancellation, $f(\pi_2+\pi_3+\pi)=X$. By Consistency and noting that $f(\pi_1+(\pi_2+\pi_3)) \cap f(\pi_2+\pi_3+\pi) \neq \emptyset$, it follows that

$$f(\pi_1+(\pi_2+\pi_3)) = f(\pi_1+(\pi_2+\pi_3)+\pi_2+\pi_3+\pi) \tag{5.3}$$

Note that $f(\pi_1+(\pi_2+\pi_3)+\pi_2+\pi_3+\pi)=f(\pi_1+\pi_2+\pi_3+(\pi_2+\pi_3)+\pi)$. Thus,

$$f(\pi_1+(\pi_2+\pi_3)) = f(\pi_1+\pi_2+\pi_3+(\pi_2+\pi_3)+\pi) \tag{5.4}$$

Therefore, (5.1) follows from (5.2) and (5.4) immediately. In particular, we note that,

$$f(\pi_2+\pi_3) = f((\pi_2+\pi_3)) \tag{5.5}$$

Take any ballot response profile $\pi \in \Pi$ and consider $\pi' \in \Pi$ such that $[\pi'(A) > 0 \Rightarrow |A|=1$ for all $A \in \mathcal{K}]$ and $[n(x,\pi)=n(x,\pi')$ for all $x \in X]$. Starting with (5.5) and by the repeated use of (5.1), it then follows that $f(\pi)=f(\pi')$.

To show that $f(\pi)=\{x \in X : n(x,\pi) \geq n(y,\pi)$ for all $y \in X\}$, we first define $m^* = \max_{x \in X} n(x,\pi)$. Since X is finite, m^* is well-defined. For each $m=0,1,\ldots,m^*$, we define

$$X_m = \{x \in X : n(x,\pi) = m\}$$

Define

$$A_0 = X_{m^*},\ A_1 = X_{m^*} \cup X_{m^*-1},\ \ldots,\ A_{m^*-1} = X_{m^*} \cup X_{m^*-1} \cup \ldots \cup X_1$$

Consider $\pi^0,\ldots,\pi^{m^*-1} \in D(\Pi)$ defined as follows: for each $m=0,1,\ldots,m^*$, $\pi_m(A_m)=1$ and $\pi_m(B)=0$ for all $B \in \mathcal{K}-\{A_m\}$. By Faithfulness, for each $m=0,\ldots,m^*$, $f(\pi^m)=A_m$. By Consistency, $f(\pi^0+\pi^1)=X_{m^*},\ldots,f(\pi^0+\cdots+\pi^{m^*})=X_{m^*}$. Note that $n(x,\pi')=n(x,\pi^0+\cdots+\pi^{m^*})$ for all $x \in X$. Therefore, $f(\pi^0+\cdots+\pi^{m^*})=f(\pi')$. Consequently, $f(\pi)=X_{m^*}$. □

The proof of Theorem 5.3.1 is based on Alós-Ferrer (2006). It may be noted that, in Fishburn (1978b), Neutrality is also used on top of the other three properties figured in the theorem. Alós-Ferrer (2006) shows that Neutrality is redundant.

Note that, in the proof of Theorem 5.3.1, Cancellation can be replaced by Disjoint equality in deriving (5.1). It is clear that approval voting satisfies Disjoint equality. Therefore, the following result follows immediately.

Theorem 5.3.2. *A ballot aggregation function f is approval voting if and only if f satisfies Consistency, Disjoint equality and Faithfulness.*

Next, we present a characterization of approval voting due to Fishburn (1978a) without giving a proof. Interested readers can consult Fishburn (1978a) for a proof.

Theorem 5.3.3 (Fishburn 1978a). *A ballot aggregation function f is approval voting if and only if f satisfies Consistency, Disjoint equality and Neutrality.*

The independence of the axioms used in each of the above results can be checked. For example, in Theorem 5.3.1, we can check that the three axioms, faithfulness, consistency and cancellation, used there are independent. For this purpose, consider first the following two ballot aggregation functions, f_1 and f_2, defined below: for all $\pi \in \Pi$,

$$f_1(\pi) = X$$
$$f_2(\pi) = \{x \in X : n(x, \pi) > 0\}$$

f_1 selects every candidate in X for any ballot response profile. f_2 selects every candidate that is chosen by at least one voter. Note that f_1 satisfies both consistency and cancellation, but fails faithfulness thus showing that faithfulness is not implied by the combination of consistency and cancellation. f_2 satisfies faithfulness and cancellation, but fails consistency showing that consistency is not implied by faithfulness and cancellation. Finally, to see that cancellation is not implied by faithfulness and consistency, we consider the following ballot aggregation function f_3: let $x \in X$ be fixed, for all $\pi \in \Pi$, $f_3(\pi) = \{x\}$ if $[n(a,\pi) = n(b,\pi) > 1$ for all $a, b \in X$ and $\pi(X) = 0]$ and $f_3(\pi)$ is given by approval voting otherwise. f_3 fails cancellation. It can also be checked that f_3 satisfies faithfulness and consistency.

The independence of the axioms figured in the other two results can be checked as well and we leave this as an exercise for interested readers.

It may be noted that, in our approach, since we do not allow the empty ballot to be casted by any voter, the abstention is represented by the full ballot X. In Fishburn (1978a), the abstention is captured by both the empty ballot and the full ballot. As a consequence, in Fishburn (1978a), the empty ballot and the full ballot X are treated equivalently by a ballot aggregation function, so that for both cases, a ballot aggregation function yields every candidate in X as a winning candidate. Sertel (1988) makes an argument that the empty ballot and the full ballot are different ways of handling abstention. When the empty ballot and the full ballot are treated differently, Sertel (1988) defines a slightly modified approval voting in which it is approval voting except when every voter casts the empty ballot, the ballot aggregation function picks no candidate as a winning candidate. In this framework, Sertel (1988) presents an axiomatization of this modified approval voting by similar axioms used in Theorem 5.3.2.

5.4 Fixed Electorate

In this section, we present an alternative axiomatization of approval voting with a fixed electorate for a given and fixed set of candidates X. It is therefore assumed that, for some positive integer n, for any ballot response profile π in the domain, $\sum_{A \in \mathcal{K}} \pi(A) = n$ and the set of participating voters is fixed. Let $D^F(\Pi)$ denote this domain of ballot response profiles.

For the purpose of axiomatization, we introduce two further axioms that are based on Baigent and Xu (1991). A ballot aggregation function f satisfies:

- *Equal treatment* if and only if, for all $\pi, \pi' \in D(\Pi)$, if $n(x,\pi) = n(x,\pi')$ for all $x \in X$, then $f(\pi) = f(\pi')$.
- *Monotonicity* if and only if, for all $\pi, \pi' \in D(\Pi)$, all $A \in \mathcal{K}$ and all $x \in X$, if $n(x,\pi') = n(x,\pi) + 1$ and $n(y,\pi') = n(y,\pi)$ for all $y \in X - \{x\}$, then $x \in f(\pi) \Rightarrow \{x\} = f(\pi')$.

Equal treatment is a variant and stronger version of Independence of symmetric substitutions introduced in Baigent and Xu (1991). It says that, for two ballot response profiles π and π', if, for each candidate $x \in X$, the number of voters who approve x under π is the same as the number of voters who approve x under π', then $f(\pi) = f(\pi')$ – the sets of winning candidates under π and π', respectively, must be identical. It essentially requires that the information about "who approves whom" should not play a significant role in figuring out winning candidates. Monotonicity requires that, if a candidate x is a winning candidate under π, and if some voter does not approve x under π but approves x under π' while all other candidates get exactly the same number of approved voters under π and under π', then x becomes the sole winning candidate under π'. Monotonicity is a stronger version of a similar property proposed in Baigent and Xu (1991) where it was called Positive responsiveness.

Theorem 5.4.1. *A ballot aggregation function f on $D^F(\Pi)$ is approval voting if and only if f satisfies Neutrality, Equal treatment and Monotonicity.*

Proof. We note that approval voting satisfies Neutrality, Equal treatment and Monotonicity. The remainder of the proof is to show that, if a ballot aggregation function f satisfies Neutrality, Equal treatment and Monotonicity, then f must be approval voting.

Let f be a ballot aggregation function satisfying Neutrality, Equal treatment and Monotonicity. We first show that, for any $\pi \in D^F(\Pi)$ and any $x, y \in X$,

$$n(x,\pi) = n(y,\pi) \Rightarrow [x \in f(\pi) \Leftrightarrow y \in f(\pi)] \tag{5.6}$$

To see (5.6) is true, consider $\pi \in D^F(\Pi)$ and any $x, y \in X$ with $n(x,\pi) = n(y,\pi)$, and $x \in f(\pi)$. Consider the following permutation σ on X: $\sigma(x) = y, \sigma(y) = x$, $\sigma(z) = z$ for all $z \in X - \{x, y\}$. For each $A \in \mathcal{K}$, let $\sigma(A) = \{\sigma(a) : a \in A\}$.

Consider π' such that $\pi'(\sigma(A)) = \pi(A)$ for all $A \in \mathcal{K}$. Note that $\pi' \in D^F(\Pi)$: whoever casted a ballot A under π casts the ballot $\sigma(A)$ under π'. Note that, by Neutrality, $y \in f(\pi')$ since $x \in f(\pi)$ and $\sigma(x) = y$. Since $n(x,\pi) = n(y,\pi)$, from the definition of the permutation σ, it follows that $n(z,\pi) = n(z,\pi')$ for all $z \in X$. By Equal treatment, we must have $f(\pi) = f(\pi')$. Note that $y \in f(\pi')$. Therefore, $y \in f(\pi)$. This completes the proof of (5.6).

Next, we show that, for any $\pi \in D^F(\Pi)$ and any $x \in X$,

$$x \in f(\pi) \Rightarrow [n(x,\pi) \geq n(y,\pi) \text{ for all } y \in X] \tag{5.7}$$

Suppose, to the contrary that $x \in f(\pi)$ and there exists $z \in X$ such that $n(x,\pi) < n(z,\pi)$. To begin with, suppose $n(x,\pi) + 1 = n(z,\pi)$. Since $n(x,\pi) < n(z,\pi)$, there must be a voter who does not approve x under π and who casts a ballot A ($x \notin A$). Consider $\pi' \in D^F(\Pi)$ such that the voter who casts the ballot A now casts the ballot $A \cup \{x\}$ and all other voters cast the same ballot under π and π'. Clearly, $\pi' \in D^F(\Pi)$. It is also clear that $n(x,\pi') = n(x,\pi) + 1$ and $n(y,\pi') = n(y,\pi)$ for all $y \in X - \{x\}$. By Monotonicity, from $x \in f(\pi)$, we must have $\{x\} = f(\pi')$. However, from (5.6), given that $n(x,\pi') = n(z,\pi')$, we must have $x, z \in f(\pi')$, a contradiction. This completes (5.7).

To complete the proof of the theorem, we establish the following: For all $\pi \in D^F(\Pi)$ and any $x \in X$,

$$[n(x,\pi) \geq n(y,\pi) \text{ for all } y \in X] \Rightarrow x \in f(\pi). \tag{5.8}$$

To prove (5.8), we need only to show that $x \notin f(\pi)$ implies that there exists $z \in X$ such that $n(x,\pi) < n(z,\pi)$. It must be the case that either $n(x,\pi) = 0$ or $n(x,\pi) > 0$. If $n(x,\pi) = 0$, then we consider any $y \in A$ such that $\pi(A) > 0$. Clearly, $n(x,\pi) < n(y,\pi)$. If $n(x,\pi) > 0$, then we consider any $y \in f(\pi)$. It follows from (5.7) that $n(y,\pi) \geq n(x,\pi)$. If $n(y,\pi) = n(x,\pi)$, then, from (5.6), $x \in f(\pi)$, a contradiction. Therefore, $n(y,\pi) > n(x,\pi)$. This completes (5.8).

Equation (5.7), together with (5.8), completes the proof of the theorem. □

It may be checked that the axioms used in Theorem 5.4.1 are independent. The ballot aggregation function f_1 defined in the last section satisfies neutrality and equal treatment, but fails to satisfy monotonicity proving that monotonicity is not implied by neutrality and equal treatment together. To see that neutrality is not implied by the combination of equal treatment and monotonicity, let us consider the following ballot aggregation function f_4 which is due to Alós-Ferrer (2006): let $X = \{x, y, z\}$ and let $f_4(\pi) = \{x\}$ if $\pi(\{x\}) = \pi(\{y\}) > 0$ and $\pi(B) = 0$ for all $B \neq \{x\}, \{y\}$ and $f_4(\pi)$ coincide with approval voting otherwise; then f_4 satisfies equal treatment and monotonicity, but fails neutrality. And finally, consider the following ballot aggregation function f_5: let i be a fixed voter who is a participating voter under any ballot response profile and let X_i^π be the ballot casted by voter i under π; define $f_5(\pi)$ as approval voting applied over X_i^π; then f_5 satisfies neutrality and monotonicity, but fails to satisfy equal treatment.

To complete this section, we introduce two axioms used by Baigent and Xu (1991) in their characterization of approval voting in the current context.

A ballot aggregation function f satisfies:

1. *Independence of symmetric substitutions* if and only if, for all $\pi, \pi' \in D^F(\Pi)$, all $A, B, C, D \in \mathcal{K}$ and all $x \in A, y \in B$ with $C = (A - \{x\}) \cup \{y\}$, $D = (B - \{y\}) \cup \{x\}$, $y \notin A, x \notin B$, if $\pi(A) = \pi'(A) - 1$, $\pi(B) = \pi'(B) - 1$, $\pi'(C) = \pi(C) + 1$ and $\pi'(D) = \pi(D) + 1$, and $\pi(E) = \pi'(E)$ for all $E \in (\mathcal{K} - \{A, B, C, D\})$, then $f(\pi) = f(\pi')$.
2. *Positive responsiveness* if and only if, for all $\pi, \pi' \in D^F(\Pi)$, all $A \in \mathcal{K}$, and all $x \in X - A$, if $\pi'(C) = \pi(C)$ for all $C \in \mathcal{K} - \{A, A \cup \{x\}\}$, $\pi'(A) = \pi(A) - 1 \geq 0$ and $\pi'(A \cup \{x\}) = \pi(A \cup \{x\}) + 1$, then $x \in f(\pi) \Rightarrow \{x\} = f(\pi')$.

It may be noted that, in Baigent and Xu (1991), the axioms are formulated in terms of choice functions. Independence of symmetric substitutions is a weaker property than equal treatment. The basic idea underlying this axiom is that it does not matter "who votes for whom." For further discussions of these two axioms, see Baigent and Xu (1991).

We now present the result which closely resembles the characterization of approval voting by Baigent and Xu (1991). Its proof is similar to the proof of Theorem 5.4.1 and we omit it.

Theorem 5.4.2. *A ballot aggregation function f is approval voting if and only if it satisfies Neutrality, Independence of symmetric substitution and Positive responsiveness.*

For the independence of the axioms figured in the above theorem, interested readers are referred to Baigent and Xu (1991).

5.5 Variable Electorate with a Single Ballot Response Profile

In this section, we consider an axiomatization of approval voting with variable electorates that are drawn from a given and fixed set $N = \{1, \ldots, m\}$ of finite number of voters and with a single ballot response profile. The approach in this section is closely related to the classical framework of aggregating a single profile of individual preferences developed in the literature of social choice theory. For this purpose and throughout this section, we assume that (1) each voter $i \in N$ casts a non-empty ballot $A_i \in \mathcal{K}$; (2) the profile of ballots $(A_1, \ldots, A_m)$ casted by voters in N is fixed; and (3) any voter i in an electorate $E \subseteq N$ casts the ballot A_i. These assumptions will put a restriction on the domain of a ballot aggregation function f. We shall denote this domain by $D^{VS}(\Pi)$ which is given by

$$\{\pi \in \Pi : \pi(A) > 0 \Rightarrow [\text{ for some } E \subseteq N, \#E = k, A = A_i \text{ for some } i \in E]\}$$

It may be noted that, for any $\pi \in D^{VS}(\Pi)$, $\sum_{A \in \mathcal{K}} \pi(A) = k$ for some $E \subseteq N$ with $\#E = k$.

Example 5.5.1. An example will be helpful in understanding the above restrictions on the domain. Let $X = \{x, y, z\}$ and a fixed electorate be given by $\{1, 2, 3\}$. There are seven ballots: $X, \{x, y\}, \{y, z\}, \{x, z\}, \{x\}, \{y\}, \{z\}$. Suppose voter 1 casts the ballot $\{x, y\}$, voter 2 casts $\{y\}$ and voter 3 casts $\{z\}$. Then, $D^{VS}(\Pi)$ contains the following 7 ballot response profiles:

$$\pi_1 : \pi_1(A) = \begin{cases} 1 & \text{if } A \in \{\{x, y\}, \{y\}, \{z\}\} \\ 0 & \text{if } A \in \{X, \{y, z\}, \{x, z\}, \{x\}\} \end{cases}$$

$$\pi_2 : \pi_2(A) = \begin{cases} 1 & \text{if } A \in \{\{x, y\}, \{y\}\} \\ 0 & \text{if } A \in \{X, \{y, z\}, \{x, z\}, \{x\}, \{z\}\} \end{cases}$$

$$\pi_3 : \pi_3(A) = \begin{cases} 1 & \text{if } A \in \{\{x, y\}, \{z\}\} \\ 0 & \text{if } A \in \{X, \{y, z\}, \{x, z\}, \{x\}, \{y\}\} \end{cases}$$

$$\pi_4 : \pi_4(A) = \begin{cases} 1 & \text{if } A \in \{\{y\}, \{z\}\} \\ 0 & \text{if } A \in \{X, \{y, z\}, \{x, z\}, \{x\}, \{x, y\}\} \end{cases}$$

$$\pi_5 : \pi_5(A) = \begin{cases} 1 & \text{if } A \in \{\{x, y\}\} \\ 0 & \text{if } A \in \{X, \{y, z\}, \{x, z\}, \{x\}, \{y\}\} \end{cases}$$

$$\pi_6 : \pi_6(A) = \begin{cases} 1 & \text{if } A \in \{\{y\}\} \\ 0 & \text{if } A \in \{X, \{y, z\}, \{x, z\}, \{x, y\}, \{z\}\} \end{cases}$$

$$\pi_7 : \pi_7(A) = \begin{cases} 1 & \text{if } A \in \{\{z\}\} \\ 0 & \text{if } A \in \{X, \{y, z\}, \{x, z\}, \{x, y\}, \{y\}\} \end{cases}$$

We now introduce three further axioms to be imposed on a ballot aggregation function f (see Xu 2008).

A ballot aggregation function f on $D^{VS}(\Pi)$ satisfies:

- *Disjoint inclusion* if and only if, for all $\pi, \pi' \in D^{VS}(\Pi)$ and all $A \in \mathcal{K}$, if $\pi'(B) = \pi(B)$ for all $B \in \mathcal{K} - \{A\}$, $\pi(A) = 0$ and $\pi'(A) = 1$, then $A \cap f(\pi) = \emptyset \Rightarrow f(\pi) \subseteq f(\pi')$.
- *Weak consistency* if and only if, for all $\pi, \pi' \in D^{VS}(\Pi)$ and all $A \in \mathcal{K}$, if $\pi'(A) = 1$, $\pi'(B) = 0$ for all $B \in \mathcal{K} - \{A\}$, and $(\pi + \pi') \in D^{VS}(\Pi)$, then $f(\pi) \cap f(\pi') \neq \emptyset \Rightarrow f(\pi + \pi') = f(\pi) \cap f(\pi')$.
- *Dual consistency* if and only if, for all $\pi, \pi' \in D^{VS}(\Pi)$ and all $x \in X$, if $x \notin f(\pi)$, $n(x, \pi') = 0$, and $(\pi + \pi') \in D^{VS}(\Pi)$, then $x \notin f(\pi + \pi')$.

Disjoint inclusion requires that, if a ballot response profile π' is enlarged from a ballot response profile π by adding a voter who is not a participant under π and who casts a ballot A, and if none of the candidates in A is a winning candidate under π, then all the winning candidates under π continue to be winning candidates under π'. Weak consistency is similar to consistency, but a bit weaker than consistency. Dual consistency stipulates that, if a candidate is not a winning candidate under π and is not approved by any voter under π', then it cannot be a winning candidate when the two ballot response profiles are joined.

Theorem 5.5.1 (Xu 2008). *A ballot aggregation function f on $D^{VS}(\Pi)$ is approval voting if and only if f satisfies Faithfulness, Disjoint inclusion, Weak consistency and Dual consistency.*

Proof. It can be checked that approval voting defined on $D^{VS}(\Pi)$ satisfies Faithfulness, Disjoint inclusion, Weak consistency and Dual consistency. It remains to be shown if a ballot aggregation function f on $D^{VS}(\Pi)$ satisfies Faithfulness, Disjoint inclusion, Weak consistency and Dual consistency, then f must be approval voting.

Let f on $D^{VS}(\Pi)$ satisfy Faithfulness, Disjoint inclusion, Weak consistency and Dual consistency. Let $\pi \in D^{VS}(\Pi)$. We shall prove the result by induction on $\sum_{A\in\mathcal{K}} \pi(A) = k$. Clearly, $1 \leq k \leq m$. By Faithfulness,

$$f(\pi) = A \text{ if } \pi(A) = 1 \text{ and } \pi(B) = 0 \text{ for all } B \in \mathcal{K} - \{A\} \tag{5.9}$$

Suppose $f(\pi)$ is approval voting for any π such that $\sum_{A\in\mathcal{K}} \pi(A) \leq k$. We now show that $f(\pi)$ is approval voting for any π such that $\sum_{A\in\mathcal{K}} \pi(A) = k+1$. Let $\pi \in D^{VS}(\Pi)$ be such that there are $k+1$ voters who cast ballots. We denote these ballots by $A_1, \ldots, A_{k+1}$. For each $p = 1, \ldots, k+1$, let $\pi_{-p} \in D^{VS}(\Pi)$ be such that π_{-p} is generated by an electorate consisting of all the participating voters under π except the voter who casts A_p, and let $\pi_p \in D^{VS}(\Pi)$ be such that it is generated by a single voter who casts the ballot A_p. Note that, by Faithfulness, $f(\pi_p) = A_p$ for each $p = 1, \ldots, k+1$. Note also that $\pi = \pi_p + \pi_{-p}$ for every $p = 1, \ldots, k+1$. If, for some $p = 1, \ldots, k+1$, $f(\pi_p) \cap f(\pi_{-p}) \neq \emptyset$, then by Weak consistency, $f(\pi) = f(\pi_p) \cap f(\pi_{-p})$. Since $f(\pi_{-p})$ and $f(\pi_p)$ are given by approval voting, it is immediately clear that $f(\pi)$ is given by approval voting as well. If, for all $p = 1, \ldots, k+1$, $f(\pi_p) \cap f(\pi_{-p}) = \emptyset$, then, by Disjoint inclusion, $f(\pi_{-p}) \subseteq f(\pi)$. That is, $\cup_{p=1,\ldots,k+1} f(\pi_{-p}) \subseteq f(\pi)$. It is easy to check that, in this case, for any $x \in \cup_{p=1,\ldots,k+1} f(\pi_{-p})$, $n(x,\pi) \geq n(y,\pi)$ for all $y \in X$. If we can show $f(\pi) = \cup_{p=1,\ldots,k+1} f(\pi_{-p})$, then we are done. Consider any $z \in X - (\cup_{p=1,\ldots,k+1} f(\pi_{-p}))$, $z \in f(\pi)$. Since $z \notin f(\pi_{-p})$ for all $p = 1, \ldots, k+1$ and each $f(\pi_{-p})$ is given by approval voting, it must be the case that, for some $p = 1, \ldots, k+1$, $z \notin A_p$. Then, by Dual consistency, $z \notin f(\pi_p + \pi_{-p}) = f(\pi)$. Therefore, in this case, $f(\pi) = \cup_{p=1,\ldots,k+1} f(\pi_{-p})$, which shows that $f(\pi)$ is given by approval voting.

From our induction hypothesis and by (5.9), it then follows that f is approval voting. □

Interested readers may want to check whether the axioms used in Theorem 5.5.1 are independent.

Acknowledgements For helpful comments on an earlier draft of the article, I am grateful to Jean-Francois Laslier and Remzi Sanver.

References

Alós-Ferrer, C. (2006). A simple characterization of approval voting. *Social Choice and Welfare, 27*, 621–625.

Baigent, N., & Xu, Y. (1991). Independent necessary and sufficient conditions for approval voting. *Mathematical Social Sciences, 21*, 21–29.

Bogomolnaia, A., Moulin, H., & Stong, R. (2005). Collective choice under dichotomous preferences. *Journal of Economic Theory, 122*, 165–184.

Brams, S., & Fishburn, P. (1978). Approval voting. *American Political Science Review, 72*, 831–847.

Chebotarev, P. Y., & Shamis, E. (1998). Characterizations of scoring methods for preference aggregation. *Annals of Operations Research, 80*, 299–332.

Fishburn, P. (1978a). Axioms for approval voting: Direct proof. *Journal of Economic Theory, 19*, 180–185.

Fishburn, P. (1978b). Symmetric and consistent aggregation with dichotomous preferences. In J. Laffont (Ed.), *Aggregation and revelation of preferences*. Amsterdam: North-Holland.

Sertel, M. (1988). Characterizing approval voting. *Journal of Economic Theory, 45*, 207–211.

Vorsatz, M. (2007). Approval voting on dichotomous preferences. *Social Choice and Welfare, 28*, 127–141.

Xu, Y. (2008). *Coalitional decision makings and approval voting* (mimeo). Department of Economics, Georgia State University.

Part III
Committees

Chapter 6
Approval Balloting for Multi-winner Elections

D. Marc Kilgour

6.1 Introduction

Approval voting is a well-known voting procedure for single-winner elections. Voters approve of as many candidates as they like, and the candidate with the most approvals wins (Brams and Fishburn 1978, 1983, 2005). But Merrill and Nagel (1987) point out that there are many ways to aggregate approval votes to determine a winner, justifying a distinction between *approval balloting*, in which each voter submits a ballot that identifies the candidates the voter approves of, and *approval voting*, the procedure of ranking the candidates according to their total numbers of approvals.

Approval balloting can also be used in a multi-winner election, where the objective is to identify a "best" subset of candidates using the ballots, i.e., the voters' approvals, as input. We discuss several different procedures for determining a subset of candidates based on a profile of approval ballots. In practice, the subset selected could be anything from a standing committee of, say, university faculty, to a constitutional convention, to an all-star team. Nonetheless, we will refer to subsets of the candidates, including the subset selected by the voters, as *committees*. We consider only systems in which every voter has an equal role; in particular, every voter must receive an identical ballot, and must have the opportunity to vote for any and all approved-of candidates. Note that these conditions rule out some multi-winner elections, such as those for national legislatures in many countries. But many elections do fit this description and, as will be seen, there are many ways to determine a winning committee in such elections.

Filling out an approval ballot is equivalent, of course, to selecting a subset of the candidates – the voter's approved subset. Thus there is a natural correspondence between a committee and an approval ballot, in that both are subsets of the set of candidates. This fundamental link is exploited by some procedures for combining

D.M. Kilgour
Wilfrid Laurier University, 75 University Avenue West, Waterloo, ON, Canada N2L 3C5
e-mail: mkilgour@wlu.ca

J.-F. Laslier and M.R. Sanver (eds.), *Handbook on Approval Voting*, Studies in Choice and Welfare, DOI 10.1007/978-3-642-02839-7_6,

the voters' preferences, as registered in a profile of approval ballots, to produce a "best" committee.

It is usual to assess voting systems by identifying and comparing their properties (Arrow et al. 2002; Brams and Fishburn 2002). Some voting systems for electing committees are studied in this way by Ratliff (2003, 2006). Though the main objective of this chapter is to collect and classify procedures that use approval balloting to elect committees, we will compare some of them according to properties that are obvious or well-established. It should be noted that all procedures mentioned here are anonymous (treat all voters fairly) and neutral (treat all candidates fairly). Nonetheless, the presentation of properties below is no doubt far from complete. This article will simply present procedures in a format suitable for comparison; some conclusions are drawn about which systems are appropriate for particular purposes, but many open questions remain.

6.2 The Setting

We assume throughout that there are $n > 1$ voters and $m > 1$ candidates. Let $[m] = \{1, 2, \ldots, m\}$. We will survey procedures to select a committee based on n approval votes with candidate set $[m]$; thus, we are conducting a multi-winner election in the sense that every member of the elected committee wins. However, we allow for ties, so there may be several winning committees. (Most of our notation and terminology is drawn from Kilgour et al. (2006) and Fishburn and Pekeč (2004).)

The set of all subsets of the set of candidates is $2^{[m]}$, and in general there are 2^m possible committees, namely the members of $2^{[m]}$. In practice, however, multi-winner elections are often conducted under a priori restrictions on the possible winning subsets. In other words, it is typical that many of the 2^m subsets of candidates *cannot* win. For example, the size of a committee is often decided in advance of the election, for example by a constitution. Rarely is it meaningful to select the entire set of candidates, $[m]$. (Why hold the election, if not to reject at least one candidate?) As well, there are often "representativeness" restrictions; the winning subset must contain at least one woman, or equal numbers of men and women, or at least one member of a predefined subset of the candidates, or representation from each of several defined subgroups. For example, a basketball all-star team requires one center, two forwards, and two guards; similarly, university committees are often required to include members from various subdivisions.

Another condition, often implicit, is that some candidate(s) must win; i.e., the winning set cannot be $\emptyset$. This condition might not be appropriate for some elections, such as for members of a "hall of fame," which must allow for the collective decision that no candidates are suitable for enshrinement. (For example, in the National Baseball Hall of Fame, ballots have about 25 candidates, of whom only a handful are elected in a typical year. Voting is by approval balloting except that voters are permitted to vote for at most 10 candidates. See National Baseball Hall of Fame [BBHOF] (2009) for details.)

For any multi-winner election, we call the allowable winning committees the *admissible committees*, denoted $\mathcal{A}$. We assume throughout that $\mathcal{A} \subseteq 2^{[m]}$ is a fixed non-empty collection of subsets of the set of all candidates, and consider it to be a "parameter" of the election, like n and m. For instance, it is common for the admissible committees to be all subsets containing exactly k candidates, where k is fixed and satisfies $1 \leq k < m$. This set of committees is denoted $\mathcal{A}_k$; if the election is to choose such a committee, we say that $\mathcal{A} = \mathcal{A}_k$. In another useful example, all non-empty committees are admissible; in this case, we say that $\mathcal{A} = \mathcal{A}_F = 2^{[m]} - \emptyset$.

Let $i = 1, 2, \ldots, n$. Then voter i's ballot is $V_i \subseteq [m]$; note that a voter is allowed to vote for no candidates, so V_i may be empty. The ballot profile is $V = (V_1, V_2, \ldots, V_n)$, and the set of all possible ballot profiles is $\mathcal{V} = \left(2^{[m]}\right)^n$. Any voting procedure is then a function, possibly multi-valued, from $\mathcal{V}$ to $\mathcal{A}$.

We now introduce several examples that will be useful to illustrate the distinctions among the procedures. The first is from Kilgour et al. (2006), and the second from Fishburn and Pekeč (2004).

Example 6.2.1. There are $n = 4$ voters named 1, 2, 3, and 4; $m = 3$ candidates named 1, 2, and 3; and all non-empty committees are admissible ($\mathcal{A} = \mathcal{A}_F$).

Voter	1	2	3	4
Ballot	1	12	13	13

Example 6.2.2. $n = 9$ voters; $m = 8$ candidates; any three-member committee admissible ($\mathcal{A} = \mathcal{A}_3$).

Voter	1	2	3	4	5	6	7	8	9
Ballot	2	12	12	13	37	45	46	47	48

Example 6.2.3. $n = 6$ voters; $m = 4$ candidates; any two-member committee admissible ($\mathcal{A} = \mathcal{A}_2$).

Voter	1	2	3	4	5	6
Ballot	12	12	24	34	13	13

Example 6.2.4. $n = 6$ voters; $m = 6$ candidates; any two-member committee admissible ($\mathcal{A} = \mathcal{A}_2$).

Voter	1	2	3	4	5	6
Ballot	1	1	234	24	235	356

Suppose that voter i's approval ballot is $V_i \subseteq [m]$ and that $S \subseteq [m]$. Then the set of candidates in S approved by voter i is $V_i \cap S$, and the number of such candidates is $App(V_i, S) = App_i(S) = |V_i \cap S|$. To illustrate, let $S = \{2, 3\}$, which we will write $S = 23$. In Example 6.2.1, for instance, $App_i(23) = 0, 1, 1, 1$ for $i = 1, 2, 3, 4$, respectively. Similarly, in Example 6.2.2, the nine values of $App_i(23)$ are, in sequence, 1, 1, 1, 1, 1, 0, 0, 0, 0. In Example 6.2.3, the sequence is $App_i(23) = 1, 1, 1, 1, 1, 1$; every voter approves of either candidate 2 or candidate 3, but never both.

6.3 Scoring Procedures

A "score" for a committee election conducted with approval balloting is a measure of the ability of a possible committee to represent the voters. A score defines a scoring procedure: an admissible committee with maximum score is elected. Multiple winning committees arise if, and only if, two or more admissible committees tie for the highest score. There are many ways to define a score using a profile of approval ballots, but before introducing them, we discuss one important property that scores may or may not possess.

Consider a multi-winner election conducted with approval ballots, in which there are m candidates, n voters, the admissible set is $\mathcal{A}$, and the ballot profile is V. A *score* is a function $f : \mathcal{A} \longrightarrow \mathbb{R}$; $f(S)$ is usually interpreted as a measure of the appropriateness of S to win the election. In fact, we will usually assume that the score is defined on all subsets, so that $f : 2^{[m]} \longrightarrow \mathbb{R}$, which is often convenient. As well, it is usually required that $f(\emptyset) = 0$.

A score $f(\cdot)$ is *additive* iff, whenever $S_1, S_2 \subseteq 2^{[m]}$, $S_1 \cap S_2 = \emptyset$, then $f(S_1 \cup S_2) = f(S_1) + f(S_2)$. In words, the score of a union of disjoint subsets is the sum of the scores of the subsets. It follows that, for an additive score, the score of a subset is equal to the sum of the scores of the members of that subset. In other words, once the scores of individual candidates are known, then the score of any possible committee can be obtained by summation.

The property of additivity makes a score easier to work with, provided that the admissible set has suitable structure, for then inspection of the scores of individual candidates can make the winning committee obvious. For example, if $\mathcal{A} = \mathcal{A}_k$, a committee is winning if and only if it contains k top-scoring candidates. In particular, there is a tie for winning committee if and only if, when the candidates are listed in decreasing order of score, there is a tie between the kth and $(k+1)$st candidates. Similarly, if $\mathcal{A} = 2^{[m]}$, a committee is winning if and only if it includes all positive-scoring candidates and excludes all negative-scoring candidates. In particular, the unique winning committee is $\emptyset$ if and only if all candidates' scores are negative. In these cases, and others, additivity of the scoring rule reduces the computational requirements substantially.

6.3.1 *(Simple) Approval*

The natural way to use approval ballots to score a committee is simply to count the total number of approvals received by the committee's members. The total number of approvals for committee S is

$$App(S) = \sum_i App(V_i, S) = \sum_i App_i(S),$$

and the (Simple) Approval rule is to select any $S \in \mathcal{A}$ that maximizes the score $App(S)$. The Approval rule is an obvious generalization of approval voting: If

$\mathcal{A} = \mathcal{A}_1$, then the Approval procedure reproduces a conventional (Single-Winner) approval-voting election.

It is easy to see that the score $App(\cdot)$ is additive. Let $S_1, S_2 \subseteq 2^{[m]}$, $S_1 \cap S_2 = \emptyset$. Then

$$App(S_1 \cup S_2) = |V_i \cap (S_1 \cup S_2)| = |(V_i \cap S_1) \cup (V_i \cap S_2)|$$
$$= |V_i \cap S_1| + |V_i \cap S_2| = App(S_1) + App(S_2),$$

since S_1 and S_2 are disjoint. It follows that, for any $S \in \mathcal{A}$, $App(S) = \sum_{j \in S} App(j)$.

For instance, in Example 6.2.2 it is easy to verify that $App(j) = 3, 3, 2, 4, 1, 1, 2, 1$ for $j = 1, 2, \ldots, 8$. The Approval Committee, the three-member committee S that maximizes $App(S)$, is therefore 124. It is easy to verify that, in Example 6.2.3, the Approval Committee is $\{12, 13\}$, i.e., the two committees 12 and 13 tie. Similarly, in Example 6.2.1, the approval committee is clearly 123, the set of all candidates. In fact, it is obvious that the Approval rule tends to select the larger committees, since the scores of individuals, $App(i)$, are never negative. In particular, Approval will select $2^{[m]}$ whenever it is admissible. This property explains why the Approval procedure is not recommended except when admissible committees are of fixed size, i.e., $\mathcal{A} \subseteq \mathcal{A}_k$ for some k satisfying $1 \leq k \leq m$.

Fishburn and Pekeč (2004, p. 6) noted that the Approval procedure is additive, and discussed the implications for efficiency. The procedure is easy to apply when there are relatively few candidates, but it is not efficient in the sense of NP-completeness (Garey and Johnson 1979). In fact, the natural algorithm based on additivity of the Approval score, $App(\cdot)$, is polynomial in $mn + |\mathcal{A}|$. Since $|\mathcal{A}| \leq 2^m$, then if the number of candidates, m, is relatively small, computational effort is reasonable even if the number of voters, n, is large.

6.3.2 *Net Approval*

The Approval rule works well, and is easy to implement, especially if the number of candidates is not large. But because of its natural bias toward larger committees, it can be recommended only in situations where the committee size is fixed in advance. A related rule that lacks this bias can be found by taking a "two-way" (or self-dual) approach to approval voting: If the score of a committee increases with the addition of an approved candidate, then it should decrease with the addition of a disapproved candidate. A natural measure of the score that voter i, whose vote was V_i, would assign to committee $S \subseteq [m]$ is its Net Approval score,

$$NApp_i(S) = |V_i \cap S| - |V_i^c \cap S|,$$

where V_i^c is the complement of V_i in $[m]$. The Net Approval procedure is to select any $S \in \mathcal{A}$ that maximizes $NApp(S) = \sum_i NApp_i(S)$.

It is easy to see that, if we define $App^+(S) = App(S) = \sum_i |V_i \cap S|$ and $App^-(S) = \sum_i |V_i^c \cap S|$, then $NApp(S) = App^+(S) - App^-(S)$. In other words, the Net Approval score of S is the Approval score of S calculated on the basis of $V = (V_1, V_2, \ldots, V_n)$, minus the Approval score of S calculated on the basis of $V^c = (V_1^c, V_2^c, \ldots, V_n^c)$. Thus, the Net Approval score of a subset equals its Approval score (candidates in S voted for) minus its Disapproval score (candidates in S voted against).

Because the Approval score, $App(S) = App^+(S)$ is additive, it is immediate that the Disapproval score, $App^-(S)$, is also additive. Also, it is easy to check that the sum or difference of two additive functions is additive, which implies that the Net Approval score, $NApp(\cdot)$, is additive. This can also be demonstrated directly. It follows that, for any $S \in \mathcal{A}$, $NApp(S) = \sum_{j \in S} NApp(j)$.

If, for example, $\mathcal{A} = \mathcal{A}_F$, i.e., any committee with at least one member may win, then additivity makes the Net Approval procedure easy to apply if at least one candidate has non-negative score. The winning committees are precisely those committees containing all candidates with positive score and no candidates with negative score. For Example 6.2.1, it is easy to check that $NApp(j) = 4, -2, 0$ for $j = 1, 2, 3$, so under Net Approval there is a tie between the committees 1 and 13. This result might be criticized for being indecisive, but it does establish that committees of different sizes can be competitive under Net Approval, which directly contrasts with the tendency of (Simple) Approval to select the largest admissible committee.

When $\mathcal{A} \subseteq \mathcal{A}_k$, the two scores $App_i(S)$ and $NApp_i(S)$ are related. Because $(V_i \cap S) \cup (V_i^c \cap S) = S$, it is easy to show that

$$NApp_i(S) = 2App_i(S) - k,$$

whenever $|S| = k$. Thus, when admissible committees are all the same size, the variation in $NApp_i(S)$ reflects only variation in $App_i(S)$, and similarly for $NApp(S)$ and $App(S)$. Net Approval scoring is valuable only when committees of different sizes are to be compared. To illustrate, consider Example 6.2.3, where the candidate approval scores are $App(j) = 4, 3, 3, 2$ for $j = 1, 2, 3, 4$, respectively, and the corresponding candidate net approval scores are $NApp(j) = 2, 0, 0, -2$. So if, as in Example 6.2.3, the admissible sets are $\mathcal{A}_2$, then the winning committees are 12 and 13 under the Net Approval procedure, exactly the same as under the Approval procedure.

One benefit of a score like Net Approval is that it can provide an absolute threshold for assessment of voter support. In particular, it gives the voters complete control over the size of the committee, which may be appropriate for some elections, such as enshrinement in a hall of fame. (In a hall of fame – see, for instance, BBHOF, 2009 – supermajority threshold approval is typically required for election.) If all candidates' scores fall below the threshold – in the majority case, if they are negative – then the best committee, if it is admissible, is $\emptyset$. If $\mathcal{A} = 2^{[m]}$, i.e., all committees are admissible, then the Net Approval procedure is equivalent to Candidate-by-Candidate Majority Voting – any candidate with more approvals than disapprovals is elected, any candidate with more disapprovals than approvals is defeated. There is a tie if and only if there are candidates with equally many approvals and disapprovals, in which case a subset is winning if and only if it contains no candidates with more disapprovals than approvals. Example 6.2.1 illustrates such a tie.

If every candidate's Net Approval score is negative and $\emptyset$ is not admissible – for example, if $\mathcal{A} = \mathcal{A}_F$ – then the winner is any non-empty admissible subset of candidates tied for the maximum score. For instance, in Example 6.2.2 the candidates' Net Approval scores are $NApp(j) = -3, -3, -5, -1, -7, -7, -5, -7$ for $j = 1, \ldots, 8$, so the winning committee would be $\emptyset$ if it were admissible. But it is not, and the winning committee is 124.

6.3.3 Satisfaction

Brams and Kilgour (2010) proposed another method for using approval ballots in multi-winner elections, called Satisfaction (Approval) Voting. This procedure is a scoring rule based on selection of the admissible committee, S, that maximizes the Satisfaction score

$$Sat(S) = \sum_i \frac{|S \cap V_i|}{|V_i|}.$$

(By convention, the fraction in the summation equals zero if voter i voted for no candidates, i.e., if $V_i = \emptyset$. Such voters have no effect on the outcome of the election using the Satisfaction procedure.) The definition of the score $Sat(S)$ reflects another view of what distinguishes a good committee. Voter i's "satisfaction" with committee S equals the proportion of candidates supported by i who belong to S. As Brams and Kilgour (2010) point out, under the Satisfaction procedure a voter pays a high price for approving of two or more candidates, so it seems likely that voters would bullet vote unless they were indifferent, or nearly so, among several candidates.

The score $Sat(\cdot)$ is additive, for if $V_i \neq \emptyset$ and if $S_1, S_2 \subseteq 2^{[m]}$, $S_1 \cap S_2 = \emptyset$, then

$$\begin{aligned} Sat(S_1 \cup S_2) &= \frac{|V_i \cap (S_1 \cup S_2)|}{|V_i|} = \frac{|(V_i \cap S_1) \cup (V_i \cap S_2)|}{|V_i|} \\ &= \frac{|V_i \cap S_1| + |V_i \cap S_2|}{|V_i|} = \frac{|V_i \cap S_1|}{|V_i|} + \frac{|V_i \cap S_2|}{|V_i|} \\ &= Sat(S_1) + Sat(S_2), \end{aligned}$$

since S_1 and S_2 are disjoint. It follows that, for any $S \in \mathcal{A}$, $Sat(S) = \sum_{j \in S} Sat(j)$.

The Satisfaction procedure can be recommended only for elections in which $\mathcal{A} \subseteq \mathcal{A}_k$ for some k, since it tends to favor larger committees over smaller. (After all, $Sat(S)$ is never negative.) For Example 6.2.4, the individual Satisfaction scores are $Sat(j) = 2, 1.17, 1, 0.83, 0.67, 0.33$ for $j = 1, \ldots, 6$, so the winning committee under the Satisfaction procedure is 12. (In contrast, under Approval, the winner is 23.) For Example 6.2.1, the candidates' Satisfaction scores are $S(j) = 2.5, 0.5, 1$ for $j = 1, 2, 3$, so the winning committee is 123 illustrating that, as for with Approval, the Satisfaction procedure is biased toward larger committees.

6.3.4 Net Satisfaction

Just as the Net Approval score can be developed from the Approval score by taking a "two-way" (or self-dual) approach, a Net Satisfaction score can be developed from the Satisfaction score. If satisfaction with a committee increases with the addition of an approved candidate, then it should decrease with the addition of an unapproved candidate. The Net Satisfaction score captures this idea; a measure of the satisfaction that voter i, whose vote was V_i, would assign to committee $S \subseteq [m]$ is

$$NSat_i(S) = \frac{|V_i \cap S|}{|V_i|} - \frac{|V_i^c \cap S|}{|V_i^c|},$$

where V_i^c is the complement of V_i in $[m]$. (By convention, any fraction with denominator 0 is taken to equal 0.) The Net Satisfaction procedure is to select any $S \in \mathcal{A}$ that maximizes $NSat(S) = \sum_i NSat_i(S)$.

Again, it is easy to see that if we define $Sat^+(S) = Sat(S) = \sum_i \frac{|V_i \cap S|}{|V_i|}$ and $Sat^-(S) = \sum_i \frac{|V_i^c \cap S|}{|V_i^c|}$, then $NSat(S) = Sat^+(S) - Sat^-(S)$. As with Net Approval, the Net Satisfaction score of S is the Satisfaction score of S calculated on the basis of $V_1, V_2, \ldots, V_n$, minus the Satisfaction score of S calculated on the basis of $V_1^c, V_2^c, \ldots, V_n^c$. We can say that the Net Satisfaction score of a subset equals its Satisfaction score (based on elected candidates voted for) minus its Dissatisfaction score (based on elected candidates voted against).

Because the Satisfaction score, $Sat(S) = Sat^+(S)$ is additive, it is again immediate that the Dissatisfaction score, $Sat^-(S)$, is also additive. Because the difference of additive functions is additive, the Net Satisfaction score, $NSat(\cdot)$, is additive. It follows that, for any $S \in \mathcal{A}$, $NSat(S) = \sum_{j \in S} NSat(j)$.

For Example 6.2.1, it is easy to check that $NSat(j) = 2.5 - 0 = 2.5, 0.5 - 2.5 = -2, 1.0 - 1.5 = -0.5$ for $j = 1, 2, 3$, so applying the Net Satisfaction procedure rule to Example 6.2.1 produces uniquely the committee 1. (Recall that Net Approval produced a tie between 1 and 13 for this example.)

Like Net Approval, Net Satisfaction can be interpreted as providing an absolute threshold for assessment of voter support. If $\mathcal{A} = 2^{[m]}$, then there can be no problem applying the Net Satisfaction with threshold zero. However, admissibility restrictions may increase the complexity of the procedure.

6.3.5 Representativeness

Monroe (1995) proposed a design principle for electoral systems: maximize proportional representation by minimizing "misrepresentation." Potthoff and Brams (1998) showed how to implement such systems using integer programming, and suggested that approval ballots made misrepresentation easy to measure. Below we show that, under approval balloting, a maximally representative subset can be determined by implementing a scoring procedure.

The fundamental idea of Monroe (1995) is that a procedure should assign a specific elected candidate to each voter; the voter is then "represented" by the assigned candidate. A condition of the assignment is that, as nearly as possible, each elected candidate should be assigned to an equal number of voters.

In the approval balloting context, it is natural to say that a candidate can represent a voter if and only if the voter approved of the candidate. Therefore, minimizing misrepresentation is equivalent to maximizing the number of voters assigned elected candidates they voted for. Based on this idea, and provided that a fixed number, k, of candidates is to be elected, i.e., that $\mathcal{A} \subseteq \mathcal{A}_k$, the integer program of Potthoff and Brams (1998) can be implemented as a Representativeness score,

$$Rep(S) = \sum_{j \in S} \sum_{i=1}^{n} x_{ij} Ind(j, V_i)$$

where $Ind(j, V_i) = 1$ if $j \in V_i$ and $Ind(j, V_i) = 0$ otherwise. (To see that this score measures representation, note that $x_{ij} = 1$ if elected candidate j is assigned to voter i, and $x_{ij} = 0$ otherwise. If the candidates in S are elected, $Rep(S)$ is thus equal to the number of times that a voter is assigned to an elected candidate that the voter approved of.)

For any $S \in \mathcal{A}_k$, define $x_j, j = 1, 2, \ldots, m$ by $x_j = 1$ if $j \in S$ and $x_j = 0$ otherwise. The 0–1 variables $x_{ij}, i = 1, 2, \ldots, n, j = 1, 2, \ldots, m$ must be chosen in accordance with the following conditions:

$$\sum_{j} x_{ij} = 1 \text{ for each } i = 1, 2, \ldots, n$$

$$-Lx_j + \sum_{i=1}^{n} x_{ij} \geq 0 \text{ for each } j = 1, 2, \ldots, m$$

$$-Ux_j + \sum_{i=1}^{n} x_{ij} \leq 0 \text{ for each } j = 1, 2, \ldots, m$$

where $L = \lfloor \frac{n}{k} \rfloor$ and $U = \lceil \frac{n}{k} \rceil$. Note that if $\frac{n}{k}$ is an integer, then $L = U$ and the last two conditions can be replaced by

$$nx_j = k \sum_{i=1}^{n} x_{ij} \text{ for each } j = 1, 2, \ldots, m.$$

To interpret the constraints, note that $x_j = 1$ indicates that candidate j is elected, and $x_{ij} = 1$ indicates that candidate j is assigned to voter i. (Each voter, i, is assigned exactly one elected candidate.) If $x_j = 0$, then $x_{ij} = 0$ for all i. If $x_j = 1$, then there are at least L and at most U voters, i, for whom $x_{ij} = 1$, i.e., who are assigned to candidate j. Note that $L = U$ if $\frac{n}{k}$ is an integer and $L = U - 1$ otherwise.

To summarize, the Representativeness score of candidate j is 0 if j is not elected; if j is elected, it equals the number of voters assigned to j who approved of j. The representativeness score of an elected candidate therefore cannot exceed the number of voters assigned to that candidate, either L or U. If there exists an admissible committee such that every voter can be assigned to a member of that committee of whom the voter approved, subject to all constraints, then every voter is represented, and that committee achieves the maximum representativeness score, n.

To see the role of the constraints, consider Example 6.2.3. Since $\frac{n}{k} = 3$, each elected candidate must be assigned to represent three voters. Candidate 1 could represent voters 1, 2, 5, or 6, but not voters 3 or 4. The only candidate who could represent both voters 3 and 4 is Candidate 4. But Candidate 4 received only two approval votes and can therefore contribute a maximum of two to the Representativeness score of any committee. Therefore *Rep*(14) = 5, and similarly *Rep*(12) = *Rep*(13) = 5, whereas *Rep*(23) = 6, the maximum score. It is easy to check that no other committees achieve Representativeness score 6, so that according to the Representativeness procedure the unique winning committee in Example 6.2.3 is 23. In contrast, most other procedures find winning committees that include Candidate 1, who received the most approval votes.

It is obvious that $Rep(S)$ is an additive score. However, the need to assign elected candidates to voters seems to cancel out any computational advantages conferred by additivity. See Potthoff and Brams (1998) or Brams (2008, Chap. 6), for a discussion of the implications of relaxing the integrality constraint, i.e., allowing two or more candidates to be assigned, fractionally, to represent a voter.

6.3.6 Proportional Approval

Another idea for scoring approval ballots for committees was suggested by Simmons (2001). Under the (Simple) Approval rule, each member of a committee voted for by a voter contributes equally to the committee's score, so that a committee that includes two of voter 1's candidates and none of voter 2's scores just as well as a committee that includes one candidate supported by each voter. The motivation for the Proportional Approval rule is that it is more important for a committee to represent more voters than to give extra representation to a voter who is already represented.

For a general presentation, set $r(0) = 0$ and let $r(1), r(2), \ldots, r(m)$ be any increasing sequence of positive numbers. The specific sequence suggested by Simmons (2001) was

$$r(k) = 1 + \frac{1}{2} + \cdots + \frac{1}{k} = \sum_{j=1}^{k} \frac{1}{j} \tag{6.1}$$

for $k = 1, 2, \ldots, m$, which matches the Hamilton method of apportionment. For any $S \subseteq [m]$, define

$$PApp(S) = \sum_i r(|S \cap V_i|)$$

The Proportional Approval procedure is to select any $S \in \mathcal{A}$ that maximizes $PApp(S)$.

For example, using the sequence (6.1), the score of a subset S is increased by 1 for each voter who votes for one member of S, by $1 + \frac{1}{2} = \frac{3}{2}$ for each voter who votes for two members of S, by $1 + \frac{1}{2} + \frac{1}{3} = \frac{11}{6}$ for each voter who votes for three members of S, etc.

A difficulty with the Proportional Approval score is that in general it is not additive. For example, using the standard scoring sequence (6.1) in Example 6.2.3 produces $PApp(1) = 4$ and $PApp(2) = 3$ but $PApp(12) = 6$. Computation with Proportional Approval is therefore more difficult. It can be shown that a Proportional Approval sequence $r(1), r(2), \ldots, r(m)$ produces an additive score if and only if it satisfies $r(k) = Ak$ for some positive A. Of course, the Approval score is captured by the score sequence $r(k) = k$. We have shown that the "diminishing marginal value of representation" property, achieved iff the sequence $r(k)$ is increasing at a decreasing rate, is incompatible with additivity.

Of course, the proportional approval score of a subset S increases as S gets larger, because the sequence $r(k)$ is increasing, and a bigger S cannot have fewer intersections with V_i (for each i). A consequence, of course, is that the Proportional Approval rule does not work well unless the size of the committee to be elected is restricted, and it is usually applied only when $\mathcal{A} \subseteq \mathcal{A}_k$ for some $k = 1, 2, \ldots, m-1$. A related rule that might be appropriate if committees of different sizes are admissible could be based on a Net Proportional Approval score, which would have the form

$$NPApp(S) = \sum_i \left[r^{+}(|S \cap V_i|) - r^{-}(|S \cap V_i^c|) \right]$$

where $r^{+}(k)$ and $r(k)$ are increasing sequences of positive numbers, possibly equal to each other, and possibly both given by (6.1).

6.3.7 *Sequential Proportional Approval*

Another computationally difficult system is the Sequential Proportional Approval procedure, proposed by Thiele (c. 1890). As noted above, the (Simple) Approval rule in an election with $\mathcal{A} \subseteq \mathcal{A}_1$ produces the usual approval voting winner.

To express the Sequential Proportional Approval procedure as a scoring rule, the single-winner Approval procedure must be executed iteratively, but with weighted voters. Suppose that voter i has weight $w_i > 0$. Then in a single-winner election (i.e., a committee of size $k = 1$ is to be elected), the Weighted Approval score of candidate $j \in [m]$ can be taken to be

$$WApp(j) = \sum_i w_i Ind(j, V_i), \tag{6.2}$$

where $Ind(j, V_i) = |j \cap V_i|$. The winning candidate would be any j that maximizes $WApp(j)$.

If $k > 1$, the Sequential Proportional Approval procedure to elect a k-member committee (i.e., with $\mathcal{A} \subseteq \mathcal{A}_k$) is an iterative procedure:

- Begin by setting $\mathcal{C}_0 = [m]$ and letting $w_i^1 = 1$ for all voters i. Apply (6.2) to obtain scores $WApp^1(j)$ for all $j \in \mathcal{C}_0$. The first candidate seated on the committee is then any candidate $j_1 \in \mathcal{C}_0$ that maximizes $WApp^1(j)$. Now set $\mathcal{C}_1 = \mathcal{C}_0 - j_1$.
- Suppose that $1 < h \leq k$ and that candidates $j_1, j_2, \ldots, j_{h-1}$ have been seated on the committee, and that the subset of remaining candidates is $\mathcal{C}_{h-1}$. Reweight the voters so that the weight of voter i is

$$w_i^h = \frac{1}{1 + |V_i \cap \{j_1, j_2, \ldots, j_{h-1}\}|}.$$

  Apply (6.2) to obtain scores $WApp^h(j)$ for all $j \in \mathcal{C}_{h-1}$. The hth candidate seated on the committee is any candidate $j_h \in \mathcal{C}_{h-1}$ that maximizes $WApp^h(j)$. If $h = k$, stop. Otherwise set $\mathcal{C}_h = \mathcal{C}_{h-1} - j_h$ and repeat.

Examples show that the Sequential Proportional Approval procedure is different from the Proportional Approval procedure. For instance, in Example 6.2.3, Sequential Proportional Approval produces a tie among 12, 13, and 14, whereas Proportional Approval produces a four-way tie: these three committees, as well as 23.

Like the Approval score, the Weighted Approval score (6.2) is additive. But the facts that candidates are elected to the committee one at a time, and that after every election each voter's weight must be recalculated, make the efficiency of the Sequential Proportional Approval procedure low, and closer to the Proportional Approval procedure than the Simple Approval or Satisfaction procedures.

6.4 Threshold Procedures

Each scoring rule is characterized by a score function that measures the "appropriateness" of each admissible subset; then any subset with maximum score wins the election. The score of a possible committee should increase as the subset becomes more similar to the voters' ballots; in general, scores reflect that overlap is good and more overlap is better, presumably because it means better representation.

Threshold methods, developed by Fishburn and Pekeč (2004), are characterized by binary (Yes or No) judgements about representativeness: either a subset has sufficient overlap with a voter's ballot to represent that voter, or it does not. The best committee is then the one that represents the most voters. In particular, insufficient overlap with a voter's ballot counts for nothing, and if the overlap exceeds a threshold, then any extra preference for that committee because of the additional overlap also counts for nothing. Thus, two subsets that meet the same thresholds for representation have the same argument for selection, even if one exceeds all thresholds while the other merely meets them.

To develop an array of threshold methods, Fishburn and Pekeč (2004) define a *threshold function* to be a function $t : \mathcal{A} \longrightarrow \mathbb{R}^+$, which maps every admissible set to a positive real number. The interpretation is that $t(S)$ is the threshold for a possible committee S to be representative: S represents any voter i for whom $|V_i \cap S| \geq t(S)$. The Threshold Approval procedure t is to select $S \in \mathcal{A}$ if and only if

$$|\{i : |V_i \cap S| \geq t(S)\}| \geq |\{i : |V_i \cap T| \geq t(T)\}| \text{ for all } T \in \mathcal{A}.$$

In other words, select S to maximize $|\{i : |V_i \cap S| \geq t(S)\}|$, the number of voters represented by S, according to the threshold embodied in $t(\cdot)$.

Note that the criterion of representativeness of a voter by a subset may depend on the subset, but it does not depend on the voter. Once the votes are in, i.e., given that $V \in \mathcal{V}$ is fixed, then whether voter i_1 is counted as represented by $S \in \mathcal{A}$ depends on $t(S)$, but it is determined in exactly the same way as whether voter i_2 is represented by S.

Note that many threshold rules are possible, depending on the choice of the threshold function $t(\cdot)$. Fishburn and Pekeč (2004) suggest many possibilities. If $\mathcal{A} \subseteq \mathcal{A}_k$, so that any committee elected must have exactly k members, then a *constant* threshold, $t(S) = \ell$, for $1 \leq \ell \leq k$ is a natural choice. Thus, a committee S is counted as representing a voter i iff $|V_i \cap S| \geq \ell$, i.e., voter i has voted for at least ℓ members of S. For Example 6.2.2, the Constant Threshold Approval rule with $t = 1$ produces the committee 234, which is approved by all voters (at this threshold). In the same example, the Constant Threshold Approval rule with $t = 2$ produces the committee 123, which is approved by three voters at this threshold, while no other committee is approved by more than two.

Since only neutral voting rules are considered here, thresholds must be *cardinal*, i.e., for all $S_1, S_2 \in \mathcal{A}$, if $|S_1| = |S_2|$, then $t(S_1) = t(S_2)$. Clearly, constant thresholds are cardinal. For situations in which committees of different sizes are admissible, constant threshold rules tend to select larger committees – exactly the same phenomenon as for scoring rules like Approval and Satisfaction. To avoid nonsensical results, Fishburn and Pekeč (2004) suggest that thresholds should be *non-decreasing*; cardinal thresholds that are non-decreasing have the property that if $|S_1| < |S_2|$, then $t(S_1) \leq t(S_2)$ for all $S_1, S_2 \in \mathcal{A}$. With a non-decreasing threshold, when a larger committee represents a voter, its overlap with the voter's ballot is not less than would be required for a smaller committee to represent the voter.

Two non-decreasing thresholds suggested by Fishburn and Pekeč (2004) are the *majority* threshold, $t(S) = \frac{|S|}{2}$ and the *strict majority* threshold, $t(S) = \frac{(|S|+1)}{2}$. In Example 6.2.1, for instance, 1, 12, and 13 tie for majority threshold committee, whereas the unique strict majority threshold committee is 1.

Finally, Fishburn and Pekeč (2004) point out that threshold methods are all NP-hard; for instance, they prove that in the special case that $|V_i| = 2$ for all voters i, the problem of determining whether there exists an $S \in \mathcal{A}_k$ that is approved by all voters is equivalent to finding a vertex cover of a graph with vertex set $[m]$, a problem known to be NP-complete (Garey and Johnson 1979). Nonetheless, if the number of candidates is relatively small, computational effort is not excessive even as the number of voters becomes large.

6.5 Centralization Procedures

Centralization procedures for committee elections with approval balloting are adaptations of an approach used in many problems: Each voter's ballot can be considered to propose a committee, so the most representative committee is the one that is "closest" to the ballots. These voting procedures can be traced to a study (Brams et al. 2004) of ways to identify a good outcome in a multi-party negotiation over many binary (Yes–No) issues. If there are m issues, a party's position can be represented as the subset of issues on which it supports the Yes side, which can be thought of as a vertex of an m-dimensional hypercube. The principle of negotiating by conducting a majority vote of the parties on each issue was demonstrated to be equivalent to finding all vertices of the hypercube – i.e., Yes–No sequences – that minimize the total distance (or average distance) to the vertices representing the positions of all negotiators. Brams et al. (2004) also adapted the Fallback procedure of Brams and Kilgour (2001) to the bargaining problem, showing that it is equivalent to finding all vertices of the hypercube that minimize the maximum distance to any vertex representing the position of a party.

These ideas were adapted to voting in multi-winner elections by Kilgour et al. (2006) and Brams et al. (2007), who also raised the possibility of forming a committee member-by-member, using majority voting. As shown below, this procedure can be expressed as a scoring rule, except that it allows no natural way to account for admissibility. A candidate whom more voters support than oppose must join the committee, and a candidate with more opponents than supporters cannot. The decision on each candidate is based on the balance of votes for that candidate only. For instance, the empty committee cannot be excluded a priori. Thus, this correspondence relies on $\mathcal{A} = 2^{[m]}$, i.e., any subset, including the empty set, is admissible.

The representation of distance between subsets used by Kilgour et al. (2006) can, however, account for admissibility in a natural way. Let $S, T \subseteq [m]$, and measure the distance between S and T as the *Hamming distance*, $d(S, T)$, defined by

$$d(S,T) = |S \,\Delta\, T| = |S - T| \cup |T - S| = |\,(S \cap T^c) \cup (S^c \cap T)\,|.$$

Thus, the distance between S and T equals the number of points (candidates) in one of S and T but not the other.

Recall that the ballot profile is called V. For any $S \in 2^{[m]}$, define $d(S, V) = \sum_i d(S, V_i)$. Then $d(S, V)$ represents the total distance from S to the collection of all ballots. Brams et al. (2004) proved that any committee $S \in 2^{[m]}$ that minimizes $d(S, V)$ must contain every candidate who is supported on more than half the ballots, and cannot contain any candidate who is supported on fewer than half the ballots. Thus, the Candidate-by-Candidate Majority Voting rule mentioned above can be implemented using the total distance minimization criterion, which was christened "Minisum." But choosing the admissible committee, S, that minimizes $d(S, V)$ does allow inadmissible committees to be excluded, so – unlike Candidate-by-Candidate Majority Voting – the Minisum procedure respects admissibility.

In fact, we have already identified a scoring procedure that is equivalent to the Minisum procedure. When all subsets are admissible, the Net Approval procedure elects any subset that contains all candidates with more approvals than disapprovals and no candidates with fewer approvals than disapprovals. Thus, the Net Approval scoring procedure, Candidate-by-Candidate Majority voting, and the Minisum procedure are all equivalent when $\mathcal{A} = 2^{[m]}$. It is not hard to see that the equivalence of Net Approval and Minisum continues when some subsets are inadmissible.

We illustrate the Minisum procedure to select any $S \in \mathcal{A}$ that minimizes $d(S, V) = \sum_i d(S, V_i)$, using Example 6.2.1. The admissible committees are $S =$ 1, 2, 3, 12, 13, 23, and 123, and the respective values of $d(S, V)$ are 3, 9, 7, 5, 3, 9, and 5. Thus for Example 6.2.1, the Minisum procedure produces a tie between committee 1 and committee 13. As noted above, the Net Approval procedure produces exactly the same result.

Clearly, then, the Minisum centralization procedure is a scoring procedure. But a second centralization procedure is not. The Minimax procedure also originated in the study of multi-party negotiation over binary issues, where the Fallback Bargaining idea was shown to result in a subset of issues that minimizes the maximum distance to the subset supported by any bargainer. The analogue for approval balloting is the Minimax procedure, a centralization system suggested by Brams et al. (2007) and presented formally in Kilgour et al. (2006). Minimax selects

$$\text{argmin}_{\,S \in \mathcal{A}} \left\{ \max_i d(S, V_i) \right\}.$$

That is, the winning committee under Minimax is any admissible committee S with the property that the maximum distance from S to any V_i is a minimum. For Example 6.2.1, the admissible committees are $S =$ 1, 2, 3, 12, 13, 23, and 123, and the respective values of $\max_i d(S, V_i)$ are 1, 3, 3, 2, 2, 3, and 2, so that the Minimax rule selects committee 1 (uniquely).

A study of examples, and of the specific properties of the Minimax procedure, led to some suggested modifications. In Brams et al. (2007), the principle of applying Minimax followed by Minisum was applied to data from a large-scale election. In Kilgour et al. (2006), the observation that the Minimax procedure ignores clones completely – so that the Results for Example 6.2.1 would be unchanged by the addition of 100 new voters, all of whom voted for 13 – led to a weighting principle. In order to make the voting system responsive to "enough" voters, weights were proposed that allow the minimax calculation to register a preponderance of voting support.

The weighted minimax system had to be applied to ballots, as opposed to voters. Now let $W = \{W_1, W_2, \ldots, W_\ell\}$ denote the set of *distinct* ballots cast by the voters, and note that $|W| = \ell$. Moreover, if $h = 1, 2, \ldots, \ell$, let n_h denote the number of voters who cast the ballot W_h. Suppose that some *weight vector* $(w_1, w_2, \ldots, w_\ell)$ is given, where w_h is the weight assigned to ballot W_h. Then the *weighted distance* from a subset $S \in 2^{[m]}$ to W_h is $w_h d(S, W_h)$. A Weighted Minimax procedure is to select the admissible subset S that minimizes $\sum_{h=1}^{\ell} w_h d(S, W_h)$. Note that weights are required to be non-negative, but there is no "normalization" condition; the sum of the weights may be any positive number. As Kilgour et al. (2006) note, weights provide only relative information, so a set of weights can be multiplied by any positive number without changing the information it contains.

The distance from a given committee to a ballot can be considered to be a weighted distance if a weight has been assigned to a ballot. (This is a benefit of working with distinct ballots rather than distinct voters; thus, the distance, $d(S, V_i)$, is replaced by the weighted distance $w_h d(S, W_h)$.) This idea extends both the Minisum and Minimax systems. Moreover, it is clear that the results of the Weighted Minisum Rule and Weighted Minimax Rule depend on the particular weights used.

One natural set of weights is *count* weights, which assign to each ballot a weight equal to the number of voters who cast it. Thus, count weights are defined by $w_h = n_h$ for $h = 1, 2, \ldots, \ell$. The results are conveniently displayed, for both Weighted Minisum and Weighted Minimax, in a table (Kilgour et al., 2006), which for Example 6.2.1 using count weights is as follows:

Ballot:	1	12	13		
Weight:	1	1	2	$\sum$	max
Subsets: 1	0	1	2	3*	2*
2	2	1	6	9	6
3	2	3	2	7	3
12	1	0	4	5	4
13	1	2	0	3*	2*
23	3	2	4	9	4
123	2	1	2	5	2*

The winning subsets for Minisum ($\sum$) and Minimax (max) are indicated by asterisks. As Kilgour et al. (2006) showed, if count weights are used, the Minisum procedure is exactly as described above using voters rather than weighted ballots. In general, the Minisum Count procedure is equivalent to Net Approval and, provided all subsets are admissible, to Candidate-by-Candidate Majority Voting. In Example 6.2.1, for instance, the Minisum outcomes using count weights remain 1 and 13. But while the direct Minisum procedure is identical to the Weighted Minisum procedure with count weights, the Minimax procedure gives different results when count weights are applied; now, the committees selected are 1, 13, 123 (tied).

Quite different results are obtained using *proximity weights*, which are defined by

$$w_h = \frac{n_h}{\sum_{r=1}^{\ell} n_r d(W_h, W_r)}.$$

Proximity weights were proposed by Kilgour et al. (2006) to give less weight to ballots cast by extreme or isolated voters, thereby reducing their influence on the final decision. To elaborate on the definition systematically, the numerator shows that w_h is proportional to n_h, the number of voters who voted for W_h. The denominator of the expression for w_h is the sum of the distances from W_h to all other ballots. (Of course, $d(W_h, W_h) = 0$, so the distance from W_h to itself does not contribute to this sum.) Thus w_h, the weight of W_h, is small when few voters approve of exactly W_h or any subset close to it. As a ballot moves closer to other ballots, it receives greater weight, either because the distances are smaller so the denominator is reduced or because the numerator is increased because it duplicates an existing ballot.

For Example 6.2.1, there are $\ell = 3$ ballots, $W_1 = 1$, $W_2 = 12$, and $W_3 = 13$, with counts $n_1 = 1, n_2 = 1$, and $n_3 = 2$. Thus, count weights are $(w_1, w_2, w_3) = (1, 1, 2)$, while proximity weights are $w_1 = \frac{1}{1{\cdot}0+1{\cdot}1+2{\cdot}1} = \frac{1}{3}, w_2 = \frac{1}{1{\cdot}1+1{\cdot}0+2{\cdot}2} = \frac{1}{5}$, and $w_3 = \frac{2}{1{\cdot}1+1{\cdot}2+2{\cdot}0} = \frac{2}{3}$. Multiplying by 15 to clear fractions gives $(w_1, w_2, w_3) = (5, 3, 10)$. For the seven admissible subsets, these proximity weights give the following table:

Ballot:	1	12	13		
Weight:	5	3	10	$\sum$	max
Subsets: 1	0	3	10	13	10
2	10	3	30	43	30
3	10	9	10	29	10
12	5	0	20	25	20
13	5	6	0	11*	6*
23	15	6	20	41	20
123	10	3	10	23	10

Thus, in Example 6.2.1 with proximity weights, both the Minisum and Minimax procedures produce (uniquely) the subset 13.

6.6 Conclusions

This chapter has surveyed methods of using approval ballots in multi-winner elections, where both the ballot cast by a voter and the election result can be considered to be subsets of the candidates. Consideration has been restricted to voting procedures that are anonymous (treat voters equally), neutral (treat candidates equally) and that permit the class of *admissible*, or potentially winning, subsets to be specified independent of the procedure. The systems studied here were classified into *scoring* procedures, in which the admissible subset with the highest total score wins, *threshold* procedures, in which the maximally representative admissible subset is selected, and *centralization* procedures, in which the admissible subset that is most central among the ballots is selected.

The table below compares the procedures discussed here in the context of the four examples. As noted above, two procedures, Net Approval and Minisum Count are identical. The table itself is proof that, except for those two, all procedures are different, since any two of them differ on at least one example. Note that two procedures, Representativeness and Sequential Proportional Approval, are defined only for some admissible sets. Specifically, Representativeness requires that $\mathcal{A} \subseteq \mathcal{A}_k$, and Sequential Proportional Approval that $\mathcal{A} = \mathcal{A}_k$. Therefore, neither procedure can be applied to Example 6.2.1, where $\mathcal{A} = \mathcal{A}_F$.

Example:	1	2	3	4
Simple Approval	123	124	12, 13	23
Net Approval	1, 13	124	12, 13	23
Satisfaction	123	124	12, 13	12
Net Satisfaction	1	124	12, 13	12
Representativeness	–	147, 234, 247	23	12, 13
Proportional Approval	123	124, 234	12, 13, 14, 23	12, 13, 23
Sequential Prop. App.	–	124, 234	12, 13, 14	12, 13, 23
Threshold – Majority	1, 12, 13	123	14, 23	12, 13
Threshold – Str. Maj.	1	123	12, 13	23, 24, 35
Minisum Count	1, 13	124	12, 13	23
Minisum Proximity	13	124	12, 13	23
Minimax Count	1, 13, 123	123, 124, 125 126, 127, 128	12, 13, 14, 23	13, 23, 25, 26 34, 35, 36, 45
Minimax Proximity	13	124	12, 13, 14, 23	12

There are many ways that voting systems can be compared, and much work remains to be done to compare these systems on grounds of theoretical properties, computational complexity, and practical utility. One large-scale comparison of procedures is reported by Brams et al. (2007) and Brams and Kilgour (2010), based on ballot data from the 2003 election by the Game Theory Society of 12 new members of council from a list of 24 candidates. The Approval, Satisfaction, Minimax and Minisum councils were compared, both when only 12-member committees were admissible, and without this restriction. None of the 12-member councils was representative, in the sense of including at least one candidate supported by each voter, even though there are subsets containing only eight candidates that represent every voter. But Brams and Kilgour (2010) also point out that, given a representative subset, there are no natural procedures for expanding or contracting it to construct a committee of predetermined size.

Acknowledgements Useful comments from Steven J. Brams, Hannu Nurmi, and Richard F. Potthoff are gratefully acknowledged. Erica Marshall provided valuable research assistance.

References

Arrow, K. J., Sen, A. K., & Suzumura, K. (Eds.) (2002). *Handbook of social choice and welfare* (Vol. 1). Amsterdam: North-Holland.

Brams, S. J. (2008). *Mathematics and democracy: Designing better voting and fair-division procedures*. Princeton, NJ: Princeton University Press.

Brams, S. J., & Fishburn, P. C. (1978). Approval voting. *American Political Science Review, 72*(3), 831–857.

Brams, S. J., & Fishburn, P. C. (1983). *Approval voting*. Cambridge, MA: Birkhäuser Boston.

Brams, S. J., & Fishburn, P. C. (2002). Voting procedures. In K. Arrow, A. Sen, & K. Suzumura (Eds.), *Handbook of social choice and welfare* (pp. 175–236). Amsterdam: Elsevier Science.

Brams, S. J., & Fishburn, P. C. (2005). Going from theory to practice: The mixed success of approval voting. *Social Choice and Welfare, 25*(2–3), 457–474.

Brams, S. J., & Kilgour, D. M. (2001). Fallback bargaining. *Group Decision and Negotiation, 10*(4), 287–316.

Brams, S. J., & Kilgour, D. M. (2010). *Satisfaction approval voting*. Preprint. New York: New York University.

Brams, S. J., Kilgour, D. M., & Sanver, M. R. (2004). A minimax procedure for negotiating multilateral treaties. In M. Wiberg (Ed.), *Reasoned choices: Essays in honor of Hannu Nurmi* (pp. 108–139). Turku, Finland: Finnish Political Science Association.

Brams, S. J., Kilgour, D. M., & Sanver, M. R. (2005). A minimax procedure for electing committees. *Public Choice, 132*(3–4), 401–420.

Fishburn, P. C., & Pekeč, A. (2004). *Approval voting for committees: Threshold approaches*. Retrieved April 14, 2009, from http://dimacs.rutgers.edu/Workshops/DecisionTheory-2/PekecFishburn04a.pdf.

Garey, M. R., & Johnson, D. S. (1979). *Computers and intractability: A guide to the theory of NP-completeness*. New York: Freeman.

Kilgour, D. M., Brams, S. J., & Sanver, M. R. (2006). How to elect a representative committee using approval balloting. In B. Simeone & F. Pukelsheim (Eds.), *Mathematics and democracy: Recent advances in voting systems and collective choice* (pp. 83–95). Heidelberg: Springer.

Merrill, S., III, & Nagel, J. H. (1987). The effect of approval balloting on strategic voting under alternative decision rules. *American Political Science Review, 81*(2), 509–524.

Monroe, B. L. (1995). Fully proportional representation. *American Political Science Review, 89*(4), 925–940.

National Baseball Hall of Fame. (2009). *Hall of fame election rules*. Retrieved August 15, 2009, from http://web.baseballhalloffame.org/hofers/rules.jsp.

Potthoff, R. F., & Brams, S. J. (1998). *Proportional representation: Broadening the options. Journal of Theoretic Politics, 10*(2), 147–178.

Ratliff, T. C. (2003). Some startling inconsistencies when electing committees. *Social Choice and Welfare, 21*, 433–454.

Ratliff, T. C. (2006). Selecting committees. *Public Choice, 126*(3–4), 343–355.

Simmons, F. (2001). *Proportional approval voting*. Retrieved May 21, 2009, from http://www.nationmaster.com/encyclopedia/Proportional-approval-voting.

Thiele, T. N. (c. 1890). *Proportional approval voting*. Retrieved May 21, 2009, from http://www.nationmaster.com/encyclopedia/Sequential-proportional-approval-voting.

Chapter 7
Does Choosing Committees from Approval Balloting Fulfill the Electorate's Will?

Gilbert Laffond and Jean Lainé

7.1 Introduction

An approval ballot is a voting ballot where voters indicate the candidates they approve among finitely many ones running for elections. We review below some recent studies of procedures that select groups of candidates, or committees, from approval ballots. Many examples can be found of collective decision-making situations where a committee, rather than a single candidate, has to be chosen: deciding about who among a class of students are the ones to be awarded, selecting a board of trustees, appointing new members of an academy, or new professors in a faculty department are all cases where a group of candidates has to be chosen by an electorate. Another example is provided by multiple referendum, where several issues are presented to the voters, who are asked issue-wise to answer by either yes or no.

We address the following question: how faithfully does the outcome of a voting rule designed from approval ballots represent the actual preferences of the voters? Approval ballots ex ante provide little information about how voters compare committees. As long as they sincerely vote, their ballot describes their most preferred outcome, and there is no obvious way to deduce from the observed votes the way they compare any two committees. Thus, some assumptions are to be made about underlined preferences. Special attention has been paid to the case of separable preferences, where the voters' position regarding each of the candidates is preferentially independent from the decision regarding any other candidate. Separability naturally calls for candidate-wise voting rules, through which the selected committee results from separate decisions, each regarding one candidate. A special candidate-wise voting rule is the (parallel) majority rule, which is typically used in multiple referendum: are elected all those candidates who are approved by a majority of voters. When preferential dependencies exist between candidates, there is little hope for a candidate-wise voting rule to perform well, since ballots provide no information

J. Lainé (✉)
Department of Economics, Istanbul Bilgi University, Dolapdere Campus, Kurtulus Deresi Cad No 47, Istanbul 34440, Turkey
e-mail: jean@bilgi.edu.tr

J.-F. Laslier and M.R. Sanver (eds.), *Handbook on Approval Voting*, Studies in Choice and Welfare, DOI 10.1007/978-3-642-02839-7_7,

about the dependencies, and also since the voting rule itself ignores them. A theoretical answer consists of asking voters to report their preferences about all possible committees. However, since the number of committees exponentially grows with the number of candidates, such a solution cannot be implemented in practice. We discuss below several proposals that have been made to overcome the difficulty.

We focus on three different notions of representativeness for a voting rule. Pareto efficiency relates to choosing consistently with a unanimous will: a voting rule is Pareto efficient if it never selects a committee that is unanimously less preferred than another one. Condorcet efficiency pertains to the majority will: a voting rule is consistent with the Condorcet winner (resp. loser) if it always selects, when it exists, the committee that is more preferred than any other one (resp. never selects a committee that is less preferred than any other one) by a majority of voters. Finally, we consider two notions of compromise, both based on the idea that voters may accept to lose in satisfaction in order to favor a large consensus on some committee.

These three approaches are investigated under the assumption of separable preferences. That a candidate-wise voting rule may fail at satisfying one of them may be considered as a voting paradox. Indeed, under sincere voting, separable preferences provide the most favorable situation for such a rule to well represent the electorate's will. We show in Sect. 7.3 that almost all these paradoxes hold, and furthermore, we study the (strong) conditions that allow for avoiding them. Finally, we briefly review in Sect. 7.4 how bad candidate-wise voting rules may behave under non-separable preferences. This calls for the design of alternative procedures of preference elicitation through ballots, such as sequential voting (where voters successively approve or disapprove a candidate knowing the result about the previous ones), or set-wise voting (where several candidates are packaged into bundles, or where the number of possible committees presented to the voters is limited). Section 7.2 is devoted to the formal presentation of voting rules from approval ballots.

7.2 Voting from Approval Ballots

We adopt the following notation: $\mathbb{N}$ stands for the set of non-zero integers, $\mathcal{N} = \{1, \ldots, n, \ldots, N\}$ stands for the set of voters, and $\mathcal{C} = \{1, \ldots, c, \ldots, C\}$ for the set of candidates, where both N, C belong to $\mathbb{N}$. Each candidate applies to a position in a committee, which can be either of a given fixed size $k \leq C$, or of any possible size.

A *committee* is a subset of $\mathcal{C}$. A committee involving k members can equivalently be described as an element $x = (x^1, \ldots, x^C) \in \{0,1\}^C$, where $k = |\{c \in \mathcal{C} : x^c = 1\}|$. We denote by $1(x)$ the set of candidates who are appointed in x: $1(x) = \{c : x^c = 1\}$. The set of all such committees is denoted by Ω_{Ck}, and $\Omega_C = \cup_{k \leq C} \Omega_{Ck}$, and $\Omega = \cup_{C \in \mathbb{N}} \Omega_C$ is the set of possible committees for a variable number of candidates and all possible committee sizes. For any subset $\mathcal{D}$ of $\mathcal{C}$ of size D and any committee $x \in \{0,1\}^C$, we denote by $x/\mathcal{D} \in \{0,1\}^D$ the restriction of

Table 7.1

	1	2	3
x_1	0	1	1
x_2	1	1	0
x_3	1	0	0

x to $\mathcal{D}$, which is defined by: $\forall c \in \mathcal{D}$, $(x/\mathcal{D})^c = x^c$. Moreover, if $\{\mathcal{D}_1, \mathcal{D}_2\}$ is a partition of $\mathcal{C}$ into two non-empty sets, then we equivalently write x and $(x/\mathcal{D}_1, x/\mathcal{D}_2)$.

Each voter n casts an *approval ballot*, by which she approves as many candidates as she wishes. Formally, an approval ballot is defined as a vector $x_n = (x_n^c)_{c=1,\ldots,C} \in \{0,1\}^C$, where $x_n^c = 1$ (resp. $x_n^c = 0$) means that n approves (resp. disapproves) candidate c. For any $x \in \{0,1\}^C$, $(-x)$ is defined by $\forall c$, $(-x)^c = 1 \Leftrightarrow x^c = 0$. A *ballot set* is a set $X = \{x_1, \ldots, x_N\} \subseteq \{0,1\}^C$. We denote by $1_X(c)$ (resp. $0_X(c)$) the number of approvals (resp. disapprovals) c receives in X (that is $1_X(c) = \sum_n x_n^c$, and $0_X(c) = N - 1_X(c)$). A ballot set can equivalently be written as a matrix $X = [x_n^c]_{n=1,\ldots,N}^{c=1,\ldots,C}$, where row n corresponds to voter n' s approval ballot x_n, and each column is associated with one specific candidate. Table 7.1 describes a ballot set involving three voters and three candidates.

A ballot set X is said to be *rich* if whenever $x \in X$, then $(-x) \in X$: for any cast ballot, one finds at least once its opposite in the ballot box. We denote by $\mathcal{X}^C$ the set of all ballot sets involving C candidates, and by $\mathcal{X} = \cup_{C \in \mathbb{N}} \mathcal{X}^C$ the set of all ballot sets for any possible number of candidates.

7.2.1 Candidate-Wise Voting Rules

A *voting rule* describes how one or several committees are selected from a ballot set. We restrict the presentation to the case where a unique committee is always chosen. Formally, a voting rule is an application V from $\mathcal{X}_V$ to Ω, such that, for any ballot set $X \in \mathcal{X}^C$ and any C, $V(X) \subseteq \Omega_C$, where $\mathcal{X}_V$ is the subset of profiles that are *admissible* for V, that is such that V is well-defined. A voting rule is said to be *simple* when the special case $C = 1$ holds: the decision is whether to elect or not one candidate. A voting rule is *anonymous* if its outcome is non-sensitive to the voters' names.[1]

Furthermore, a voting rule V is *candidate-wise* if, for any $X \in \mathcal{X}_V$, one can write

$$V(X) = (v_1(X), \ldots, v_c(X), \ldots, v_C(X))$$

where $v_1, \ldots, v_C$ are simple voting rules: the collective choice is defined as a set of separate decisions, each regarding one single candidate.

[1] Formally, for any $X = \{x_1, \ldots, x_N\} \in \mathcal{X}_V$, for any permutation σ of $\{1, \ldots, N\}$, $V(X) = V(X^\sigma)$, where $X^\sigma = \{x_1^\sigma, \ldots, x_N^\sigma\}$ is defined by: $\forall n$, $x_n^\sigma = x_{\sigma^{-1}(n)}$.

We focus below on two specific classes of candidate-wise voting rules (CWVR), namely the parallel and the sequential rules.

A CWVR V is *parallel* if one can write for any $X \in \mathcal{X}_V$,

$$V(X) = (v_1(X/\{1\}), \ldots, v_c(X/\{c\}), \ldots, v_C(X/\{C\}))$$

where $v_1, \ldots, v_C$ are simple, and where $X/\{c\}$ is the restriction of X to candidate c (that is the cth column of X). A parallel rule is a referendum-type rule, where the decision to appoint a candidate only depends on the voters' positions about this candidate. Hence, a parallel rule decomposes the election into simultaneous mutually independent choices, each dealing with one single candidate.

If V is parallel and anonymous, then at any admissible ballot set X and for any candidate c, one can write $v_c(X/\{c\}) = f_c(1_X(c), 0_X(c))$: the collective position regarding c only depends on the number of approvals and disapprovals given to c.

An example of parallel and anonymous voting rule is the *candidate-wise simple majority rule Maj*, under which candidates are appointed if they gather more approvals than disapprovals: for any ballot set X involving an odd number of voters, $Maj(X) = (m(X/\{1\}), \ldots, m(X/\{C\}))$, where, for all c, $m(X/\{c\}) = 1$ if and only if $1_X(c) > 0_X(c)$.

Maj is a particular example of a threshold rule: a candidate is elected if she receives a given minimum percentage of approvals.

When the committee size is restricted, a threshold rule may fail to select a committee having the relevant size k (even if there are exactly k approvals per ballot). A natural adaptation of *Maj* is the k-plurality rule $Plur_k$, which selects the k best candidates in terms of number of approvals. Since several different committees can be chosen through $Plur_k$, we adopt a simple tie-breaking rule which ensures a unique choice. Formally, $Plur_k$ is the CWVR that is defined as follows. For $k \leq C$, for $X \in \mathcal{X}^C$, the linear order $\succ_X$ of $\mathcal{C}$ is defined by: for any two $c, c' \in \mathcal{C}$, $c \succ_X c'$ if either $[1_X(c) > 1_X(c')]$ or $[1_X(c) = 1_X(c')$ and $c < c']$. Then $Plur_k(X)$ selects the first k best elements of $\mathcal{C}$ for $\succ_X$.

Instead of decomposing the election in parallel, a *sequential* CWVR sequentially composes simple voting rules according to an exogenous order ρ of candidates. Denote by $\rho(r)$ the candidate having rank r in ρ. Then successive elections take place, through which the social decision $v_{\rho(1)}(X^1)$ is made upon candidate $\rho(1)$ from a ballot set $X^1 \in \{0,1\}^N$, and then upon $\rho(2)$ from $X^2 \in \{0,1\}^N$ with common knowledge of $v_{\rho(1)}(X^1) \in \{0,1\}$, and so on, the choice about candidate $\rho(r)$ being made given the preceding choices about candidates $\rho(1), \ldots, \rho(r-1)$. Organizing successive elections through approval balloting generates a sequence of ballot sets $\{X^r, r = 1, \ldots, C\}$, where in each X^r, voters indicate their position regarding the candidate having rank r in ρ, knowing the results of all successive past votes $v_{\rho(1)}(X^1), \ldots, v_{\rho(r-1)}(X^{r-1})$. Given a set of C simple rules $v_1, \ldots, v_C$ and an order ρ of $\mathcal{C}$, this defines the sequential rule $Seq(\rho, v_1, \ldots, v_C)$. A typical example is given by the sequential majority rule *SeqMaj*, where the successive simple rules $v_1, \ldots, v_C$ all coincide with the simple majority rule m.

7.2.2 Preferences Over Committees

In order to evaluate how well the outcome of alternative CWVR depicts the will of the electorate, we have to make assumptions about how voters compare committees. Since approval ballots are the only observed data, preferences over committees have to be elicited from the ballots, through some consistency properties which relate the voting behavior and the underlined preferences over possible choices.

Ignoring for a while any committee size constraint, we assume that voters' preferences over committees are complete preorders of $\{0,1\}^C$. A preference profile is a N-tuple of complete preorders $\pi = (R_1, \ldots, R_N)$.[2] We denote by Π^C the set of all profiles with C candidates, and $\Pi = \cup_{C \in \mathbb{N}} \Pi^C$ is the set of all possible profiles when C varies.

The construction of preferences is made by mapping ballot sets to preference profiles, using a *preference extension rule* ε, defined as an application from $\mathcal{X}$ to Π, which maps each ballot set $X = \{x_1, \ldots, x_N\} \in \mathcal{X}^C$ to a profile $\pi(X) = (R_1, \ldots, R_N)$ in Π^C.

A specific preference extension rule is the *Hamming rule*, through which voters compare committees according to the symmetric distance to their ballot. The Hamming distance between any two vectors $x = (x^1, \ldots, x^C)$ and $y = (y^1, \ldots, y^C)$ in $\{0,1\}^C$ is defined by $d(x, y) = \mid \{c : x^c \neq y^c\} \mid$. The *Hamming extension rule* ε^H is then defined by: $\forall X \in \mathcal{X}^C$, $\varepsilon^H(X) = (R_1^H, \ldots, R_N^H)$, where for all n and $y, z \in \Omega_C$, $d(x_n, y) < d(x_n, z) \Leftrightarrow y \; P_n^H \; z$ and $d(x, y) = d(x, z) \Leftrightarrow y \; I_n^H \; z$.

Consider the ballot set described in Table 7.1. The Hamming extension rule produces the preference profile on $\{0,1\}^C$ (Table 7.2).[3]

The Hamming extension rule shares several interesting properties. First, it is *top-consistent*: every voter's ballot is her unique most preferred committee. In other words, each vote is assumed to be sincere. In addition, ε^H shares the property of *separability*: each candidate c is preferentially independent from $\mathcal{C} - \{c\}$, in the sense that a voter's position about a candidate does not depend on the choice to be made about other candidates. Put differently, preferences over committees produce a clear ranking of individual candidates. Separability is formally defined as follows: for any candidate c and any two committees $x = (x^1, \ldots, x^C)$, $y = (y^1, \ldots, y^C) \in \{0,1\}^C$, let $(y_{-c}, x^c) = (y^1, \ldots, y^{c-1}, x^c, y^{c+1}, \ldots, y^C)$; the preference R_n of

Table 7.2

Voter 1	Voter 2	Voter 3
(0, 1, 1)	(1, 1, 0)	(1, 0, 0)
(1, 1, 1), (0, 0, 1), (0, 1, 0)	(1, 1, 1), (0, 1, 0), (1, 0, 0)	(1, 1, 0), (0, 0, 0), (1, 0, 1)
(1, 0, 1), (1, 1, 0), (0, 0, 0)	(0, 0, 1), (0, 1, 1), (1, 0, 1)	(0, 1, 0), (0, 0, 1), (1, 1, 1)
(1, 0, 0)	(0, 0, 1)	(0, 1, 1)

[2] The asymmetric counterpart of R_n is denoted by P_n, while I_n stands for its indifference part.

[3] Decreasing preference is to be read downwards in each column, and several committees figuring in the same cell are indifferent.

Table 7.3

	1	2	3	4	5
x_1	0	1	1	1	0
x_2	1	1	1	1	1

voter n is separable if, for any three committees x, y, z, any candidate c, (y_{-c}, x^c) P_n $(y_{-c}, -x^c)$ implies (z_{-c}, x^c) P_n $(z_{-c}, -x^c)$, and (y_{-c}, x^c) I_n $(y_{-c}, -x^c)$ implies (z_{-c}, x^c) I_n $(z_{-c}, -x^c)$.

Top-consistency is not problematic as long as separability prevails too. Indeed, it is already known (Lacy and Niou 2000) that, when *Maj* (or *SeqMaj*) is the prevailing rule, each voter with separable preferences casts a ballot which is her most preferred committee.

When there is a size constraint k, the Hamming extension rule obviously provides a complete preorder of the k-member committees. However, since ballots may contain more than k approvals, the Hamming rule is no longer top-consistent. For instance, in Table 7.2, voter 3 has two most preferred 2-member committees. This motivates the following *top-k-consistency property*: when a voter approves at most k candidates, she prefers any committee that includes those candidates than any other one, and if, when she approves more than k candidates, she would prefer any committee all members of which she approves than any other one.[4]

Let us suppose that a two-member committee has to be chosen from the ballot set depicted in Table 7.3.

Consider the two-member committees $x = (0, 0, 1, 1, 0)$ and $y = (0, 1, 1, 0, 0)$. The Hamming extension rule makes both voters indifferent between x and y. The reason is that voters' positions regarding candidates who are appointed neither in x nor in y do not matter when comparing x and y. This leads to the following *k-independence property*. The extension rule ε is k-independent if, for any $C \geq k$, for any ballot set X in $\mathcal{X}^C$, for any two k-member committees y and z, for any two ballots x_1 and x_2 who coincide on $1(y) \cup 1(z)$, one has $[y\ I_1\ z] \Leftrightarrow [y\ I_2\ z]$ and $[y\ P_1\ z] \Leftrightarrow [y\ P_2\ z]$.

Many further properties can be retained for preference extension rules. In fact, comparing committees in this setting is equivalent to extending dichotomous preferences over individuals to preferences over sets of individuals.[5]

Among the properties described above, the most controversial is certainly separability. Indeed, many examples can be found where complementarities or spill-over effects prevail in the comparison of committees. Among them is the case of sport teams: some player might be judged worthwhile to get appointed under the condition she plays with another specific one. Similarly, voters may be reluctant to face an elected assembly which over-represents a party.

---

[4] Formally, ε is top-k-consistent if, for any $C \geq k$, for any ballot set $X = \{x_1, \ldots, x_N\} \in \mathcal{X}^C$, for any n, one has (1) $[|\ 1(x_n)\ | \leq k \Rightarrow y\ P_n\ z]$ for all $y, z \in \Omega_k$ with $1(x_n) \subseteq 1(y)$ and $1(x_n) \subsetneq 1(z)$, and (2)

$[|\ 1(x_n)\ | > k \Rightarrow y\ P_n\ z]$ for all $y, z \in \Omega_k$ with $1(y) \subseteq 1(x_n)$ and $1(z) \subsetneq 1(x_n)$.

[5] The reader can refer to Barbera et al. (2001) for reviewing how to design preferences over sets.

A way to relax separability is suggested in (Lang and Xia 2009), consisting on inducing linear orders of committees from conditional preference networks, or *CP-net preferences*. It appears that this preference domain allows for a well-defined sequential voting behavior. Consider a case where there are C candidates. Write $\{0,1\}^C = D_1 \times \ldots \times D_C$, where $D_c = \{0,1\}$ for all c is called a dimension.[6] Furthermore, let $\succ$ be a linear order of the dimensions, say $D_1 \succ D_2 \succ \ldots \succ D_C$. Moreover, suppose that a voter n has preferences over committees represented by the linear order P_n such that, for any two $z, z' \in D_2 \times \ldots \times D_C$, one has $(1,z)P_n(0,z) \Leftrightarrow (1,z')P_n(0,z')$ and $(0,z)P_n(1,z) \Leftrightarrow (0,z')P_n(1,z')$. Thus, n's position about candidate 1 does not depend on the decisions to be taken about the other candidates. Similarly, we assume that for any $a \in \{0,1\}$ and any two $z, z' \in D_3 \times \ldots \times D_C$, one has $(a,1,z)P_n(a,0,z) \Leftrightarrow (a,1,z')P_n(a,0,z')$ and $(a,0,z)P_n(a,1,z) \Leftrightarrow (a,0,z')P_n(a,1,z')$. More generally, there is a preferential independence between any candidate c (or dimension D_c) and the choice made about the subsequent candidates $c+1,\ldots,C$, that is the choice made in $D_{c+1} \times \ldots \times D_C$. Note however that preferences within each dimension D_c depend on the previous choices in $D_1 \times \ldots \times D_{c-1}$. It is obviously seen that sincere voting in a sequential CWVR is well-defined in that case, since the best decision about a candidate is independent from the future decisions to be made. In fact, Lang and Xia (2009) show that this is still the case when $\succ$ is replaced with any acyclic binary relation over the dimensions. Formally, one defines a directed graph $\mathcal{G} = (\mathcal{C}, E)$ having $\mathcal{C}$ as set of vertices, and E as set of edges, where, for a candidate c, the set of edges $E(c)$ to c is the set of all candidates the position about c is preferentially dependent from. Given the graph $\mathcal{G}$, one also defines a set $CP(\mathcal{G})$ of conditional preferences, that describes how the preferred position regarding each of the candidates c depends from the decision taken about her 'parents' $E(c)$. It is easily shown that $CP(\mathcal{G})$ generates a partial preference relation on committees, which is finally extended to a linear order. Consider the next example: $\mathcal{C} = \{1,2,3\}$, $\mathcal{G} = \{(1,2),(2,3),(1,3)\}$, and $CP(\mathcal{G})$ is given in Table 7.4.[7]

Table 7.4

$c=1$	$c=2$	$c=3$
$1 \succ 0$	$x^1 = 1 \Rightarrow 1 \succ 0$	$(x^1, x^2) = (1,1) \Rightarrow 1 \succ 0$
	$x^1 = 0 \Rightarrow 0 \succ 1$	$(x^1, x^2) = (1,0) \Rightarrow 1 \succ 0$
		$(x^1, x^2) = (0,1) \Rightarrow 1 \succ 0$
		$(x^1, x^2) = (0,0) \Rightarrow 0 \succ 1$

[6] Lang and Xia (2009) consider the more general case where each dimension is a finite set. The interpretation is then that a committee is defined as a set of seats, each seat receiving a finite number of candidates. This leads to seat-wise instead of candidate-wise voting rules. We focus here on the case of approval ballots, which is equivalent to seat-wise designation by voters of the best among exactly two candidates.

[7] Table 7.4 is to be read as follows: in column 3, approving is better than disapproving candidate 3, if both candidates 1 and 2 are elected, but the reverse holds if neither 1 nor 2 is elected.

$CP(\mathcal{G})$ induces the following partial preference relation $\succ^*$on $\{0,1\}^3$:

- $(1,1,1) \succ^* (0,1,1), (1,1,0) \succ^* (0,1,0), (1,0,1) \succ^* (0,0,1), (1,0,0) \succ^* (0,0,0)$
- $(1,1,1) \succ^* (1,0,1), (1,1,0) \succ^* (1,0,0), (0,0,1) \succ^* (0,1,1), (0,0,0) \succ^* (0,1,0)$
- $(1,1,1) \succ^* (1,1,0), (1,0,1) \succ^* (1,0,0), (0,1,1) \succ^* (0,1,0), (0,0,0) \succ^* (0,0,1)$

The transitive closure $\succ^T$of $\succ^*$ is defined by: $(1,1,1) \succ^T (1,0,1), (1,1,0) \succ^T (1,0,0) \succ^T (0,0,0) \succ^T (0,0,1) \succ^T (0,1,1) \succ^T (0,1,0)$. which finally allows for two possible linear orders that are consistent with $\succ^T$.

Lang and Xia (2009) show that, as long as all voters agree on the same acyclic graph $\mathcal{G}$, then sequential voting behavior is well-defined. That all voters must follow the same preferential dependencies among candidates is a strong assumption, which is still weaker than separability.

7.3 Representativeness of Voting Rules Under Separable Preferences

Do both parallel and sequential CWVR select a committee that can be assessed as a 'satisfactory representative' of voters' preferences? Since approval ballots offer a very incomplete information about the latter, we may anticipate a rather negative answer, unless strong assumptions are made about the way ballots depict preferences. In particular, CWVR ignore complementarities in preferences. Hence, ruling out these complementarities, by assuming separability, brings the most favorable situation for CWVR to perform well. Moreover, assuming that votes are sincere, parallel and sequential CWVR always select the same committee.

We distinguish two broad types of representativeness properties. The first relates to the size of the largest fraction of voters who prefer at least one non-chosen committee instead of the chosen one. More precisely, we focus on the Pareto efficiency and several Condorcet-consistency properties of voting rules. A voting rule V is *Pareto-efficient* if it always selects a committee which is not less preferred than another one by all voters. Moreover, V is *Condorcet-winner* (resp. *Condorcet-loser*) *consistent* if it always selects the Condorcet winner whenever it exists, that is a committee which is more preferred than any other one by a majority of voters (resp. never selects a Condorcet loser, that is a committee which is less preferred than any other one by a majority of voters).[8] The second type of representativeness property rests on the idea that the elected committee should appear as a *compromise* which would arise when voters directly bargain over the outcome. Examples of compromises are the majoritarian compromise, and the majority approval.

Situations where a parallel CWVR fails at satisfying each of these properties are generally presented as *voting paradoxes*, since the assumption of separable

[8] For alternative definitions of Condorcet winning committees in settings where approval ballots are not the premises, see Fishburn (1981), Gehrlein (1985), Ratliff (2003), and Kaymak and Sanver (2003).

preferences naturally calls for separate candidate-wise voting. We briefly review below recent results obtained on these voting paradoxes.

7.3.1 Pareto Efficiency

It is already known that candidate-wise majority voting is Pareto-efficient under Hamming preferences, both in the variable and restricted committee size cases (see Brams et al. 1997; Brams et al. 2004, 2007). Indeed, for any ballot set $X = \{x_1, \ldots, x_N\}$, *Maj* always selects the committee which minimizes the total Hamming distance $\sum_{n \in \mathcal{N}} d(x_n, x)$ in the set of all possible committees with variable size. Since the total Hamming distance can be interpreted as the sum of dis-utilities, then *Maj* selects the committee which fulfills the utilitarian criterion.[9]

However, it is worth mentioning that *Maj* can produce a committee that is almost Pareto dominated. More precisely, for $\alpha \in]0, 1[$, say that the CWVR V is *α-efficient* for the extension rule ε if, at any ballot set, no fraction representing a proportion strictly more than α of the electorate can agree on another committee. Note that weak Pareto efficiency is equivalent to the 1-efficiency limit case. Then, Cuhadaroglu and Lainé (2009) describe a ballot set X such that *Maj*(X) is not α-efficient for *any* $\alpha \in]0, 1[$.

More importantly, *Maj* is no longer Pareto efficient when the Hamming rule is replaced with another separable one (Kadane 1972). For instance, consider the ballot set X depicted in Table 7.5.

Then *Maj*$(X) = (1, 1, 1)$. Let ε be the extension rule leading to the preference profile (Table 7.6).

It is easily checked that ε is separable, while voters unanimously prefer $(0, 0, 0)$ than *Maj*(X).

However, separability ensures that *Maj* never selects a universally Pareto dominated committee, that is less preferred than any other committee by all voters (Lacy and Niou 2000).

This example illustrates the following general result by Benoit and Kornhauser (2006).[10] Define dictatorship as the voting rule which identifies the collective decision with a specific voter's ballot (formally, the dictatorship $Dict_{n^*}$ for voter n^* is defined by: $\forall x_{n^*} \in \{0, 1\}^C$, $\forall X = \{x_1, \ldots, x_{n^*}, \ldots, x_N\} \in \mathcal{X}^C$, $Dict_{n^*}(X) = x_{n^*}$).

Proposition 7.3.1 (Benoit and Kornhauser 2006). *If one allows for any separable preference extension rule that leads to a profile of linear orders over committees,*

[9] With C candidates, ε^H produces from any ballot set $X = \{x_1, \ldots, x_N\}$ a profiles $(P_1, \ldots, P_N)$ of preorders where each P_n is represented by the utility function U_n defined on $\{0, 1\}^C$ by $U_n(x) = (C - \sum_c \mid x_n^c - x^c \mid) = C - d(x_n, x)$. Hence, $\sum_n d(x_n, x) = NC - \sum_n U_n(x)$.

[10] We state here a specific version of the result consistent with approval balloting. In fact, the result is more general, since it deals with seat-wise voting rules when finitely many candidates apply for each seat in a committee.

Table 7.5

c	1	2	3
x_1	1	1	0
x_2	0	1	1
x_3	1	0	1

Table 7.6

R_1	R_2	R_3
$(1,1,0)$	$(0,1,1)$	$(1,0,1)$
$(0,1,0)$	$(0,1,0)$	$(1,0,0)$
$(1,0,0)$	$(0,0,1)$	$(0,0,1)$
$(0,0,0)$	$(0,0,0)$	$(0,0,0)$
$(1,1,1)$	$(1,1,1)$	$(1,1,1)$
$(0,1,1)$	$(1,1,0)$	$(1,1,0)$
$(1,0,1)$	$(1,0,1)$	$(0,1,1)$
$(0,0,1)$	$(1,0,0)$	$(0,1,0)$

and if there are at least three candidates, dictatorship is the unique Pareto-efficient candidate-wise voting rule.[11]

In the two-candidate case, Özkal-Sanver and Sanver (2006) have proven that *Maj* is the unique anonymous Pareto efficient CWVR if any separable extension rule is allowed. In fact, *Maj* is Pareto efficient for any top-consistent extension rule. To see why, consider the next four possible ballots $x = (1,1)$, $y = (1,0)$, $z = (0,1)$, and $w = (0,0)$. Denote by N_t the number of ballots $t = x, y, z, w$ in the ballot set X. Now suppose that $Maj(X) = w$ (this entails no loss of generality, since a relevant relabelling of ballots can be done to ensure it), and that w is Pareto-dominated. Top-consistency ensures that $N_w = 0$. Moreover, one must have $N_z > N_x + N_y$. Hence, $Maj(X) = z$, a contradiction.

Since *Maj* is Pareto-efficient with Hamming preferences extension, Proposition 7.3.1 raises the following question: what is the largest domain (for inclusion) of separable extension rules for which *Maj* is Pareto efficient? It is already known from Benoit and Kornhauser (1994) that if any separable and top-lexicographic extension rule is allowed, then *Maj* is Pareto-efficient. Top-lexicographic means that all voters agree on the order of importance of candidates. For instance, if this order is $1 > \cdots > C$, then what should matter the most for all voters is whether the choice regarding candidate 1 fulfills their wish. Note that the extension rule that brings the profile in Table 7.7 is not lexicographic: R_2 calls for ranking the candidates in the order $1 > 2 > 3$, but R_1 disagrees with such an order.

[11] This proposition extends a result by Özkal-Sanver and Sanver (2006), which states that, as long as there are at least three candidates and any top-consistent and separable extension rule is allowed, then there is no anonymous Pareto-efficient CWVR. Note however that Özkal-Sanver and Sanver allow for indifference between committees. Moreover, Lang and Xia (2009) have shown that Proposition 7.3.1 remains true for voting correspondences.

Table 7.7

c	1	2	3
x_1	1	1	1
x_2	1	1	0
x_3	1	0	1
x_4	0	1	1
x_5, x_6, x_7	1	0	0
x_8, x_9, x_{10}	0	1	0
x_{11}, x_{12}, x_{13}	0	0	1

A partial answer is given in Cuhadaroglu and Lainé (2009), where attention is restricted to *neutral* domains of separable extension rules.

Proposition 7.3.2 (Cuhadaroglu and Lainé 2009). *The largest neutral domain of top-consistent and separable preference extension rules for which candidate-wise majority voting is Pareto-efficient is the domain of Hamming-consistent extension rules.*

A domain of preference extension rules $\mathcal{E}$ is neutral if neither the names of the candidates, nor the labelling of the ballots provide valuable information about preferences. Formally, the two following properties must hold:

(1) Consider a ballot set X where ballots x_1 and x_2 contain the same number of approvals. Then, x_2 can be deduced from x_1 by means of a permutation σ of the candidates. Let $R_t(\varepsilon)$ the preference obtained from x_t, where $t = 1, 2$, through some extension rule $\varepsilon \in \mathcal{E}$, and suppose that $yR_1(\varepsilon)z$. Then, there exists $\varepsilon' \in \mathcal{E}$ for which $yR_1(\varepsilon')z$ and $y^\sigma R_2(\varepsilon')z^\sigma$, where y^σ and z^σ are obtained from y and z by running σ.

(2) Moreover, suppose now that $x_2 = (x_1/\mathcal{D}_1, -x_1/\mathcal{D}_2)$, where $\{\mathcal{D}_1, \mathcal{D}_2\}$ is a partition of $\mathcal{C}$ into two non-empty sets, and suppose that $yR_1(\varepsilon)z$ for some $\varepsilon \in \mathcal{E}$. Then, there exists $\varepsilon' \in \mathcal{E}$ for which $yR_1(\varepsilon')z$ and $(y/\mathcal{D}_1, -y/\mathcal{D}_2)R_1(\varepsilon')(z/\mathcal{D}_1, -z/\mathcal{D}_2)$.

Furthermore, we say that a preference extension rule is *Hamming-consistent* if it extends the Hamming preferences to the case where indifference may be cut within its indifference classes.[12]

In the case where the committee size is fixed to k, then k-plurality also minimizes the total Hamming distance in the set of all committees with size k (Brams et al. 2007).[13] Furthermore, k-plurality voting shares an even stronger property: the Hamming preference extension rule is the unique strongly neutral and top-consistent rule for which k-plurality voting is Pareto-efficient for any non-zero integer k, where strong neutrality essentially means that committees are compared by only considering their respective numbers of approved members (Cuhadaroglu and Lainé (2009)).

[12] Formally, ε is Hamming-consistent if for any $X = \{x_1, \ldots, x_N\} \in \mathcal{X}^C$, for any voter n and any two committees y and z, $d(x_n, y) < d(x_n, z) \Rightarrow y\ P_n\ z$.

[13] Note that, similarly to the case of unrestricted size, $Plur_k$ may select a α-inefficient committee for any $\alpha \in]0, 1[$.

As long as votes are assumed to be sincere, a stronger requirement than Pareto-efficiency is that the chosen committee is the one which is the ideal outcome of the highest number of voters. The paradox of multiple elections occurs whenever the committee chosen by *Maj* receive the fewest votes. Brams et al. (1998) show that this paradox generalizes the paradox of voting, and provide an extensive study of the three-candidate case. Consider the ballot set depicted in Table 7.7 involving three candidates and 13 voters.

Then, $Maj(X) = (0, 0, 0)$ does not belong to X, hence the paradox. Scarsini (1998) introduced a stronger version of the paradox, where not only $Maj(X)$ but also all the committees sufficiently close to it, receive zero votes. Sufficiently close means at a Hamming distance strictly less than the smaller integer larger than $\frac{C+1}{2}$.

Avoiding this strong paradox clearly ensures that, under top consistency, *Maj* is Pareto-efficient. A property of ballot sets that provides a sufficient condition for avoiding the strong paradox is stated below. Let $B \subset \{0,1\}^C$ be a set of at least three different ballots. Say that B is stable if $Maj(X) \in B$ for any ballot set X such that $x_n \in B$ for all n.

Proposition 7.3.3 (Laffond and Lainé 2009d). *Let $B \subset \{0,1\}^C$ be a set of at least three different ballots. Then B is stable if and only if $Maj(\{x, y, z\}) \in B$ for any triple $\{x, y, z\}$ of different ballots in B.*[14]

It follows that if B is stable, and if all ballots in B are cast in X, then $Maj(X)$ is a Pareto efficient committee.

7.3.2 Condorcet Properties

7.3.2.1 Voting Paradoxes

A Pareto efficient voting rule never chooses a committee against a unanimous will. More demanding is not to choose against a majority will. Kadane (1972) proved that *Maj* is Condorcet-winner consistent for any separable extension rule (see also Schwartz 1977). Moreover, it is easy to show that separability ensures the Condorcet-loser consistency of *Maj*.

However, a Condorcet winner may not exist, even under Hamming preferences. Equivalently, under separable preferences, *Maj* may produce a committee that is majority defeated. Consider the ballot set X depicted in Table 7.8.

Then $Maj(X) = (0, 0, 0)$, while the majority formed by the first 3 voters less prefer $Maj(X)$ than its opposite. It is easily checked that every committee is defeated by another one under the majority rule. Benoit and Kornhauser (1994) prove that *Maj* selects the Condorcet winner if voters' preferences share a very demanding lexicographic property, which in particular implies that all voters agree on the relative importance of each of the candidates.

[14] Furthermore, *Maj* is the unique Pareto-efficient, neutral and anonymous CWVR for which this property holds.

Table 7.8

	1	2	3
x_1	1	0	1
x_2	1	1	0
x_3	0	1	1
x_4, x_5	0	0	0

Hence, the possibility for *Maj*(X) to be majority defeated relates to the well-known Condorcet paradox adapted to a specific setting. There is a close relationship between such a possibility and the Ostrogorski paradox (Rae and Daudt 1976; Bezembinder and Van Acker 1985; Deb and Kelsey 1987; Nurmi 1998, 1999), although they are not equivalent. The Ostrogorski paradox occurs when, interpreting 0 and 1 as two competing political parties having opposite views on issues $c = 1, \ldots, C$, and assuming that voters vote for the party they agree with on a majority of issues, the winning party get issue-wise a minority of supports.[15]

Another related compound-majority voting paradox is the Anscombe paradox (Anscombe 1976), which refers to ballot sets X where a majority of voters disagree with *Maj*(X) on a majority of candidates. Table 7.8 also illustrates this paradox: three voters among five disagree with *Maj*(X) on two among the three candidates. In fact, the Anscombe paradox is equivalent to *Maj*(X) being majority defeated by its opposite (see Laffond and Lainé 2009c).

For a set X involving an odd number of ballots, and for the preference extension rule ε, the majority tournament $T(X, \varepsilon)$ is the complete and reflexive binary relation defined on $\{0,1\}^C \times \{0,1\}^C$ by $(y, z) \in T(X, \varepsilon)$ if the number of voters in $\varepsilon(X)$ who prefer y than z is strictly larger than the number of those who prefer z than y.[16]

We say that a ballot set X faces the *(resp. strict) majoritarian paradox for* ε if there exists a committee y such that $(y, Maj(X)) \in T(X, \varepsilon)$ (resp. $(-Maj(X), Maj(X)) \in T(X, \varepsilon)$). Note that under Hamming preferences, a committee is non-defeated in $T(X, \varepsilon^H)$ if and only if it is the Condorcet winner of X (Laffond and Lainé 2009a).[17] Since *Maj*(X) is the Condorcet winner of X whenever it exists, then the majoritarian paradox prevails if X has no Condorcet winner.

A paradox stronger than the majoritarian paradox has been introduced in Laffond and Lainé (2009a). A tournament solution S is a correspondence from

---

[15] Table 7.8 depicts an Ostrogorski paradox, but the following ballot set

	1	2	3
x_1, x_2	0	1	0
x_3	1	0	0
x_4, x_5	1	1	1

also depicts an Ostrogorski paradox where *Maj* elects a Condorcet winner.

[16] Under Hamming preferences, $(y, z) \in T(X, \varepsilon^H)$ if $\mid \{n : d(x_n, y) < d(x_n, z)\} \mid > \mid \{n : d(x_n, z) < d(x_n, y)\} \mid$.

[17] Recall that we assume an odd number of voters. When the number of voters is even, this equivalence disappears: add up one voter to the set in Table 7.7, with ballot *Maj*(X). One get that $Maj(X \cup \{Maj(X)\}) = Maj(X)$. However, both *Maj*$(X)$ and $-Maj(X)$ are non-defeated in $T(X, \varepsilon^H)$.

the set of all complete and reflexive binary relations on $\{0,1\}^C \times \{0,1\}^C$ to the set of non-empty subsets of $\{0,1\}^C$, such that, for any set X and any extension rule ε, $S(T(X,\varepsilon)) = \{x^*\}$ whenever the Condorcet winner x^* of X exists.[18] Hence, S either selects from $T(X,\varepsilon)$ the unique Condorcet winner if it exists, or a subset of committees otherwise. The *S-paradox for* ε occurs at the ballot set X if $Maj(X) \notin S(T(X,\varepsilon))$.

The *Top-Cycle* TC (Schwartz 1972) and the *Uncovered Set* UC (Miller 1977) are well-known tournament solutions. A committee x belongs to $TC(T(X,\varepsilon))$ if x defeats either directly or indirectly all other committees: formally, $x \in TC(T(X,\varepsilon))$ if, for any $y \in \{0,1\}^C$, there exists a sequence $\{y_h\}_{0 \leq h \leq H}$ of committees such that $y_0 = x$, $y_H = y$, and $(y_h, y_{h+1}) \in T(X,\varepsilon)$ for all $h = 0,\ldots,H-1$. Furthermore, x belongs to $UC(T(X,\varepsilon))$ if x defeats either directly or indirectly in two steps all other committees. An equivalent definition is: $x \in UC(T(X,\varepsilon))$ if there is no other committee y such that $(y,x) \in T(X,\varepsilon)$ and for any committee z, $(x,z) \in T(X,\varepsilon)$ implies $(y,z) \in T(X,\varepsilon)$.

The next example proves that the outcome of *Maj*(X) may not belong to the Uncovered Set, when $\varepsilon = \varepsilon^H$. Since most tournament solutions are refinements of UC, then the UC-paradox essentially states that *Maj* is inconsistent with Condorcet-type choice from Hamming profiles.

In the case of five candidates, consider the ballot set X containing $N = \alpha + 10\beta$ ballots, where α voters cast $(0,0,0,0,0)$, and where each ballot with three approvals is cast by β voters. Denoting by Ω_t the subset of ballots with t approvals, one get that:

- $(-Maj(X), Maj(X)) \in T(X,\varepsilon^H)$ if $\alpha < 10\beta$
- $(Maj(X), x) \in T(X,\varepsilon^H)$ where $x \in \Omega_1 \cup \Omega_2$ if $\alpha > 2\beta$
- $(Maj(X), x) \in T(X,\varepsilon^H)$ where $x \in \Omega_3 \cup \Omega_4$ if $\alpha > 4\beta$
- $(-Maj(X), x) \in T(X,\varepsilon^H)$ where $x \in \Omega_1 \cup \Omega_2$ if $\alpha < 4\beta$
- $(-Maj(X), x) \in T(X,\varepsilon^H)$ where $x \in \Omega_3 \cup \Omega_4$ if $\alpha < 2\beta$

Thus, if $2 < \frac{\alpha}{\beta} < 4$, then $(-Maj(X), x) \in T(X,\varepsilon^H)$ whenever $(Maj(X), x) \in T(X,\varepsilon^H)$, which implies that $Maj(X) \notin UC(T(X,\varepsilon^H))$.

However, for separable preference extension rules, *Maj* always selects within the Top-Cycle (Laffond and Lainé (2009a)).

7.3.2.2 Avoiding Paradoxes

Which restrictions upon ballot sets are sufficient to avoid the majoritarian and S-paradoxes? A possible route deals with the level of the candidate-wise majority size. First, under the Hamming extension rule, the majoritarian paradox cannot

[18] Since Hamming preferences allow for indifference, $T(X)$ is a weak tournament. See Peris and Subiza (1999) for an analysis of weak tournament solutions, and Laslier (1997) for a complete review of tournament solutions.

occur when all such sizes are greater that 75% (see Wagner 1983, 1984 about the Anscombe paradox). Furthermore, this 'three-fourth' rule draws also a benchmark for the UC-paradox. For the ballot set X and candidate c, we denote by $m(c)$ the candidate $c's$ size, that is $m(c, X) = \frac{|\{n: Maj(X/\{c\}) = x_n^c|}{N}$.

Proposition 7.3.4 (Laffond and Lainé 2009a). *Let X be a ballot set with N ballots such that $m(c, X) \geq \frac{3}{4}$ for all candidates c. Then, under the Hamming extension rule, Maj(X) is a Condorcet winner of X. Moreover, for any $\alpha > 0$, there exists a ballot set X such that $m(c, X) > (\frac{3}{4} - \alpha)$ for all c and $Maj(X) \notin UC(T(X, \varepsilon^H))$.*

Another approach, proposed in Laffond and Lainé (2009b), is to relate the existence of a paradox to the level of proximity between ballots. The underlying intuition is that the mutually closer are the ballots, the more unlikely is the possibility of a paradox. One measure of proximity is the maximal Hamming distance one can find between two cast ballots. For a ballot set $X = \{x_1, \ldots, x_N\}$ with C candidates, the *proximity level* of X is defined by $\delta(X) = Max\{\frac{d(x_n, x_m)}{C}, x_n, x_m \in X\}$, that is the maximal fraction of candidates whom two ballots disagree on.

Proposition 7.3.5 (Laffond and Lainé 2009b). *If the electorate is large enough, then the strict majoritarian (or Anscombe) paradox never prevails at X if $\delta(X) \leq \sqrt{2} - 1$. Moreover, for any $\alpha > 0$, there exists a ballot set X with $\delta(X) < \alpha + \sqrt{2} - 1$ at which the paradox holds.*

Proposition 7.3.5 can be generalized. Let $\theta(X)$ denote the proportion of voters who prefer *Maj*(X) than its opposite. Hence, the Anscombe paradox holds at X whenever $\theta(X) < \frac{1}{2}$.

Proposition 7.3.6 (Laffond and Lainé 2009b). *Let $\eta \in [0, 1]$. Let r_η be the minimal number in $[0, 1]$ for which there exists a ballot set X such that $\delta(X) = r_\eta$ and $\theta(X) = \eta$. If the electorate is large enough, then $r_\eta = \frac{1-\eta}{\eta}.(\sqrt{\frac{1}{1-\eta}} - 1)$.*

Proposition 7.3.6 is to be read as follows. Suppose that one aims at *Maj*(X) being supported by 75% of the electorate against $-Maj(X)$. Thus, $\eta = \frac{3}{4}$ and $r_{0.75} = \frac{1}{3}$. This means that we reach the target for any ballot set where if any two ballots differ on less than a third of the candidates, and also that, for any $r > \frac{1}{3}$, there exists a ballot set X' such that the maximal distance between two ballots is r and more than a fourth of the electorate less prefer *Maj*(X') than its opposite.

Laffond and Lainé (2006) note that a ballot set with at most three different ballots never faces the majoritarian paradox. They also establish a necessary and sufficient condition for avoiding the majoritarian paradox. We first proceed with several preliminary definitions.

A ballot set $Y \subseteq \{0, 1\}^C$ is said to be *simple* if all its elements are mutually different. Given a ballot set X, we call *range* of X the simple ballot set $Y(X)$ whose all elements are ballots in X. For any subset $\mathcal{D} \subseteq \mathcal{C}$ of candidates, a $\mathcal{D}$-relabelling of the simple voting set Y is obtained by reversing in each ballot approvals and disapprovals regarding all the candidates in $\mathcal{D}$. Furthermore, for any permutation σ

of $\mathcal{C}$, a σ-permutation of Y is the simple ballot set obtained by reshuffling the set of candidates (i.e., columns of Y) without modifying the voters' positions regarding each of them.[19]

Two simple ballot sets Y and Y' are *equivalent* if there exist a subset $\mathcal{D} \subseteq \mathcal{C}$ of candidates and a permutation σ of $\mathcal{C}$ such that Y' is obtained from Y through a $\mathcal{D}$-relabelling together with a σ-permutation.

A simple ballot set $Y = \{y_1, \ldots, y_N\}$ has a *single-switch representation* if in each ballot y_n, there exists at most one candidate $1 \leq c(n) \leq C - 1$ such that $y_n^{c(n)} \neq y_n^{c(n)+1}$. Moreover, Y is said to be *single-switch* if it is equivalent to a ballot set having a single-switch representation.[20]

The single-switchness property characterizes the rich ballot sets that are not exposed to the majoritarian paradox.[21]

Proposition 7.3.7 (Laffond and Lainé 2006). *If there are at least three voters, a rich ballot set with at least three different pairs of opposite ballots cannot face the majoritarian paradox if and only if its range is single-switch.*

Single-switchness is a strong restriction which rules out a large number of ballots. It relates to some consistency across ballots: indeed, a simple ballot set is single-switch if and only if, for any two voters n and m, the set of voters approved by n either contains the set of those approved by m, or contains the set of those disapproved by m.

7.3.3 *Compromising Through Majority Voting*

We now define the representativeness of a CWVR as its ability to reach a compromise. A compromise relates to the outcome of some non-specified negotiation among voters through which some (hopefully large) fraction of them agree on a

---

[19] The $\mathcal{D}$*–relabelling* of $Y = [y_n^c]_{n=1,\ldots,N}^{c=1,\ldots,C}$ is the ballot set $V^{\mathcal{D}} = [y_n'^c]_n^c$ defined by: $\forall c \in \mathcal{D}$, $\forall n$, $y_n^c = 1 \Leftrightarrow y_n'^c = 0$, and $\forall c \notin \mathcal{D}$, $\forall n$, $y_n^c = y_n'^c$. The σ-permutation of Y is a ballot set $Y^\sigma = [x_n^{\sigma c}]_n^c$ defined by: $\forall c$, $\forall n$, $x_n^{\sigma c} = x_{\sigma^{-1}(n)}^c$.

[20] For instance, the simple ballot set

	1	2	3	4
x_1	1	0	1	0
x_2	1	1	1	0
x_3	0	1	1	1
x_4	0	0	0	1

is single switch. To see why, the $\{1\}$-relabelling of Y gives

	1	2	3	4
x_1	0	0	1	0
x_2	0	1	1	0
x_3	1	1	1	1
x_4	1	0	0	1

, while

	4	1	2	3
x_1	0	0	0	1
x_2	0	0	1	1
x_3	1	1	1	1
x_4	1	1	0	0

has a single-switch representation.

[21] Remember that a ballot set is rich if it contains only pairs of opposite ballots.

single outcome. Compromising refers to the acceptance by individuals of some decrease in their satisfaction in order to reach an agreement. We categorize below two types of compromise solutions from preference profiles over committees.

For a top-consistent preference extension rule ε, a committee x is a *Fallback α-bargaining committee* in the ballot set X if it fulfills three conditions: (1) it is supported by a fraction α of the voters at some maximal loss of h ranks in their preference given in $\varepsilon(X)$, (2) no other committee can be supported by a fraction α of the electorate whose members suffer from a lower loss in satisfaction, and (3) no other committee get a larger support under the maximal loss h. Fallback α-bargaining is thus a negotiation procedure under which voters begin by indicating their preference ranking over all committees. They then fall back, in lockstep, to less and less preferred committees - starting with first choices, then adding second choices, and so on - until one is found on which a fraction α of the voters agree.

Attention has been paid to the cases where either $\alpha = \frac{1}{2}$ or $\alpha = 1$. The former is known as the Majoritarian Compromise, or MC (Sertel 1987; Sertel and Yılmaz 1998; Giritligil Kara and Sertel 2005), while the latter is the Fallback Bargaining solution, or FB (Brams and Kilgour 2001; Brams et al. 2004, 2007; Kilgour et al. 2006).[22] We respectively denote by $MC(X)$ and $FB(X)$ the set of majoritarian and fallback bargaining committees at X given the extension rule ε^H.

Assuming Hamming preferences, we will use the following notations and definitions: for a committee x, a ballot set X with N voters, and $k \in \mathbb{N}$, the $k-$support of x in X is the integer $Supp_k(x, X)$ equal to the number of voters who place x at worst at the kth rank in their preferences. Moreover, for $\alpha \in [0, 1]$, let $k^*(X, \alpha) = Min\{k \in \mathbb{N} : \exists x \in \{0, 1\}^Q$ such that $Supp_k(x, X) \geq \alpha.N\}$, that is the minimal loss in satisfaction to be accepted for a fraction α of the voters to agree on some committee. A committee x is a Fallback α-bargaining committee at X if, for any other committee y, $Supp_{k^*(X,\alpha)}(x, X) \geq Supp_{k^*(X,\alpha)}(y, X)$. It is obviously checked that there always exist one or maybe several α-bargaining committees for any value of α.

Consider the ballot set X depicted in Table 7.9.

Under the Hamming extension rule, one get the following profile over committees (Table 7.10).

The reader will check that $k^*(X, \frac{1}{2}) = 2$, which leads to $MC(X) = \{(0, 0, 0, 0), (1, 0, 0, 0)\}$. Furthermore, $k^*(X, 1) = 3$, and one get $FB(X) = \{(1, 1, 0, 0), (1, 0, 1, 0), (1, 0, 0, 1)\}$.

Table 7.9

	1	2	3	4
x_1	1	1	1	1
x_2	1	0	0	0
x_3	0	0	0	0

[22] Brams et al. (2007) introduce two notions of weighted FB we will not discuss here.

Table 7.10

Voter 1	Voter 2	Voter 3
1111	1000	0000
1110,1101,1011,0111	1100,1010,1001,0000	1000,0100,0010,0001
0011,0101,0110,1001,1010,1100	0100,0010,0001,1110,1101,1011	1100,1010,1001,0110,0101,0011
1000,0100,0010,0001	0110,0101,0011,1111	1110,1101,1011,0111
0000	0111	1111

Table 7.11

x_1	1	1	1	0
x_2	1	1	0	1
x_3	1	0	1	1
x_4	0	1	1	1
x_5, x_6, x_7	0	0	0	0

The second type of compromise solution rests upon a dual approach. Given a ballot set X and the extension rule ε^H, the number of indifference classes generated by ε^H in the profile $\varepsilon(X)$ is equal to $C + 1$. Say that a committee x is β-acceptable for voter n if the rank given to x in R_n is less than $\beta.(C + 1)$. The *β-compromise* of X is the set of all committees that are β-acceptable for a maximal number of voters. We focus here on the case $\beta = \frac{1}{2}$: a committee x is $\frac{1}{2}$-acceptable for voter n if n agrees with x on more than half of the candidates. We denote by $C(X)$ the $\frac{1}{2}$-compromise at X. It is easily checked that $C(X) = MC(X)$ when X is defined in Table 7.9 (hence $C(X) \cap FB(X) = \varnothing$). But it is easy to find cases where $C(X) \cap MC(X) = \varnothing$.

Consider again Table 7.9. Since $Maj(X) = (1, 0, 0, 0)$, then $Maj(X) \in MC(X)$ while $Maj(X) \notin FB(X)$. How well does *Maj* perform as a way to reach the *Maj*oritarian, the Fallback Bargaining and the $\frac{1}{2}$-compromise is examined in Brams et al. (2004, 2007) and in Laffond and Lainé (2009c). Since *Maj* minimizes the sum of distances, while FB minimizes the maximal distance to the preference profile, it is not a surprise that *Maj* may not select a Fallback Bargaining committee, unless restrictions are made about ballot sets.

Table 7.11 gives a case where the outcome of *Maj* is not a majoritarian compromise, while its opposite is.

One get $Maj(X) = x_5$. Since no ballot gathers a majority of votes, then $k^*(X, \frac{1}{2}) > 1$. Moreover, note that $Supp_2(-x_5, X) = 4 > \frac{N}{2}$. Suppose that $(-x_5) \notin MC(X)$. Let $y \in MC(X)$. Then it must be true that the majority supporting y involves $5, 6$ and 7. Hence y must contains at least three disapprovals. But this implies that, for $n = 1, 2, 3, 4$, $d(y, x_h) > 1$, so that $5, 6$ and 7 must represent more than half of the voters, a contradiction. Thus, $(-x_5) \in MC(X)$, which clearly implies that $x_5 \notin MC(X)$.

Table 7.12 illustrates the same problem with the $\frac{1}{2}$-compromise.

$Maj(X) = w = (0, 0, 0, 0, 0)$ and the 45 voters from 51 to 95 find that w is $\frac{1}{2}$-acceptable, while $-w$ is $\frac{1}{2}$-acceptable for the first 50 voters. Moreover, no voter among the last 45 ones find $\frac{1}{2}$-acceptable any three-member or four-member

Table 7.12

$x_1, \ldots, x_{10}$	1	1	1	0	0
$x_{11}, \ldots, x_{20}$	0	1	1	1	0
$x_{21}, \ldots, x_{30}$	0	0	1	1	1
$x_{31}, \ldots, x_{40}$	1	0	0	1	1
$x_{41}, \ldots, x_{50}$	1	1	0	0	1
$x_{51}, \ldots, x_{59}$	1	0	0	0	0
$x_{60}, \ldots, x_{68}$	0	1	0	0	0
$x_{69}, \ldots, x_{77}$	0	0	1	0	0
$x_{78}, \ldots, x_{86}$	0	0	0	1	0
$x_{87}, \ldots, x_{95}$	0	0	0	0	1

committee. One also check that each one-member committee is found $\frac{1}{2}$-acceptable by 48 voters, and that each two-member committee is found $\frac{1}{2}$-acceptable by 38 voters. Thus, $C(X) = \{-w\}$, so that $Maj(X) \notin C(X)$.[23]

The three propositions below summarize conditions for *Maj* to select within each of the three compromise concepts. The first points out that the three-fourth rule is of little help.

Proposition 7.3.8 (Laffond and Lainé 2009c). *Let X be a ballot set with at most three different ballots such that $m(c, X) \geq \frac{3}{4}$ for all candidates c. Then, $Maj(X)$ belongs to $MC(X)$. However, there exists a ballot set X with six different ballots such that $m(c, X) \geq \frac{3}{4}$ and $Maj(X) \notin MC(X)$. Finally, there is no ballot set X such that $m(c, X) \geq \frac{3}{4}$ and $-Maj(X) \in MC(X)$.*

Actually, no condition dealing with the level candidate-wise majority size can secure a compromise through *Maj*: indeed, *Maj* may choose a committee that is not a majoritarian compromise (or a $\frac{1}{2}$-compromise, or a Fallback bargaining) even when voters are almost unanimous candidate-wise.

Proposition 7.3.9 (Laffond and Lainé 2009c). *For any $\alpha > 0$, there exist three ballot sets X, X', X'' with respectively N, N', N'' voters such that $m(c, X) > (1 - \alpha)$ and $Maj(X) \notin MC(X)$, $m(c, X') > (1 - \alpha)$ and $Maj(X') \notin C(X')$, and $m(c, X'') > (1 - \alpha)$ and $Maj(X'') \notin FB(X'')$.*

Furthermore, looking for the equivalent of the single-switch condition leads to an almost impossibility result.

Proposition 7.3.10 (Laffond and Lainé 2009c). *A rich ballot set X cannot be such that $Maj(X) \notin MC(X)$ if and only if its range contains either one unique pair of opposite ballots, or two pairs $\{x, -x\}$ and $\{y, -y\}$ of opposite ballots such that $d(x, y) \leq 1$. Moreover, for any two distinct pairs of opposite ballots $\{x, -x\}$ and $\{y, -y\}$, there exists a ballot set X with range $\{\{x, -x\}, \{y, -y\}\}$ such that $Maj(X) \notin C(X)$.*

[23] It is shown in Laffond and Lainé (2009c) that having $-Maj(X) \in MC(X)$ requires at least four candidates, whereas having $-Maj(X) \in C(X)$ requires at least five candidates. Finally, having $[Maj(X) \notin MC(X)]$ or $Maj(X) \notin C(X)]$ requires at least three candidates.

Since *Maj* always selects the Condorcet winner when it exists, a by-product of Proposition 7.3.10 is that no Condorcet choice voting rule always selects either in the Majoritarian Compromise or in the approval $\frac{1}{2}$-compromise. To conclude, it is obvious that both Propositions 7.3.8 and 7.3.9 remain valid if any separable preference rule is allowed beyond the Hamming rule. And Proposition 7.3.10 becomes even more negative: there is no set of potential ballots where *Maj* selects a majoritarian compromise committees whatever the distribution of votes among ballots.

Finally, most results dealing with Condorcet properties and Compromise solutions still hold when the size of committee is restricted.

7.4 Non-separable Preferences

We have shown that candidate-wise voting, and in particular *Maj*, may poorly represent the voters' preferences about committees, even when these preferences are assumed to be separable. We also showed that only strong restrictions, beyond separability, upon either preferences or ballot sets, can overcome this lack of representativeness. Not surprisingly, allowing for non-separable preferences exposes *Maj* to even worst difficulties, and referring to a voting paradox is no longer appropriate.

Preferential dependencies are likely in many real-life elections, and this has been demonstrated in several studies: among them, Lacy and Niou (2000) use data from elections via the Internet, and Ratliff (2006) analyzes the ballots of two elections at Wheaton College in Massachusetts. Hodge and Schwallier (2006) study, by means of randomly generated preference profiles, how non-separability influences the representativeness of multiple referendum, where the measure of representativeness is based on the Borda score.

Under non-separable preferences, *Maj* is no longer Condorcet-loser consistent, and its outcome may even be Pareto-dominated by all other committees (Lacy and Niou 2000). Consider the two ballot sets depicted in Table 7.13.

If preferences over committees are not separable, then nothing precludes that both voters 1 and 2 uniquely rank $Maj(X) = (0, 0)$ last in $\varepsilon(X)$, and that all three voters uniquely rank $Maj(X') = (1, 1, 1)$ last in $\varepsilon(X')$.[24]

By construction, CWVR like *Maj* ignore some essential features which drive the preferences behind the ballots. So, separating the overall profile of preferences into a set of candidate-wise (binary) profiles, and designing vote from the latter, exposes the choice procedure to a (maybe strong) lack of representativeness.[25]

[24] Lacy and Niou (2000) prove that, when preferences are separable, *Maj* selects a committee that cannot be unanimously less preferred than any other one.

[25] This simple argument is also put ahead in Saari and Sieberg (2001), "these paradoxical behaviors arise because the separation of inputs into disconnected parts can cause a concomitant loss of available and crucial information."

Table 7.13

	1	2
x_1	1	0
x_2	0	1
x_3	0	0

	1	2	3
x_1	1	1	0
x_2	1	0	1
x_3	0	1	1

Table 7.14

	1	2
x_1	0	1
x_2	1	0
x_3	0	0

Table 7.15

Voter 1	Voter 2	Voter 3
(0, 1)	(1, 0)	(0, 0)
(1, 1)	(1, 1)	(1, 1)
(0, 0)	(0, 0)	(0, 1)
(1, 0)	(0, 1)	(1, 0)

There is another important consequence of non-separability: votes may not be sincere. For instance, by casting the ballot (0, 1) in Table 7.13, voter 1 ensures the election of the candidate 2 only, and hence may increase her level ofs satisfaction. More generally, since candidate-wise positions are not clear when preferences are non-separable, it may somehow be difficult to forecast the voters' behavior when facing either a parallel or a sequential CWVR: how would voter 1 decide about candidate 1 in a three-candidate case if willing 1 to get elected also depend on the decision to be taken, either simultaneously, or in the future, about another candidate? Another way to look at strategic voting is to expect it could help at escaping from the major problems met by *Maj*. For instance, we already know that, under separability, strategic voting leads to sincere votes, and *Maj* always select the Condorcet winner when it exists.[26] Unfortunately, this is no longer true when some voter has non-separable preferences. Consider the ballot set X defined in Lacy and Niou (2000) by (Table 7.14) and suppose that the extension rule ε defines the following profile $\varepsilon(X) = (P_1, P_2, P_3)$ of orders (Table 7.15).

Suppose that *Maj* is used. Both voters 1 and 2 having separable preferences, a dominant strategy is for them to vote sincerely. It follows that voter 3 is pivotal on each candidate, so that his best response to the other strategies is voting also sincere. It follows that $Maj(X) = (0, 0)$. Finally, $(1, 1)$ is the Condorcet winner of $T(X, \varepsilon)$. Hence, strategic voting cannot secure the choice of the Condorcet winner when it exists.

We are left with the following statement: when voters are asked to cast approval ballots, and when a parallel CWVR is used, there is little hope to always avoid choosing a committee that is poorly considered by a significant fraction of the

[26] See Kramer (1972) for the same result in a more general setting.

voters. Even under the most favorable case, namely separable preferences, strong additional assumptions are required in order to escape from the different voting paradoxes. And when preferential dependencies among candidates prevail, the best route to be followed is to design voting procedures which better elicit voters' preferences than CWVR do.

Asking voters to report their entire preference relation has only a theoretical value, and must be ruled out from any practically implementable voting procedure, unless the number of candidates is very small. Indeed the exponential number of possible committees should preclude to defend this option. We are left with three alternative methods:[27]

- Choosing among a subset of possible committees (Kilgour and Bradley 1998; Brams et al. 1997)
- Designing procedures based on partial reports of preferences (Brams et al. 1997; Ratliff 2006)
- Sequential voting (Lacy and Niou 2000; Lang and Xia 2009)

As argued in Lang and Xia (2007, 2009), and already recognized in Brams et al. (1997), the first option, that is 'packaging the candidates' in order to reduce the number of committees to be compared, is hardly feasible in a systematic way, and might be an actual option in very specific contexts. The second option consists of asking voters either to report their first k best committees (where k is given a reasonably small value), and to apply a choice function on this restricted profile, or to indicate which committees they approve and select the plurality winner(s). In both methods, voters cast several instead of a single approval ballot.[28] To our knowledge, a complete analysis of the representativeness properties of both options 1 and 2 remain to be done.

While options 1 and 2 follow the 'global' way (voters vote for bundles of candidates), sequential voting follows the 'local' way of candidate-wise choice, although non-separability does not naturally call for it. The intuition is that successive votes may help at taking into account at least part of the preferential dependencies. Still, the actual voting behavior cannot be specified without assumptions on voters' expectations about the future candidates. Lacy and Niou (2000) assume that voters are optimistic, in the sense that they always vote for the candidate according to their most preferred committee given the past decisions. Under this assumption, the sequential majority voting is Condorcet-loser consistent. However, as illustrated by the next example (Table 7.16) with three candidates (Lang and Xia 2009), the selected committee may be a 'nearly' Condorcet loser:

[27] The list is not exhaustive. An alternative method has been proposed by Bock et al. (1998), in which the size of the committee is determined from votes cast on single candidates.

[28] Brams et al. (1997) describe two refinements of approval voting on committees which deal with the possibility to abstain over certain candidates, and compare them with approval voting using the data of a specific multiple referendum election. Ratliff (2006) shows how voting from partially reported preferences may help in a specific real-life election.

Table 7.16

voter 1	voter 2	voter 3
$(1, 1, 0)$	$(1, 0, 1)$	$(0, 1, 1)$
$(0, 0, 0)$	$(0, 0, 0)$	$(0, 0, 0)$
...	...	$(0, 1, 0)$
...	...	$(0, 0, 1)$
...	...	$(1, 1, 1)$
...	...	$(1, 1, 0)$
$(1, 0, 1)$	$(1, 1, 1)$	$(1, 0, 0)$
$(1, 1, 1)$	$(1, 1, 0)$	$(1, 0, 1)$

Table 7.17

voter 1	voter 2	voter 3
$(1, 1)$	$(1, 0)$	$(0, 1)$
$(0, 1)$	$(1, 1)$	$(0, 0)$
$(1, 0)$	$(0, 1)$	$(1, 0)$
$(0, 0)$	$(0, 0)$	$(1, 1)$

Suppose that the decisions are made first about candidate 1, then candidate 2, then 3. Then 1 is elected, leading voters 1 and 3 to elect candidate 2, and finally voters 2 and 3 approve candidate 3. The resulting committee $(1, 1, 1)$ is majority defeated by all committees but $(1, 1, 0)$.

Another positive property of sequential majority voting is that it never selects a universally Pareto dominated committee, although it can be Pareto dominated (Lacy and Niou 2000).

We argue that optimistic voting behavior is unlikely to prevail. Indeed, *SeqMaj* is manipulable, while optimistic voting is equivalent to sincere voting. In the example with two candidates given in Table 7.17, we suppose that candidate 1 is chosen first. If voter 1 is sincere at the first vote, then $(1, 0)$ is chosen, while disapproving candidate 1 leads to the preferred committee $(0, 1)$.

However, Lacy and Niou (2000) show that sophisticated voting[29] always results in a Condorcet winner whenever it exists.[30] The example in Table 7.16 illustrates the result: $(0, 0, 0)$ is the Condorcet winner. At the last vote, if $(0, 0)$ is the current situation, then clearly $(0, 0, 0)$ will be elected against $(0, 0, 1)$. Consider backwards the vote about candidate 2. Then $(0, 0)$ will be elected against $(0, 1)$ since voters anticipate the last vote, Similarly, candidate 1 will be not elected, voters anticipating the subsequent sequence of results. Hence, strategic voting preserves Condorcet-winning consistency with non-separable preferences.

Finally, Lang and Xia (2007, 2009) offer a general study of sequential voting when voters have CP-net preferences. We summarize below the properties dealing with representativeness:[31]

[29] Sophisticated voting prevails when voters rule our their dominated strategies, and vote candidate-wise according to the most preferred outcome their choice is likely to produce.

[30] This extends a result by Farquharson (1969) to multiple referendum. See also Kramer (1972).

[31] Lang and Xia (2007, 2009) essentially study whether the sequential composition of simple voting rules inherits a given property satisfied by all the simple rules. They also address the

Proposition 7.4.1 (Lang and Xia 2009). *Suppose that all voters have CP-net preferences. Then,*

(1) SeqMaj is Condorcet-winner consistent.[32]

(2) SeqMaj is Condorcet-loser consistent.

The main interest of Proposition 7.4.1 is to generalize the Condorcet consistency properties established for separable preferences (Kadane 1972) to a strictly larger domain. However, since Pareto efficiency is violated under separable preferences, it remains so in this larger domain.

As a conclusion, how to design committee choice procedures from approval balloting remains a widely unsolved question. Both parallel majority voting *Maj* with separable preferences, and sequential majority voting *SeqMaj* with CP-net preferences are Condorcet (winner and loser) consistent. However, *Maj* (and thus *SeqMaj*) does not share other appealing representativeness properties, unless very strong restrictions are made on separable preferences. And there is no hope to move out this dead end when preferential dependencies are allowed. Several suggestions, like driving a richer information about preferences through the provision of more than one ballot, or presenting bundles of candidates, seem to allow for some promising benefits in specific real-life elections. Their general formalization is a challenging question of economic design.

References

Anscombe, G. E. M. (1976). On the frustration of the majority by fulfillment of the majority's will. *Analysis, 36*, 161–168.

Barbera, S., Bossert, W., & Pattanaik, P. (2001). Ranking sets of objects. In S. Barbera, P. J. Hammond, & Ch. Seidl (Eds.), *Handbook of utility theory* (Vol. 2). Boston: Kluwer.

Benoit, J.-P., & Kornhauser, L. A. (1994). Social choice in a representative democracy. *American Political Science Review, 88*(1), 185–192.

Benoit, J.-P., & Kornhauser, L. A. (2006). *Only a dictatorship is efficient or neutral* (New York University Public Law and Legal Theory working papers).

Bezembinder, T., & Van Acker, P. (1985). The Ostrogorski paradox and its relation to non-transitive choice. *Journal of Mathematical Psychology, 11*, 131–158.

Bock, H. H., Day, W. H. E., & McMorris, F. R. (1998). Consensus rule for committee elections. *Mathematical Social Sciences, 35*(3), 219–232.

Brams, S. J., & Kilgour, D. M. (2001). Fallback bargaining. *Group Decision and Negotiation, 10*(4), 287–316.

Brams, S. J., Kilgour, D. M., & Zwicker, W. S. (1997). Voting on referenda: The separability problem and possible solutions. *Electoral Studies, 16*(3), 359–377.

Brams, S. J., Kilgour, D. M., & Zwicker, W. S. (1998). The paradox of multiple elections. *Social Choice and Welfare, 15*, 211–236.

reverse question, namely: do the candidate-wise rules inherit a given property from their sequential composition?

[32] Lang and Xia also extend the result of Benoit and Kornhauser (1994) to CP-net preferences: *SeqMaj* always selects the Condorcet winner if voters' preferences are conditionally lexicographic.

Brams, S. J., Kilgour, D. M., & Sanver, M. R. (2004). A minimax procedure for negotiating multilateral treaties. In M. Wiberg (Ed.), *Reasoned choices: Essays in honor of Hannu Nurmi*. Turku, Finland: Finnish Political Science Association.

Brams, S. J., Kilgour, D. M., & Sanver, M. R. (2007). A minimax procedure for electing committees. *Public Choice, 132*, 401–420.

Cuhadaroglu, T., & Lainé, J. (2009). *Pareto efficient voting with approval balloting* (Istanbul Bilgi University, working paper).

Deb, R., & Kelsey, D. (1987). On constructing a generalized Ostrogorski paradox: Necessary and sufficient conditions. *Mathematical Social Sciences, 14*, 161–174.

Farquharson, R. (1969). *The Theory of Voting*, Basil Blackwell.

Fishburn, P. C. (1981). An analysis of simple voting systems for electing committees. *SIAM Journal on Applied Mathematics, 41*(3), 499–502.

Gehrlein, W. V. (1985). The Condorcet criterion and committee selection. *Mathematical Social Sciences, 10*(3), 199–209.

Giritligil Kara, A. E., & Sertel, M. R. (2005). Does majoritarian approval matters in selecting a social choice rule? An explanatory panel study. *Social Choice and Welfare, 25*, 43–73.

Hodge, J. K., & Schwallier, P. (2006). How separability affects the desirability of referendum elections outcome. *Theory and Decision, 61*, 251–276.

Kadane, J. (1972). On division of the question. *Public Choice, 13*, 47–54.

Kaymak, B., & Sanver, M. R. (2003). Sets of alternatives as Condorcet winners. *Social Choice and Welfare, 20*(3), 477–494.

Kilgour, D. M., & Bradley, W. J. (1998). *Nonseparable preferences and simultaneous elections*. Boston: American Political Science Association.

Kilgour, D. M., Brams, S. J., & Sanver, M. R. (2006). How to elect a representative committee using approval balloting. In F. Pukelsheim & B. Simeone (Eds.), *Mathematics and democracy: Voting systems and collective choice* (pp. 83–95). Berlin: Springer.

Kramer, G. H. (1972). Sophisticated voting over multidimensional choice spaces. *Journal of Mathematical Sociology, 2*, 165–180.

Lacy, D., & Niou, E. M. S. (2000). A problem with referendums. *Journal of Theoretical Politics, 12*(1), 5–31.

Laffond, G., & Lainé, J. (2006). Single-switch preferences and the Ostrogorski paradox. *Mathematical Social Sciences, 52*(1), 49–66.

Laffond, G., & Lainé, J. (2009a). Condorcet choice and the Ostrogorski paradox. *Social Choice and Welfare, 32*, 317–333.

Laffond, G., & Lainé, J. (2009b). *Avoiding the Anscombe paradox with mutually close ballots* (Istanbul Bilgi University, working paper).

Laffond, G., & Lainé, J, (2009c). *Choosing a committee: Approval balloting and the search for a compromise* (Istanbul Bilgi University, working paper).

Laffond, G., & Lainé, J. (2009d). *Triple-consistent social choice and the majority rule* (Istanbul Bilgi University, working paper).

Lang, J., & Xia, L. (2007). Sequential voting rules and multiple elections paradoxes. In *Proceedings of the eleventh conference on theoretical aspects of rationality and knowledge*, pp. 279–288.

Lang, J., & Xia, L. (2009). Sequential composition of voting rules in multi-issue domains. *Mathematical Social Sciences, 57*, 304–324.

Laslier, J. F. (1997). *Tournament solutions and majority voting*. Berlin: Springer.

Nurmi, H. (1998). Voting paradoxes and referenda. *Social Choice and Welfare, 15*, 333–350.

Nurmi, H. (1999). *Voting paradoxes and how to deal with them*. Heidelberg: Springer.

Miller, N. R. (1977). Graph-theoretical approaches to the theory of voting. *American Journal of Political Science, 24*, 769–803.

Özkal-Sanver, I., & Sanver, M. R. (2006). Ensuring Pareto optimality by referendum voting. *Social Choice and Welfare, 27*(1), 211–219.

Peris, J. E., & Subiza, B. (1999). Condorcet choice correspondences for weak tournaments. *Social Choice and Welfare, 16*, 217–231.

Rae, D., & Daudt, H. (1976). The Ostrogorski paradox: A peculiarity of compound majority decisions. *European Journal of Political Research, 4*, 391–398.

Ratliff, T. C. (2003). Some startling inconsistencies when electing committees. *Social Choice and Welfare, 21*(3), 433–454.

Ratliff, T. C. (2006). Selecting committees. *Public Choice, 126*, 343–355.

Saari, D. G., & Sieberg, K. (2001). The sum of the parts can violate the whole. *American Political Science Review, 95*(2), 415–433.

Scarsini, M. (1998). A strong paradox of multiple elections. *Social Choice and Welfare, 15*, 237–238.

Schwartz, T. (1972). Rationality and the myth of the maximum. *Noûs, 6*, 97–117.

Schwartz, T. (1977). Collective choice, separation of issues and vote trading. *American Political Science Review, 71*, 999–1010.

Sertel, M. R. (1987). A non-dictatorial compromise. *Social Choice and Welfare, 4*(1), 1–11.

Sertel, M. R., & Yılmaz, B. (1998). The majoritarian compromise is majoritarian optimal and subgame-perfect implementable. *Social Choice and Welfare, 16*, 615–627.

Wagner, C. (1983). The Anscombe paradox and the rule of three-fourth. *Theory and Decision, 15*, 303–308.

Wagner, C. (1984). Avoiding the Anscombe paradox. *Theory and Decision, 16*, 233–238.

Part IV
Strategic Voting

Chapter 8
The Basic Approval Voting Game

Jean-François Laslier and M. Remzi Sanver

8.1 Introduction

There is a vast literature which conceives Approval Voting as a mechanism where the approval of voters is a mere strategic action with no intrinsic meaning. As usual, a group of voters who have preferences over a set candidates is considered. Every voter announces the list of candidates which he approves of and the winners are the candidates which receive the highest number of approvals. Assuming that voters take simultaneous and strategic actions, we are confronted to a normal form game whose analysis dates back to Brams and Fishburn (1983). This chapter surveys the main results of this literature.

The problem with this approach is that the main conceptual tool of game theory – Nash equilibrium – is of little help for understanding Approval Voting and most voting rules. By definition, an equilibrium is a vote profile in which no voter can, by changing her vote only, change the outcome of the game in such a way that the new outcome is strictly better for her. In a world where voters are only interested in who wins the election (instrumental and consequentialist voting, opposed to expressive voting), the outcome of the game is just the identity of the elected candidate, or candidates in case of a tie. Then it is almost always the case that no voter can, by changing her vote only, change the outcome of the game. With Approval Voting, as well as with most voting rules, this will happen as soon as one candidate is winning the election with a margin of more than two votes. Therefore, apart cases where several candidates tie or almost tie, almost everything is a Nash equilibrium. In particular, except in some very degenerated cases, any candidate is winning at some Nash equilibrium.

The game-theoretical literature on voting, and in particular on Approval Voting, has therefore focused on the possibility of using more powerful tools than Nash equilibrium in order either to predict the outcome of a voting game or at least to

---

J.-F. Laslier (✉)
Laboratoire d'Econométrie, École Polytechnique, 91128 Palaiseau, France
e-mail: jean-francois.laslier@polytechnique.edu

J.-F. Laslier and M.R. Sanver (eds.), *Handbook on Approval Voting*, Studies in Choice and Welfare, DOI 10.1007/978-3-642-02839-7_8,

narrow down the set of possible outcomes. To this aim, several routes have been followed.

The first route is to restrict the set of voting strategies that a voter is supposed to possibly use. The natural idea, from the game theoretic perspective, is to suppose that voters do not use dominated strategies. Although this idea reveals very powerful in solving sequential voting games (Farquharson 1969; Moulin 1979; Moulin 1983; Banks 1985; Bag et al. 2009) this is not the case for simultaneous voting games defined by the usual voting rules (Dhillon and Lockwood 2004; Buenrostro and Dhillon 2003; Dellis and Oak Dellis and Oak). For Approval Voting, undominated strategies are often called "admissible strategies" and can be characterized: If the voter's preference is strict, she approves her preferred candidate, she does not approve her worse candidate, and no constraint is imposed as to the other, intermediate candidates (for a precise statement, see Proposition 8.3.1). For Approval Voting, another meaningful restriction on the set of strategies is the sincerity requirement, which imposes that when the voter approves a candidate, she also approves all the candidates she strictly to prefers to this one. Brams and Sanver (2006) have described the set of possible outcomes when voters use only undominated ("admissible") and sincere strategies. It turns out that, except in some degenerated situations, all candidates pass this test (see Sect. 8.4.1).

The second route is to come back to a notion of equilibrium and to refine the notion of Nash equilibrium according to the usual concepts of game theory. (See Myerson 1991 or Van Damme 1991 for the general theory and De Sinopoli 2000 for an application to plurality voting.) In comparison with the previous approach, this amounts to give up the idea that the voter's behavior can be restricted a priori and to instead consider that each voter is reacting to what she believes are the voting intentions of the other voters. Remark that among the plethora of Nash equilibria of the voting games, most of them are degenerated from the strategic point of view in the sense that no player has any incentive *not* to deviate. In fact, unless she is "pivotal," the voter's choice has indeed no consequence at all on the outcome. This is a clear case for the refinement of equilibrium. One could hope that statements of the kind "A Condorcet loser cannot be elected at equilibrium under Approval Voting" or "Voters vote sincerely at equilibrium under Approval Voting" could be demonstrated when the notion of equilibrium is properly defined. This hope is justified if one allows not only for individual deviations but also for group deviations – hence considers strong equilibrium as the game-theoretic solution concept. (See Proposition 8.4.2 about Condorcet-consistency.) But the notion of strong equilibrium has a major drawback as a predictive tool since, in many cases, there is no such equilibrium. On the other hand, for different refinements of Nash equilibrium that yield non-empty predictions in finite normal-form games, De Sinopoli et al. (2006) have provided counter-examples (reproduced in Sect. 8.4.2) that kill the hope to make these statements true for any of the classical refinements of Nash equilibrium through concepts such as "perfection," "properness" or "stability."

The third route is to refine the concept of equilibrium following non-standard ideas that would be specific to the voting context. In politics, voting situations often involves large number of players, a fact that raises new difficulties but also new possibilities. This avenue, pioneered by Myerson and Weber (1993) and Myerson (2002) is the object of the survey of Nunez (2010) in this book and is out of the scope of the present chapter.

Section 8.2 presents the basic notation and concepts. Section 8.3 deals with undominated and sincere individual strategies. Section 8.4 deals with the aggregate outcome of the vote. Section 8.5 concludes.

8.2 The Normal Form Game

We denote by $\mathbf{I}$ the finite set *voters* (sometime called *individuals* or *players*) and by $\mathbf{X}$ the finite set of *candidates* (sometimes called *alternatives*). We assume $\#\mathbf{I} \geq 2$ and $\#\mathbf{X} \geq 2$. Every voter i has a *preference* over $\mathbf{X}$, expressed by a utility function $u_i : \mathbf{X} \longrightarrow \mathbb{R}$. So given two candidates $x, y \in \mathbf{X}$, voter i finds x at least as good as y iff $u_i(x) \geq u_i(y)$. A candidate x is *high* in u_i iff $u_i(x) \geq u_i(y)$ for all $y \in \mathbf{X}$. We say that x is *low* in u_i iff $u_i(y) \geq u_i(x)$ for all $y \in \mathbf{X}$. We call u_i *null* whenever i is indifferent among all alternatives, i.e., $u_i(x) \geq u_i(y)$ for every $x, y \in \mathbf{X}$. If u_i is null then every candidate is both low and high in u_i. If u_i is not null then the candidates which are high in u_i and those which are low in u_i form disjoint sets.

A *ballot* is any subset of the set of candidates; we denote by $2^{\mathbf{X}}$ the set of ballots. When voter i casts ballot B_i, we say that i *approves* the candidates in B_i. We let $B = (B_i)_{i \in I} \in \left(2^{\mathbf{X}}\right)^{\mathbf{I}}$ stand for a *ballot profile* and write $B = (B_i, B_{-i})$ with $B_{-i} = \left(B_j\right)_{j \in \mathbf{I} \setminus \{i\}}$, whenever we wish to highlight the dependency of B with respect to i's ballot. We refer to B_{-i} as *a ballot profile without i*.

Given a ballot profile B, the *score* of candidate x is

$$s(x; B) = \#\{i \in \mathbf{I} : x \in B_i\}$$

and the (non-empty) set of *winning candidates* (under Approval Voting) is

$$W(B) = \{x \in \mathbf{X} : s(x; B) \geq s(y; B) \forall y \in \mathbf{X}\}.$$

Similarly, we write $s(x; B_{-i}) = \#\{j \in \mathbf{I} \setminus \{i\} : x \in B_j\}$.

We suppose that voters vote simultaneously by casting a ballot which is some set of candidates while Approval Voting is used as the outcome function. So we consider a normal form game where the strategy set for any voter i is the set $2^{\mathbf{X}}$ of possible ballots. Hence a ballot profile B is also a strategy profile and the outcome is the set of winning candidates $W(B)$.

As $W(B)$ may contain more than one candidate, our strategic analysis requires the knowledge of voters' preferences over non-empty subsets of $\mathbf{X}$. We assume that ties over outcomes are broken by fair lotteries and that voters evaluate outcomes by expected Von-Neumann Morgenstern utilities. So the utility that voter i attaches to a set S of winning candidates is

$$u_i(S) = \frac{1}{\#S} \sum_{x \in S} u_i(x).$$

Note that we abuse notation and allow u_i to have arguments which are both elements and non-empty subsets of $\mathbf{X}$.

8.3 Admissibility and Sincerity

8.3.1 Admissible Strategies

Following the game-theoretical vocabulary, for any voter i with preference u_i, we say that the ballot B_i (*weakly*) *dominates* the ballot B_i' if and only if $u_i(W(B_i, B_{-i})) \geq u_i\ (W(B_i', B_{-i}))$ for all B_{-i} and $u_i(W(B_i, B_{-i})) > u_i\ (W(B_i', B_{-i}))$ for some B_{-i}. A ballot is *undominated* if and only if it is dominated by no ballot. Following Brams and Fishburn (1983), we qualify undominated ballots as *admissible* and use either words. The following proposition characterizes admissible ballots.

Proposition 8.3.1.

(*i*) *If u_i is null then all ballots are admissible for voter i.*

(*ii*) *Let the number of voters be at least three. If u_i is not null then the ballot B_i is admissible for voter i if and only if B_i contains every candidate who is high in u_i and no candidate who is low in u_i.*

Proof. (i) directly follows from the definitions. To show the "only if" part of (ii), consider a ballot B_i which fails to contain a candidate y who is high in u_i. Let $B_i' = B_i \cup \{y\}$. We will prove that B_i' dominates B_i.

Given any B_{-i}, all candidates except y have the same score at $(B_i; B_{-i})$ and $(B_i'; B_{-i})$ while the score of y is raised by one unit at the latter ballot profile. Therefore, regarding the sets of winning candidates $Y = W(B_i; B_{-i})$ and $Y' = W(B_i'; B_{-i})$, the following three cases are exhaustive:

1. $y \notin Y$ and $Y' = Y$.
2. $y \notin Y$ and $Y' = Y \cup \{y\}$.
3. $y \in Y$ and $Y' = \{y\}$.

In all three cases, $u_i(Y') \geq u_i(Y)$. Now fix some $k \in \mathbf{I}\backslash\{i\}$ and consider B_{-i} where $B_j = \emptyset$ for all $j \in \mathbf{I}\backslash\{i, k\}$ and $B_k = \{z\}$ for some candidate z who is not high in u_i. If $z \notin B_i$ then $W(B_i; B_{-i}) = B_i \cup \{z\}$ and $W(B_i'; B_{-i}) = B_i' \cup \{z\} = B_i \cup \{y, z\}$, hence $u_i(W(B_i'; B_{-i})) > u_i(W(B_i; B_{-i}))$. If $z \in B_i$, then $W(B_i; B_{-i}) = \{z\}$, $W(B_i'; B_{-i}) = \{y, z\}$ and we have $u_i(W(B_i'; B_{-i})) > u_i(W(B_i; B_{-i}))$. This proves that B_i' dominates B_i, and we conclude that an undominated ballot must contain all candidates who are high in u_i. Similar arguments show that an undominated ballot cannot contain a candidate who is low in u_i.

We now show the "if" part of (ii). Consider a ballot B_i that contains every candidate high in u_i and no candidate low in u_i. In order to show that B_i is undominated, we consider any distinct ballot B_i' and establish the existence of some B_{-i} where $u_i(W(B_i; B_{-i})) > u_i(W(B_i'; B_{-i}))$.

First let B_i' contain a candidate y low in u_i. Let B_{-i} be such that $B_j = \{y\}$ for some voter $j \in \mathbf{I}\backslash\{i\}$ and $B_k = \emptyset$ for every voter $k \in \mathbf{I}\backslash\{i, j\}$. So $W(B_i; B_{-i}) = B_i \cup \{y\}$, $W(B_i'; B_{-i}) = \{y\}$ and $u_i(W(B_i; B_{-i})) > u_i(W(B_i'; B_{-i}))$.

Now let B_i' fail to contain all candidates high in u_i. So the set Y of candidates in $B_i \backslash B_i'$ who are high in u_i is non-empty. Let L be the set of candidates who are low in u_i. Let B_{-i} be such that $B_j = Y \cup L$ for some voter $j \in \mathbf{I}\backslash\{i\}$ and $B_k = \emptyset$ for every voter $k \in \mathbf{I}\backslash\{i, j\}$. So $W(B_i; B_{-i}) = Y$ and at $(B_i'; B_{-i})$, the score of every candidate who is high in u_i is at most one and the score of some candidates who are low in u_i is one. Thus, $W(B_i'; B_{-i})$ contains candidates who are not high in u_i. Hence $u_i(W(B_i; B_{-i})) > u_i(W(B_i'; B_{-i}))$.

Finally let B_i' contain every candidate high in u_i and no candidate low in u_i. First consider the case where there exists a candidate y in B_i not in B_i'. Let B_{-i} be such that for two (distinct) voters $j, k \in \mathbf{I}\backslash\{i\}$ we have $B_j = B_k = \{y, z\}$ where z is low in u_i and $B_l = \emptyset$ for every voter $l \in \mathbf{I}\backslash\{i, j, k\}$. So $W(B_i; B_{-i}) = \{y\}$, $W(B_i'; B_{-i}) = \{y, z\}$ and $u_i(W(B_i; B_{-i})) > u_i(W(B_i'; B_{-i}))$. Now consider the case where B_i is a proper subset of B_i'. Take some $y \in B_i' \setminus B_i$. Note that y is not high in u_i. Take some candidate z high in u_i and let B_{-i} be such that for two (distinct) voters $j, k \in \mathbf{I}\backslash\{i\}$ we have $B_j = B_k = \{y, z\}$ and $B_l = \emptyset$ for every voter $l \in \mathbf{I}\backslash\{i, j, k\}$. So $W(B_i; B_{-i}) = \{y, z\}$, $W(B_i'; B_{-i}) = \{y\}$ and $u_i(W(B_i; B_{-i})) > u_i(W(B_i'; B_{-i}))$.

8.3.2 Sincerity

Following Brams and Fishburn (1983), a strategy (or ballot) B_i of voter i with preference P_i is said to be *sincere* iff for all candidates x, $y \in \mathbf{X}$,

$$y \in B_i \text{ and } u_i(x) > u_i(y) \Rightarrow x \in B_i.$$

So under a sincere strategy B_i, if i approves of a candidate y then she also approves of any candidate x which she strictly prefers to y. With K candidates, if voter i is never indifferent between two distinct candidates, she has at her disposal $K + 1$ sincere strategies, including the *full ballot* $B_i = \mathbf{X}$ which consists of approving of all candidates, and the *empty ballot* $B_i = \emptyset$ which consists of approving of none.

Proposition 8.3.1 does not make any statement about candidates who are neither high nor low. In fact, for a voter i with preference u_i, every non-sincere ballot that contains every candidate high in u_i and no candidate which is low in u_i. is an undominated strategy for i. So admissible ballots need not be sincere, nor sincere ballots have to be admissible.[1] On the other hand, sincere and non-sincere ballots can be discriminated through the fact that every ballot profile B_{-i} without i admits at least one sincere ballot B_i as a best-response of i. In other words, the set of best responses of i to B_{-i} cannot consist of insincere ballots only.

Proposition 8.3.2. *Given any voter i with preference u_i and any ballot profile B_{-i} without i, there exists a sincere ballot $B_i \in 2^{\mathbf{X}}$ such that $u_i(W(B_i; B_{-i})) \geq u_i(W(B_i'; B_{-i}))$ for every ballot $B_i' \in 2^{\mathbf{X}}$.*

Proof. Take any voter i with preference u_i and any ballot profile B_{-i} without i. Let Y be the (non-empty) set of candidates who receive the highest number of approvals at B_{-i}. Let Z be the (possibly empty) set of candidates who receive at B_{-i} precisely one approval less than the highest number of approvals. The outcome set $W(B_i, B_{-i})$ when B_i vary can take two forms: if $B_i \cap Y \neq \emptyset$, then $W(B_i, B_{-i}) = B_i \cap Y$, and if $B_i \cap Y = \emptyset$, then $W(B_i, B_{-i}) = Y \cup Z'$, for $Z' = B_i \cap Z$. Denote by u_i^* the maximum utility obtained by i. Then $u_i^* \geq \max_{y \in Y} u_i(y)$, and $u_i^* \geq \max_{Z' \subseteq Z} u_i(Y \cup Z')$, with one of these two inequalities being an equality. Let $y^* \in Y$ be such that $u_i(y^*) = \max_{y \in Y} u_i(y)$. Let $B_i^1 = \{x \in X : u_i(x) \geq u_i(y^*)\}$. This is a sincere ballot, so if B_i^1 is a best response, we are done.

Notice that B_i^1 brings at least the level of utility $u_i(y^*)$; so if B_i^1 is a not best response, it must be the case that $u_i(y^*) < u_i^*$ and that $u_i^* = \max_{Z' \subseteq Z} u_i(Y \cup Z')$. In that case, let $Z^* = \{z \in Z : u_i(z) \geq u_i(Y)\}$. Recall that the utility for a subset is the average utility of its elements; as one can easily check, it follows that $u_i(Y \cup Z^*) = \max_{Z' \subseteq Z} u_i(Y \cup Z')$. Let $B_i^2 = \{x \in X : u_i(x) \geq u_i(Y \cup Z^*)\}$ This is again a sincere ballot. Moreover, in that case, $B_i^2 \cap Y = \emptyset$ so that the ballot B_i^2 brings the utility $u_i(Y \cup (B_i^2 \cap Z))$. Here, $B_i^2 \cap Z = \{z \in Z : u_i(z) \geq u_i(Y \cup Z^*)\}$ and $u_i(z) \geq u_i(Y \cup Z^*)$ if and only if $u_i(z) \geq u_i(Y)$, so that $B_i^2 \cap Z = Z^*$, and B_i^2 brings the maximal utility u_i^*. We again found a sincere best response.

Proposition 8.3.1 slightly differs from the existing results of the literature regarding the way preferences over sets are handled. In fact, it makes the same statement as Corollary 2.1 in Brams and Fishburn (2007) which is shown under more general assumptions for extending preferences over sets. On the other hand, the result announced by Proposition 8.3.2 has no analogous in Brams and Fishburn (1983, 2007), as it fails to hold under these more general assumptions.[2]

---

[1] Nevertheless, if there are precisely three candidates, then every admissible ballot is sincere.

[2] To see this, let voter i have the preference $u_i(x_1) > u_i(x_2) > u_i(x_3) > u_i(x_4) > u_i(x)$ $\forall x \in \mathbf{X} \backslash \{x_1, x_2, x_3, x_4\}$ and let B_{-i} be such that $s(x_2; B_{-i}) = s(x_4; B_{-i}) > s(x_1; B_{-i}) = s(x_3; B_{-i}) > s(x; B_{-i}) \forall x \in \mathbf{X} \backslash \{x_1, x_2, x_3, x_4\}$ while $s(x_2; B_{-i}) - s(x_1; B_{-i}) = 1$. The ballot $B_i = \{x_1, x_3\}$ which yields $\{x_1, x_2, x_3, x_4\}$ can be a best-response under the Brams and Fishburn (1983, 2007) assumptions while there is no sincere ballot for voter i which yields the same outcome. Endriss (2009) identifies the assumptions on preferences over sets which rule out incentives to vote insincerely.

Proposition 8.3.2 has no analogous for insincere ballots. In other words, the best response of i to B_{-i} can consist of sincere ballots only.[3] As a result, one may be tempted to assume – as we do in Sect. 8.4.1 – that voters restrict their strategies to those which are admissible and insincere. On the other hand, in Sect. 8.4.2, we see that such an assumption is not totally innocuous.

8.4 Approval Voting Outcomes

8.4.1 Admissible and Sincere Outcomes

Brams and Sanver (2006) study the set of candidates which are chosen under Approval Voting at a given preference profile, assuming that voters use admissible and sincere strategies. For a formal expression of their findings, let $u = (u_i)_{i \in \mathbf{I}}$ be a preference profile. Write

$$\alpha(u) = \left\{B \in \left(2^{\mathbf{X}}\right)^{\mathbf{I}} : \forall\, i \in \mathbf{I},\ B_i \text{ is admissible and sincere with respect to } u_i\right\}.$$

We define

$$AV(u) = \{x \in X : x \in W(B) \text{ for some } B \in \alpha(u)\}$$

as the set of (admissible and sincere) Approval Voting outcomes at u. So candidate x is an Approval Voting outcome at u if and only if there exists a profile of sincere and admissible strategies B where x is a (possibly tied) winning candidate under Approval Voting.

Note that a voter who strictly ranks K candidates has exactly $K - 1$ admissible and sincere strategies which consist of approving her first $k \in \{1, \ldots, K-1\}$ best candidates. This is a drastic reduction of a voter's strategy space which originally contained 2^K strategies. Nevertheless, this does not restrict much the size of $AV(u)$ which Brams and Sanver (2006) characterize, assuming that voters are never indifferent between any two candidates, i.e., $u_i(x) \neq u_i(y)\ \forall i \in \mathbf{I}, \forall x, y \in \mathbf{X}$.

Proposition 8.4.1. *Given a preference profile u with no indifferences, a candidate x is not in $AV(u)$ if and only if there exists a candidate $y \in \mathbf{X} \backslash \{x\}$ such that according to u, the number of voters who rank y as the best and x as the worst candidate exceeds the number of voters who prefer x to y.*

Based on Proposition 8.4.1, $AV(u)$ may contain Pareto dominated alternatives[4] as well as Condorcet losers. Moreover, at every preference profile u, a Condorcet

---

[3] Consider four voters and four candidates where each of voters 2, 3 and 4 approve of precisely one candidate; say x, y and z respectively. Let the fourth candidate w be ranked last in the preference of voter 1 whose unique admissible best response is to approve of the candidate he prefers the most.

[4] In the environment we consider, if a Pareto dominates b and $b \in AV(u)$, then $a \in AV(u)$ as well.

winner (whenever it exists); all scoring rule outcomes; the Majoritarian Compromise winner; the Single Transferable Vote winner are always in $AV(u)$. We refer the reader to Brams and Sanver (2006) for a more detailed and formal expression of these results. Nevertheless, we can right away conclude that, in our game theoretic framework, assuming that voters restrict their strategies to those which are admissible and sincere does not suffice to have a fine prediction of the election result under Approval Voting.

8.4.2 Equilibrium Outcomes

The model can be more predictive, when admissible and sincere strategy profiles are required to pass certain stability tests. A profile of sincere and admissible strategies B is *strongly stable* at preference profile u iff given any other profile of admissible and sincere strategies B', there exists a voter i with $B_i \neq B'_i$ while $u_i(W(B)) \geq u_i(W(B'))$. So B is strongly stable at u iff there exists no coalition of voters whose members can all be better-off by switching their strategies to another admissible and sincere one (which may differ among the members of the coalition). Let $AV^*(u) = \{x \in X : x \in W(B)$ for some $B \in \alpha(u)$ which is strongly stable$\}$ be the set of strongly stable AV outcomes at u. Clearly, $AV^*(u)$ refines $AV(u)$ and the reduction is indeed dramatic:

Proposition 8.4.2. *Given a preference profile u, a candidate x is strongly stable at u if and only if x is a weak Condorcet winner at u.*

Note that the definition of a Condorcet winner is a weak one: x is a weak Condorcet winner at u iff given any other candidate y, the number of voters who prefer x to y is at least as much as the number of voters who prefer y to x. So, in some cases, u may admit more than one weak Condorcet winner. Of course, u may admit no weak Condorcet winner, hence no strongly stable profile of admissible and sincere strategies. This last observation is not surprising, as strong stability – which corresponds to strong Nash equilibrium – is a rather demanding condition. The interested reader can see Sertel and Sanver (2004) for a more general treatment of strong equilibrium outcomes of voting games.

The complete proof of Proposition 8.4.2 can be found in Brams and Sanver (2006). However, we wish to give a simple and instructive description of the proof. If an outcome x is not a weak Condorcet winner, it means that there exists another outcome y which is preferred to x by some majoritarian coalition of voters which can block any strategy profile which yields x as the Approval Voting outcome. If x is a weak Condorcet winner, then no coalition can block the strategy profile where voters for whom x is not low approve x but do not approve anything below x and voters for whom x is low approve only their high candidate.[5]

[5] We take the occasion to claim that Proposition 8.4.2 remains valid when strong stability is further strengthened so as to allow non-admissible and non-sincere strategies.

We now present two results from De Sinopoli et al. (2006) which advise caution in interpreting Propositions 8.4.2 and 8.3.2:

1. There may exist non-trivial equilibria where a Condorcet winner obtains no vote.
2. There may exist non-trivial equilibria with some voters voting non-sincerely.

Example 8.4.1 (Condorcet in-consistency). There are four candidates, $\mathbf{X} = \{a, b, c, d\}$ and three voters $\{1, 2, 3\}$ with utility:

$$u_1(a) = 10,\ u_1(b) = 0,\ u_1(c) = 1,\ u_1(d) = 3,$$
$$u_2(a) = 0,\ u_2(b) = 10,\ u_2(c) = 1,\ u_2(d) = 3,$$
$$u_3(a) = 1,\ u_3(b) = 0,\ u_3(c) = 10,\ u_3(d) = 3.$$

Candidate d is the Condorcet winner of this utility profile. Consider the following strategy profile:

- Voter 1 votes $\{a\}$.
- Voter 2 votes $\{b\}$.
- Voter 3 votes $\{c\}$.

In such a situation there is a tie among the candidates a, b, and c, so that the payoff to each player is $11/3$. Starting from this situation each player is playing a unique best response: any other choice would lead to a strictly lower payoff. In this strict equilibrium, the Condorcet winner receives no vote.

The question of sincerity is raised by considering the possibility that players use mixed strategies. A mixed strategy is a probability distribution over the set of pure strategies. Here the set of mixed strategies is thus the simplex $\Delta(2^X)$ with 2^K vertices, that is an affine space of dimension $2^K - 1$. We denote by

$$\sigma_i \in \Delta(2^X)$$

a mixed strategy of voter i and by σ_{-i} a profile of mixed strategies for the other voters. Payoffs are defined in the usual ways as expected values. For a mixed strategy profile σ, $\sigma(B)$ is the probability of the pure-strategy profile B under σ. Players are supposed to randomize independently the ones from the others so that;

$$\sigma(B) = \prod_{i \in \mathbf{I}} \sigma_i(B_i)$$

and

$$u_i(\sigma) = \sum_{B \in (2^X)^{\mathbf{I}}} u_i(B)\sigma(B) = \sum_{B \in (2^X)^{\mathbf{I}}} \frac{1}{\#W(B)} \sum_{x \in W(B)} u_i(x)\sigma(B).$$

Example 8.4.2 (A non-sincere equilibrium). There are four candidates, $\mathbf{X} = \{a, b, c, d\}$ and three voters $\{1, 2, 3\}$ with utility:

$$u_1(a) = 1000,\ u_1(b) = 867,\ u_1(c) = 866,\ u_1(d) = 0,$$
$$u_2(a) = 115,\ u_2(b) = 1000,\ u_2(c) = 0,\ u_2(d) = 35,$$
$$u_3(a) = 0,\ u_3(b) = 35,\ u_3(c) = 115,\ u_3(d) = 1000.$$

Candidate d is the Condorcet winner of this utility profile. Consider the following strategy profile:

- Voter 1 votes $\{a, c\}$.
- Voter 2 votes $\{b\}$ with probability 1/4 and $\{a, b\}$ with probability 3/4.
- Voter 3 votes $\{d\}$ with probability 1/4 and $\{c, d\}$ with probability 3/4.

Note that voter 1 is not voting sincerely. Nevertheless, this strategy profile is an equilibrium and De Sinopoli et al. (2006) show that it forms a singleton-stable set, an important refinement of Nash equilibrium.

8.5 Conclusion

The analysis above raises three issues:

- An a priori restriction of voters' strategies based on a reasonable intuition such as undominated and sincere voting is not sufficient to restrict the set of possible outcomes of an Approval Voting election.
- Many refinements of Nash equilibrium, when applied to Approval Voting games, ensure the existence of equilibrium but the outcome of these equilibria do not seem to behave particularly well with respect to social choice requirements.
- Strong Nash equilibrium predicts Condorcet winners as the only Approval Voting outcomes but equilibrium fails to exist when there is no Condorcet winner.

These essentially negative theoretical results call for developing a finer understanding of how a voter chooses a ballot under Approval Voting. This analysis could rely on some general, game-theoretic principles such as the ones just described, but should probably also embody some elements specific to real voting situations such as the large size of the electorate, the specific structures of Approval Voting strategies, or the specificities of political information.

References

Bag, P. K., Sabourian, H., & Winter, E. (2009). Multi-stage voting, sequential elimination and Condorcet consistency. *Journal of Economic Theory, 144*(3), 1278–1299.

Banks, J. S. (1985). Sophisticated voting outcomes and agenda control. *Social Choice and Welfare, 2*, 295–306.

Brams, S. J., & Fishburn, P. C. (1983). *Approval Voting*. Boston: Birkhäuser.

Brams, S. J., & Fishburn, P. C. (2007). *Approval Voting* (2nd ed.). Berlin: Springer.

Brams, S. J., & Sanver, M. R. (2006). Critical strategies under Approval Voting: Who gets ruled in and ruled out. *Electoral Studies, 25*(2), 287–305.

Buenrostro, L., & Dhillon, A. (2003). *Scoring rule voting games and dominance solvability* (Warwick Economic Research Papers No. 698).

De Sinopoli, F. (2000). Sophisticated voting and equilibrium refinements under Plurality Rule. *Social Choice and Welfare, 17*, 655–672.

De Sinopoli, F., Dutta, B., & Laslier, J. F. (2006). Approval Voting: Three examples. *International Journal of Game Theory, 35*, 27–38.

Dellis, A., & Oak, M. (2007). Policy convergence under Approval and Plurality Voting: The role of policy commitment. *Social Choice and Welfare, 29*, 229–245.

Dhillon, A., & Lockwood, B. (2004). When are Plurality Rule voting games dominance-solvable? *Games and Economic Behavior, 46*, 55–75.

Endriss, U. (2009). Sincerity and manipulation under Approval Voting. working paper, ILLC, University of Amsterdam.

Farquharson, R. (1969). *Theory of voting*. New Haven: Yale University Press.

Moulin, H. (1979). Dominance-solvable voting schemes. *Econometrica, 47*, 1337–1351.

Moulin, H. (1983). *The strategy of social choice*. Amsterdam: North-Holland.

Myerson, R. B. (1991). *Game theory: Analysis of conflict*. Cambridge: Harvard University Press.

Myerson, R. B. (2002). Comparison of scoring rules in Poisson voting games. *Journal of Economic Theory, 103*, 219–251.

Myerson, R. B., & Weber, R. J. (1993). A theory of voting equilibria. *American Political Science Review, 87*, 102–114.

Nunez, M. (2010). Approval Voting in large electorates. In J. F. Laslier & M. R. Sanver (Eds.), *Handbook on Approval Voting*. Heildelberg: Springer-Verlag.

Sertel, M. R., & Sanver, M. R. (2004). Strong equilibrium outcomes of voting games are the generalized Condorcet winners. *Social Choice and Welfare, 22*, 331–347.

Van Damme, E. (1991). *Stability and perfection of Nash equilibria* (2nd ed.). Heidelberg: Springer.

Chapter 9
Approval Voting in Large Electorates

Matías Núñez

9.1 Introduction

The strategic analysis of voting rules has given some insight into the understanding of their properties. However, one can assert that these analyses are "too rich" in the sense that they show that a plethora of equilibria can arise under most voting rules. In particular, there is a controversy over Approval voting or AV, a voting rule which has been called "the electoral reform of the twentieth century." This voting rule allows the voter to vote for as many candidates as he wishes and the candidate who gets the most votes wins the election. Its detractors claim that this kind of method enhances strategic voting when compared for instance to Plurality voting (henceforth PV), whereas its proponents consider that it has several advantages as far as strategic voting is concerned. For an extensive discussion on this controversy over AV, the reader can refer to Brams (2008) and Weber (1995).

One important feature of AV was characterized by Brams and Fishburn (1981). They show that if a Condorcet Winner exists then the AV game has a Nash equilibrium in undominated strategies that selects the Condorcet Winner. The Condorcet Winner – the candidate who beats all other candidates on pairwise contests – has often been considered to be a good equilibrium solution in voting games. The robustness of the previous result has been weakened by De Sinopoli et al. (2006). To do so, they apply Nash equilibrium refinements such as the perfect equilibrium solution to Approval games. Using these techniques, they prove that there may exist equilibria in which the Condorcet Loser and Condorcet Winner are selected with the same probability or even in which the Condorcet Winner gets no vote at all. Therefore, AV does not guarantee what is called Condorcet consistency: the Winner of the election does not always coincide with the Condorcet Winner.

M. Núñez
CNRS, THEMA, Université Cergy-Pontoise, 95011 Cergy-Pontoise, France
e-mail: matias.nunez@polytechnique.edu

J.-F. Laslier and M.R. Sanver (eds.), *Handbook on Approval Voting*, Studies in Choice and Welfare, DOI 10.1007/978-3-642-02839-7_9,

However, the previous works were performed in a basic game theoretical framework.[1] Such a framework faces some criticisms when dealing with elections with a large number of voters. Indeed, it is no longer realistic to assume that voters have no uncertainty over the scores of the candidates.[2] The existence of candidates with almost no chance of winning the election might affect voters' behavior as a voter might not vote for such a candidate. The introduction of *commonly shared prior beliefs* over the outcome of the election is the main objective of models with large electorates. To our knowledge, there exists three main models dealing with elections with a large number of strategic voters: the Myerson–Weber framework (Myerson and Weber 1993), the Score Uncertainty model (Laslier 2009) and the Population Uncertainty model (Myerson 1998, 2000, 2002).

The so-called Myerson–Weber framework (Myerson and Weber 1993) introduces the idea that in a voting equilibrium, voters behave in accordance to their preferences and with respect to their anticipations regarding the relative chances of different pairs of candidates of being in contention for victory. The Myerson–Weber framework skips the main technical difficulties and introduces in an exogenous manner the pivot probabilities, i.e., of changing the winner of the election from one candidate to another. To keep things simple, it is assumed that these pivot probabilities will be common knowledge for voters in the election and that they respect some ordering condition (in some sense, candidates' expected scores and pivot probabilities will be correlated in an intuitive way). The authors draw a positive conclusion over the properties of AV when compared with PV and the Borda Count.

The remaining models (Score and Population Uncertainty model) set up formal game-theory models in which the pivot probabilities are neither exogenously introduced nor assumed to be common knowledge for all voters.

Laslier's (2009) Score Uncertainty model is performed in a standard game theoretical framework where uncertainty is introduced by assuming that there is some small but strictly positive probability that each vote is erased. Under this approach, Laslier (2009) shows that AV leads to equilibria with desirable properties such as Condorcet Consistency and sincerity of voters' best responses. These positive results are a consequence of the properties of pivot probabilities in such a setting. In the Score Uncertainty model, pivot probabilities are ordered in such a manner that voters' unique best responses satisfy a simple rule. If we let a denote the candidate who is considered to be the most likely winner, a voter will approve of any candidate he prefers to candidate a. Besides, he will never approve of a candidate he prefers

[1] See the chapter on this book (Laslier and Sanver 2010a) that presents a detailed account of the main results concerning strategic approval voting in the classic framework.

[2] Whereas in a Nash equilibrium, voters perfectly know candidates' scores, this is not the case in a trembling-hand perfect equilibrium. Indeed, such an equilibrium concept is the limit of a sequence of completely mixed strategies equilibria, in which a mixed strategy represents voters' mistakes (voters have a trembling hand). Within each of these mixed strategy equilibria, voters are uncertain about candidates' scores. However, as will be shown the advantage of the models with large electorates, is that all voters share the same prior probability distribution over their probability of affecting the outcome of the election. Hence, these models provide a simpler way of computing voters' best responses.

candidate a to. Finally, to decide whether to vote for candidate a, the voter compares a to the second most likely winner. This simple rule will not be satisfied in the last model addressed within this work, the Population Uncertainty model. As far as the information is concerned, pivot probabilities will not be equally shared by voters in the Score Uncertainty model as assumed in the Myerson–Weber framework. However, as the electorate becomes large the differences between the pivot probabilities are greatly reduced so that voters' best responses are not affected.

The third model with a large number of strategic voters is Myerson's Population Uncertainty framework, also known as Large Poisson Games.[3] Myerson (1998, 2000, 2002) introduces an uncertainty over the total number of voters in the election. To do so, it is assumed that the total number of voters in the game is not constant and is drawn from a Poisson distribution of a given parameter n, the expected size of the population. Due to the Poisson uncertainty, Myerson (1998) shows that pivot probabilities are common knowledge in *any* Poisson game (independently of the size of the electorate). Besides, Myerson (2002) draws a positive conclusion over the properties of AV when compared to other voting rules by analyzing some simple voting situations. This conclusion is drawn by showing that AV does not have the undesirable properties of other one-shot voting rules such as PV or the Borda Count. However, Myerson (2002) does not provide a full characterization of the voting equilibria that remain under AV. In order to address such an issue, Nuñez (2009, 2010) shows that AV need not correctly aggregate preferences. Nuñez (2010) constructs a simple voting situation where a candidate who is ranked first by more than half of the population (and thus the Condorcet Winner) is not the Winner of the election in equilibrium. In equilibrium, voters anticipate that the Condorcet Winner is not included in the most probable pivot outcome. This information concerning the probability of affecting the outcome of the election makes the majority of the voters vote for their preferred *and* for their second preferred candidate and this leads to the election of the latter. The existence of such an equilibrium is a consequence of the non-intuitive ordering of pivot probabilities that arise in Poisson games. This example shows that the refinement of the set of Nash equilibria on Large Poisson Games is limited.

However, in the previously mentioned situation, there also exist equilibria where the Condorcet Winner wins the election. As argued by Schelling (1960) and Myerson and Weber (1993) the multiplicity of equilibria has a political significance. A large set of equilibria in an electoral situation implies that informational issues have a great influence when determining the result of the election. In order to address this multiplicity of equilibria, Nuñez (2009) shows that it can be the case that, with three candidates, the Condorcet Winner is not the winner of the election in *any* of the equilibria of the game. Hence, AV can lead to worse preference aggregation than PV in Large Poisson Games. In addition to the Condorcet Consistency of AV, Nuñez (2009) investigates whether this voting rule leads to sincere best responses.

---

[3] Large Poisson Games are a novel field of research. Among the few works dealing with these games, the reader can refer to Bouton and Castanheira (2008), Goertz and Maniquet (2008), Krishna and Morgan (2008), Nuñez (2009, 2010), and De Sinopoli and Gonzalez Pimienta (ming).

Indeed, the proponents of AV often suggest that this voting rule enhances sincere voting as voters are allowed to vote for as many candidates as they wish. As Nuñez (2009) shows, this is not the case on Large Poisson Games. Indeed, Nuñez (2009) provides an example in which voters' best responses are not sincere and such that the Condorcet Winner gets no vote under AV in equilibrium.

The present work is structured as follows. Section 9.2 introduces the Myerson–Weber framework, Sect. 9.3 presents the Score Uncertainty model and Sect. 9.4 describes in detail Large Poisson Games. Section 9.5 concludes.

9.2 The Myerson–Weber Framework

There are n voters in the election. Each voter has a type t that determines his preferences over the set of candidates $K = \{k, l, \ldots\}$. The preferences of a voter with a type t is denoted by $u_t = (u_t(k))_{k \in K}$. Thus, for a given t, $u_t(l) > u_t(k)$ implies that t-voters strictly prefer candidate l to candidate k. Each type t belongs to the finite set of types T.

Each voter's type is drawn from T according to the *distribution of types* denoted by $r = (r(t))_{t \in T}$.[4] In other words, $r(t)$ represents the probability that a voter randomly drawn from the population has type t.

For any pair of candidates $k, l \in K$, let $T_{k,l} = \{t \in T \mid u_t(k) > u_t(l)\}$ be the set of preference types where candidate k is strictly preferred to candidate l. The Condorcet Winner ($C.W.$) of the election is defined as:

Definition 9.2.1. A candidate k is called the Condorcet Winner of the election if

$$\sum_{t \in T_{k,l}} r(t) > 1/2 \ \forall\, l \in K, l \neq k.$$

Similarly, the Condorcet Loser of the election is a candidate k such that $\sum_{t \in T_{k,l}} r(t) < 1/2 \ \forall\, l \in K, l \neq k$.

Each voter i must choose a ballot c from a finite set of possible ballots denoted by C. Within this work, we stick to the comparison of Plurality and Approval voting.

Definition 9.2.2 (One Man, One Vote). A Plurality voting ballot (PV) specifies the candidate the voter approves of.

Definition 9.2.3 (One Man, Many Votes). An Approval voting ballot (AV) specifies the subset of candidates the voter approves of.

Formally, an AV ballot consists of a vector of length K that lists whether a candidate has been approved or not (whenever candidate k is approved there is a one in the kth coordinate, whereas the lack of approval is represented by a zero). A

[4] The distribution of types satisfies $r(t) > 0 \ \forall\, t \in T$ and $\sum_{t \in T} r(t) = 1$.

PV ballot is a vector of length K in which every coordinate equals zero but the one corresponding to the approved candidate that is denoted by one. Hence, in order to unify both notations we refer generally to the set of available ballots as C.

We assume that each voter maximizes his expected utility to determine which ballot in the set C he will cast. In this model, his vote has an impact in his payoff if it changes the winner of the election. Therefore, a voter needs to estimate the probability of these situations: the pivot outcomes.

We say that two candidates are tied if their vote totals are equal. Furthermore, let H denote the set of all unordered pairs of distinct candidates. We denote a pair $\{k,l\}$ in H as kl with $kl = lk$.

For each pair of candidates k and l, the kl-pivot probability p_{kl} is the probability of the outcome perceived by the voters that candidates k and l will be tied for first place in the election. Furthermore, we assume that the probability of candidates k and l being tied for first place is the same than the probability of candidate k being in first place one vote ahead candidate l (and both candidates above the rest of the candidates), which is in turn the same one than the probability of candidate l being in first place one vote ahead candidate k.[5]

A vector that lists the pivot probabilities for all pairs of candidates is denoted by $p = (p_{kl})_{kl \in H}$. This vector p is assumed to be the same one for all voters in the election. A voter with kl-pivot probability p_{kl} anticipates that submitting the ballot c has the impact P_{kl} on his expected utility with

$$P_{kl} = \begin{cases} p_{kl} & \text{if he approves candidate } k \text{ and does not approve candidate } l \\ -p_{kl} & \text{if he approves candidate } l \text{ and does not approve candidate } k \\ 0 & \text{elsewhere} \end{cases}$$

Let $EU_t[c]$ denote the expected utility by a voter of type t from casting ballot c when p is the common vector of pivot probabilities. It follows that

$$EU_t[c] = \sum_{kl \in H} P_{kl}[u_t(k) - u_t(l)].$$

A strategy function is a probability distribution σ over the set C that summarizes the voting behavior of voters of each type. For any $c \in C$ and any $t \in T$, $\sigma(c \mid t)$ is the probability that a voter with type t casts ballot c. Therefore,

$$\tau(c) = \sum_{t \in T} r(t)\sigma(c \mid t),$$

[5] Myerson and Weber (1993) justify this assumption by arguing that it seems reasonable when the electorate is large enough. However, Large Poisson Games (Myerson 1998, 2000, 2002) do not respect this intuition. It can be the case that the probability of candidates k and l being tied for first place becomes infinitely more likely than the probability of candidate k being in first place one vote ahead candidate l as the electorate becomes large enough. For an example of these divergent probabilities, see the voting game analyzed by Sect. 9.4.5 within this chapter.

is the share of the electorate who cast ballot c. Given a vote distribution τ, the expected score of candidate k is

$$\rho(k) = \sum_{c \in C_k} \tau(c),$$

in which C_k consists of the subset of ballots in which candidate k is approved. A Winner of the election is a candidate whose expected score is maximal.

9.2.1 Voting Equilibrium in the Myerson–Weber Framework

One substantive assumption of the Myerson–Weber framework is what we will refer to as the *ordering condition*. As will be shown throughout, the main differences between three models with a large number of voters lie on this type of conditions over the pivot probabilities. Myerson and Weber (1993) assumes that voters expect candidates with lower expected scores are less likely serious contenders for first place than candidates with higher expected scores. In other words, if the expected score for some candidate k is strictly less than the expected score for some candidate l, then the voters would perceive that candidate k's being tied with any third candidate m is much less likely than candidate l's being tied for first place with candidate m.

Definition 9.2.4 (Ordering condition). Given a strategy function σ and any $0 \leq \varepsilon < 1$, a pivot probability vector p satisfies the ordering condition for ε (with respect to σ) if, for every three distinct candidates k, l and m, if $\rho(k) < \rho(l)$, then $p_{km} \leq \varepsilon p_{lm}$.

Besides, Myerson and Weber (1993) assumes that the probability of three (or more) candidates being tied for first place is infinitesimal in comparison to the probability of two-candidate tie.

Definition 9.2.5 (Equilibrium in the Myerson–Weber framework). We refer to σ as an equilibrium of the game if and only if, for every positive number ε, there exists some vector p of positive pivot probabilities that satisfies the ordering condition and such that, for each $c \in C$ and for each $t \in T$,

$$\sigma(c \mid t) > 0 \implies c \in \arg\max_{d \in C} EU_t[d].$$

It can be shown that the set of voting equilibria is non-empty given the existence of the ordering condition.[6] In order to ensure the existence of equilibrium, Myerson and Weber (1993) assume that the pivot probability vector is a probability distribution over the set H of unordered pairs of candidates so that $\sum_{kl \in H} p_{kl} = 1$.

[6] See Theorem 1, p. 105 in Myerson and Weber (1993).

9.2.2 Comparison of AV and PV in the Myerson–Weber Framework

Given the previous simple framework, Myerson and Weber (1993) draws a positive conclusion over the properties of AV when compared with PV. The current subsection presents a brief outline of their results. Let us consider a Myerson–Weber voting game where there are three candidates $K = \{a, b, c\}$ and three different types $T = \{t_1, t_2, t_3\}$ such that:

t_1	t_2	t_3
a	b	c
b	a	a
c	c	b

in which the utility of the voters satisfies $u_{t_1}(a) = 10 > u_{t_1}(b) = 9 > u_{t_1}(c) = 0$; $u_{t_2}(b) = 10 > u_{t_2}(a) = 9 > u_{t_2}(c) = 0$ and $u_{t_3}(c) = 10 > u_{t_3}(a) = u_{t_3}(b) = 0$. Besides, the distribution of types satisfies

$$r(t_1) = 0.3, \quad r(t_2) = 0.3 \text{ and } r(t_3) = 0.4.$$

Given this distribution, candidate c is the Condorcet Loser as $r(t_3) < r(t_1) + r(t_2)$.

Proposition 9.2.1. *In the previous example, PV can implement the Condorcet loser as the unique Winner of the election.*

Proposition 9.2.2. *In the previous example, AV does not implement the Condorcet loser as the unique Winner of the election.*

Proof. The present situation is the typical case of a divided majority election.[7] There is a majority of the electorate that prefers candidates a and b to candidate c. However, this majority is divided in two symmetric groups: one of which strictly prefers candidate a to candidate b and the other that prefers candidate b to candidate a.

Under PV, there are three voting equilibria: two equilibria on which voters on the majority coordinate and make either candidate a or candidate b to be elected and a third equilibrium on which voters with type t_1 and t_2 split their votes and candidate c is the expected Winner. The latter equilibrium is such that

$$\sigma(a \mid t_1) = \sigma(b \mid t_2) = \sigma(c \mid t_3) = 1,$$

which implies that

$$\rho(a) = 0.3, \ \ \rho(b) = 0.3, \ \ \rho(c) = 0.4,$$

[7] See the chapter on this book (Laslier and Sanver 2010b) that presents a detailed account on voting experiments dealing with the classical case of a divided majority election.

so that candidate c is the Winner of the election. This equilibrium exists whenever the pivot probability vector p satisfies $p_{ab} = 0$, $9/19 \leq p_{ac} \leq 10/19$ and $p_{bc} = 1 - p_{ac}$. Since candidates a and b have similar probabilities of being in contention for victory with candidate c, voters with type t_1 and t_2 fail to coordinate.

Under AV, there are also three voting equilibria. In two of them, voters on the majority coordinate and make either candidate a or b to be elected in a similar manner to the one with PV. However there does not exist an equilibrium under which candidate c has the strictly highest expected score. In the third equilibrium is such that the three candidates get the same expected score. Indeed, such an equilibrium satisfies

$$\sigma(a \mid t_1) = 2/3, \quad \sigma(a,b \mid t_1) = 1/3,$$
$$\sigma(b \mid t_2) = 2/3, \quad \sigma(a,b \mid t_2) = 1/3, \text{ and } \sigma(c \mid t_3) = 1,$$

which implies that

$$\rho(a) = r(t_1)\big(\sigma(a \mid t_1) + \sigma(a,b \mid t_1)\big) + r(t_2)\sigma(a,b \mid t_2) = 0.4,$$

and similarly $\rho(b) = \rho(c) = 0.4$, so that the three candidates get the same expected score. This equilibrium exists whenever the pivot probability vector p satisfies $p_{ab} = 9/11$, $p_{ac} = p_{bc} = 1/11$. In this equilibrium, none of the pivot probabilities is negligible with respect to the others but the probability of a pivot between candidates a and b is nine times probable than the other two candidate pivot outcomes.

A Change on the Type Distribution

In order to prove that the positive conclusion over AV drawn on the previous example lies on the particular distribution of types, Myerson and Weber (1993) modify the distribution of types so that

$$r(t_1) = 0.49, \quad r(t_2) = 0.49 \text{ and } r(t_3) = 0.02.$$

In this modified version of the example, AV uniquely leads to a unique equilibrium in which everyone for his most preferred candidate. Such an equilibrium satisfies

$$\sigma(a \mid t_1) = \sigma(b \mid t_2) = \sigma(c \mid t_3) = 1,$$

which implies that

$$\rho(a) = \rho(b) = 0.49, \quad \rho(c) = 0.02,$$

so that both candidates a and b get the same expected score. To have such an equilibrium, it suffices to specify a pivot probability vector p such that $p_{ab} = 1$ and $p_{ac} = p_{bc} = 0$. The set of voting equilibria under PV is much larger in this example and can lead voters from both types t_1 and t_2 to vote for candidate a (whenever the pivot probability vector p satisfies $p_{ac} = 1$ and $p_{ab} = p_{bc} = 0$).

The results of the Myerson–Weber framework suggest that AV leads to a better preference aggregation than other simple one-shot voting rules such as PV. However the previous analysis is performed in a setting in which strong assumptions are made over the information available to voters. The remaining models in this work try to escape from these assumptions and analyse elections building on the introduction of trembles, in a similar spirit to the trembling-hand perfect equilibrium of Selten (1975).

9.3 The Score Uncertainty Model

The Score Uncertainty model introduced by Laslier (2009) is based on the introduction on some strictly positive probability that every vote is erased. This erasing probability creates the uncertainty faced by voters and generates the pivot probability vectors, that were exogenously introduced on the previously described Myerson–Weber setting. Therefore, all the notations will remain the same unless otherwise specified.

There are n voters in the election. Each voter has a type t that determines his preferences over the set of candidates $K = \{k, l, \ldots\}$. The preferences of a voter with a type t is denoted by $u_t = (u_t(k))_{k \in K}$. Thus, for a given t, $u_t(j) > u_t(k)$ implies that t-voters strictly prefer candidate j to candidate k. Each type t belongs to the finite set of types T.

Each voter's type is drawn from T according to the distribution of types denoted by $r = (r(t))_{t \in T}$. Each voter i must choose a ballot c from a finite set of possible ballots denoted by C. The possible set of ballots we focus on (AV and PV) have already been defined in the previous section

We assume that each voter maximizes his expected utility to determine which ballot in the set C he will cast. Similarly to the previous model, his vote has an impact in his payoff if it changes the winner of the election. Therefore, a voter needs to estimate the probability of these situations: the pivot outcomes. The main difference between the Myerson–Weber framework and the Score uncertainty model is the way of introducing uncertainty in the model. Whereas Myerson and Weber (1993) introduces it in a exogenous way, Laslier (2009) introduces a small probability $\mathbf{q}$ that each vote for each candidate is erased. Voters have some uncertainty over the total scores of candidates which comes from this small but strictly positive probability that their vote is erased.

Formally, Laslier (2009) considers a large electorate. To do so, the electorate with n voters is replicated ν times as follows. By assumption, we know that $r(t)$ stands for the share of the electorate with type t with $\sum_{t \in T} r(t) = 1$. In the ν-fold replicate economy the number of type-t voters is $n\nu r(t)$ and the total number of voters is equal to $n\nu$.

Furthermore, we assume that for any voter and each candidate approved by this voter there is a probability $\mathbf{q} > 0$ that this vote is not recorded. This probability is supposed to be small, with $\mathbf{q} < 1/n$ (independently of ν). These mistakes occur independently of the voter, of the candidate, and of the voter approving or not other candidates.

For any $c \in C$ and any $t \in T$, the strategy function $\sigma(c \mid t)$ stands for the probability that a voter with type t casts the ballot c. Therefore,

$$\tau(c) = \sum_{t \in T} r(t)\sigma(c \mid t),$$

is the share of the electorate who cast ballot c. The maximal score of candidate k is

$$\rho(k) = \sum_{i \in C_k} \tau(c),$$

in which C_k consists of the subset of ballots in which candidate k is approved. However, given the erasing probability the realized score of candidate k differs from the maximal one. For any candidate k and any voter i, let $\eta_{i,k}$ denote the random variable such that

$$\eta_{i,k} = \begin{cases} 1 \text{ with probability } \mathbf{q} \\ 0 \text{ with probability } \mathbf{1} - \mathbf{q}. \end{cases}$$

When the maximal number of votes for candidate k equals $n\nu\rho(k)$, the realized number of votes for candidate k is a random variable $s(k)$. If we let $AV(k)$ denote the set of voters who approve candidate k, the random variable $s(k)$ satisfies

$$s(k) = \sum_{i \in AV(k)} (1 - \eta_{i,k}).$$

The *score profile* $s = (s(k))_{k \in K}$ is a vector that describes the realized number of votes each candidate gets. There are at most $n\nu\rho(k)$ voters who approve of candidate k so that the score $s(k)$ of candidate k is a binomial random variable with expected value and variance:

$$E[s(k)] = (1 - \mathbf{q})n\nu\rho(k)$$
$$V[s(k)] = \mathbf{q}(1 - \mathbf{q})n\nu\rho(k).$$

A Winner of the election is a candidate k whose score $\rho(k)$ satisfies $\rho(k) = \max_{l \in K} \rho(l)$. It is important to emphasize that given the score distribution $\rho(k)$, the scores of candidates $s(k)$ are independent random variables whereas this will not be the case in Large Poisson Games.

Given the score profile s, an outcome of the election is a pivot between a non-empty subset of candidates Y if and only if:

$$\forall\, y \in Y \quad s(y) \geq \max_{k \in K} s(k) - 1$$
$$\forall\, y \notin Y, \quad s(y) < \max_{k \in K} s(k) - 2.$$

A pivot between the pair of candidates k and l will be denoted by *pivot*(k, l) and its probability will be represented by p_{kl}. These pivot probabilities for all pairs of candidates are summarized by a vector $p = (p_{kl})_{kl \in H}$, in which H stands for the set of unordered pair of candidates. This vector is not anymore assumed to be common for all voters: it is indeed generated by the erasing probability. A voter with kl-pivot probability p_{kl} anticipates that submitting the ballot c has the impact P_{kl} on his expected utility with

$$P_{kl} = \begin{cases} p_{kl} & \text{if he approves candidate } k \text{ and does not approve candidate } l \\ -p_{kl} & \text{if he approves candidate } l \text{ and does not approve candidate } k \\ 0 & \text{elsewhere} \end{cases}$$

Let $EU_t[c]$ denote the expected utility by a voter of type t from casting ballot c when p is the common vector of pivot probabilities. It follows that

$$EU_t[c] = \sum_{kl \in H} P_{kl}[u_t(k) - u_t(l)].$$

in which P_{kl} is defined as previously. Indeed, the Score Uncertainty model, as the Myerson–Weber setting, assumes that the probability of three (or more) candidates being tied for first place is infinitesimal in comparison to the probability of two-candidate tie which allows us to write the previous simple expression for the expected utility of voters.

9.3.1 Voting Equilibrium in the Score Uncertainty Model

Laslier (2009) does not assume the ordering condition which was an important property of the Myerson–Weber framework. Given the erasing probability $\mathbf{q}$, it is shown that *any* pivot probability vector satisfies the *limit ordering condition.*

Definition 9.3.1 (Limit Ordering condition). Given a strategy function σ, a pivot probability vector p satisfies the limit ordering condition if, for every three distinct candidates k, l and m, if $\rho(k) < \rho(l)$, then

$$\lim_{\nu \to \infty} \frac{p_{km}}{p_{lm}} = 0.$$

Proposition 9.3.1. *Given that there are no ties in the score distribution, any pivot probability vector satisfies the limit ordering condition in the Score Uncertainty model.*

Proposition 9.3.2. *The pivot probability vectors are not equal for all the voters. However, whenever the electorate is large, the differences between the pivot probability vectors do not affect voters' best responses.*

Definition 9.3.2 (Equilibrium in the Score Uncertainty Model). We refer to σ as an equilibrium of the game if and only if for each ballot $c \in C$ and each $t \in T$,

$$\sigma(c \mid t) > 0 \implies c \in \arg\max_{d \in C} EU_t[d].$$

9.3.2 Approval Voting on the Score Uncertainty Model

Once we have properly defined the Score Uncertainty model and the equilibrium of the voting game, we introduce the two results that summarize Laslier's (2009) conclusions over AV. These results are very positive for AV in a large electorate. Indeed, both sincerity and Condorcet Consistency are satisfied by AV at equilibrium.

Definition 9.3.3 (Sincerity). An AV ballot is sincere if, given the lowest-preferred candidate k that a voter approves of, he also approves of all candidates he prefers to k.

Theorem 9.3.1. *For a large electorate, in the absence of a tie in the score distribution, best responses are sincere under AV.*

Theorem 9.3.2. *For a large electorate, in the absence of a tie in the score distribution, AV uniquely selects the Condorcet Winner (whenever it exists) as the Winner of the election. If the preference profile admits a Condorcet Winner and the Condorcet Winner has a unique best contender then the game has a unique equilibrium. In this equilibrium, the Condorcet Winner is elected.*

The underlying rationale for both theorems is the limit ordering condition. Under the limit ordering condition, we can say that pivot probabilities are "well ordered." For instance, let us pick three candidates a, b and c such that $\rho(a) > \rho(b) > \rho(c)$ (the expected score of candidate a is higher than the expected score of candidate b and so on). Whenever the electorate is large, we know that every voter in the election anticipates that the pivot between candidates a and b is the most probable one and that as the size of the electorate becomes larger, the pivot probabilities p_{ac} and p_{bc} become negligible when compared with the pivot probability p_{ab}. This, in turn, implies that voters vote according to the following rule: for every voter with type t, the unique best-response ballot R_t is such that

$$\text{a voter with type } t \text{ s.t. } u_t(a) > u_t(b) \implies R_t = \{k \in K : u_t(k) \geq u_t(a)\},$$
$$\text{a voter with type } t \text{ s.t. } u_t(a) < u_t(b) \implies R_t = \{k \in K : u_t(k) > u_t(a)\},$$

Indeed, given that voters are expected-utility maximizers, every voter will voter for either candidate a or candidate b and no voter will vote for both. When a voter has voted for either candidate a or candidate b, he still needs to decide whether he will give his approval to candidate c. However, this decision is quite easy given the limit ordering condition. Let us suppose that a voter has approved of candidate a. Whenever he prefers candidate c to candidate a, he will approve of candidate c as the most probable pivot in which candidate c is involved is against candidate a (due to the limit ordering condition). In such an outcome, the expected utility of the voter increases by approving of candidate c. Similarly, if the voter prefers candidate a to candidate c, he will not approve of candidate c as the most probable pivot in which candidate c is included is against candidate a. Similar arguments show that the unique best-response ballot satisfies the previous claim for a finite number of candidates.

The fact that the limit ordering condition implies a unique best response ballot has different consequences. First of all, it is simple to see that Theorem 9.3.1 is a direct consequence. Indeed, a sincere ballot under AV is a ballot such that whenever you give your approval to some given candidate a, you approve any candidate that you prefer to candidate a. The best response ballot R_t satisfies this definition and thus every voter is sincere at equilibrium.

The second implication of the limit ordering condition is that the score of the first-ranked candidate in equilibrium equals the share of the electorate who prefers the first-ranked candidate to the second-ranked candidate. And the score of any other candidate equals the share of the electorate who prefers such a candidate to the first-ranked candidate. Therefore, the Condorcet Winner is the only possible Winner of the election in equilibrium as the Condorcet Winner is the candidate who is preferred in pairwise comparisons to the rest of the candidates in the election.

As will be shown in the remaining chapter, the limit ordering condition is *not* satisfied by Large Poisson Games and this will be the source of the failure of preference aggregation under AV in such a setting. Indeed, given three candidates a, b and c such that $\rho(a) > \rho(b) > \rho(c)$, it could be the case that the pivot probability p_{ac} becomes infinitely larger than any other pivot probability as the size of the electorate becomes large.

9.4 Large Poisson Games

A Poisson random variable $\mathcal{P}(n)$ is a discrete probability distribution that depends on a unique parameter which represents its mean. The probability that a Poisson random variable of parameter n takes the value v, being v a nonnegative integer is equal to

$$e^{-n}\frac{n^v}{v!}.$$

A Poisson voting Game of expected size n is a game such that the actual number of voters taking part in the election is a random variable drawn from a Poisson distribution with mean n. This assumption represents the uncertainty faced by voters w.r.t. the number of voters that show up the day of the election. The probability distribution and its parameter n are common knowledge.

Each voter's type is independently drawn from T according to the *distribution of types* denoted by $r = (r(t))_{t \in T}$.

A Poisson game of expected size n is then represented by (K, T, n, r, u). The expression "large Poisson game" is used to describe the asymptotic behavior of a sequence of Poisson games of expected size n when n is large enough.

In order to completely determine an election in a Poisson voting game, the voting rule remains to be specified. A Poisson voting game will be represented by (K, T, C, n, r, u) in which C stands for the set of available ballots. The set of ballots we focus on (AV and PV) have already been defined in the description of the Myerson–Weber framework (Sect. 9.2).

As shown by Myerson (1998), assuming a Poisson population has two main advantages: common public information and independence of actions.

As usual, voters' actions depend on their type (private information) and on the actions of other voters. In such a probabilistic framework, there exists a probability distribution over the different possible outcomes that might arise in the election. When we refer to common public information, we mean that this probability distribution does not depend on the type t. Indeed, each voter in the election fully knows the probability distribution over the different outcomes independently of t. This is not the case when using solution concepts such as the perfect equilibrium of Selten (1975). In a perfect equilibrium, strategic voters have some prior beliefs over the expected scores of the candidates. However, in such an equilibrium, there is an asymmetry of information that makes more difficult the analysis of the game. This common public information property of Poisson Games entails that voters' actions uniquely depend on their private information t on this type of games in equilibrium.

The second main advantage is usually referred as the independence of actions. Indeed, the number of voters who choose a given ballot is independent from the number of voters who choose another ballot. This is not the case if we assume for instance a binomial distribution. Let us assume that a binomial random variable represents the number of voters in the election. A binomial distribution is characterized by two parameters n and p. Whereas p represents the probability of taking part in the election, the parameter n stands for the maximal size of the population. This upper-bound for the number of voters implies that voters' actions are correlated.[8] This is not the case in a Poisson voting game as there is not an upper-bound for the number of voters in the election. These two properties substantially simplify the analysis of the voting game and are unique to the Poisson games as shown by Myerson (1998).

[8] To see this correlation, it suffices to understand that under the binomial assumption, whenever a voter does not vote for a candidate, there is a most $n-1$ voters that can do it.

We represent voters' actions by the *strategy function* $\sigma(c|\, t)$[9] which is a function from T into $\Delta(C)$ the set of probability distributions over C. Formally, we write

$$\sigma : \begin{cases} T \longrightarrow \Delta(C) \\ t \longmapsto \sigma(. \mid t). \end{cases}$$

A voter with type t chooses ballot c with probability $\sigma(c \mid t)$. Then, taking into account the distribution of types r and the strategy function $\sigma(. \mid t)$, the *vote distribution* $\tau = (\tau(c))_{c \in C}$ can be determined as follows. For each $c \in C$, we define

$$\tau(c) = \sum_{t \in T} r(t)\sigma(c \mid t).$$

The vote distribution τ represents the share of votes each ballot gets. We denote by $x(c)$ the Poisson random variable with parameter $n\tau(c)$ that describes the number of voters $x(c)$ who choose ballot c. Furthermore the *vote profile* $x = (x(c))_{c \in C}$ is a vector of length C of independent random variables (due to the independent actions property).

We denote by b a vector of length C of non-negative integer numbers. Each component $b(c)$ of vector b accounts for the number of voters who vote for ballot c. The set of electoral *outcomes*[10] given ballot set C is denoted by $\mathscr{B}$, where

$$\mathscr{B} = \{b \in \mathbb{R}^C \mid b(c) \text{ is a non-negative integer for all } c \in C\}$$

The subsets of $\mathscr{B}$ will be denoted by capital letters $B \subset \mathscr{B}$.

Given the vote profile x, the (common knowledge) probability that the outcome is equal to a vector $b \in \mathscr{B}$ is such that

$$\begin{aligned} P[x = b \mid n\tau] &= P[\bigcap_{c \in C} x(c) = b(c) \mid n\tau] \\ &= \prod_{c \in C} P[x(c) = b(c) \mid n\tau] \\ &= \prod_{c \in C} \left(\frac{e^{-n\tau(c)}(n\tau(c))^{b(c)}}{b(c)!} \right). \end{aligned}$$

For ease of notation, we refer to $P[x = b \mid n\tau]$ by $P[x = b]$. We will be mainly interested in computing the probabilities of subsets of $\mathscr{B}$ rather than probabilities of vectors themselves, as for instance the probability of two given candidates getting the same number of votes. Given the vote profile x, we write that the probability of

---

[9] The strategy function satisfies $\sigma(c \mid t) \geq 0 \;\forall\, c \in C$ and $\sum_{d \in C} \sigma(d \mid t) = 1$.

[10] In probabilistic terminology, an electoral outcome is usually referred as an event or realization of a random variable, i.e., the value that is actually observed (what actually happened). For ease of notation, we will refer to them simply as outcomes.

the outcome $B \subset \mathscr{B}$ is equal to

$$P[x \in B] = \sum_{b \in B} P[x = b].$$

Let C_k denote the set of ballots in which candidate k is approved. Given the vote profile x, the *score distribution* $\rho = (\rho(k))_{k \in K}$ describes the share of votes that each candidate gets. For each $k \in K$,

$$\rho(k) = \sum_{c \in C_k} \tau(c).$$

It follows that the number of voters that vote for a candidate k is drawn from a Poisson random variable with mean $n\rho(k)$. Given the score distribution, we define the *score profile* $s = (s(k))_{k \in K}$ describes the number of voters who vote for each candidate k with

$$s(k) = \sum_{c \in C_k} x(c) \sim \mathcal{P}(n\rho(k)).$$

Given that under AV voters can vote for several candidates, it is not true in general that the score profile s is a vector of independent random variables. As will be shown this lack of independence is an important property of AV on Poisson games. Indeed, due to this correlation between the candidate scores, counterintuitive situations might arise.

Given an outcome $B \subset \mathscr{B}$, let $M(B) = \arg\max_{j \in K} \rho(j)$ denote the set of candidates with the most points. We say that candidate a is the Winner of the election whenever candidate a is the unique candidate in the set $M(B)$. Assuming a fair toss of a coin, the probability of candidate k winning the election given the vector $B \subset \mathscr{B}$ is

$$Q[k \mid B] = \begin{cases} 1/\#(M(B)) & \text{if } k \in M(B) \\ 0 & \text{if } k \notin M(B). \end{cases}$$

9.4.1 Voting Equilibrium on Large Poisson Games

For any outcome $B \subset \mathscr{B}$ and any ballot $c \in C$, we let $B + \{c\}$ denote the outcome such that one ballot c is added. That is, we write that the outcome $D \subset \mathscr{B}$ is such that

$$D = B + \{c\} = \{d \in D \mid d = b + c \text{ for any } b \in B, c \in C\}.$$

in which the sum of vectors b and c is componentwise. Thus, given the vote profile x, a voter with type t casts the ballot c that maximizes his expected utility

$$EU_t[c \mid n\tau] = \sum_{B \subset \mathscr{B}} P[x \in B] \sum_{k \in K} Q[k \mid B + \{c\}] u_t(k).$$

Again, for ease of notation, we write $EU_t[c]$ for $EU_t[c \mid n\tau]$.

Definition 9.4.1 (Equilibrium of a Poisson game). We refer to σ as an *equilibrium* of the Poisson voting game (K, T, C, n, r, u) if for each $c \in C$ and each $t \in T$, given the vote distribution τ,

$$\sigma(c \mid t) > 0 \Longrightarrow c \in \arg\max_{d \in C} EU_t[d].$$

As the focus of this work is on elections with a large number of voters, one shall look at the limits of equilibria as the expected number of voters n tends to infinity. Thus, we refer to a *large equilibrium sequence* of (K, T, C, r, u) to denote any equilibria sequence $\{\sigma_n\}_{n=1}^{\infty}$ of the voting games (K, T, C, n, r, u) such that the vectors σ_n are convergent to some limit σ as $n \to \infty$ in the sequence. We refer to this limit σ as a *large equilibrium* of (K, T, C, r, u). Furthermore, we refer to a sequence of outcomes in $\mathscr{B}$ by $\{B_n\}_{n=1}^{\infty}$. The limit B of a sequence of outcomes $\{B_n\}_{n=1}^{\infty}$ in $\mathscr{B}$ is an outcome and so it is a subset of $\mathscr{B}$.

9.4.2 The Decision Process

As previously stated, we assume that each voter determines which ballot he casts by maximizing his expected utility. As voters are instrumentally motivated, they care only about the influence of their own vote in determining the Winner's identity. As usual in voting environments with a large number of voters, a voter's action has a negligible impact on the outcome of the election. Indeed, it has some impact only if there is some set of candidates involved in a close race for first place where one ballot could pivotally change the result of the election: a *pivot*.

Definition 9.4.2. Given the score profile s and a subset Y of the set of candidates K, an outcome $B \subset \mathscr{B}$ is a *pivot*(Y) if and only if:

$$\forall\, y \in Y,\ \ s(y) \geq \max_{k \in K} s(k) - 1$$
$$\forall\, k \notin Y,\ \ s(k) < \max_{k \in K} s(k) - 2.$$

The set of all pivot outcomes is denoted by $\Sigma(C) \subset \mathscr{B}$, where

$$\Sigma(C) = \{B \subset \mathscr{B} \mid \exists\, Y \subset K,\ B = pivot(Y)\}.$$

Besides, the set of all pivot outcomes in which candidate k is involved is denoted by $\Sigma(C, k) \subset \Sigma(C)$, where

$$\Sigma(C, k) = \{B \in \Sigma(C) \mid \exists\, Y \subset K \text{ s.t. } k \in Y \text{ and } B = pivot(Y)\}.$$

The vector $p = (p_{kl})_{kl \in H}$ summarizes the pivot probabilities for all pairs of candidates in which H stands for the set of unordered pairs of candidates. Similarly to the previous models, the vector p deserves special attention. However, in Large Poisson

Games, there are no restrictions over the probabilities of pivot outcomes involving three (or more) candidates.

Thus, given the vote profile τ, the expected utility for a voter with type t of casting ballot c is such that

$$\begin{aligned} EU_t[c] &= \sum_{B \subset \mathscr{B}} P[x \in B] \sum_{k \in K} Q[k \mid x + \{c\}] u_t(k) \\ &= \sum_{B \subset \Sigma(C)} P[x \in B] \sum_{k \in K} Q[k \mid x + \{c\}] u_t(k). \end{aligned}$$

The probability of any pivot outcome generally tends to zero as the expected population n becomes large. However, we can still compare their likelihood by comparing the rates at which their probabilities tend to zero. These rates can be measured by a concept of magnitude, defined as follows.

Given a large equilibrium sequence $\{\sigma_n\}_{n=1}^{\infty}$, the magnitude $\mu[B]$ of the limit B of a sequence of outcomes $\{B_n\}_{n=1}^{\infty} \subset \mathscr{B}$ is such that

$$\mu[B] = \lim_{n \to \infty} \frac{1}{n} \log P[x \in B \mid n\tau] = \lim_{n \to \infty} \frac{1}{n} \log P[x \in B].$$

Notice that the magnitude of an outcome must be inferior or equal to zero, since the logarithm of a probability is never positive. The main advantage of using magnitudes is to have an analytical way to compare likelihoods of outcomes rather than estimations, as the following example shows.

Example 9.4.1. Probabilities and Magnitudes in a Poisson voting game.

Let (K, T, C, n, r, u) be a Poisson voting game. The vote profile x describes the number of voters who cast a given ballot. For two given ballots c and c', we write

$$x(c) \sim \mathcal{P}(n\tau(c)) \text{ and } x(c') \sim \mathcal{P}(n\tau(c')).$$

Given the independent actions property, both $x(c)$ and $x(c')$ are independent random variables. Let us denote by $\{B_n\}_{n=1}^{\infty} \subset \mathscr{B}$ the sequence of outcomes in which there is the same number of voters that choose ballot c and ballot c' for each expected size of the electorate n. We denote the limit of the sequence of outcomes $\{B_n\}_{n=1}^{\infty}$ by B. For a given n, each outcome B_n is formally defined by

$$B_n = \{b \in \mathscr{B} \mid b(c) = b(c')\},$$

The definition of the probability of an outcome implies

$$P[x \in B_n] = \sum_{b \in B_n} P[x = b] = \sum_{k=0}^{\infty} P[x(c) = k \cap x(c') = k].$$

Therefore, the independence of actions property entails that $P[x(c) = k \cap x(c^{'}) = k] = P[x(c) = k]P[x(c^{'}) = k]$ so that

$$P[x \in B_n] = e^{-n(\tau(c)+\tau(c^{'}))} \sum_{k=0}^{\infty} \frac{(n^2\tau(c)\tau(c^{'}))^k}{(k!)^2}$$
$$= e^{-n(\tau(c)+\tau(c^{'}))} I_0 \left(2n\sqrt{\tau(c)\tau(c^{'})}\right),$$

where I_0 is a modified Bessel function.[11] Hence, the magnitude of the limit outcome $B \subset \mathscr{B}$ is such that:

$$\mu[B] = \lim_{n\to\infty} \frac{1}{n} \log P[x \in B_n]$$
$$= \lim_{n\to\infty} \frac{1}{n} \log e^{-n(\tau(c)+\tau(c^{'}))} I_0 \left(2n\sqrt{\tau(c)\tau(c^{'})}\right)$$
$$= 2\sqrt{\tau(c)\tau(c^{'})} - (\tau(c) + \tau(c^{'}))$$
$$= -(\sqrt{\tau(c)} - \sqrt{\tau(c^{'})})^2,$$

which gives an explicit rate of convergence towards zero.

If one can show that a pivot between one pair of candidates has a magnitude that is strictly greater than the magnitude of a pivot between another pair of candidates, then the latter becomes infinitely less likely as the expected number of voters goes to infinity. That is to say, given two subsets Y and $Y^{'}$ of the set of candidates K, for any pair of outcomes *pivot*(Y) and *pivot*($Y^{'}$) $\subset \mathscr{B}$, if

$$\mu[pivot(Y)] > \mu[pivot(Y^{'})],$$

then we know that the pivot outcome between candidates in Y is infinitely more likely than the pivot outcome between candidates in $Y^{'}$, i.e.

$$\lim_{n\to\infty} \frac{P[x \in pivot(Y)^{'}]}{P[x \in pivot(Y)]} = 0.$$

We now move to the description of the decision process of voters. Let k be a candidate. Let c and c' be two ballots that only differ by an extra candidate k: $c' = c \cup k$. In order to evaluate which of the ballots the type-t voter casts, he computes the sign of the following expression

[11] See Formula (9.6.10) in Abramowitz and Stegun (1965). A modified Bessel function I_0 satisfies $\lim_{n\to\infty} \frac{1}{n} \log I_0(n\alpha) = \alpha$.

$$\begin{aligned}\Delta &= EU_t[c'] - EU_t[c] \\ &= \sum_{B \subset \Sigma(C)} P[x \in B] \sum_{k \in K} \left[\, Q[k \mid x \in B + \{c'\}] - Q[k \mid x \in B + \{c\}] \, \right] u_t(k).\end{aligned}$$

The sum Δ simply represents the effect of adding candidate k to his ballot in his expected utility. However, adding this extra candidate to his ballot can only have an impact in the cases where this candidate is involved in a pivot. Therefore, Δ can be rewritten as follows:

$$\sum_{B \subset \Sigma(C,k)} P[x \in B] \sum_{k \in K} \left[\, Q[k \mid x \in B + \{c'\}] - Q[k \mid x \in B + \{c\}] \, \right] u_t(k).$$

Then, if there exists a $pivot(Y) \subset \Sigma(C,k)$ where candidate k is involved which probability becomes infinitely more likely as n tends towards infinity than every other pivot $B \subset \Sigma(C,k)$, one can factor out by this pivot. Indeed, let us assume that every pivot B where candidate k is involved becomes infinitely less likely than $pivot(Y)$ as the expected number of voters n tends towards infinity,

$$\lim_{n \to \infty} \frac{P[x \in B]}{P[x \in pivot(Y)]} = 0 \qquad \text{for all } B \in \Sigma(C,k).$$

Given this focalisation of voters' attention on the outcome $pivot(Y)$, a voter's decision (the sign of Δ) is reduced to evaluating which ballot maximizes his expected utility in case of a $pivot(Y)$,

$$\text{sign}(\Delta) = \text{sign}\Big(\sum_{k \in K} \left[\, Q[k \mid x \in pivot(Y) + \{c'\}] - Q[k \mid x \in pivot(Y) + \{c\}] \, \right] u_t(k)\Big).$$

Repeating the previous procedure, one can deduce the best response for every voter in the election. Therefore, if given the vote profile x, there exists a strict ordering of the magnitudes of the pivot outcomes, we can ensure the existence of a *unique* best response, in a similar manner to the best response sets R_t described for the Score Uncertainty model.

9.4.3 Computing Magnitudes

This section introduces the main technical tools for the computation of the magnitudes in Poisson games. A reader mainly interested in the strategic properties of the voting rules can skip this section.

As previously defined, the magnitude of an outcome represents the speed of convergence towards zero of the probability of such an outcome. The *magnitude* theorem (Myerson 2000) states that a magnitude can be computed as the solution of a maximization problem with a concave and smooth objective function. The *dual magnitude* theorem or *DMT* (Myerson 2002) gives a method to compute magnitudes of outcomes that can be defined by linear inequalities involving the vote profile $x = (x(c))_{c \in C}$. Finally, as a pivot outcome cannot be defined with such linear inequalities, the Magnitude equivalence theorem or MET (Nuñez 2010) sets up a method of computing magnitudes of pivot outcomes by using the DMT.

In order to formally introduce the results, we first give the definition of offset ratio of an outcome that will be necessary throughout.

For any outcome $B \subset \mathscr{B}$ and any ballot $c \in C$, the ratio $B(c)/n\tau(c)$ is called the c-offset ratio of B when $n\tau$ is the vote distribution. That is, the c-offset is a ratio which describes the number of players who vote for ballot c as a fraction of the expected number of voters who were supposed to cast ballot c.

For any ballot $c \in C$, we say that $\alpha(c)$ is the limit of c-offsets in the sequence of outcomes $\{B_n\}_{n=1}^{\infty}$ iff $\{B_n\}_{n=1}^{\infty}$ has a finite magnitude and, for every major sequence[12] of points $\{b_n\}_{n=1}^{\infty}$ in $\{B_n\}_{n=1}^{\infty}$, we have

$$\alpha(c) = \lim_{n\to\infty} \frac{b_n(c)}{n\tau_n(c)} = \frac{b(c)}{n\tau(c)} \text{ with } \tau(c) = \lim_{n\to\infty} \tau_n(c) \text{ and } b(c) = \lim_{n\to\infty} b_n(c).$$

Theorem 9.4.1 (Magnitude Theorem, Myerson 2000). *Let $\{B_n\}_{n=1}^{\infty}$ be a sequence of outcomes in $\mathscr{B}$. Then*

$$\begin{aligned}\lim_{n\to\infty} \log P[x \in B_n]/n &= \lim_{n\to\infty} \max_{b_n \in B_n} \log P[x = b_n]/n \\ &= \lim_{n\to\infty} \max_{b_n \in B_n} \sum_{c\in C} \tau_n(c)\psi\left(\frac{b_n(c)}{n\tau_n(c)}\right).\end{aligned}$$

in which $\psi(x) = x(1-\log(x)) - 1$ whenever $x > 0$ and $\psi(0) = -1$.

Theorem 9.4.2 (Dual Magnitude Theorem, Myerson 2002). *Let $B \subset \mathscr{B}$ be an outcome defined by*

$$B = \left\{ \sum_{c\in C} a_k(c)x(c) \geq 0 \ \forall\, k \in J \right\},$$

in which J is a finite set and parameters $a_k(c)$ are given for every $k \in J$ and $c \in C$. Suppose that $\lambda \in \mathbb{R}^C$ is an optimal solution to the problem

$$\min_{\lambda} \sum_{c\in C} \tau(c)(\exp(\sum_k \lambda_k a_k(c)) - 1) \quad \text{s.t. } \lambda_k \geq 0,\ \forall\, k \in J. \quad (F)$$

[12] A sequence $\{b_n\}_{n=1}^{\infty}$ is a major sequence of points in the sequence of outcomes $\{B_n\}_{n=1}^{\infty}$ iff each b_n is a point in B_n and the sequence of points $\{b_n\}_{n=1}^{\infty}$ has a magnitude that is equal to the greatest magnitude of any sequence that can be selected from the outcomes B_n. Formally, $b_n \in B_n\ \forall\, n$ and $\lim_{n\to\infty} \log(P[x = b_n])/n = \lim_{n\to\infty} \max_{y_n \in B_n} \log(P[x = y_n])/n$. See Sect. 3 in Myerson (2000) for a more detailed account of sequences of outcomes in Large Poisson Games.

Then the optimal value of the objective function (F) *coincides with the magnitude* $\mu[B]$ *of the outcome* $B \subset \mathscr{B}$ *and the limits of the c-offset ratios associated are such that*

$$\alpha(c) = \exp(\sum_k \lambda_k a_k(c)), \text{ for all } c \in C.$$

This theorem states a simple technique to compute magnitudes of outcomes are defined w.r.t. to a finite series of inequalities.

Example 9.4.1 (continued). Probabilities and Magnitudes in a Poisson voting game.

Let us apply the DMT to compute the probability of two ballots c and c' getting the same number of votes. Indeed, we can represent the limit outcome $B \subset \mathscr{B}$ as

$$B = \left\{[x(c) - x(c') \geq 0] \cap [x(c') - x(c) \geq 0]\right\}.$$

Then, by the DMT, the magnitude $\mu[B]$ of $B \subset \mathscr{B}$ is such that

$$\mu[B] = \min_{\lambda} \tau(c)\exp(\lambda_1 - \lambda_2) + \tau(c')\exp(-\lambda_1 + \lambda_2) - \tau(c) - \tau(c'),$$

s.t. $\lambda_i \geq 0 \; \forall\, i$. Solving this minimization problem yields to

$$\alpha(c) = \exp(\lambda_1 - \lambda_2) = \sqrt{\frac{\tau(c')}{\tau(c)}} \text{ and } \alpha(c') = \exp(-\lambda_1 + \lambda_2) = \sqrt{\frac{\tau(c)}{\tau(c')}},$$

and to

$$\mu[B] = 2\sqrt{\tau(c)\tau(c')} - \tau(c) - \tau(c') = -(\sqrt{\tau(c)} - \sqrt{\tau(c')})^2.$$

which coincides with the magnitude of the limit outcome B previously computed.

The Magnitude Equivalence Theorem or *MET* (Nuñez 2010) substantially reduces the computations of the magnitude of a pivot outcome: it allows us to use directly the *DMT* to compute magnitudes of pivot outcomes.

The DMT is conceived to compute the magnitude of outcomes defined by a series of inequalities involving the vote profile $x = (x(c))_{c \in C}$. Formally, using the DMT we compute the magnitude of an outcome $B \subset \mathscr{B}$ defined by

$$B = \left\{\sum_{c \in C} a_k(c)x(c) \geq 0 \; \forall\, k \in J\right\}.$$

However, a pivot outcome does not have this geometrical structure, i.e., for some $Y \subset K$, an outcome $pivot(Y)$ is defined by

$$\forall\, y \in Y, \; s(y) \geq \max_{k \in K} s(k) - 1$$
$$\forall\, k \notin Y, \; s(k) \leq \max_{k \in K} s(k) - 2.$$

Given that the components $s(k)$ of the score profile s are sums of the components $x(c)$ of the vote profile x, i.e., $s(k) = \sum_{c \in C_k} x(c)$, we cannot express a pivot outcome only using linear inequalities involving x.

The MET shows that the magnitude of a pivot outcome coincides with the magnitude of an outcome than can be defined uniquely using this type of inequalities.

Theorem 9.4.3 (Magnitude Equivalence Theorem, Nuñez 2010). *Let Y be a subset of the set of candidates K and pivot(Y) be its associated pivot outcome. Given a large equilibrium sequence $\{\sigma_n\}_{n=1}^{\infty}$, we can write*

$$\mu[pivot(Y)] = \mu[D],$$

for some outcome $D \subset \mathcal{B}$ defined by

$$D = \{s(k) = s(l)\ \forall\ k, l \in Y\} \cap \{s(k) \geq s(l)\ \forall\ k \in Y \textit{and}\ l \in K \setminus Y\}.$$

This result shows that there exists an outcome, defined by a series of inequalities depending on the vote profile x, which magnitude coincides with the magnitude of the pivot outcome. Indeed, the outcome D defined by Theorem 9.4.3 can be written down as:

$$D = \left\{ \sum_{c \in C} a_k(c)x(c) \geq 0\ \forall\ k \in J \right\},$$

for some parameters a_k as, by definition,

$$s(k) = \sum_{c \in C_k} x(c).$$

Thus, one can directly the DMT to compute the magnitude of pivot outcomes, solving a simple constrained maximization problem.

9.4.4 Approval Voting and Plurality Voting on Large Poisson Games

This section presents an example, due to Myerson (2002), where in equilibrium AV leads to better preference aggregation than PV. There are two types of voters and three candidates, one of which is unanimously preferred. Due to the flexibility of AV, every voter votes the unanimously preferred candidate in the unique equilibrium of game. However, this is not the case under PV, which is one of the major flaws of PV in this framework. Indeed, whenever voters anticipate that a pair of candidates is the most likely one to be in contention for victory, then one of the candidates included in the pair is the Winner of the election at equilibrium. Hence, PV is too vulnerable to the information manipulation (information concerning the expected scores of the candidates) whereas AV is more robust as it allows voters more flexibility.

Let us consider a Large Poisson voting game where there are three candidates $K = \{a, b, c\}$ and three different types $T = \{t_1, t_2\}$ such that:

t_1	t_2
a	a
b	c
c	b

in which the utility of t_1-voters satisfies $u_{t_1}(a) > u_{t_1}(b) > u_{t_1}(c)$ and so on. This example does not lie on the utility levels but rather on the preference orderings. Besides, the distribution of types satisfies

$$r(t_1) = p \text{ and } r(t_2) = 1 - p \quad \text{for some } 0 < p < 1.$$

Proposition 9.4.1. *On Large Poisson Games, a unanimously preferred candidate is the unique Winner of the election under AV.*

Proposition 9.4.2. *On Large Poisson Games, a unanimously preferred candidate need not be the Winner of the election under PV.*

Proof. We claim that there is a large equilibrium σ of the game (K, T, C, r, u) in which candidate a is not the Winner of the election under PV. In this large equilibrium, the strategy function satisfies

$$\sigma(b|\, t_1) = \sigma(c|\, t_2) = 1,$$

and the vote distribution is such that

$$\tau(b) = r(t_1) \text{ and } \tau(c) = r(t_2).$$

Given the vote distribution, the vote profile $x = (x(c))_{c \in C}$ is the following vector

$$x(b) \sim \mathcal{P}(pn) \quad \text{and} \quad x(c) \sim \mathcal{P}((1-p)n).$$

In such an equilibrium, the score distribution $\rho = (\rho(k))_{k \in K}$ is such that

$$\rho(b) = p \text{ and } \rho(c) = 1 - p.$$

Given this score distribution, the Winner of the election is either candidate b or candidate c. Finally, given the score distribution, the score profile $s = (s(k))_{k \in K}$ is such that

$$s(b) = x(b) \sim \mathcal{P}(pn), \quad \text{and} \quad s(c) = x(c) \sim \mathcal{P}((1-p)n).$$

Let us now show why σ is indeed a large equilibrium of this Poisson Approval voting game. The aim is to prove that the pair σ induces a probability distribution over the set of pivot outcomes such that σ is still a best response. The solved minimization problems are included in the appendix. In this example, there are three possible pivot outcomes involving two candidates *pivot*(a,b), *pivot*(a,c) and *pivot*(b,c) and one pivot outcome in which the three candidates are involved.

Given the strategy function σ, the MET implies that the magnitude of the outcome *pivot*(b,c) is equal to the magnitude of the outcome $\{s(b) = s(c) \geq s(a)\}$. Formally, we write

$$\mu[pivot(b,c)] = \mu[\{s(b) = s(c) \geq s(a)\}].$$

The outcome $\{s(b) = s(c) \geq s(a)\}$ can be defined as

$$\{[x(b) \geq x(c)] \cap [x(c) \geq x(b)] \cap [x(b) \geq 0]\} \Longleftrightarrow \{[x(b) = x(c)] \cap [x(b) \geq 0]\}.$$

According to the *DMT*, we know that the magnitude of *pivot*(b,c) is equal to the solution of the following optimisation problem.

$$\tau(b)\exp[\lambda_1 - \lambda_2] + \tau(c)\exp[-\lambda_1 + \lambda_2] - \tau(b) - \tau(c),$$

such that $\lambda_i \geq 0 \ \forall \ i$. Thus, the magnitude of this pivot outcome is such that

$$\mu[pivot(b,c)] = 2\tau(b)\tau(c) - \tau(b) - \tau(c) = -(\sqrt{\tau(b)} - \sqrt{\tau(c)})^2,$$

which implies that $\mu[pivot(b,c)] > -1$ as $0 < p < 1$.

Similarly, combining the MET and the DMT, the magnitude of a pivot between candidates a and b is equal to

$$\mu[pivot(a,b)] = \mu[\{s(a) = s(c) \geq s(b)\}] = \mu[\{x(b) = 0\} \cap \{x(c) = 0\}] = -1,$$

and the magnitude of a pivot between candidates a and c is equal to

$$\mu[pivot(b,c)] = \mu[\{s(b) = s(c) \geq s(a)\}] = -1.$$

Moreover, the magnitude of the pivot between candidates a, b and c is equal to the magnitude of the pivot between candidates b and c, i.e.

$$\mu[pivot(a,b,c)] = -1.$$

Therefore, the magnitudes of the pivot outcomes are ordered as follows:

$$\mu[pivot(b,c)] > \mu[pivot(a,b)] = \mu[pivot(a,c)] = \mu[pivot(a,b,c)] \quad \textbf{(A)}.$$

Inequality (**A**) can be rewritten in terms of the pivot probabilities p_{kl} as follows

$$\lim_{n\to\infty} \frac{p_{kl}}{p_{bc}} = 0 \quad \forall k,\ l \in K,$$

which implies that in this equilibrium the limit ordering condition is satisfied (as both candidates b and c have a strictly higher expected score than candidate c).

Taking into account the ordering of the magnitudes (**A**), one can determine the ballot that each voter of a given type chooses. Under PV, it is clear that voter's best responses are such that

$$\sigma(b|\, t_1) = \sigma(c|\, t_2) = 1,$$

showing that σ is an equilibrium. This equilibrium simply represents the wasted-vote effect in a Large Poisson game: that is, no voting for a candidate you prefer as you anticipate he has no chance of winning the election.

However, this is not the case under AV. Indeed, even if any of the pivot outcomes in which candidate a is involved is far less probable than the pivot outcome between candidates b and c, the pivot outcomes involving candidate b arise with *strictly positive probability*. Then, as with AV a voter can approve as many candidates as he wishes, approving candidate a strictly increases his expected utility. Therefore given inequality (**A**), the strategy function satisfies under AV

$$\sigma(a, b|\, t_1) = \sigma(a, c|\, t_2) = 1,$$

showing that σ is not a large equilibrium of the game (K, T, C, r, u).

Similar claims show that the unique Winner in equilibrium is candidate a. Indeed, as there is always a strictly positive alone of no voter showing up the day of the election, a voter always approves of his preferred candidate under AV. Therefore, a unanimously preferred candidate must the Winner of the election under AV.

It is clear through the arguments presented in this proof that the properties hold independently of the example. In other words, a unanimously preferred candidate will always be the Winner of the election under AV whereas this need not be the case under PV.

9.4.5 AV Does Not Satisfy Condorcet Consistency on Large Poisson Games

In this section, an example from Nuñez (2010) is provided where, in equilibrium, the Winner of the election does not coincide with the Condorcet Winner. Moreover, in this equilibrium a candidate preferred by more than half of the voters is not elected. The majority of voters (t_2-voters) would prefer to vote just for their preferred candidate, candidate b. However, they vote for their second preferred candidate a to prevent candidate c from winning the election, as the most probable pivot outcome in which candidate a is involved is against candidate c. It is a pure coordination problem which the Poisson uncertainty does not remove. This equilibrium is characterized by a failure in preference aggregation: it is due to the correlation between the scores of the candidates that naturally arise in Large Poisson Games when a voting rule allows to vote for several candidates.

Let us consider a Large Poisson Approval voting game where there are three candidates $K = \{a, b, c\}$ and three different types $T = \{t_1, t_2, t_3\}$ such that:

t_1	t_2	t_3
a	b	c
b	a	a
c	c	b

in which the utility of t_1-voters satisfies $u_{t_1}(a) > u_{t_1}(b) > u_{t_1}(c)$ and so on. This example does not lie on the utility levels but rather on the preference orderings. Besides, the distribution of types satisfies

$$r(t_1) = 0.1, \quad r(t_2) = 0.6 \text{ and } r(t_3) = 0.3.$$

Given this distribution, candidate b is the $C.W.$ as

$$r(t_2) > r(t_1) + r(t_3)$$
$$r(t_1) + r(t_2) > r(t_3),$$

Furthermore, candidate b is more than simply a Condorcet Winner. There is more than the expected half of voters that rank him first.

Proposition 9.4.3. *On Large Poisson Games, a candidate who is ranked first by more than the expected half of voters need not be the Winner of the election under* AV.

Proof. We claim that there is a large equilibrium σ of the game (K, T, C, r, u) in which candidate b is not the Winner of the election. In this large equilibrium, the strategy function satisfies

$$\sigma(a|\, t_1) = \sigma(a, b|\, t_2) = \sigma(c|\, t_3) = 1,$$

and the vote distribution is such that

$$\tau(a) = r(t_1), \quad \tau(a, b) = r(t_2), \quad \tau(c) = r(t_3).$$

Given the vote distribution, the vote profile $x = (x(c))_{c \in C}$ is the following vector

$$x(a) \sim \mathcal{P}(0.1n), \quad x(a, b) \sim \mathcal{P}(0.6n) \quad \text{and } x(c) \sim \mathcal{P}(0.3n).$$

In such an equilibrium, the score distribution $\rho = (\rho(k))_{k \in K}$ is such that

$$\rho(a) = \tau(a) + \tau(a,b) = 0.7, \;\; \rho(b) = \tau(a,b) = 0.6 \;\text{ and }\; \rho(c) = 0.3.$$

Given this score distribution, the Winner of the election is candidate a which therefore implies that AV is not Condorcet Consistent in Poisson Games. Finally, given the score distribution, the score profile $s = (s(k))_{k \in K}$ is such that

$$s(a) = x(a) + x(a,b) \sim \mathcal{P}(0.7n), \;\; s(b) = x(b) \sim \mathcal{P}(0.6n) \;\text{ and }\; s(c) = x(c) \sim \mathcal{P}(0.3n).$$

Let us now show why σ is indeed a large equilibrium of this Poisson Approval voting game. The aim is to prove that σ induces a probability distribution over the set of pivot outcomes such that σ is still a best response for voters. The solved minimization problems are included in the appendix. In this example, there are three possible pivot outcomes involving two candidates *pivot*(a,b), *pivot*(a,c) and *pivot*(b,c) and one pivot outcome in which the three candidates are involved.

Given the strategy function σ, the MET implies that the magnitude of the outcome *pivot*(a,b) is equal to the magnitude of the outcome $\{s(a) = s(b) \geq s(c)\}$. Formally, we write

$$\mu[pivot(a,b)] = \mu[\{s(a) = s(b) \geq s(c)\}].$$

The outcome $\{s(a) = s(b) \geq s(c)\}$ can be defined as

$$\{[x(a) \geq 0] \cap [-x(a) \geq 0] \cap [x(a) + x(a,b) - x(c) \geq 0]\}.$$

According to the *DMT*, we know that the magnitude of *pivot*(a,b) is equal to the solution of the following optimisation problem.

$$\tau(a)\exp[\lambda_1 - \lambda_2 + \lambda_3] + \tau(a,b)\exp[\lambda_3] + \tau(c)\exp[-\lambda_3] - 1,$$

such that $\lambda_i \geq 0 \; \forall \; i$. Thus, the magnitude of this pivot outcome is such that

$$\mu[pivot(a,b)] = -0.1.$$

Similarly, combining the MET and the DMT, the magnitude of a pivot between candidates a and c is equal to

$$\mu[pivot(a,c)] = \mu[\{s(a) = s(c) \geq s(b)\}] = -0.0834849,$$

and the magnitude of a pivot between candidates b and c is equal to

$$\mu[pivot(b,c)] = \mu[\{s(b) = s(c) \geq s(a)\}] = -0.151472.$$

Moreover, the magnitude of the pivot between candidates a, b and c is equal to the magnitude of the pivot between candidates b and c, i.e.

$$\mu[pivot(a,b,c)] = -0.151472 = \mu[pivot(b,c)].$$

Therefore, the magnitudes of the pivot outcomes are ordered as follows:

$$\mu[pivot(a,c)] > \mu[pivot(a,b)] > \mu[pivot(b,c)] = \mu[pivot(a,b,c)] \quad \textbf{(B)}.$$

Taking into account inequality **(B)**, one can determine the ballot that each voter of a given type chooses. As previously argued, a voter votes for a candidate k iff the pivot outcome with the highest magnitude involving candidate k is against a less preferred candidate. In this case, the magnitudes of the pivot outcomes are strictly ordered so that voters' best responses immediately follow. Therefore, the strategy function satisfies

$$\sigma(a|\, t_1) = \sigma(a,b|\, t_2) = \sigma(c|\, t_3) = 1,$$

and the vote distribution is such that

$$\tau(a) = r(t_1), \quad \tau(a,b) = r(t_2), \quad \tau(c) = r(t_3),$$

showing that σ is a large equilibrium of the game (K, T, C, r, u).

On the Limit Ordering Condition

It is important to emphasize that in the previous example the limit ordering condition is violated. Indeed, candidates a and b have the highest expected scores but the most probable pivot outcome in which candidate a is involved is $pivot(a,c)$. In terms of the pivot probabilities p_{kl} that only involve pairs of candidates we can write this violation of the limit ordering condition as follows. The expected scores of candidates b and c satisfy $\rho(c) < \rho(b)$ so that the limit ordering condition would imply that the pivot probability p_{ab} becomes far more likely than p_{ac} as the expected number of voters becomes large. However, given inequality **(B)**, we can write that

$$\lim_{n\to\infty} \frac{p_{ab}}{p_{ac}} = 0 \quad \text{with} \quad \rho(c) < \rho(b).$$

This lack of ordering is the source of the bad preference aggregation that arises in equilibrium, preventing the arguments presented by Laslier (2009) from remaining valid in Large Poisson Games.

Single-Peaked Preferences

One cannot escape from this type of bad equilibria by artificially restraining voters' preferences. This example can be extended to a situation in which preferences are single-peaked. Let us we assume that there are four different types $T = \{t_0, t_1, t_2, t_3\}$ such that

t_0	t_1	t_2	t_3
a	a	b	c
c	b	a	a
b	c	c	b

in which the distribution of types r satisfies

$$r(t_0) = \varepsilon, \quad r(t_2) = 0.1 - \varepsilon, \ r(t_3) = 0.6 \text{ and } r(t_4) = 0.3.$$

for some small $\varepsilon > 0$. With such a slight alteration, the large equilibrium in which candidate a is the Winner of the election still exists and the preference profile satisfies single-peakedness.

The Equilibrium Is Not Unique

It is important to emphasize that in this game there is another large equilibrium in which the $C.W.$ coincides with the Winner of the election. In such a large equilibrium, the strategy function $\sigma(. \mid t)$ satisfies

$$\sigma(a \mid t_1) = \sigma(b \mid t_2) = \sigma(a, c \mid t_3) = 1,$$

and the vote distribution is such that

$$\tau(a) = 0.1, \quad \tau(b) = 0.6, \quad \tau(a, c) = 0.3.$$

In this alternative equilibrium, the Winner of the election is candidate b. Indeed, in such an equilibrium, the outcome *pivot*(a, b) becomes infinitely more probable than any other pivot outcome $B \subset \Sigma(C)$ as n tends towards infinity. Voters with type t_1 and t_2 vote for their preferred candidate and the t_3-voters vote for candidate a to prevent candidate b in the event of an outcome *pivot*(a, b).

Nevertheless, AV can uniquely lead to equilibria in which Condorcet Consistency is violated. Indeed, Nuñez (2009) constructs an example with three candidates in the Condorcet Winner is not the Winner of the election at *any* of the equilibria of the game.

9.5 Conclusion

This work analyses the properties of AV on models with a large number of voters. The Myerson–Weber framework (Myerson and Weber 1993) has the virtue of being simple and at the same time setting up some simple comparisons between one-shot voting rules. In such a framework, AV leads to better preference aggregation than PV in some situations. However, its simplicity is due to the lack of a formal game-theory model that raises questions about the assumptions concerning pivot probabilities. Both Score Uncertainty model and Large Poisson Games address these technical problems and give positive answers: it is indeed possible by means of a formal model to obtain that pivot probabilities are common knowledge (as far as voters' best responses are concerned, this is true for both models) and that pivot probabilities are "well ordered" (this is only correct in the Score Uncertainty model).

Large Poisson Games possess several advantages such as the independent actions or the environmental equivalence property that simplify the analysis of the voting equilibria. Using these games, Myerson (2002) shows that AV is more robust to information manipulation than other one-shot voting rules such as PV in some simple voting games. However, AV does not preclude paradoxical situations from arising as a consequence of the independent actions property as shown by Nuñez (2009, 2010). When the voting rule allows to vote for more than one candidate, the fact that the number of voters who cast a given ballot is independent of the number of voters who cast another one (independent actions property) naturally implies that the scores of the candidates are correlated. This correlation implies that the limit ordering condition of the pivot probabilities is violated. As a consequence of this non-intuitive ordering, the Winner of the election does not always coincide with the Condorcet Winner. Whenever the voters anticipate that the Condorcet Winner is not included in the most probable pivot outcome, he need not be the Winner of the election in equilibrium. This fact limits the reduction of Nash equilibria that arises in Large Poisson Games.

In the Score Uncertainty model (Laslier 2009) candidates' scores are independent random variables. With such an independence, the pivot probabilities satisfy the limit ordering condition. Hence, AV ensures that voters' best responses are sincere and the Condorcet Winner wins the election whenever it exists, provided that every candidate gets a strictly positive share of votes.

Appendix

This appendix provides the constrained minimization problems used to compute the magnitudes of the pivot outcomes in Sect. 9.4.5, in the large equilibrium in which the Condorcet Winner does not coincide with the Winner of the election.

Magnitude of a Pivot Between Candidates a and b

$$\mu[\{s(a) = s(b) \geq s(c)\}] = \min_{\lambda} \tau(a)\exp[\lambda_1 - \lambda_2 + \lambda_3] + \tau(a,b)\exp[\lambda_3] + \tau(c)\exp[-\lambda_3] - 1,$$

such that $\lambda_i \geq 0 \ \forall \ i$. The solution to this problem yields

$$\mu[\{s(a) = s(b) \geq s(c)\}] = \mu[\{x(a) = 0\}] = -r(t_1) \text{ as } r(t_2) > r(t_3).$$

Magnitude of a Pivot Between Candidates a and c

$$\mu[\{s(a) = s(c) \geq s(b)\}] = \min_{\lambda} \tau(a)\exp[\lambda_1 - \lambda_2 + \lambda_3] + \tau(a,b)\exp[\lambda_1 - \lambda_2] + \tau(c)\exp[-\lambda_1 + \lambda_2] - 1,$$

such that $\lambda_i \geq 0 \ \forall \ i$. Therefore,

$$\begin{aligned} \mu[pivot(a,c)] &= \mu[\{s(a) = s(c) \geq s(b)\}] \\ &= -(\sqrt{r(t_1) + r(t_2)} - \sqrt{r(t_3)})^2 = \mu[x(a) + x(a,b) = x(c)]. \end{aligned}$$

Magnitude of a Pivot Between Candidates b and c

$$\mu[\{s(b) = s(c) \geq s(a)\}] = \min_{\lambda} \tau(a)\exp[-\lambda_3] + \tau(a,b)\exp[\lambda_1 - \lambda_2] + \tau(c)\exp[-\lambda_1 + \lambda_2] - 1,$$

such that $\lambda_i \geq 0 \ \forall \ i$. Therefore,

$$\begin{aligned} \mu[pivot(b,c)] = \mu[\{s(b) = s(c) \geq s(a)\}] &= -r(t_1) - (\sqrt{r(t_2)} - \sqrt{r(t_3)})^2 \\ &= -r(t_1) + \mu[x(a,b) = x(c)] \\ &= \mu[pivot(a,b,c)]. \end{aligned}$$

References

Abramowitz, M., & Stegun, I. (1965). *Handbook of mathematical tables*. New York: Dover.

Bouton, L., & Castanheira, M. (2008). *One person, many votes: Divided majority and information aggregation* (mimeo). ECARES.

Brams, S. (2008). *Mathematics and democracy: Designing better voting and fair-division procedures*. Princeton, NJ: Princeton University Press.

Brams, S., & Fishburn, P. (1981). Approval voting, Condorcet's principle and runoff elections. *Public Choice, 36*, 89–114.

De Sinopoli, F., Dutta, B., & Laslier, J.-F. (2006). Approval voting: Three examples. *International Journal of Game Theory, 38*, 27–38.

De Sinopoli, F., & Gonzalez Pimienta, C. (2009). Undominated (and) perfect equilibria in Poisson games. *Games and Economic Behaviour, 66*(2), 775–784.

Goertz, J., & Maniquet, F. (2008). *On the informational efficiency of approval voting* (mimeo). CORE.

Krishna, V., & Morgan, J. (2008). *Voluntary voting: Costs and benefits* (mimeo). Penn State University.

Laslier, J. (2009). The leader rule: A model of strategic approval voting in a large electorate. *Journal of Theoretical Politics, 21*, 113–136.

Laslier, J., & Sanver, R. (2010a). The basic approval voting game. In J. Laslier & R. Sanver (Eds.), *Handbook on approval voting*. Heildelberg: Springer-Verlag.

Laslier, J., & Sanver, R. (2010b). Laboratory experiments on approval voting. In J. Laslier & R. Sanver (Eds.), *Handbook on approval voting*. Heildelberg: Springer-Verlag.

Myerson, R. (1998). Population uncertainty and Poisson games. *International Journal of Game Theory, 27*, 375–392.

Myerson, R. (2000). Large Poisson games. *Journal of Economic Theory, 94*, 7–45.

Myerson, R. (2002). Comparison of scoring rules in Poisson voting games. *Journal of Economic Theory, 103*, 219–251.

Myerson, R., & Weber, R. (1993). A theory of voting equilibria. *American Political Science Review, 87*, 102–114.

Nuñez, M. (2009). *Two examples of strategic approval voting* (mimeo).

Nuñez, M. (2010). Condorcet consistency of approval voting: A counter example on large Poisson games. *Journal of Theoretical Politics, 22*(1), 64–84.

Schelling, T. (1960). *The strategy of conflict*. Harvard: Cambrige University Press.

Selten, R. (1975). A reexamination of the perfectness concept for equilibrium points in extensive games. *International Journal of Game Theory, 4*, 25–55.

Weber, R. (1995). Approval voting. *Journal of Economic Perspectives, 9*, 39–49.

Chapter 10
Computational Aspects of Approval Voting

Dorothea Baumeister, Gábor Erdélyi, Edith Hemaspaandra, Lane A. Hemaspaandra, and Jörg Rothe

10.1 Introduction

"Yes, we can!" – Barack Obama's campaign slogan inspired enough of his supporters to go to the polls and give him their "yes" votes that he won the 2008 U.S. presidential election. And this happened notwithstanding the fact that many other voters said "no" when pollsters asked if they viewed Barack Obama as qualified for the office. "Yes" and "no" are perhaps the most basic ways for us, as voters, to express our preferences about candidates, and "yes" and "no" are what approval voting is all about.

In approval voting, every voter either approves or disapproves of each candidate. At the end of the day, all approvals are counted and whoever is approved of by the most voters wins. Since Brams and Fishburn (1978) proposed this system three decades ago, it has been studied intensely in social choice theory (see, e.g., Brams 1980; Brams and Fishburn 1981, 1983, 2002, 2005; Brams et al. 2004, 2007a; Brams and Sanver 2006, 2009; Dutta et al. 2006; Kilgour et al. 2006), it has been adopted by numerous scientific and engineering societies (such as the IEEE),[1] and it has even been dubbed "the electoral reform of the twentieth century" by its proponents (see Dutta et al. 2006). This chapter focuses on the computational aspects of approval voting.

Why should one bother to study the computational aspects of approval voting? Isn't this just a matter of summing up the approvals each candidate receives and

[1]Notwithstanding the success of approval voting in many such societies, Brams and Fishburn (2005) also note that approval voting "is not currently used in any public elections, despite efforts to institute it, so its success should be judged as mixed." For example, the U.S. presidential election, which involves the U.S. electoral college and has aspects of both plurality and majority voting, is not held under approval voting.

J. Rothe (✉)
Institut für Informatik, Heinrich-Heine-Universität Düsseldorf, 40225 Düsseldorf, Germany
e-mail: rothe@cs.uni-duesseldorf.de

J.-F. Laslier and M.R. Sanver (eds.), *Handbook on Approval Voting*, Studies in Choice and Welfare, DOI 10.1007/978-3-642-02839-7_10,

comparing the resulting approval scores? Sure enough, adding and comparing are quite easy computational tasks, and this is one important point in favor of approval voting: It is easy to determine the winners of an approval election. However, "computational aspects" here is meant in a much broader and deeper sense than with regard to mere winner determination. In particular, for approval voting (AV, for short) and one of its variants (SP-AV), this chapter discusses computational issues that model attempts to tamper with the outcome of an election in various ways, and we will pay particular attention to the complexity results known for these computational problems.

For example, the issue of voters tampering with election outcomes by reporting insincere preferences, commonly referred to as strategic voting, has been studied intensely in the social choice literature since the seminal work of Gibbard (1973) and Satterthwaite (1975) (see also Duggan and Schwartz 2000; Everaere et al. 2007). The celebrated Gibbard–Satterthwaite Theorem states that essentially all natural voting systems are in some sense manipulable by a strategic voter.

This is bad news. But there is hope! Imagine a voting system that, though manipulable in principle, has the property that it would confront a strategic voter with a really hard problem to solve when attempting to find a successful manipulative preference to reveal – or even to tell if in the given election such a manipulation is possible. Such a voting system would still be susceptible to manipulation in theory, but one could reasonably hope that due to the complexity of the associated problem no manipulation attempts would ever occur in real elections (or, if they did occur, that they wouldn't be successful).

Employing computational complexity as a means to protect voting systems from manipulation and other attacks known as "control" attacks was first proposed by Bartholdi et al. (1989a, 1992) and Bartholdi and Orlin (1991) and has since been studied for a wide range of concrete systems (see, e.g., Conitzer and Sandholm 2003, 2006; Elkind and Lipmaa 2005; Faliszewski et al. 2006, 2009a; Faliszewski et al. 2008b; Conitzer et al. 2007; Hemaspaandra and Hemaspaandra 2007; Procaccia and Rosenschein 2007; Meir et al. 2008; Zuckerman et al. 2008; Brelsford et al. 2008 for results on manipulation, and see Sect. 10.3 for the literature regarding control attacks). Some systems have been shown to be "resistant" to manipulation (i.e., informally speaking, their manipulation problem is computationally hard) and some systems have turned out to be "vulnerable" to manipulation (i.e., their manipulation problem is easy to solve).

What does it mean for a problem to be "computationally hard" or "easy" to solve? Complexity theory provides the notions and techniques needed to answer these questions. Two important tasks in complexity theory are to classify problems in terms of their algorithmic complexity and to compare the complexities of two given problems via reductions. This chapter will present numerous concrete reductions that establish the computational hardness of problems related to approval voting. For more background on complexity theory, we refer to the textbooks Papadimitriou (1994), Hemaspaandra and Ogihara (2002), and Rothe (2005).

For approval voting and manipulation, however, the bad news persists: Approval voting is vulnerable to manipulation, even if we allow voters to have "weights" (for example, one approval of a voter of weight five counts as much as five weight-one approvals) and even if we allow a coalition of strategic voters to coordinate their manipulative efforts (as opposed to having a single strategic voter). That is, even in such more general settings, there is a polynomial-time algorithm that solves the manipulation problem for AV. Let us state this problem in the form that will be used for representing computational problems throughout this chapter:

Name: AV-MANIPULATION.
Given: A set C of candidates, a list V of ballots over C (that already have been cast by the nonstrategic voters, where each ballot gives that voter's approvals/disapprovals for each $c \in C$) along with these voters' weights, a list S with the weights of the strategic voters (whose ballots haven't been cast yet), and a distinguished candidate $p \in C$.
Question: Is it possible for the strategic weighted voters to cast their ballots in such a way that p becomes an approval winner of the resulting election?

As noted by Faliszewski et al. (2006, 2009a),[2] one can use the following simple greedy strategy to solve this problem in polynomial time: The strategic voters simply approve of their favorite candidate p and disapprove of all other candidates. If this manipulation makes p an approval winner, they have reached their goal (and the polynomial-time algorithm can check whether this happens because, as mentioned earlier, winner determination is easy for AV, and so the algorithm accepts its input in this case). But if p still is not an approval winner after this manipulation, then no strategy whatsoever can turn p into an approval winner (and so, in this case, the algorithm can safely reject its input).

So, as there is no hope for approval voting to computationally resist manipulation, it's time for a change! Let's change what is being changed in the tampering attempts and how it is being changed – that is, let's change the tampering scenario from manipulation to either bribery or control. And let's also change who performs these changes in the election: Both these scenarios differ from manipulation in that they model situations where external actors do seek to affect the outcome.

The model for the complexity-theoretic study of bribery was introduced by Faliszewski et al. (2006, 2009a). In bribery settings, the "briber" seeks to influence the outcome of an election via bribing certain voters to make them change their preferences, without exceeding the briber's budget. The specific bribery scenarios we will consider involve, for example, weighted and unweighted voters, voters with and without price tags, changing one complete ballot (dubbed "bribery") as opposed to changing just one approval/disapproval in a ballot at unit cost (dubbed "microbribery"), and we will present the associated computational problems and their complexities.

[2] In a different, slightly more flexible scenario, Meir et al. (2008) showed that approval voting is vulnerable to manipulation by a single strategic voter.

The model for the complexity-theoretic study of control was introduced by Bartholdi et al. (1992). In control settings, the election's "chair" seeks to influence the outcome of an election via modifying its structure, namely, via actions such as adding/deleting/partitioning either candidates or voters. These scenarios loosely model activities that we often encounter in political elections, such as get-out-the-vote drives ("adding voters"), disenfranchisement and other means of voter suppression ("deleting voters"), introducing new "spoiler" candidates into an election ("adding candidates"), and so on. This chapter will present the complexity of AV and SP-AV in each of the common 22 control scenarios, which results in a total of 44 complexity results in this section.

Unfortunately, there is again bad news for AV: Approval voting is vulnerable to many (and is computationally resistant to only four) of these 22 control attacks. However, it will also turn out that SP-AV, which stands for "sincere-strategy preference-based approval voting," displays broad resistance to control: SP-AV is resistant to 19 of the 22 control attacks. That is, among natural voting systems with a polynomial-time winner-determination procedure, SP-AV possesses the broadest resistance to control currently known to hold (see Sect. 10.3 for a more detailed discussion that compares SP-AV with other systems displaying an exceptionally broad resistance to control – voting systems that belong to the Copeland/Llull complex of systems, Faliszewski et al. 2007, 2008a, 2009b).

Unlike many natural and widely used voting systems that are defined in terms of rankings (i.e., strict linear orderings) of the candidates, approval voting merely distinguishes between each voter's acceptable and unacceptable candidates, yet completely ignores the preference rankings that voters may have about the candidates. To overcome this shortcoming, Brams and Sanver (2006) proposed SP-AV as a voting system that combines preference-based voting with approval voting. In their definition, they require each voter to have an approval strategy[3] that is both "sincere" and "admissible." An *approval strategy* of a voter is simply a partition of the candidates into approved and disapproved candidates. It is *sincere* if there are no "gaps" (with respect to this voter's preference ranking of the candidates), i.e., if this voter approves of some candidate then he or she also approves of each candidate ranked higher. A voter's approval strategy is said to be *admissible* if the voter approves of his or her top candidate and disapproves of his or her bottom candidate. Note that in a one-candidate election no voter can have an admissible approval strategy.

Sincerity and admissibility are quite natural notions to require. For example, sincerity makes sure that there is no conflict between a voter's preference ranking and approval strategy. Admissibility in particular prevents approval strategies from being trivial: It is not admissible for a voter to either approve or disapprove of every candidate in an election. Brams and Sanver (2006) point out that an admissible approval strategy is not dominated in a game-theoretic sense (see Brams and

[3] To avoid confusion, we stress that the "strategy" in "approval strategy" or "AV strategy" is not meant in the sense of strategic voting, but rather simply refers to which candidates are approved of and which are disapproved of by a voter.

Fishburn 1978 and also, e.g., Dutta et al. 2006). Obviously, if voters are not allowed to either approve or disapprove of everybody then sincere approval strategies are always admissible. Following Erdélyi et al. (2008b,c), we require that all voters must cast only sincere votes and, when there are at least two candidates, voters must cast only sincere, admissible votes (we will call this *Convention 1*).

Within AV, ballots are given as 0–1 (or, equivalently no–yes) approval vectors, where "1" represents approval and "0" disapproval. Within SP-AV, ballots contain more information than this, as they also provide a preference ranking of the candidates. We represent such rankings from left to right (so the leftmost candidate is the most preferred one), and we additionally indicate the approvals/disapprovals by inserting an "approval line" into the ranking, where all candidates to the left of this line are approved and all candidates to the right of this line are disapproved. For example, "$a \ b \ c \ | \ d$" means that this voter approves of a, b, and c, but disapproves of d. Since we require approval strategies (with at least two candidates) to be admissible, the approval line will never be to the left or to the right of all candidates.

There is a central point regarding admissibility that we need to discuss with respect to our control scenarios. In cases of control by deleting candidates or by partition of candidates or voters (the formal definitions of which will be presented in Sect. 10.3), it may happen that admissibility will be lost due to the control action. That is, an originally admissible approval strategy might be changed into an inadmissible one by, for example, deletion of candidates. To prevent this from happening (and to obey our convention), we handle such cases by applying the following rule (that we will call *Rule 1*): "If during or after a control action an election with at least two candidates is obtained such that some voter approves of either all candidates or no candidates, then the approval line in each such voter's ballot is moved so as to respectively disapprove of that voter's bottom candidate or approve of that voter's top candidate." This rule was introduced by Erdélyi et al. (2008c, see also Erdélyi et al. 2008b) in order to preserve (or re-enforce) admissibility under such control actions. So, for example, if candidate d has been deleted from an election containing the ballot $a \ b \ c \ | \ d$ then, according to this rule, this ballot is being changed into $a \ b \ | \ c$.

Now, coming back to our original question of why one should bother to study the computational aspects of approval voting, it is important to note that this study is motivated not just from the social choice and political science point of view. Indeed, decision-making and preference aggregation are central tasks in many areas of computer science, and voting procedures are far from being confined to political elections in human societies. They have many applications in computer science, ranging from multiagent systems within artificial intelligence (see, e.g., Ephrati and Rosenschein 1997) to the development of recommender systems (see Ghosh et al. 1999) to aggregating the web-page rankings from various search engines (see Dwork et al. 2001), to name just a few. These are topics at the heart of the emerging area of *computational social choice*, which is at the interface of social choice theory and computer science and is developing rapidly (see Endriss and Lang 2006 and Endriss and Goldberg 2008).

Of course, the computational aspects of voting systems and their associated problems comprise more than just proving these problems hard or easy to solve. Certainly, if some problem has been shown to be computationally hard, the actors involved (be it strategic voters, the chair, or the briber) will seek to find ways to circumvent this obstacle. Fortunately, computer scientists have developed many approaches for how to cope with the computational hardness of problems in practical applications, and some of these apply here. In particular, we will present an approximation algorithm and some local search heuristics for "minimax approval voting," an interesting variant of approval voting that was proposed by Brams et al. (2004, see also Brams et al. 2007a,b) for the purpose of electing a committee of fixed size.

This chapter is organized as follows. Section 10.2 is a detailed discussion of worries about the model, and presents what we feel is the ideal framing. In Sect. 10.3, we will present the 44 control results for AV and SP-AV mentioned above. We will also describe the needed complexity-theoretic notions in a way accessible to readers not familiar with complexity theory. In particular, for each of the reductions constructed, we will give comprehensible examples, and also the problems from which we reduce will be illustrated via examples. Section 10.4 presents the complexity results for bribery. Finally, in Sect. 10.5, we will present local search heuristics for minimax approval voting. (Since this is a survey chapter, by "present" we do not mean that the results are due to this chapter, and in each case we will point the reader to the source papers in the technical literature.)

10.2 Discussion of Models for Control of Domain-Sensitive Rules

We now come to a particularly important discussion – of the model and approach to SP-AV under control. The previous section describes the "rule," Rule 1, that Erdélyi et al. (2008c see also Erdélyi et al. 2008b) creates to handle the fact that unlike any other election system whose control properties have been studied, SP-AV has a domain restriction that has the property that some (in fact, six) of the 22 common control types can turn legal (i.e., in the domain of SP-AV) inputs into inputs that are not in the domain of SP-AV. On its surface, thus, control analysis simply conflicts with SP-AV elections.

Erdélyi et al. (2008c, see also Erdélyi et al. 2008b) approaches this with a rule, Rule 1, that within the control framework readjusts preferences, plus keep in mind also Erdélyi et al.'s (2008c) Convention 1, which itself blocks certain votes from even being legal to cast at all. However, this approach is arguably unsatisfying and may lead readers to think, incorrectly, that they are seeing control results about SP-AV. Treating a preference rewrite rule as if it is part of control is unconvincing since control itself is not about doing anything other than what its definition embraces, and in various settings various rewrite rules could be proposed, all in ways whose justification is not about control but hinges on one's own subjective notion of what the "natural" correction is to move one back onto the domain of the election systems. After all, the Erdélyi et al. (2008b,c) idea that (no, no, no, no) and (yes, yes, yes, yes), each of which show complete equality among the candidates, should be rewritten to, respectively, (yes, no, no, no) and (yes, yes, yes, no), is a matter of taste and this rewrite is not a part of the notion of SP-AV as Brams and Sanver defined it – (no, no, yes, yes) is for example another possible rewrite.

In fact, the Brams and Sanver notion of admissibility was not designed with control in mind and in the context of control has restrictive, grotesque effects. In control by adding candidates, if one dislikes the two initial candidates but loves all the spoiler candidates, this approach would force one to cast a vote that approved one of the two disliked candidates, and one would be bound under control to have that approval of a disliked candidate remain – along with new approvals of liked candidates – after spoiler candidates were added.[4] Again, we stress that Brams and Sanver were not anticipating control to be spliced on top of their system, so in mentioning this we are not criticizing their system. However, we will soon give a system, SP-AV-CTA, that we suggest is the natural way to merge control with the general flavor of SP-AV, and we mention now that our approach will remove the effect just noted above; one will be able to cast a vote that says "I will approve my favorite candidate from among those that end up as active candidates in this control by adding candidates election."

Of course, returning to the Erdélyi et al. (2008b,c) preference rewrite rule, if the preference rewrite rule is of minor importance to the control results one might think there is no issue here. But in fact, the broadened resistance that Erdélyi et al. (2008b,c) prove for "SP-AV" is due not at all to SP-AV itself, but rather is completely due to the preference rewrite rule Erdélyi et al. (2008b,c) introduce. For this entire chapter, for consistency with their work, we will echo their view that modifies control with the preference rewrite rule/convention and speaks of the results as if they were about SP-AV.

However, for posterity, we point out what we feel is a more natural approach. In particular, if our process is going to rewrite preferences by some ad hoc rule, the most intellectually frank way to do that is to openly admit that one's election system is not SP-AV. Rather, consider the following election system, SP-AV-CTA (*C*oerce *T*o *A*dmissible). The system's domain will be the same as that of SP-AV except

[4] This assumes that regarding Convention 1 one's notion of the "initial election" in control by adding candidates is the basic candidate set C with no spoiler candidates yet added. This is the natural approach and is consistent with how most papers define this control type: Voters indeed have preferences over $C \cup S$, with S the potential spoiler set, but the base election is with C and then one adds in some candidates (call this view of addition of candidates the Base-Is-C view). Read absolutely literally, Erdélyi et al. (2008b,c), by diverging from the standard definition, seem to hint at the approach that the initial election in terms of Convention 1 is over $C \cup S$, and after enforcing Convention 1 on that, one deletes some of S. Call this view of addition of candidates the Base-Is-($C \cup S$) view. We mention that this unusual view would avoid the issue mentioned above, but at the cost of taking a quite counterintuitive view of what the original ("before control") election is in the addition of candidates case.

it allows inadmissible votes. And its action on any in-domain input is to treat the admissible votes as themselves, to coerce each inadmissible vote into an admissible vote using precisely the preference rewrite rule of Erdélyi et al. (2008b,c), and to then act precisely as SP-AV would on that input.

SP-AV-CTA is in effect precisely what Erdélyi et al. (2008b,c) are feeding their post-control votes into. We suggest one bring SP-AV-CTA out of the closet, and simply directly study it and its control properties. (Of course, SP-AV-CTA may lose some of the nice game-theoretic or other properties of SP-AV. But if so, that too should be openly faced.) Note that SP-AV-CTA sidesteps the control by adding candidates "two disliked candidates initially" example mentioned earlier, as in SP-AV-CTA, a vote of (no, no, ..., no) plus one's preference order over all initial and spoiler candidates has the effect of approving precisely one's favorite candidate (if there are at least two candidates left in, of course).

As a final comment, we mention that control of SP-AV-CTA is not precisely what the Erdélyi et al. (2008b,c) results formally speak to. There is a very subtle difference in that in SP-AV-CTA the initial votes can be inadmissible (but after control all then-inadmissible votes are interpreted as admissible votes). In contrast, in the Erdélyi et al. (2008b,c) model, the initial votes must already be admissible (because our system is SP-AV and admissibility is part of its domain requirement),[5] plus after the control action they in effect feed the post-control election to the SP-AV-CTA system. However, this subtle distinction in this particular case seems unlikely to remove any resistances, and so we believe SP-AV-CTA is almost certain to retain 19 resistances. But, in any case, as a model, directly studying SP-AV-CTA, with no restrictions, seems cleaner and crisper than claiming to study the known system SP-AV while in effect really studying effects related precisely to SP-AV-CTA's departures from SP-AV. Indeed, the attractiveness of our SP-AV-CTA approach is sufficiently compelling that Erdélyi, Nowak, and Rothe, in light of the present chapter, adopted the SP-AV-CTA model in their final version Erdélyi et al. 2009c.

[5] Even this doesn't mean as much as it seems. Since, for example, control by adding candidates always takes as its input preferences over the union of the initial and the spoiler candidate set, and Erdélyi et al. (2008b,c) diverge from most papers in that they explicitly state that the input election *is* $(C \cup S, V)$ seems to actually in that control case allow, under what we earlier called the Base-is-$(C \cup S)$ view, inputs that are inadmissible with respect to the original election's candidate set. Presumably, their rule would coerce the ballots that didn't via the addition of candidates end up admissible. Under the base-is-$(C \cup S)$ view, their approach would be extremely close or quite possibly even semantically identical to looking at – with no special rules or conventions – control for SP-AV-CTA. However, if one really wants to enforce Convention 1, it seems to us that the Base-is-C view is far more compelling. However, our suggestion is that Convention 1 and Rule 1 be discarded – along with the convoluted structure they form – and one simply define stand-alone election systems, such as SP-AV-CTA, that can be analyzed under the utterly standard, long-studied notion of control. In particular, preference coercions should be internalized into the election system.

10.3 The Control Complexity of Approval Voting and Sincere-Strategy Preference-Based Approval Voting

10.3.1 Introduction, Background, and Discussion

In this section, we will present the control complexity of AV and SP-AV for each of the ten classic types of procedural control, and also for an 11th type that has recently been suggested as the "right" version of one of those ten types. Remarkably, each of these complexities – and there are 44 problems involved here – is known. Our goal here is not to prove each of these 44 results, but rather to make clear what each of the results means, and to provide representative proofs that give the flavor of how one obtains such results.

Recall from the Introduction that by studying control, we are asking how hard it is to determine whether a favored candidate can be made the unique winner (or, in the destructive cases, whether a despised candidate can be precluded from being a unique winner) via a given type of alteration of the structure of the election by the election's "chair." The 11 types of control are abstracting many types of actions that occur in the real world, such as voter suppression, get-out-the-vote efforts, candidate recruitment, etc. Ideally, we would hope that our election systems block all such attacks – that each such attack either never succeeds (so-called immunity) or that it is computationally intractable for the chair to find such a successful alteration (indeed, even to tell if one exists – so-called resistance). The case we want to avoid – so-called vulnerability – is that the chair's task is easy: The chair in polynomial time can determine whether he/she can make a successful alteration.

One might worry that such ideal hardness cannot possibly be achieved. For example, perhaps there is an Arrow-like "impossibility" theorem showing that no election system satisfying some natural, attractive, broadly-satisfied-by-typical-election-systems axioms is, for each of our 11 control attacks, resistant or immune to the attack. The existence of such an impossibility theorem is, on its face, at least plausible. However, recently it was shown that for control such an impossibility claim is itself impossible to obtain – because it is untrue. In particular, Hemaspaandra et al. (2007b, 2009, see also Faliszewski et al. 2009b regarding how to include also the 11th control type) proved that there is an election system – indeed, an election system whose winner-determination problem is computationally feasible – that is resistant to all of our control types.

Given that result, one might naturally ask why the results of this section – on AV and SP-AV – are even worth doing. After all, Hemaspaandra et al. (2009) provides a "perfect" election system. But one must think about that in context. That election system indeed is "perfect" in terms of avoiding our control manipulations. But that election system is not AV or SP-AV, and in the real world, approval voting is very attractive and often used, and so the properties of approval voting are richly worth studying – thus, this book! In fact, the election system of Hemaspaandra et al. (2009) not only is not AV or SP-AV but indeed is a quite complicated, unnatural construct. What that paper does is to show how to hybridize election systems together in such a way that if any one of the constituent systems is immune or resistant to a given control type, then the hybridized system is immune or resistant to that type. Simply put, the process unions the advantages of all the constituent systems. But the price of doing this is a bulky, complicated election system that – while sufficient to prove the impossibility of obtaining an impossibility theorem – would not be attractive for everyday use.

For this reason, it is natural to study already existing, attractive election systems to determine whether they are highly resistant to control. Our hope, of course, is that the systems we study will be highly resistant (or immune) to control. What we will see in this section, however, is that approval voting is vulnerable to many types of control, as shown by Hemaspaandra et al. (2007a). This is disappointing, and is one weakness of approval voting. On the other hand, approval voting is quite attractive in many ways, and so we certainly are not suggesting that AV's vulnerability to control means AV should never be used. Rather, in selecting an election system, one must weigh the strengths and weaknesses of the systems one is considering, and then must choose the one that is most appropriate for the use to which one will be putting the election system.

However, if one is a fan both of approval voting and of resistance to control, one will find good news in this section. SP-AV, which itself combines aspects of approval voting with aspects of plurality voting, has been shown by Erdélyi et al. (2008c, see also Erdélyi et al. 2008b) to have precisely those resistances possessed by either of those systems. As such, it has a higher number of control resistances than any previously studied natural system. The natural systems that previously had the most proven control resistances were from the Copeland/Llull election system complex (Faliszewski et al. 2008a, 2007). (Those systems remain the natural systems with the most proven resistances among voting systems in which, unlike AV and SP-AV, votes are simply a strict ordering over the candidates.)

The rest of this section is organized as follows. Section 10.3.2 quickly presents the election model, the needed notions from complexity theory, and the notions used to classify the hardness of control. Section 10.3.3, for each control type, describes the motivation for that control type and the results that hold for that control type for AV and SP-AV elections. Section 10.3.4 gives some conclusions for control issues related to AV and SP-AV.

10.3.2 Notions: Elections and Complexity

Our elections will have finite numbers of candidates and voters, but the numbers can vary from input to input. So an election is a pair (C, V), where C is the candidate set and V is a list of ballots (votes) expressing preferences over C. For AV, each ballot will be a length $\|C\|$ bit-vector, expressing approval or disapproval for each candidate. For SP-AV, as discussed in the Introduction, a ballot will be such a vector along with a strict ordering over C. We will use the terms ballot, vote, and voter relatively interchangeably. Within both AV and SP-AV, for each election (C, V) and for each candidate $c \in C$, we use $score_{(C,V)}(c)$ to denote the number of approvals c receives from voters in V. In the case of weighted voters, each approval of a voter of weight k counts as k weight-one approvals.

In this section, we'll be discussing the complexity of control problems. The control types will be presented in Sect. 10.3.3, but for now it will suffice to say that for each control-type problem we study, the input to the problem will be an election, (C, V), a specified candidate, p, and in some cases some information specific to the control problem.

For each control type, we will study two different problems, the constructive problem (trying to make p the unique winner) and the destructive problem (trying to ensure that p ends up not being a unique winner).

To make this a bit more concrete, we now define explicitly two of our problems: Constructive Control by Adding (a Limited Number of) Candidates (CCAC, for short) and Destructive Control by Adding (a Limited Number of) Candidates (DCAC, for short). For these and all other control types, we will in order to ensure cross-comparability with earlier work take the control-type definitions essentially word-for-word from Faliszewski et al. (2009b), which itself for historical consistency is in general following even earlier papers (except Faliszewski et al. 2009b studies both the model in which we seek to make/preclude a winner and in which we seek to make/preclude a unique winner, but in this chapter, like the papers we are most focused on Hemaspaandra et al. 2007a, Erdélyi et al. 2008c,b, we focus on the unique-winner model). "$\mathcal{E}$" denotes the election system's name (AV or SP-AV in our case).

Control by Adding a Limited Number of Candidates

Name: $\mathcal{E}$-CCAC and $\mathcal{E}$-DCAC.

Given: Disjoint sets C and D of candidates, a collection V of voters represented via their ballots over the candidates in the set $C \cup D$, a distinguished candidate $p \in C$, and a nonnegative integer k.

Question ($\mathcal{E}$-CCAC): Is there a subset E of D such that $\|E\| \leq k$ and p is the unique winner of the $\mathcal{E}$ election $(C \cup E, V)$?

Question ($\mathcal{E}$-DCAC): Is there a subset E of D such that $\|E\| \leq k$ and p is not a unique winner of the $\mathcal{E}$ election $(C \cup E, V)$?

Now, we are ready to define the notions that are used to describe the difficulty (and possibility) of a given type of control for a given election system. If for a given control type and election system the "chair" – the actor exerting the given type of control – can never, on any input, change p from not being a unique winner to being the unique winner (or, for the destructive case, change p from being the unique winner to not being a unique winner), we say the system is *immune* to the given type of control. One might think that immunity never occurs, but in fact it does. For example, in AV, if p already is losing to (or tieing for winner with) some other candidate q, then adding additional candidates certainly won't make p become the unique winner. Thus we have our first theorem: AV is immune to constructive control by adding candidates (Hemaspaandra et al. 2007a). If immunity does not hold for a given control type, we say the election system is *susceptible* to this type of control.

Regarding election systems susceptible to a type of control, there are two cases that will particularly interest us. If the given problem has a polynomial-time algorithm to tell whether on a given input there exists a control action of the problem's sort that achieves the chair's goal for p, we say the election system is *vulnerable* to the given type of control. (In each case in this chapter where vulnerability is asserted, something even stronger in fact holds, as noted by the original papers, namely, there is a polynomial-time algorithm that tells whether a successful control action by the chair exists, and if so *produces such an action*.) However, if a given control problem is NP-hard (a central notion in complexity theory to be defined in the next paragraph), we say that the election system is *resistant* to the given type of control. Informally put, resistance to some control type means that the corresponding control problem is impracticable (although potentially not impossible) to solve.

Some readers of this chapter may not be familiar with the notion of being "NP-hard," so we briefly discuss the concept. NP, nondeterministic polynomial time, is the class of all problems that can be solved in polynomial time on a nondeterministic Turing machine. However, an elegant way to describe the class without having to introduce nondeterminism or Turing machines is that NP is the class of all languages L for which there exist a polynomial q and a polynomial-time computable boolean predicate R such that $L = \{x \mid (\exists y)[|y| \leq q(|x|) \text{ and } R(x, y)]\}$. NP is a tremendously important class – it captures the complexity of thousands of crucial problems, ranging from satisfiability of boolean formulas to the traveling salesperson problem. Of course, some NP problems are computationally simple, e.g., the empty set is in NP. The problems that have the property that each NP problem can be rephrased in terms of them are called the NP-*hard* problems. Formally, a problem A is NP-hard if $(\forall B \in \mathrm{NP})[B \leq_m^p A]$. $B \leq_m^p A$ (B *many-one polynomial-time reduces to* A) by definition means there exists a polynomial-time computable function f such that $(\forall x)[x \in B \iff f(x) \in A]$. A problem is called NP-*complete* exactly if it is in NP and is NP-hard. Each of the problems stated in this chapter as being resistant in fact is not only NP-hard, but also happens to be NP-complete.

The concepts of immunity, susceptibility, vulnerability, and resistance for control were introduced by Bartholdi et al. (1992), and the above definitions are theirs except we follow the more logical, now more common approach, introduced in Hemaspaandra et al. (2007b, 2009), of defining resistant as being "susceptible, NP-hard" problems (Bartholdi et al. 1992 defined it as "susceptible, NP-complete" problems).

10.3.3 44 Control Results

In this section, for each of the 11 commonly discussed types of control, and for each of the constructive and destructive cases, we present what results hold for AV and

Table 10.1 Overview of results. Key: I means immune, R means resistant, V means vulnerable, TE means ties-eliminate, and TP means ties-promote. Results for SP-AV are due to Erdélyi et al. (2008c); their proofs are either due to Erdélyi et al. (2008c) or draw on proofs from Hemaspaandra et al. (2007a). Results for AV are due to Hemaspaandra et al. (2007a). (The results for control by adding a limited number of candidates for AV, though not stated explicitly in Hemaspaandra et al. 2007a, follow immediately from the proofs of the corresponding results for the "unlimited" variant of the problem.)

	SP-AV		AV	
Control by	Constructive	Destructive	Constructive	Destructive
Adding an unlimited number of candidates	R	R	I	V
Adding a limited number of candidates	R	R	I	V
Deleting candidates	R	R	V	I
Partition of candidates	TE: R TP: R	TE: R TP: R	TE:V TP: I	TE:I TP: I
Run-off partition of candidates	TE: R TP: R	TE: R TP: R	TE: V TP: I	TE: I TP: I
Adding voters	R	V	R	V
Deleting voters	R	V	R	V
Partition of voters	TE: R TP: R	TE: V TP: R	TE: R TP: R	TE: V TP: V

SP-AV – 44 results in all. We'll do so by going right through the types (sometimes in groups), mentioning the intuition/motivation for the type (when doing so we will – unless we explicitly mention otherwise – be giving the intuition for the constructive case, but from that one can naturally see the intuition for the destructive case), and then will state what results hold. For a handful of cases – enough to give the reader the flavor of how immunity, susceptibility, vulnerability, and resistance are proven – we will include proofs. (Readers not interested in how such results are proven will want to skip over such proofs.)

To collect and have the results all in one place, Table 10.1 presents all 44 results.

10.3.3.1 Adding and Deleting Candidates

We already defined control by adding a limited number of candidates. We now define the remaining two types of control by candidate addition/deletion.

Control by Adding an Unlimited Number of Candidates

Name: $\mathcal{E}$-CCAC$_\text{u}$ and $\mathcal{E}$-DCAC$_\text{u}$.

Given: Disjoint sets C and D of candidates, a collection V of voters represented via their ballots over the candidates in the set $C \cup D$, and a distinguished candidate $p \in C$.

Question ($\mathcal{E}$-CCAC$_\mathrm{u}$): Is there a subset E of D such that p is the unique winner of the $\mathcal{E}$ election $(C \cup E, V)$?
Question ($\mathcal{E}$-DCAC$_\mathrm{u}$): Is there a subset E of D such that p is not a unique winner of the $\mathcal{E}$ election $(C \cup E, V)$?

Control by Deleting Candidates

Name: $\mathcal{E}$-CCDC and $\mathcal{E}$-DCDC.
Given: A set C of candidates, a collection V of voters represented via their ballots over C, a distinguished candidate $p \in C$, and a nonnegative integer k.
Question ($\mathcal{E}$-CCDC): Is it possible to by deleting at most k candidates ensure that p is the unique winner of the resulting $\mathcal{E}$ election?
Question ($\mathcal{E}$-DCDC): Is it possible to by deleting at most k candidates other than p ensure that p is not a unique winner of the resulting $\mathcal{E}$ election?

Following Faliszewski et al. (2009b), we consider "limited" candidate addition the more natural addition notion, but we mention that "unlimited" candidate addition is what (along with the same other nine notions we will present here – except for the handling of ties in the subelections of the control-by-partition cases, see Sect. 10.3.3.2) was used by Bartholdi et al. (1992).

The three add/delete candidate control types model candidate recruitment and forcing candidates out of the race. For example, Ralph Nader joining the U.S. presidential contest may have helped George W. Bush in his race against Al Gore. Similarly, if once in the race Ralph Nader could have been persuaded to drop out, that might have been helpful to Al Gore.

The results that hold for these control types are as follows. AV is immune to constructive control by both limited and unlimited addition of candidates, and the argument we gave about this earlier in this chapter suffices to prove both those cases. Essentially the same reasoning shows that AV is immune to destructive control by deletion of candidates – if p is already the unique winner under AV, and so is approved by more voters than is any other candidate, then deleting candidates (other than p) will clearly leave p still the unique winner, since p still will have more approvals. So, we have shown the following.

Theorem 10.3.1 (Hemaspaandra et al. 2007a). *AV is immune to constructive control both by adding a limited and by adding an unlimited number of candidates, and to destructive control by deleting candidates.*

However, if one considers the same control cases as in Theorem 10.3.1 but with "constructive" and "destructive" being swapped, one obtains vulnerability rather than immunity results.

Theorem 10.3.2 (Hemaspaandra et al. 2007a). *AV is vulnerable to destructive control both by adding a limited and by adding an unlimited number of candidates, and to constructive control by deleting candidates.*

In contrast, SP-AV is resistant to control in these six cases. As an example of proving susceptibility and indeed resistance, we give a proof of the result of Erdélyi et al. (2008c) that SP-AV is resistant to constructive control by deleting candidates.

Theorem 10.3.3 (Erdélyi et al. 2008c). *SP-AV is resistant to constructive control by deleting candidates.*

Proof. The reason why SP-AV is susceptible to constructive control by deleting candidates is quite simple: SP-AV is a "voiced" voting system (i.e., SP-AV has the property that in each single-candidate election, this candidate – even with zero approvals – wins), and it is known that every voiced voting system is susceptible to constructive control by deleting candidates (Hemaspaandra et al. 2007a). Indeed, consider an election with exactly two candidates, p and q, in which p and q tie for winner, i.e., they have the same number of approvals. So p is not a unique winner in this election, but deleting q from it makes p the unique SP-AV winner of the resulting election.

To prove that SP-AV is resistant to constructive control by deleting candidates, we need to show that the corresponding control problem, SP-AV-CCDC, is NP-hard (which, as mentioned in Sect. 10.3.2, means that each NP problem is $\leq_m^p$-reducible to SP-AV-CCDC). Fortunately, we do not have to give a $\leq_m^p$-reduction from every single NP problem to this problem; rather, it is enough to reduce just one NP-*complete* problem to SP-AV-CCDC.[6]

We choose Hitting Set as our NP-complete problem (see, e.g., Garey and Johnson 1979) from which to reduce. Before we formally define this problem, we give an illustrative explanation from everyday life. Consider a set, B, of m students. Each student has chosen to sign up for some subset of n courses, where we assume that courses for which no students signed up have already been cancelled. So, the ith course, $1 \leq i \leq n$, is attended by some nonempty subset of the students, say $S_i \subseteq B$. Of course, students may have signed up for different courses and some lazy students may have chosen to sign up for no course at all. The professors who teach these courses are exceedingly busy doing research, which is why they want to pay a student in each of the n courses to help them produce a course report for that course (to be given to the Dean). However, since the departmental budget is limited, they can afford to pay only $k < m$ students. Also, the limit on the departmental budget is the reason why students will get one and the same amount of money for their help, no matter how many course reports they help to write. We further assume that the students in B (except the lazy ones, of course, who haven't signed up for any course) are so eager to help that they would volunteer to do the job even if they had to write all n course reports. The task the professors face is to determine whether there is some set of at most k students in B such that each course is attended by at least one of them – in other words, they need to determine whether there exists a

[6] This follows by transitivity of the $\leq_m^p$-reducibility: Each NP problem can be $\leq_m^p$-reduced to any NP-complete problem, so if we find a $\leq_m^p$-reduction from some NP-complete problem to SP-AV-CCDC, then each NP problem $\leq_m^p$-reduces to SP-AV-CCDC.

hitting set of size at most k for $\mathcal{S} = \{S_1, S_2, \ldots, S_n\}$. A bit more formally phrased, this problem is defined as follows:

Name: HITTING SET.
Given: A set $B = \{b_1, b_2, \ldots, b_m\}$, a collection $\mathcal{S} = \{S_1, S_2, \ldots, S_n\}$ of nonempty subsets S_i of B, and a positive integer $k < m$.
Question: Does there exist a hitting set of size at most k for $\mathcal{S}$, i.e., does there exist a subset $B' \subseteq B$ such that $\|B'\| \leq k$ and for each i, $1 \leq i \leq n$, we have $S_i \cap B' \neq \emptyset$?

Let $(B, \mathcal{S}, k)$ be a given instance of HITTING SET, where $B = \{b_1, b_2, \ldots, b_m\}$, $\mathcal{S} = \{S_1, S_2, \ldots, S_n\}$ with $\emptyset \neq S_i \subseteq B$ for each i, $1 \leq i \leq n$, and $k < m$ is a positive integer.[7] To prove HITTING SET $\leq_m^p$ SP-AV-CCDC, we now define a reduction that in polynomial time transforms $(B, \mathcal{S}, k)$ into an instance $(C, V, p, m-k)$ of SP-AV-CCDC such that the HITTING SET question for $(B, \mathcal{S}, k)$ is answered "yes" if and only if the SP-AV-CCDC question for $(C, V, p, m-k)$ is answered "yes."

Our instance $(C, V, p, m-k)$ of SP-AV-CCDC is constructed from the given HITTING SET instance $(B, \mathcal{S}, k)$ as follows. The candidate set is $C = B \cup \{p\}$. There are $4n(k+1)+4m-2k+3$ voters in V whose ballots belong to the following five groups:

1. The first group contains, for each i with $1 \leq i \leq n$, $2(k+1)$ ballots of the following type (this notation will be explained more clearly below):

$$S_i \mid (B - S_i)\ p.$$

2. The second group contains, for each i with $1 \leq i \leq n$, $2(k+1)$ ballots of the following type:

$$(B - S_i)\ p \mid S_i.$$

3. The third group contains, for each j with $1 \leq j \leq m$, two ballots of the following type:

$$b_j \mid p\ (B - \{b_j\}).$$

4. The fourth group contains $2(m-k)$ ballots of the following type:

$$B \mid p.$$

5. The fifth group contains three ballots of the following type:

$$p \mid B.$$

[7] HITTING SET is sometimes defined with "$k \leq m$" instead of "$k < m$." However, since B is always a hitting set of size k if $m = k$ (due to $\mathcal{S}$ containing no empty sets), we may require $k < m$ in the problem instance.

In the above notation, each time a set is written in a ballot it is a shorthand for having its elements listed in the same order that they occur in the list $(b_1, b_2, \dots, b_m)$. For example, if $B = \{b_1, b_2, \dots, b_6\}$ contains six members and $S_3 = \{b_2, b_3, b_5\}$ contains three, then "$S_3|(B - S_3)p$" means that each of the $2(k+1)$ ballots corresponding to S_3 has the form: $b_2\ b_3\ b_5 \mid b_1\ b_4\ b_6\ p$.

It is obvious that $(C, V, p, m-k)$ can be computed from $(B, \mathcal{S}, k)$ in polynomial time.

Note that, regardless of whether $(B, \mathcal{S}, k)$ is a "yes" or a "no" instance of Hitting Set, p is not the unique SP-AV winner (and not even *an* SP-AV winner) of the election (C, V) constructed. In particular, note that p is approved by $2n(k+1)+3$ voters in (C, V), but every $b_i \in B$ has $2n(k+1)+2+2(m-k)$ approvals, and since $k < m$, every $b_i \in B$ is better off than p.

On the other hand, we will show that if $\mathcal{S}$ has a size-k hitting set then p can be made the unique SP-AV winner by deleting at most $m-k$ candidates, and if $\mathcal{S}$ does not have a hitting set of size k then this is not possible. But before we proceed with the proof, let us first look at an illustrative example of the construction.

Example 10.3.1. Suppose we are given the Hitting Set instance $(B, \mathcal{S}, 2)$, where $B = \{b_1, b_2, b_3, b_4\}$ and $\mathcal{S} = \{S_1, S_2, S_3\}$ with $S_1 = \{b_1, b_2\}$, $S_2 = \{b_1, b_4\}$, and $S_3 = \{b_3, b_4\}$. Our reduction then yields the instance $(C, V, p, 2)$ of SP-AV-CCDC with candidate set $C = \{b_1, b_2, b_3, b_4, p\}$ and with V consisting of the 51 ballots shown in Table 10.2.

The ballots in the left column correspond to the first group, those in the middle column to the second group, and the ballots in the right column correspond

Table 10.2 Fifty-one ballots generated from a "yes" instance of Hitting Set

$b_1\ b_2 \mid b_3\ b_4\ p$	$b_3\ b_4\ p \mid b_1\ b_2$	$b_1 \mid p\ b_2\ b_3\ b_4$
$b_1\ b_2 \mid b_3\ b_4\ p$	$b_3\ b_4\ p \mid b_1\ b_2$	$b_1 \mid p\ b_2\ b_3\ b_4$
$b_1\ b_2 \mid b_3\ b_4\ p$	$b_3\ b_4\ p \mid b_1\ b_2$	
$b_1\ b_2 \mid b_3\ b_4\ p$	$b_3\ b_4\ p \mid b_1\ b_2$	$b_2 \mid p\ b_1\ b_3\ b_4$
$b_1\ b_2 \mid b_3\ b_4\ p$	$b_3\ b_4\ p \mid b_1\ b_2$	$b_2 \mid p\ b_1\ b_3\ b_4$
$b_1\ b_2 \mid b_3\ b_4\ p$	$b_3\ b_4\ p \mid b_1\ b_2$	
		$b_3 \mid p\ b_1\ b_2\ b_4$
$b_1\ b_4 \mid b_2\ b_3\ p$	$b_2\ b_3\ p \mid b_1\ b_4$	$b_3 \mid p\ b_1\ b_2\ b_4$
$b_1\ b_4 \mid b_2\ b_3\ p$	$b_2\ b_3\ p \mid b_1\ b_4$	
$b_1\ b_4 \mid b_2\ b_3\ p$	$b_2\ b_3\ p \mid b_1\ b_4$	$b_4 \mid p\ b_1\ b_2\ b_3$
$b_1\ b_4 \mid b_2\ b_3\ p$	$b_2\ b_3\ p \mid b_1\ b_4$	$b_4 \mid p\ b_1\ b_2\ b_3$
$b_1\ b_4 \mid b_2\ b_3\ p$	$b_2\ b_3\ p \mid b_1\ b_4$	
$b_1\ b_4 \mid b_2\ b_3\ p$	$b_2\ b_3\ p \mid b_1\ b_4$	$b_1\ b_2\ b_3\ b_4 \mid p$
		$b_1\ b_2\ b_3\ b_4 \mid p$
$b_3\ b_4 \mid b_1\ b_2\ p$	$b_1\ b_2\ p \mid b_3\ b_4$	$b_1\ b_2\ b_3\ b_4 \mid p$
$b_3\ b_4 \mid b_1\ b_2\ p$	$b_1\ b_2\ p \mid b_3\ b_4$	$b_1\ b_2\ b_3\ b_4 \mid p$
$b_3\ b_4 \mid b_1\ b_2\ p$	$b_1\ b_2\ p \mid b_3\ b_4$	
$b_3\ b_4 \mid b_1\ b_2\ p$	$b_1\ b_2\ p \mid b_3\ b_4$	$p \mid b_1\ b_2\ b_3\ b_4$
$b_3\ b_4 \mid b_1\ b_2\ p$	$b_1\ b_2\ p \mid b_3\ b_4$	$p \mid b_1\ b_2\ b_3\ b_4$
$b_3\ b_4 \mid b_1\ b_2\ p$	$b_1\ b_2\ p \mid b_3\ b_4$	$p \mid b_1\ b_2\ b_3\ b_4$

to the third, fourth, and fifth groups. Note that $(B, \mathcal{S}, 2)$ is a "yes" instance of HITTING SET: Choose, for example, the two-element set $B' = \{b_1, b_4\}$ and note that B' has a nonempty intersection with each of S_1, S_2, and S_3, so B' is a hitting set of size two for $\mathcal{S}$.

Our construction ensures that $(C, V, p, 2)$ is also a "yes" instance of SP-AV-CCDC. In (C, V), p is approved by 21 voters and each $b_i \in B$ is approved by 24 voters, so p does not win. However, p can be made the unique SP-AV winner by deleting two candidates from C. In particular, if we delete b_2 and b_3 from B (i.e., if we delete all candidates not in the hitting set B' mentioned above), we obtain the election with candidate set $B' \cup \{p\} = \{b_1, b_4, p\}$ and voters in V restricted to $B' \cup \{p\}$. By the rule (Rule 1) about moving the approval line to ensure admissibility, p gains four approvals in this election, namely, p gains two approvals by changing the two ballots of the form $b_2 \mid p\ b_1\ b_3\ b_4$ into $p \mid b_1\ b_4$ and two approvals by changing the two ballots of the form $b_3 \mid p\ b_1\ b_2\ b_4$ into $p \mid b_1\ b_4$. Since p doesn't lose any of the approvals it already had before the deletion, p now has 25 approvals. However, both b_1 and b_4 still have 24 approvals, so p is now the unique SP-AV winner.

Now, consider the same HITTING SET instance as above except with the parameter k decreased by one: $(B, \mathcal{S}, 1)$. Note that $(B, \mathcal{S}, 1)$ is a "no" instance of HITTING SET: No one-element subset of B hits each of S_1, S_2, and S_3, i.e., for no i, $1 \leq i \leq 4$, does $\{b_i\}$ have a nonempty intersection with each of S_1, S_2, and S_3.

Starting from $(B, \mathcal{S}, 1)$, our reduction yields the SP-AV-CCDC instance $(C, V', p, 3)$ with the same candidate set as above, $C = \{b_1, b_2, b_3, b_4, p\}$, and with the new V' consisting of the 41 ballots shown in Table 10.3.

To see that $(C, V', p, 3)$ is a "no" instance of SP-AV-CCDC, first note that p has 15 approvals in (C, V'), yet each $b_i \in B$ is approved by 20 voters, so p does not win. Moreover, it can be checked that deleting any subset of at most three candidates from B will leave at least one $b_i \in B$ in the race that has more approvals than p.

It is not a coincidence that, as in Example 10.3.1, "yes" (respectively, "no") instances of HITTING SET yield "yes" (respectively, "no") instances of SP-AV-CCDC. Indeed, continuing with the proof of Theorem 10.3.3, we will now show that, in general, B has a hitting set of size k for $\mathcal{S}$ if and only if p can be made the unique SP-AV winner by deleting at most $m - k$ candidates from the election (C, V).

Suppose there is a hitting set B' of size k for $\mathcal{S}$. Since deleting the $m - k$ candidates in $B - B'$ from C will move the approval line – due to the rule, Rule 1, of re-enforcing admissibility – for $2(m - k)$ voters of the third group, p will gain these approvals of the candidates in $B - B'$. Thus in $(B' \cup \{p\}, V)$, p is now approved by $2n(k + 1) + 2(m - k) + 3$ voters and each member of B' still has $2n(k + 1) + 2 + 2(m - k)$ approvals, so p is the unique SP-AV winner of the election $(B' \cup \{p\}, V)$.

Conversely, suppose that p can be made the unique SP-AV winner by deleting no more than $m - k$ candidates from C. Of course, p cannot have been deleted, so the deleted candidates all belong to B. Let B' be the set of candidates from B

Table 10.3 Forty-one ballots generated from a "no" instance of HITTING SET

$b_1\ b_2 \mid b_3\ b_4\ p$	$b_3\ b_4\ p \mid b_1\ b_2$	$b_1 \mid p\ b_2\ b_3\ b_4$
$b_1\ b_2 \mid b_3\ b_4\ p$	$b_3\ b_4\ p \mid b_1\ b_2$	$b_1 \mid p\ b_2\ b_3\ b_4$
$b_1\ b_2 \mid b_3\ b_4\ p$	$b_3\ b_4\ p \mid b_1\ b_2$	$b_2 \mid p\ b_1\ b_3\ b_4$
$b_1\ b_2 \mid b_3\ b_4\ p$	$b_3\ b_4\ p \mid b_1\ b_2$	$b_2 \mid p\ b_1\ b_3\ b_4$
		$b_3 \mid p\ b_1\ b_2\ b_4$
		$b_3 \mid p\ b_1\ b_2\ b_4$
		$b_4 \mid p\ b_1\ b_2\ b_3$
		$b_4 \mid p\ b_1\ b_2\ b_3$
$b_1\ b_4 \mid b_2\ b_3\ p$	$b_2\ b_3\ p \mid b_1\ b_4$	
$b_1\ b_4 \mid b_2\ b_3\ p$	$b_2\ b_3\ p \mid b_1\ b_4$	
$b_1\ b_4 \mid b_2\ b_3\ p$	$b_2\ b_3\ p \mid b_1\ b_4$	$b_1\ b_2\ b_3\ b_4 \mid p$
$b_1\ b_4 \mid b_2\ b_3\ p$	$b_2\ b_3\ p \mid b_1\ b_4$	$b_1\ b_2\ b_3\ b_4 \mid p$
		$b_1\ b_2\ b_3\ b_4 \mid p$
		$b_1\ b_2\ b_3\ b_4 \mid p$
		$b_1\ b_2\ b_3\ b_4 \mid p$
		$b_1\ b_2\ b_3\ b_4 \mid p$
$b_3\ b_4 \mid b_1\ b_2\ p$	$b_1\ b_2\ p \mid b_3\ b_4$	
$b_3\ b_4 \mid b_1\ b_2\ p$	$b_1\ b_2\ p \mid b_3\ b_4$	$p \mid b_1\ b_2\ b_3\ b_4$
$b_3\ b_4 \mid b_1\ b_2\ p$	$b_1\ b_2\ p \mid b_3\ b_4$	$p \mid b_1\ b_2\ b_3\ b_4$
$b_3\ b_4 \mid b_1\ b_2\ p$	$b_1\ b_2\ p \mid b_3\ b_4$	$p \mid b_1\ b_2\ b_3\ b_4$

remaining in the race after deletion. Since $\|B - B'\| \le m - k$, we have $\|B'\| \ge k$. We will show that B' is a hitting set of size exactly k for $\mathcal{S}$. Let h be the number of sets S_i in $\mathcal{S}$ not hit by B', i.e., $h = \|\{i \mid S_i \in \mathcal{S} \text{ and } B' \cap S_i = \emptyset\}\|$.

In $(B' \cup \{p\}, V)$, p is approved by

$$2(n - h)(k + 1) + 2(m - \|B'\|) + 3 \tag{10.1}$$

voters. From our assumption that all candidates occur in the same ordering in each of the $2n(k + 1)$ first-group voters (those of the form $S_i \mid (B - S_i)\ p$, where $1 \le i \le n$), it follows that for each S_i not hit by B', one and the same candidate in B' (namely, the candidate occurring first in the order of B') receives an additional $2(k+1)$ approvals – due to moving the approval line – thanks to the $2(k+1)$ voters corresponding to S_i after deletion of the candidates in $B - B'$. Summing up, this candidate will end up having exactly

$$2(n + h)(k + 1) + 2 + 2(m - k) \tag{10.2}$$

approvals after deletion of $B - B'$. Since p has been made the unique SP-AV winner by deleting the candidates in $B - B'$ from C, p must have more approvals in $(B' \cup \{p\}, V)$ than any candidate in B'. So by (10.1) and (10.2), we have

$$2(n - h)(k + 1) + 2(m - \|B'\|) + 3 > 2(n + h)(k + 1) + 2 + 2(m - k)$$

or, equivalently,

$$1 + 2(k - \|B'\|) > 4h(k + 1).$$

However, since h is a nonnegative integer satisfying

$$1 = \frac{4(k+1)}{4(k+1)} > \frac{1 + 2(k - \|B'\|)}{4(k+1)} > h,$$

we have $h = 0$, which implies that B' is a hitting set for $\mathcal{S}$. Moreover, $h = 0$ also implies that $1 + 2(k - \|B'\|)$ must be positive, so $\|B'\| \leq k$. Summing up, B' is a hitting set of size k for $\mathcal{S}$. □

10.3.3.2 The Partition Cases

We now come to a particularly interesting and challenging collection of cases: those cases having to do with partitioning into subelections. In these cases, the election becomes a two-round process. There is a first round, containing one or two subelections depending on the case, and then there is a (final) second round.

Since there is a first round, an issue arises here that we did not have in Sect. 10.3.3.1. Namely, what should we do if in a first-round election some candidates *tie* as winners? There are two natural approaches, both introduced in Hemaspaandra et al. (2007a), and since people seem to have strong feelings as to one or the other being more natural – but with no broad agreement among people – many papers take the cautious approach of giving results for both approaches to the issue of whom to move forward from tied first-round subelections. The two approaches are *ties promote* (in Table 10.1, "TP"), in which all subelection winners move forward to the final round, and *ties eliminate* (in Table 10.1, "TE"), in which one moves forward from a subelection exactly if one is the one and only winner of that subelection. It is worth remarking that Table 10.1 contains three cases in which these two rules yield different results. Dealing with ties truly is part of a partition-based control model, and not some unimportant detail that never affects one's study.

Among the two models, we consider the TE model by far the more consistent with the rest of the framework. The reason is that in control problems (and so, in the second-round elections), we are asking whether candidate p can be made the (or be precluded from being a) *unique* winner. Although control problems are most typically studied in the unique-winner model, sometimes they are studied in the model in which one just asks whether the given candidate is (this is the constructive case) or is not (this is the destructive case) *a* winner. For example, Faliszewski et al. (2007, 2009b) study both models. We mention that, analogously to the comments just made, if one's model is the nonunique-winner model, then our feeling is that the more natural approach to ties in partition-related subelections is the TP model.

We now state the three partition-based control types.

Control by Run-Off Partition and Partition of Candidates

Name: $\mathcal{E}$-CCRPC and $\mathcal{E}$-DCRPC (Control by Run-Off Partition of Candidates).
Given: A set C of candidates, a collection V of voters represented via ballots over C, and a distinguished candidate $p \in C$.
Question ($\mathcal{E}$-CCRPC): Is there a partition of C into C_1 and C_2 such that p is the unique winner of the two-stage election where the winners of subelection (C_1, V) that survive the tie-handling rule compete against the winners of subelection (C_2, V) that survive the tie-handling rule? Each subelection (in both stages) is conducted using election system $\mathcal{E}$.
Question ($\mathcal{E}$-DCRPC): Is there a partition of C into C_1 and C_2 such that p is not a unique winner of the two-stage election where the winners of subelection (C_1, V) that survive the tie-handling rule compete against the winners of subelection (C_2, V) that survive the tie-handling rule? Each subelection (in both stages) is conducted using election system $\mathcal{E}$.

The above description defines four computational problems for a given election system $\mathcal{E}$: $\mathcal{E}$-CCRPC-TE, $\mathcal{E}$-CCRPC-TP, $\mathcal{E}$-DCRPC-TE, and $\mathcal{E}$-DCRPC-TP.

Name: $\mathcal{E}$-CCPC and $\mathcal{E}$-DCPC (Control by Partition of Candidates).
Given: A set C of candidates, a collection V of voters represented via ballots over C, and a distinguished candidate $p \in C$.
Question ($\mathcal{E}$-CCPC): Is there a partition of C into C_1 and C_2 such that p is the unique winner of the two-stage election where the winners of subelection (C_1, V) that survive the tie-handling rule compete against all candidates in C_2? Each subelection (in both stages) is conducted using election system $\mathcal{E}$.
Question ($\mathcal{E}$-DCPC): Is there a partition of C into C_1 and C_2 such that p is not a unique winner of the two-stage election where the winners of subelection (C_1, V) that survive the tie-handling rule compete against all candidates in C_2? Each subelection (in both stages) is conducted using election system $\mathcal{E}$.

This description defines four computational problems for a given election system $\mathcal{E}$: $\mathcal{E}$-CCPC-TE, $\mathcal{E}$-CCPC-TP, $\mathcal{E}$-DCPC-TE, and $\mathcal{E}$-DCPC-TP.

Control by Partition of Voters

Name: $\mathcal{E}$-CCPV and $\mathcal{E}$-DCPV.
Given: A set C of candidates, a collection V of voters represented via ballots over C, and a distinguished candidate $p \in C$.
Question ($\mathcal{E}$-CCPV): Is there a partition of V into V_1 and V_2 such that p is the unique winner of the two-stage election where the winners of subelection (C, V_1) that survive the tie-handling rule compete against the winners of subelection (C, V_2) that survive the tie-handling rule? Each subelection (in both stages) is conducted using election system $\mathcal{E}$.

Question ($\mathcal{E}$-DCPV): Is there a partition of V into V_1 and V_2 such that p is not a unique winner of the two-stage election where the winners of subelection (C, V_1) that survive the tie-handling rule compete against the winners of subelection (C, V_2) that survive the tie-handling rule? Each subelection (in both stages) is conducted using election system $\mathcal{E}$.

The motivation for run-off partition of candidates is any process, perhaps a legislative process, in which a given committee considers two batches of alternatives, votes on each separately, and then considers only the winners. For example, a physics department's faculty search might result in many experimental and many theoretical candidates, and the department chair might try to clarify the decision process by having the entire department faculty vote separately among the experimental candidates and among the theoretical candidates, and after that limit the discussion and final-round vote to only those candidates that survived the first-round vote.

In contrast, (non-run-off) partition of candidates models systems in which there is a qualifying election that some candidates are exempted from. For example, in some tournaments, the host country may be given such an exemption from qualification.

Finally, partition of voters models a primary system in which the electorate is divided into (two) groups, each group votes over the candidates, and the winners move forward to a final election in which everyone votes. In our previous example, this would be the case if the department had the theoreticians hold an election over all candidates and separately had the experimentalists hold an election over all candidates – yes, one can make some guesses about who might win in each subelection! – and then had both groups jointly vote over those candidates surviving the first round.

The results that hold for the partition cases are as follows. For AV, resistance holds for constructive control by partition of voters (both TP and TE), vulnerability holds for the TE cases of constructive partition of candidates, constructive run-off partition of candidates, and destructive partition of voters, vulnerability also holds for the TP case of destructive partition of voters, and immunity holds in the remaining six cases. For SP-AV, vulnerability holds for the TE case of destructive control by partition of voters, and resistance holds for the other 11 cases.

To illustrate vulnerability proofs, we prove the following result.

Theorem 10.3.4 (Hemaspaandra et al. 2007a). *Approval voting is vulnerable to destructive control by partition of voters in the TE model.*

Proof. First, we give an example showing that approval voting is susceptible to destructive control by partition of voters in the TE model. Consider the election (C, V) with candidate set $C = \{a, b, c\}$, distinguished candidate c, and the following collection $V = \{v_1, v_2, \dots, v_{17}\}$ of ballots, each being represented as an approval vector for abc in $\{0, 1\}^3$: $v_1 = \cdots = v_5 = 100$, $v_6 = 110$, $v_7 = \cdots = v_{10} = 010$, and $v_{11} = \cdots = v_{17} = 001$. Clearly, c is the unique approval winner in (C, V). However, if we partition V into $V_1 = \{v_1, \dots, v_5, v_{11}, \dots, v_{14}\}$ and $V_2 = V - V_1$, then candidate a is the unique approval winner of the first-round subelection (C, V_1) and candidate b is the unique approval winner of the other first-round subelection, (C, V_2). Since c doesn't proceed to the final round, c's victory has been successfully blocked by voter partition. Thus approval voting is susceptible to destructive control by partition of voters in the TE model.

To prove vulnerability, we now describe a polynomial-time algorithm for the control problem at hand. The algorithm takes as input an election (C, V) and a distinguished candidate $p \in C$. The output of the algorithm either will be a successful voter partition (V_1, V_2) (i.e., a partition such that p won't be a unique approval winner in the final round of the corresponding two-stage election), or will be "control impossible" (this will be the case exactly if there is no way for the chair to exert a successful control action of this type).

In a nutshell, the basic idea in the algorithm (after having handled certain trivial cases including the case where p is not a unique approval winner of (C, V) and the cases with $\|C\| \leq 2$ – so we now have that $\|C\| \geq 3$ and that p is the unique approval winner of (C, V)) is to check whether it is possible to find two candidates (other than the distinguished candidate p) who can prevent p from proceeding to the final round of the two-stage election induced by some partition of voters. More precisely, if two such candidates indeed can be found, they will preclude p from the final round by tieing or defeating p in each of the two first-round subelections. And if they cannot be found, it is impossible to block p's final-round victory in any partition of voters.

Before describing the algorithm, let us establish some useful notation. Consider (for elections having at least three candidates) any two distinct candidates $a, b \in C - \{p\}$. Our notation will focus on only the approvals/disapprovals of a, b, and p, and will not care about which of the other candidates are approved of. For $x \in \{a, b\}$, define the following six sets:

$$
\begin{aligned}
S_x &= \{v \in V \mid v \text{ approves of } x \text{ and disapproves of both } p \text{ and the candidate in } \{a, b\} - \{x\}\};\\
S_{xp} &= \{v \in V \mid v \text{ approves of both } x \text{ and } p \text{ and disapproves of the candidate in } \{a, b\} - \{x\}\};\\
W_p &= \{v \in V \mid v \text{ approves of } p \text{ and disapproves of both } a \text{ and } b\};\\
L_p &= \{v \in V \mid v \text{ disapproves of } p \text{ and approves of both } a \text{ and } b\}.
\end{aligned}
$$

We now describe our algorithm. The algorithm first checks the following trivial cases:

1. If C contains p alone, then output "control impossible" and stop. (There is no other candidate who could possibly prevent p from winning.)
2. Else if p is not a unique approval winner, then output the successful partition $(V, \emptyset)$ and stop.
3. Else if $\|C\| = 2$, then output "control impossible" and stop. (In this case, p wins at least one of the subelections – in whatever partition of voters the chair chooses – and proceeds to the final stage, where it basks in glory as the unique approval winner.)

Now, if none of the trivial cases apply, we know that p currently is the unique approval winner in (C, V) and C has at least two members other than p. To determine whether p can be dethroned as the unique approval winner, the algorithm enters a loop to check for each pair of distinct candidates $a, b \in C - \{p\}$ whether

$$\|W_p\| > \|S_a\| + \|S_b\| + \|L_p\| \tag{10.3}$$

is satisfied. If so, p has so many more approvals than this a and b that this a and b are helpless against p in any voter partition, and the algorithm enters the next loop iteration to check the next a and b. As soon as some a and b are found for which $\|W_p\| \leq \|S_a\| + \|S_b\| + \|L_p\|$ is satisfied (i.e., (10.3) does not hold), it outputs the successful partition (V_1, V_2) and stops, where $V_1 = S_a \cup S_{ap} \cup W'_p$ with W'_p containing the first $\min(\|S_a\|, \|W_p\|)$ voters of W_p, and where V_2 is $V - V_1$. If all possible pairs of candidates (other than p) have been checked and none has produced a successful partition of V, the algorithm outputs "control impossible" and stops.

Why does this algorithm correctly decide whether destructive control by partition of voters in model TE is possible for AV? On the one hand, if (10.3) is satisfied for all pairs of candidates a and b then, for whatever voter partition is chosen, no pair of candidates can tie or defeat p in *both* subelections, so p will be the unique approval winner of at least one subelection and will thus proceed to the final round, which it alone will win. On the other hand, if there exists a candidate pair a and b for which (10.3) fails to hold then the algorithm partitions V into (V_1, V_2) as stated above. It follows that, in (C, V_1), we have

$$score_{(C,V_1)}(a) - score_{(C,V_1)}(p) = \|S_a\| - \min(\|S_a\|, \|W_p\|) \geq 0,$$

which means that a ties or defeats p in (C, V_1). To show that b also ties or defeats p in the other subelection, (C, V_2), we have to show that

$$score_{(C,V_2)}(b) - score_{(C,V_2)}(p) = \|S_b\| + \|L_p\| - (\|W_p\| - \min(\|S_a\|, \|W_p\|)) \geq 0.$$

Transform this inequality into the form

$$\|W_p\| - \|L_p\| \leq \min(\|S_a\|, \|W_p\|) + \|S_b\|.$$

If $\|W_p\| < \|S_a\|$, we have $\|W_p\| - \|L_p\| \leq \|W_p\| + \|S_b\|$, which is always true because both $\|L_p\|$ and $\|S_b\|$ are nonnegative. If $\|W_p\| \geq \|S_a\|$, however, we have $\|W_p\| \leq \|S_a\| + \|S_b\| + \|L_p\|$, which is true because (10.3) does not hold in the current case. Thus this algorithm correctly decides AV-DCPV for model TE in polynomial time. □

As an example of a resistance proof based on a reduction from the "X3C" problem (which will be defined in a minute), we now prove the following result.

Theorem 10.3.5 (Hemaspaandra et al. 2007a). *Approval voting is resistant to constructive control by partition of voters in the TP model.*

Proof. To prove susceptibility, consider the same example as in the proof of Theorem 10.3.4, except that now a instead of c is the distinguished candidate.

Initially, a is not a unique approval winner of (C, V), but the same partition of V that was described in the proof of Theorem 10.3.4 does make a the unique approval winner. Thus, approval voting is susceptible to constructive control by partition of voters in model TP.

To prove resistance, we reduce from the following problem.

Name: EXACT COVER BY THREE-SETS (X3C, for short).
Given: A set $B = \{b_1, b_2, \ldots, b_{3m}\}, m \geq 1$, and a collection $\mathcal{S} = \{S_1, S_2, \ldots, S_n\}$ of sets such that for each i, $1 \leq i \leq n$, it holds that $S_i \subseteq B$ and $\|S_i\| = 3$.
Question: Does there exist an exact cover for B, i.e., does there exist a set $A \subseteq \{1, 2, \ldots, n\}$, $\|A\| = m$, such that $\bigcup_{i \in A} S_i = B$?

To give an illustrative (even though, admittedly, not really everyday) example of this problem, we mention the infamous theft of the splendid glass mosaic covering the interior of the Cathedral of Monreale on Sicily, which was created by artists from Sicily and Constantinople in the twelfth century. One morning, the bishop entered the cathedral and, to his utter dismay, every single one of the 15,000,000 mosaic pieces, which had covered about 6,340 square meters, was gone! Let us enumerate the stolen mosaic pieces and call the mosaic $B = \{b_1, b_2, \ldots, b_{15{,}000{,}000}\}$, so $m = 5{,}000{,}000$ in this example. The police started an intense search for the evildoers and the lost treasure. Soon thereafter, a number of pieces of this invaluable mosaic – surprisingly always in batches of three – turned up on black markets for historical art all over the world. It soon became clear, however, that many of these pieces were not the original ones but faked. For example, the first four of the size-three batches of mosaic pieces found were $S_1 = \{b_{17}, b_{3{,}471}, b_{4{,}946{,}071}\}$, $S_2 = \{b_{17}, b_{463}, b_{94{,}228}\}$, $S_3 = \{b_{231}, b_{56{,}463}, b_{12{,}094{,}578}\}$, and $S_4 = \{b_{17}, b_{3{,}471}, b_{94{,}228}\}$ – obviously, these three-element subsets of B weren't disjoint and so some of them contained faked mosaic pieces. In total, a collection $\mathcal{S} = \{S_1, S_2, \ldots, S_n\}$ of $n = 55{,}557{,}390$ size-three batches of mosaic pieces (each S_i being offered for only one dollar) were spotted on black markets world-wide. "I do not care if they are original or faked," the bishop was quoted as saying, "I just want to have one (real or fake) copy of each piece of my mosaic. Look at my cathedral: It is *naked* inside!" So he started collecting money to buy enough size-three batches of mosaic pieces to be able to rebuild the complete mosaic in the cathedral, and eventually he had collected a total of exactly five million dollars. The problem he was facing now is the X3C problem: Is it possible to find five million sets S_i that exactly cover B?[8]

It is not hard to believe that (large enough) instances of the X3C problem are computationally intractable, and it indeed is known that X3C is NP-complete (see, e.g., Garey and Johnson 1979).

Turning back to the proof, let an instance $(B, \mathcal{S})$ of X3C be given, where $B = \{b_1, b_2, \ldots, b_{3m}\}$, $m \geq 1$, $\mathcal{S} = \{S_1, S_2, \ldots, S_n\}$, and for each i, $1 \leq i \leq n$, we have that $S_i \subseteq B$ and $\|S_i\| = 3$. Define the value $h_j = \|\{S_i \in \mathcal{S} \mid b_j \in S_i\}\|$ for each j, $1 \leq j \leq 3m$. Define the election (C, V), where $C = B \cup \{p, y, z\}$ is

[8] This is a fictional example to illustrate X3C.

the set of candidates with the distinguished candidate p (i.e., p is the candidate the chair wishes to make the unique winner of the election), and where V is defined to consist of the following $2n + 2m + 3$ ballots:

1. For each i, $1 \le i \le n$, there is one voter who approves of the candidates in $\{z\} \cup S_i$ and who disapproves of the candidates in $\{p, y\} \cup (B - S_i)$.
2. For each i, $1 \le i \le n$, there is one voter who approves of the candidates in $\{p\} \cup B_i$, where $B_i = \{b_j \in B \mid i + h_j \le n\}$, and who disapproves of the candidates in $\{y, z\} \cup (B - B_i)$.
3. There are $m + 1$ voters who each approve of only y and disapprove of the candidates in $\{p, z\} \cup B$.
4. There are $m + 2$ voters who each approve of the candidates in $\{p, z\} \cup B$ and disapprove of only y.

Note that $score_{(C,V)}(y) = m + 1$ and $score_{(C,V)}(p) = score_{(C,V)}(z) = score_{(C,V)}(b_j) = n + m + 2$ for all j, $1 \le j \le 3m$.

The following example illustrates the construction.

Example 10.3.2. Suppose we are given the X3C instance $(B, \mathcal{S})$, where $B = \{b_1, b_2, \ldots, b_6\}$ and $\mathcal{S} = \{S_1, S_2, S_3\}$ with $S_1 = \{b_1, b_3, b_5\}$, $S_2 = \{b_2, b_4, b_6\}$, and $S_3 = \{b_1, b_4, b_6\}$. Our reduction then yields the instance (C, V, p) of AV-CCPV (in model TP) with candidate set $C = \{p, y, z\} \cup B$ and with V consisting of the 13 ballots shown in Table 10.4.

It is easy to see that this X3C instance has an exact cover: $S_1 \cup S_2 = B$. Our reduction has the property that p can be made the unique approval winner by partition of voters in model TP: Partition V into (V_1, V_2), where V_1 contains the two voters of the first group that correspond to the cover (these voters are given in boldface in Table 10.4) and all voters of the third group, and where V_2 contains the remaining voters. The resulting subelections are shown in Table 10.5. Since y is the unique approval winner of subelection (C, V_1) and p is the unique approval winner of subelection (C, V_2), they are the only candidates proceeding to the final round, where p beats y by 7 to 3.

Table 10.4 Thirteen ballots generated from a "yes" instance of X3C

C	1st group			2nd group			3rd group			4th group				Score
p	**0**	**0**	0	1	1	1	0	0	0	1	1	1	1	7
y	**0**	**0**	0	0	0	0	1	1	1	0	0	0	0	3
z	**1**	**1**	1	0	0	0	0	0	0	1	1	1	1	7
b_1	**1**	**0**	1	1	0	0	0	0	0	1	1	1	1	7
b_2	**0**	**1**	0	1	1	0	0	0	0	1	1	1	1	7
b_3	**1**	**0**	0	1	1	0	0	0	0	1	1	1	1	7
b_4	**0**	**1**	1	1	0	0	0	0	0	1	1	1	1	7
b_5	**1**	**0**	0	1	1	0	0	0	0	1	1	1	1	7
b_6	**0**	**1**	1	1	0	0	0	0	0	1	1	1	1	7

Table 10.5 Subelections (C, V_1) and (C, V_2) obtained by voter partition from (C, V) in Example 10.3.2

Subelection (C, V_1)						
C	1st		3rd			Score
p	0	0	0	0	0	0
y	0	0	1	1	1	3
z	1	1	0	0	0	2
b_1	1	0	0	0	0	1
b_2	0	1	0	0	0	1
b_3	1	0	0	0	0	1
b_4	0	1	0	0	0	1
b_5	1	0	0	0	0	1
b_6	0	1	0	0	0	1

Subelection (C, V_2)									
C	1st	2nd			4th				Score
p	0	1	1	1	1	1	1	1	7
y	0	0	0	0	0	0	0	0	0
z	1	0	0	0	1	1	1	1	5
b_1	1	1	0	0	1	1	1	1	6
b_2	0	1	1	0	1	1	1	1	6
b_3	0	1	1	0	1	1	1	1	6
b_4	1	1	0	0	1	1	1	1	6
b_5	0	1	1	0	1	1	1	1	6
b_6	1	1	0	0	1	1	1	1	6

Turning back to the proof of Theorem 10.3.5, we will now show that, in general, $\mathcal{S}$ has an exact cover for B if and only if p can be made the unique approval winner by partition of voters in model TP.

From left to right: Suppose $\mathcal{S}$ contains an exact cover for B. Then the following partition of V into two subsets makes p the unique approval winner. Let V_1 contain the $m + 1$ voters of the third group (i.e., those voters approving of only y) and the m first-group voters corresponding to the cover, and let V_2 be $V - V_1$. It is easy to see that y is the unique approval winner of subelection (C, V_1) and p is the unique approval winner of subelection (C, V_2). And when p and y meet in the second (and final) stage of the election, p will be the unique approval winner.

From right to left: Suppose p can be made the unique approval winner by partition of voters in model TP. So, in particular, p must win the final round. Recall, however, that everybody in $B \cup \{p, z\}$ has $n + m + 2$ approvals with respect to the whole electorate V, so each of the candidates in $B \cup \{z\}$ would tie for winner with p in the final round. Thus y is the only candidate who can run against p in the final round. Since we are in the TP model, it follows that each of the two subelections must have a unique approval winner: One must have y and the other one must have p.

Let (V_1, V_2) be such a partition of V where y is the unique approval winner of, say, subelection (C, V_1), and p is the unique approval winner of the other subelection, (C, V_2). Without loss of generality, we may assume that none of the voters in V_1 approve of p, so V_1 contains only voters of the first and the third group.

For p to become the unique approval winner of subelection (C, V_2), p must in particular defeat each $b_j \in B$. For each $b_j \in B$, all voters of the third and fourth group approve of b_j exactly if they approve of p, so they don't help p to get an advantage over b_j and may thus be disregarded. If all the remaining voters (those of the first and second group) were in V_2, each b_j would tie p in (C, V_2). But since every $b_j \in B$ must lose at least one approval against p in (C, V_2) (and since all second-group voters approve of p and so, by our assumption, stay in V_2), the first-group voters that are not in V_2 (i.e., those in V_1) must form a cover for B.

However, since y is the unique approval winner of (C, V_1) and since y has $m+1$ approvals overall, z can have no more than m approvals in (C, V_1). However, among the voters in V_1 (to which, as noted above, only first- and third-group voters belong), z is approved by only first-group voters, so it can have no more than m approvals from these voters. Thus, V_1 contains no more than m (and thus exactly m) first-group voters, which represent an exact cover for B.

10.3.3.3 Adding and Deleting Voters

We have saved for last those cases whose motivations probably have the most emotional resonance for each of us, as voters. These cases are those related to adding and deleting voters.

We start by stating these control types.

Control by Adding Voters

Name: $\mathcal{E}$-CCAV and $\mathcal{E}$-DCAV.
Given: A set C of candidates, two disjoint collections of voters, V and W, represented via ballots over C, a distinguished candidate $p \in C$, and a nonnegative integer k. (V is sometimes called the registered voter set, and W, the pool of voters available for adding, is sometimes called the unregistered voter set.)
Question ($\mathcal{E}$-CCAV): Is there a subset Q, $\|Q\| \leq k$, of voters in W such that the voters in $V \cup Q$ elect p as the unique winner according to system $\mathcal{E}$?
Question ($\mathcal{E}$-DCAV): Is there a subset Q, $\|Q\| \leq k$, of voters in W such that the voters in $V \cup Q$ do not elect p as a unique winner according to system $\mathcal{E}$?

Control by Deleting Voters

Name: $\mathcal{E}$-CCDV and $\mathcal{E}$-DCDV.
Given: A set C of candidates, a collection V of voters represented via ballots over C, a distinguished candidate $p \in C$, and a nonnegative integer k.
Question ($\mathcal{E}$-CCDV): Is it possible to by deleting at most k voters ensure that p is the unique winner of the resulting $\mathcal{E}$ election?
Question ($\mathcal{E}$-DCDV): Is it possible to by deleting at most k voters ensure that p is not a unique winner of the resulting $\mathcal{E}$ election?

The reason we said above that these are the control types with the most emotional resonance is that these types loosely model activities that go to the heart of people's participation in elections. For example, deleting voters models voter suppression, where the chair's budget is enough to (by whatever means – visits, intimidation, spreading rumors, buying ads, making phone calls) keep any choice of k voters from voting. Similarly, adding voters can be viewed as loosely modeling get-out-the-vote drives, e.g., the chair's budget can pay to drive the chair's choice of k people to the polls.

With respect to these control types, AV and SP-AV behave basically the same, and so it is not surprising that the results here are identical for AV and SP-AV: Both are vulnerable in the destructive cases and resistant in the constructive cases. As our final sample proof of this section, we prove the following result.

Theorem 10.3.6 (Erdélyi et al. 2008c). *SP-AV is vulnerable to destructive control by deleting voters and is resistant to constructive control by adding voters.*

Proof. In both cases, susceptibility immediately follows from the fact that the proof of susceptibility for destructive control by partition of voters from Theorem 10.3.4 also shows that SP-AV is susceptible to destructive control by partition of voters and the results of Hemaspaandra et al. (2007a) that establish links between the susceptibility claims for certain types of control. In particular, it is known that every voiced voting system that is susceptible to destructive control by partition of voters in model TE or TP is also susceptible to destructive control by deleting voters (Hemaspaandra et al. 2007a, Theorem 4.3). Susceptibility to destructive control by deleting voters in turn is equivalent to susceptibility to constructive control by adding voters (Hemaspaandra et al. 2007a, Theorem 4.1).

For concreteness, here is a quite simple example: Consider an election with two candidates, c and d, having two ballots of the form $c \mid d$ and one ballot of the form $d \mid c$. So c is the unique SP-AV winner. However, deleting one ballot of the form $c \mid d$ yields a tie between c and d. Thus SP-AV is susceptible to destructive control by deleting voters. On the other hand, if we view one ballot of the form $c \mid d$ as that of an unregistered voter, and the remaining two ballots, $c \mid d$ and $d \mid c$, as those of registered voters, then we can turn c from not being a unique SP-AV winner into the unique SP-AV winner by adding the originally unregistered ballot. Thus SP-AV is also susceptible to constructive control by adding voters.

To prove that SP-AV is vulnerable to destructive control by deleting voters, we present an algorithm for solving the problem SP-AV-DCDV in polynomial time. The algorithm takes as input an election (C, V) (that fulfills the requirements of SP-AV), a distinguished candidate p (whom the chair seeks to prevent from being a unique winner), and a nonnegative integer k (the maximum number of voters allowed to be removed from the election). The output of the algorithm either will be "control impossible" (if it is not possible to via deleting at most k votes prevent p from being a unique SP-AV winner), or it will be a subset $V' \subseteq V$ with $\|V'\| \leq k$ such that p is not a unique winner of the election $(C, V - V')$.

The algorithm first checks the following trivial cases:

1. If C contains p alone, then output "control impossible" and stop. (There is no other candidate who could possibly prevent p from being the unique SP-AV winner.)
2. Else if p already is not a unique SP-AV winner, then output the empty set as the set V' of voters to be deleted. (There is no need for the chair to intervene in this case.)

If none of the trivial cases applies, we know that $\|C\| \geq 2$ and p has more approvals than any other candidate in C. For each $c \in C$, let $surplus_{(C,V)}(p,c) = score_{(C,V)}(p) - score_{(C,V)}(c)$. Note that $surplus_{(C,V)}(p,c)$ is positive for each candidate $c \neq p$ in C. Now, the algorithm determines some candidate $q \neq p$ in C with smallest $surplus_{(C,V)}(p,q)$, and if $surplus_{(C,V)}(p,q) > k$ then it outputs "control impossible" and stops. In this case, deleting any choice of at most k voters will not dethrone p as the unique SP-AV winner. Otherwise (i.e., if $surplus_{(C,V)}(p,q) \leq k$), the algorithm outputs, as the set V' of voters to be deleted, $surplus_{(C,V)}(p,q)$ voters who approve of p and disapprove of q,[9] and stops.

To prove that SP-AV is resistant to constructive control by adding voters, we again give a reduction from the NP-complete problem X3C, which was defined in the proof of Theorem 10.3.5. Let an instance $(B, \mathcal{S})$ of X3C be given, where $B = \{b_1, b_2, \ldots, b_{3m}\}$ (we assume $m > 1$, which is possible because the thus modified problem is still NP-complete), $\mathcal{S} = \{S_1, S_2, \ldots, S_n\}$, and $S_i \subseteq B$ with $\|S_i\| = 3$ for each i, $1 \leq i \leq n$.

Given $(B, \mathcal{S})$, construct an instance (C, V, W, p, m) of SP-AV-CCAV as follows. Let $C = B \cup \{p\}$, where p is the distinguished candidate. Let V contain $m - 2$ registered voters of the form $B \mid p$, and let W consist of the following n unregistered voters: For each i, $1 \leq i \leq n$, there is one voter of the form $p \; S_i \mid (B - S_i)$ in W. Clearly, p is not a unique SP-AV winner of the election (C, V), since $score_{(C,V)}(p) = 0$ and $score_{(C,V)}(b_j) = m - 2$ for each j, $1 \leq j \leq 3m$.[10]

Before proceeding with the proof, we present a small example to illustrate the construction.

Example 10.3.3. Consider the "yes" instance $(B, \mathcal{S})$ of X3C that is defined by $B = \{b_1, b_2, \ldots, b_9\}$ (so $m = 3$) and $\mathcal{S} = \{S_1, S_2, S_3, S_4\}$ with $S_1 = \{b_1, b_3, b_5\}$, $S_2 = \{b_2, b_4, b_6\}$, $S_3 = \{b_1, b_4, b_6\}$, and $S_4 = \{b_7, b_8, b_9\}$. From this instance we construct the SP-AV-CCAV instance $(C, V, W, p, 3)$ with candidate set $C = \{p\} \cup B$, distinguished candidate p, and with registered and unregistered voters, V and W, as shown in Table 10.6.

Clearly, p is not a unique SP-AV winner of the election (C, V), as p has zero approvals and each $b_j \in B$ has one approval. However, adding to V the first two and the last of the unregistered voters of W (which correspond to an exact cover for B) makes p the unique SP-AV winner of the resulting election, as in this election p has three approvals, but each $b_j \in B$ has only two.

For the sake of contrast, consider the "no" instance $(B, \mathcal{S}')$ of X3C that is obtained from $(B, \mathcal{S})$ by modifying only the fourth set in $\mathcal{S}$ from S_4 to

[9] It is easy to see that, by definition of $surplus_{(C,V)}(p,q)$, such voters do exist.

[10] It is worth discussing the boundary case of $m = 2$ here, which is the smallest m possible. In this case, the election (C, V) has seven candidates, but none of these candidates is approved by any voter, simply because there are no voters (due to $\|V\| = m - 2 = 0$). By the definition of *SP-AV winner* (which, recall, is any candidate with the largest number of approvals), each of these seven candidates in C is a SP-AV winner, so it indeed is true that p is not a *unique* SP-AV winner of (C, V), even if $m = 2$.

Table 10.6 Registered and unregistered voters generated from a "yes" and from a "no" instance of X3C

V: one registered voter
b_1 b_2 b_3 b_4 b_5 b_6 b_7 b_8 b_9 | p

W: four unregistered voters	W': four unregistered voters
p b_1 b_3 b_5 | b_2 b_4 b_6 b_7 b_8 b_9	p b_1 b_3 b_5 | b_2 b_4 b_6 b_7 b_8 b_9
p b_2 b_4 b_6 | b_1 b_3 b_5 b_7 b_8 b_9	p b_2 b_4 b_6 | b_1 b_3 b_5 b_7 b_8 b_9
p b_1 b_4 b_6 | b_2 b_3 b_5 b_7 b_8 b_9	p b_1 b_4 b_6 | b_2 b_3 b_5 b_7 b_8 b_9
p b_7 b_8 b_9 | b_1 b_2 b_3 b_4 b_5 b_6	p b_2 b_5 b_8 | b_1 b_3 b_4 b_6 b_7 b_9

$S'_4 = \{b_2, b_5, b_8\}$, so $\mathcal{S}' = \{S_1, S_2, S_3, S'_4\}$. Now, this instance is transformed by our reduction into the SP-AV-CCAV instance $(C, V, W', p, 3)$, where W' can also be found in Table 10.6. Clearly, $(C, V, W', p, 3)$ is a "no" instance of SP-AV-CCAV, since adding any subset W'' of at most three voters from W' to V will fail to make p a unique SP-AV winner of the election $(C, V \cup W'')$. For example, if we added the first two and the last of the unregistered voters of W' to V then p would tie for winner with both b_2 and b_5, with each having three approvals.

We now prove that $\mathcal{S}$ has an exact cover for B if and only if p can be made the unique SP-AV winner by adding at most m voters.

If $\mathcal{S}$ contains an exact cover for B, then let $W' \subseteq W$ be the set of voters corresponding to this cover and add W' to V. Since $score_{(C,V \cup W')}(p) = m$ and $score_{(C,V \cup W')}(b_j) = m - 1$ for each j, $1 \leq j \leq 3m$, p is the unique SP-AV winner of the election $(C, V \cup W')$.

Conversely, suppose that p can be made the unique SP-AV winner by adding at most m voters from W to V. Every candidate b_j in B has $m - 2$ approvals more than p in (C, V), and adding any voter from W will give both p and three members of B one more approval. Thus for p to become the unique SP-AV winner we need to add exactly m voters from W, while making sure that no candidate from B gains more than one additional approval. It follows that the m voters added correspond to an exact cover for B. □

10.3.4 Conclusions

Table 10.1 already summarized the control results. However, some comments are in order. We mention that Table 10.1 reflects the fact that each of the 44 cases' complexity is known. None of the cases remain open. We also note that SP-AV (see however the discussion and caveats of Sect. 10.2) has a very large number of resistances. This is clearly a strong point in its favor, although of course one's choice of an election system for any particular task will depend on the task, and maximizing control resistances will rarely be the only – or even the most important – factor one weighs in choosing one's system. Other factors may include simplicity (for the voter), perceived fairness, acceptability to the electorate, resistance to voter manipulation (see Bartholdi et al. 1992; Faliszewski et al. 2009c), etc. And even if control resistances were what one cared about, the Llull/Copeland complex of systems, while having fewer resistances in number, has some resistances that SP-AV lacks, and so again one would judge by what resistances are most needed for the task at hand – what attacks we most want to be protected from. Nonetheless, it is clear that SP-AV is an interesting system having a quite large number of control resistances, and as such is worth being at least seriously considered – weighing its overall strengths and weaknesses – when one is choosing an election system for a task.

10.4 The Complexity of Bribery for Approval Voting

In this section, we are concerned with the complexity of bribery for approval voting. The notion of bribery was introduced by Faliszewski et al. (2006, 2009a), who studied the hardness of bribery for voting systems as diverse as approval voting, scoring protocols (an important class of voting systems including in particular Borda count, plurality, veto, and k-approval), and Dodgson voting (Dodgson 1876, see Hemaspaandra et al. 1997; Homan and Hemaspaandra 2006, 2009; Caragiannis et al. 2009 for results on the computational aspects of Dodgson's system). Bribery has been subsequently explored for a variety of other voting systems (see, e.g., Faliszewski et al. 2007, 2008a, 2009b; Faliszewski 2008). Here we will focus on the complexity results for bribery with respect to approval voting established by Faliszewski et al. (2006, 2009a).

As mentioned in the Introduction, bribery models scenarios in which an external agent, the "briber," seeks to make his or her favorite candidate win (i.e., we in this section focus on the constructive case only) by bribing some voters to change their votes. As such, bribery can be seen as sharing some aspects with control (namely, that an external actor seeks to change the electoral outcome) and some with manipulation (namely, that the voters' ballots may be changed). Unlike the chair in control settings, the briber doesn't alter the procedure of an election but rather alters the voters' ballots, and unlike the strategic voters in manipulation settings, it here is the briber (and not some of the voters) who seeks to do something bad.

For each of the bribery problems we will define, the briber's budget will be limited and we will consider the following variants of how the budget can be spent in order to reach the briber's goal:

1. Voters can be "weighted" or "unweighted." For example, if a voter of weight three approves of some candidate then this candidate walks off with the equivalent of three weight-one approvals from just this voter. Weighted voting occurs in many real-life settings. In voting on referenda, stockholders' votes are typically weighted by the number of shares they own.
2. Voters may or may not have price tags. The priced-voters case again is modeling a natural situation, namely, it reflects the fact that some voters may be more expensive to bribe than others.

3. For both the weighted-voters case and the priced-voters case, one can distinguish between whether weights/prices are represented in binary or in unary. Clearly, for the computational complexity of the problems thus defined the representation of the problem instances can – and we will show that in some cases it does – make a difference.
4. Finally, in addition to bribery where a voter's ballot either is or is not bought by the briber, we will also consider a more fine-tuned, local approach, called "microbribery," where the briber pays for each bit-flip.[11] This model reflects the fact that voters naturally may have stronger feelings about some candidates than about others, and so a voter's approval or disapproval of some candidate may be for sale but the same voter may not be willing to change his or her approval/disapproval of some other candidate. If it comes to convictions, voters may have preferences that money can't buy.

We start by formally defining the most basic variant of bribery for approval voting. Unlike in Sect. 10.3, we define all bribery problems in the nonunique-winner model, since that is the core model taken by Faliszewski et al. (2006, 2009a) (although do note that essentially all their results also hold in the nonunique winner mode) and the other papers on bribery, and it makes sense to keep the studies of a given type of attack as uniform as possible in their model. We mention in passing that some papers study bribery, manipulation, or control in both the unique-winner and the nonunique-winner models. For example, Faliszewski et al. (2009b) do so for bribery and control regarding the Llull/Copeland complex of systems, and they obtain the same complexity results for both models in each case. Similarly, Hemaspaandra et al. (2006) prove that the complexity of winner determination, which originally was shown in the nonunique-winner model for Dodgson (Hemaspaandra et al. 1997), Young (Rothe et al. 2003), and Kemeny elections (Hemaspaandra et al. 2005), remains the same in the unique-winner model. To the best of our knowledge, the only complexity results where the unique-winner model parts company with the nonunique-winner model are due to Faliszewski et al. (2008b) and are related to manipulating Copeland elections.

Name: AV-BRIBERY.
Given: A set C of candidates, a collection V of voters represented via ballots over C, a distinguished candidate p, and a nonnegative integer k (the "budget").
Question: Is it possible to change at most k voters' ballots so that p is an approval winner of the resulting election?

In an AV-BRIBERY instance, all voters are both unweighted and unpriced. The corresponding bribery problem for weighted but unpriced voters is denoted by AV-WEIGHTED-BRIBERY, that for unweighted but priced voters by

[11] The study of microbribery was initiated in Faliszewski et al. (2006), where it however was notated bribery′, and was also studied in Faliszewski et al. (2007). The term "microbribery" was first used for this concept in Faliszewski et al. (2008a), which studies (among other things) microbribery of the Llull/Copeland election systems.

AV-$BRIBERY, and if the voters both are weighted and have price tags, we write AV-WEIGHTED-$BRIBERY.

The first result in this section is that even AV-BRIBERY, the simplest of these four problems, is NP-complete. Since this problem is a special case of each of the other three problems just defined, they each immediately inherit the NP-hardness lower bound of AV-BRIBERY. On the other hand, it is easy to see that each of these four problems is contained in NP: Given an instance (C, V, p, k), nondeterministically guess a bribery action involving at most k voters and verify deterministically that p wins the resulting election. Obviously, this can be done in polynomial time, even if weights and prices are given in binary.[12]

Theorem 10.4.1 (Faliszewski et al. 2006, 2009a). AV-BRIBERY *is* NP-*complete.*

Proof. Membership in NP was justified above. It remains to prove that AV-BRIBERY is NP-hard. To this end, we now describe a reduction from the NP-complete problem X3C (which was defined in the proof of Theorem 10.3.5). Let an instance $(B, \mathcal{S})$ of X3C be given, where $B = \{b_1, b_2, \dots, b_{3m}\}$, $m \geq 1$, $\mathcal{S} = \{S_1, S_2, \dots, S_n\}$, and for each i, $1 \leq i \leq n$, we have that $S_i \subseteq B$ and $\|S_i\| = 3$. Without loss of generality, we may assume that $n \geq m$, since otherwise $\mathcal{S}$ wouldn't have an exact cover for B, and mapping such a "no" instance of X3C to some fixed "no" instance of AV-BRIBERY would correctly handle this case. We again need the values h_j, which were defined in the proof of Theorem 10.3.5 as $h_j = \|\{S_i \in \mathcal{S} \mid b_j \in S_i\}\|$ for each j, $1 \leq j \leq 3m$.

Define the election (C, V), where $C = B \cup \{p\}$ is the set of candidates with the distinguished candidate p, and where V is defined to consist of the following $(3m + 2)n + 2m - \sum_{j=1}^{3m} h_j = 3mn + 2m - n$ ballots:

1. For each i, $1 \leq i \leq n$, there is one voter who approves of the three candidates in S_i and who disapproves of all other candidates.
2. For each j, $1 \leq j \leq 3m$, there are $n - h_j + 1$ voters who each approve of only b_j and who disapprove of all other candidates.
3. There are $n - m$ voters who each approve of only p and who disapprove of the candidates in B.

Note that $score_{(C,V)}(p) = n - m$ and $score_{(C,V)}(b_j) = n + 1$ for each j, $1 \leq j \leq 3m$.

Define our AV-BRIBERY instance to be (C, V, p, m). As in the previous reductions presented in Sect. 10.3, we give an example to illustrate this construction of (C, V, p, m) from the given X3C instance $(B, \mathcal{S})$.

Example 10.4.1. Suppose we are given the X3C instance $(B, \mathcal{S})$, where $B = \{b_1, b_2, \dots, b_6\}$ and $\mathcal{S} = \{S_1, S_2, S_3, S_4\}$ with $S_1 = \{b_1, b_2, b_3\}$, $S_2 = \{b_2, b_4, b_6\}$, $S_3 = \{b_1, b_3, b_5\}$, and $S_4 = \{b_1, b_2, b_6\}$. Our reduction then yields the instance

[12] Clearly, representing weights or prices in unary provides a less succinct input than using the binary representation. This implies that a problem using unary encoding is at most as hard, computationally, as the corresponding problem using binary encoding.

Table 10.7 Twenty-four ballots generated from a "yes" instance of X3C

C	1st				2nd																		3rd		Score
p	0	**0**	**0**	0	0	0	0	0	0	0	0	0	0	0	0	0	0	0	0	0	0	0	1	1	2
b_1	1	**0**	**1**	1	1	1	0	0	0	0	0	0	0	0	0	0	0	0	0	0	0	0	0	0	5
b_2	1	**1**	**0**	1	0	0	1	1	0	0	0	0	0	0	0	0	0	0	0	0	0	0	0	0	5
b_3	1	**0**	**1**	0	0	0	0	0	1	1	1	0	0	0	0	0	0	0	0	0	0	0	0	0	5
b_4	0	**1**	**0**	0	0	0	0	0	0	0	0	1	1	1	1	0	0	0	0	0	0	0	0	0	5
b_5	0	**0**	**1**	0	0	0	0	0	0	0	0	0	0	0	0	1	1	1	1	0	0	0	0	0	5
b_6	0	**1**	**0**	1	0	0	0	0	0	0	0	0	0	0	0	0	0	0	0	1	1	1	0	0	5

$(C, V, p, 2)$ of AV-BRIBERY with candidate set $C = \{p\} \cup B$ and with V consisting of the 24 ballots shown in Table 10.7.

Since $S_2 \cup S_3 = B$, $\mathcal{S}$ has an exact cover for B. Our reduction ensures that p can be made an approval winner by bribing the two voters of the first group that correspond to this exact cover (i.e., the two boldfaced first-group voters in Table 10.7): If their ballots are changed such that they both approve of only p and disapprove of all candidates in B, then all candidates are winners with score 4 in the resulting election.

We now prove that $\mathcal{S}$ has an exact cover for B if and only if p can be made an approval winner by bribing at most m voters.

From left to right: Suppose that $\mathcal{S}$ has an exact cover for B. Simply bribe the voters corresponding to this exact cover by changing their ballots such that each bribed voter approves of p and disapproves of the candidates in B. Then p gains m approvals (i.e., $score_{(C,V')}(p) = n$, where V' denotes the collection of ballots after the bribery) and every $b_j \in B$ loses one approval (i.e., $score_{(C,V')}(b_j) = n$). Thus p is an approval winner in (C, V').

From right to left: Suppose that p can be made an approval winner by bribing at most m voters. Let (C, V') denote the election that results from any such bribery. By bribing at most m voters, p can gain no more than m additional approvals. Thus $score_{(C,V')}(p) \leq (n - m) + m = n$. On the other hand, all candidates in B have $n + 1$ approvals initially, so for p to become an approval winner by bribing at most m voters, each $b_j \in B$ has to lose at least one approval by the bribery. It follows that exactly m voters must have been bribed to yield (C, V'), and these m voters correspond to an exact cover for B. □

Corollary 10.4.1 (Faliszewski et al. 2006, 2009a). *Each of the problems* AV-WEIGHTED-BRIBERY, AV-\$BRIBERY, *and* AV-WEIGHTED-\$BRIBERY *is* NP-*complete.*

We now turn to the formal definition of the microbribery problem for approval voting. Again, we start with the most basic case where voters are unweighted and don't have price tags.

Name: AV-MICROBRIBERY.

Given: A set C of candidates, a collection V of voters represented via ballots over C, a distinguished candidate p, and a nonnegative integer k.

Question: Is it possible to flip at most k entries in the ballots (to be clear: not k entries per ballot but k entries overall, in all the ballots) so that p is an approval winner of the resulting election?

As with bribery, AV-WEIGHTED-MICROBRIBERY denotes the microbribery problem for weighted but unpriced voters within approval voting, AV-\$MICROBRIBERY denotes this problem for unweighted but priced voters (each bit flip may have a different price), and we write AV-WEIGHTED-\$MICROBRIBERY if the voters both are weighted and have price tags. Here, it will also make sense to distinguish between binary and unary representations for weights or prices. The names of the above-defined four microbribery problems for AV refer to the more succinct binary representation of, respectively, weights and prices, which is the standard way of representing numbers. To indicate that either weights or prices or both are given in unary, we use the subscript "unary" at the corresponding place in the problem name. For example, AV-WEIGHTED$_{\text{unary}}$-\$MICROBRIBERY denotes the same problem as AV-WEIGHTED-\$MICROBRIBERY but with weights given in unary and prices given in binary. It is known that the microbribery problem for AV is easy to solve even if both weights and prices are given, provided that at least one of them is represented in unary. We omit the proof.

Theorem 10.4.2 (Faliszewski et al. 2006, 2009a). AV-MICROBRIBERY, AV-WEIGHTED$_{\text{unary}}$-\$MICROBRIBERY$_{\text{unary}}$, AV-WEIGHTED$_{\text{unary}}$-\$MICROBRIBERY, *and* AV-WEIGHTED-\$MICROBRIBERY$_{\text{unary}}$ *can each be solved in polynomial time.*

In contrast, if both weights and prices are encoded in binary, the microbribery problem for AV is NP-complete.

Theorem 10.4.3 (Faliszewski et al. 2006, 2009a). AV-WEIGHTED-\$MICROBRIBERY *is* NP-*complete.*

Proof. Again, it is clear that AV-WEIGHTED-\$MICROBRIBERY is in NP. Thus, it remains to prove that AV-WEIGHTED-\$MICROBRIBERY is NP-hard. To this end, we reduce from the following famous problem:

Name: PARTITION.
Given: A nonempty sequence $s_1, s_2, \ldots, s_n$ of positive integers such that $\sum_{i=1}^{n} s_i$ is an even number.
Question: Does there exist a subset $A \subseteq \{1, 2, \ldots, n\}$ such that $\sum_{i \in A} s_i = \sum_{i \in \{1,2,\ldots,n\}-A} s_i$?

Before we proceed with the proof, let us illustrate this problem by an example that continues the example for the X3C problem presented in the proof of Theorem 10.3.5. Recall from that previous example that a magnificent mosaic had been stolen from the Cathedral of Monreale on Sicily and then was sold – piece by piece, along with faked duplicates, and always in batches of three – on black markets world-wide. Now, Giuseppe Greco and Salvatore Inzerillo, who had planned, organized, and executed this coup, were meeting in a bar in the port district of

Palermo to divide what they had earned, namely \$55,557,390. To hide their tracks, they had changed this amount of money into 1,235 bars of gold with different sizes, $s_1, s_2, \ldots, s_{1{,}235}$, each s_i being the integer amount of dollars the ith bar of gold was worth.

"Now let me get this straight," said Giuseppe who knew Salvatore all too well from other deals. "This time we do it fifty-fifty!"

"Yes, fine with me," replied Salvatore craftily. "So tell me, my friend: Which of the gold bars are for me?"

Giuseppe set down and started thinking. He was counting the gold bars, playing with them, weighing them in his hands, and comparing their dollar amounts. He was scratching his head. Meanwhile, Salvatore was watching him silently. Giuseppe was thinking for more than an hour. Then he said: "No! You tell me! Which of the gold bars are mine?"

"Well," said Salvatore slowly, "all I know about this, and these were my father's last words before he died, so it must be true: This is a very hard problem, and you cannot solve it in a trice. So it just won't be possible for us to make it fifty-fifty. You can try to solve this problem by brute force, but this won't be possible in a reasonable amount of time!"

Salvatore was right in one regard: PARTITION is an NP-complete problem (see, e.g., Garey and Johnson 1979) and thus, indeed, is very hard to solve. But he wasn't right in another regard: Giuseppe did solve the problem he was facing by brute force before dawn. Salvatore was found the next day, floating face-down in the Mediterranean near the port of Palermo. And Giuseppe and all the gold were gone.

Now, back to the proof, let an instance $(s_1, s_2, \ldots, s_n)$ of PARTITION be given, where we have $\sum_{i=1}^{n} s_i = 2S$ and where S and each s_i is a positive integer. Define the election (C, V), where $C = \{p, x\}$ is the set of candidates with p being the distinguished candidate, and where V consists of the following $n + 1$ ballots:

1. There is one voter of weight S who approves of p and disapproves of x, and the price for flipping any entry in this voter's ballot is $2S + 1$.
2. For each i, $1 \leq i \leq n$, there is one voter of weight s_i who approves of x and disapproves of p, where the price for flipping this voter's p-entry is s_i and the price for flipping this voter's x-entry is $2S + 1$.

Our AV-WEIGHTED-\$MICROBRIBERY instance then is (C, V, p, S). Note that $score_{(C,V)}(p) = S$ and $score_{(C,V)}(x) = 2S$. Thus x is the unique approval winner of the election (C, V).

The following example illustrates the construction.

Example 10.4.2. Suppose we are given the PARTITION instance $(s_1, s_2, \ldots, s_6)$, where $s_1 = 9$, $s_2 = 2$, $s_3 = 3$, $s_4 = 4$, $s_5 = 1$, and $s_6 = 1$. This is a "yes" instance of PARTITION, and a possible partition of $\{1, 2, \ldots, 6\}$ is $\{1, 5\}$ and $\{2, 3, 4, 6\}$, since $s_1 + s_5 = 10 = s_2 + s_3 + s_4 + s_6$.

Our reduction then yields the instance $(C, V, p, 10)$ of AV-WEIGHTED-\$MICROBRIBERY with candidate set $C = \{p, x\}$ and with V consisting of the seven ballots shown in Table 10.8.

Table 10.8 Seven ballots generated from a "yes" instance of PARTITION

V	Weight	Price		Ballot	
		p	x	p	x
v_0	10	21	21	1	0
v_1	9	9	21	0	1
v_2	2	2	21	0	1
v_3	3	3	21	0	1
v_4	4	4	21	0	1
v_5	1	1	21	0	1
v_6	1	1	21	0	1
Score				10	20

Clearly, x is the unique winner of (C, V), as x has 20 approvals and p has only 10. However, p can be made an approval winner by microbribing the two voters that correspond to the $\{1, 5\}$ part of the partition given above. Namely, flipping the p-entries in the ballots of voters v_1 and v_5 costs exactly ten units of bribery currency (which is just as much as the briber can afford). After the bribery, the voters v_1 and v_5 approve of both p and x, so both candidates, p and x, are approval winners with score 20.

Turning back to the proof, we now show that, in general, there is a set $A \subseteq \{1, 2, \ldots, n\}$ such that $\sum_{i \in A} s_i = S$ if and only if p can be made an approval winner by a microbribery of cost at most S.

From left to right: Suppose there is a set $A \subseteq \{1, 2, \ldots, n\}$ such that $\sum_{i \in A} s_i = S$. Change the disapprovals of p into approvals in the ballots of the voters corresponding to A. Clearly, this will cost $\sum_{i \in A} s_i = S$, and after that both p and x are approval winners with score $2S$.

From right to left: Suppose p can be made an approval winner by a microbribery of cost at most S. The given amount S allows only flips from 0 to 1 in the p-entries of the second voter group. Thus p can gain (at most) S additional approvals, whereas x's score will still be $2S$ after the microbribery. Since for the voters whose ballots were changed to make p an approval winner, the weights and the costs for flipping their p-entries are the same, p must have gained exactly S approvals and the weights of these voters must also be exactly S. Thus there is a set $A \subseteq \{1, 2, \ldots, n\}$ such that $\sum_{i \in A} s_i = S$. □

10.5 Local Search Heuristics for Minimax Approval Voting

10.5.1 Minisum and Minimax Approval Voting

In the present section, we will focus on the complexity of electing a committee of fixed size via another variant of approval voting. (For more on the complexity issues in electing committees, we commend to the reader the work of Meir et al. 2008.)

Table 10.9 Electing a committee of size two via minisum and minimax evaluation for approval voting

Committee	Ballots			Evaluation	
	(3×) 1110	(2×) 0101	(3×) 1010	sum	max
1100	3	4	6	13	**6**
1010	3	8	0	**11**	8
1001	9	4	6	19	9
0110	3	4	6	13	**6**
0101	9	0	12	21	12
0011	9	4	6	19	9

The standard way of evaluating an approval election to obtain a committee of k candidates is to sum up the approvals for each candidate and to select k candidates with the highest number of approvals (where some tie-breaking rule can be used when there is more than one such size-k committee). This method is called the *minisum* procedure, since the outcome minimizes the sum of the Hamming distances of the winning committee to all ballots. Recall that the Hamming distance between two binary vectors is the minimum number of bit-flips needed to transform one vector into the other.

Brams et al. (2004, see also Brams et al. 2007a,b) proposed a new evaluation method for approval elections to determine committees of fixed size, which is called the *minimax* procedure, since it minimizes the maximum of the Hamming distances between the winning committee and the ballots. The minisum procedure seeks to find an outcome that is close to many ballots, i.e., it minimizes the *total* dissatisfaction of the electorate, whereas the minimax procedure seeks to minimize the dissatisfaction of the most dissatisfied voter, even if this results in a higher total (or, equivalently, average) dissatisfaction.

Example 10.5.1. Table 10.9 gives an example of an election that aims to find a committee of size $k = 2$ by approval voting. There are four candidates and eight voters in this election, where three voters approve of the first three candidates, two voters approve of the second and the fourth candidate, and three voters approve of the first and the third candidate. In the first column, the possible committees with two candidates each are listed. The next three columns give the Hamming distances between each such committee and each of the three distinct ballots times their multiplicities, the "sum" column gives the sum of these values, and the "max" column gives their maximum.

A committee with a smallest entry in the "sum" column is a *minisum (winner) committee*, and in this example we happen to have a unique minisum committee, namely 1010. The two voters with ballots 0101, however, will be completely dissatisfied with the minisum outcome, since none of their candidates is in the minisum committee and all candidates they disapprove of are in this committee.

A committee with a smallest entry in the "max" column is a *minimax (winner) committee*, and in this example there are two minimax committees, 1100 and 0110.

Table 10.10 Electing a size-two committee via minisum and minimax approval voting with proximity weights

Committee	Ballots with proximity weights			Evaluation	
	(3×) 1110 $\frac{1}{3}$	(2×) 0101 $\frac{2}{21}$	(3×) 1010 $\frac{3}{11}$	sum	max
1100	693	264	1,134	2,091	1,134
1010	693	528	0	**1,221**	**693**
1001	2,079	264	1,134	3,477	2,079
0110	693	264	1,134	2,091	1,134
0101	2,079	0	2,268	4,347	2,268
0011	2,079	264	1,134	3,477	2,079

Both the minisum and the minimax evaluation presented in Example 10.5.1 take only the number of identical ballots into account. A different approach, also proposed by Brams et al. (2007a), is to base these evaluations on "proximity weights." Instead of merely counting how often each ballot occurs, the "proximity" of any ballot to all other ballots is thus taken into account. Formally, the *proximity weight* of a given ballot v_i is defined by

$$\frac{m_i}{\sum_{j=1}^{t} m_j \cdot H(v_i, v_j)}, \tag{10.4}$$

where $t > 1$ is the total number of distinct ballots (say, $v_1, v_2, \ldots, v_t$ with $v_i \neq v_j$ for $i \neq j$), m_i is the multiplicity of v_i, and $H(v_i, v_j)$ is the Hamming distance between ballots v_i and v_j. Note that, to ensure that the denominator in (10.4) isn't zero, we assume that there are at least two distinct ballots.

Example 10.5.2. Table 10.10 shows the same election as in Example 10.5.1, except that the evaluation is made by calculating the Hamming distance of each committee to each distinct ballot times this ballot's proximity weight (instead of multiplying this Hamming distance with this ballot's multiplicity). To avoid fractions, all results are multiplied by the least common multiple of the distinct voters' denominators in (10.4), which is 693 in this example. This is reasonable, since the ratios between the single alternative committees remain the same, see Kilgour et al. (2006).

Both the winning committee obtained via the minisum evaluation with proximity weights and that obtained via the minimax evaluation with proximity weights happen to coincide with the minisum committee calculated without proximity weights in Example 10.5.1. However, Brams et al. (2007a) mention that minisum outcomes not based on proximity weights and minimax outcomes based on proximity weights can diverge maximally, i.e., all bits in the minisum outcome are flipped in the minimax outcome.

10.5.2 *NP-Hardness and Approximability of Fixed-Size Minimax Approval Voting*

Given an election with m candidates, n ballots, and a fixed committee size of k, the minisum evaluation (with or without proximity weights) is easy and can be done in polynomial time. If proximity weights are not considered, simply add the number of ones in all ballots for each candidate, and declare k candidates with the highest approval scores as winners to join the committee. If ties occur (e.g., if there are more than k candidates each having the highest approval score), the winning committee of size k may be selected among the possible size-k minisum committees by using some fixed, computationally simple tie-breaking rule. If proximity weights are to be considered in a minisum evaluation, the algorithm is similarly efficient, since all proximity weights involved can be determined in polynomial time.

In a minimax evaluation, however, all $\binom{m}{k}$ committees in concept may need be considered, and although for each committee the Hamming distances to all ballots can be computed easily, this may be just too many committees. Indeed, LeGrand (2004) showed that the following problem is NP-complete.

Name: FIXED-SIZE MINIMAX APPROVAL VOTING (FSM-AV, for short).
Given: A set C of candidates, a collection V of voters represented via ballots over C, a nonnegative integer $k \leq \|C\|$, and a nonnegative integer $d \leq \|C\|$.
Question: Does there exist a vector $v \in \{0,1\}^{\|C\|}$ such that v has exactly k ones and $H(v, v_i) \leq d$ for all $v_i \in V$, i.e., is there a committee of size k whose maximum Hamming distance to the ballots in V is at most d?

Theorem 10.5.1 (LeGrand 2004). FSM-AV *is* NP-*complete.*

Proof. FSM-AV is easily seen to be in NP, so it remains to prove FSM-AV NP-hardness. LeGrand (2004) gave a reduction to FSM-AV from the NP-complete problem VERTEX COVER (see, e.g., Garey and Johnson 1979), which is formally defined as follows.

Name: VERTEX COVER.
Given: An undirected graph G with vertex set $V(G) = \{v_1, v_2, \ldots, v_n\}$ and edge set $E(G)$, and a nonnegative integer $k \leq n$.
Question: Does G have a vertex cover of size at most k, i.e., does there exist a subset $V' \subseteq V(G)$ with $\|V'\| \leq k$ such that for each edge $\{x, y\} \in E(G)$, we have that $\{x, y\} \cap V' \neq \emptyset$?

Does this problem look familiar to you? If it does, you've been carefully reading the proofs in Sect. 10.3! VERTEX COVER may be viewed as a special case of the problem HITTING SET, which was defined in the proof of Theorem 10.3.3. In fact, HITTING SET is nothing other than the vertex cover problem for hypergraphs (note: the hyperedges of a hypergraph may involve more than two vertices). So, any given instance (G, k) of VERTEX COVER may be viewed as the instance $(V(G), E(G), k)$ of HITTING SET. To see this, recall the intuitive explanation of HITTING SET from the proof of Theorem 10.3.3: Students signing up for courses and professors seeking to find a hitting set of size at most k among the students such that each course contains at least one hitting-set student. In this interpretation, VERTEX COVER is the special case of HITTING SET in which every course is attended by exactly two students, and if we draw a graph the vertices of which are the students and that has an edge between two vertices exactly if the corresponding students are jointly attending the same course, then determining a vertex cover of size at most k in the graph is just the same as asking for a hitting set of size at most k. (This observation immediately provides a reduction from VERTEX COVER to HITTING SET.)

We now describe a reduction from VERTEX COVER to FSM-AV.[13] Given an instance (G, k) of VERTEX COVER, we will define an instance (C, V, k, k) of FSM-AV such that (C, V, k, k) is a "yes" instance of FSM-AV if and only if (G, k) is a "yes" instance of VERTEX COVER. Note that the maximum size of the vertex cover allowed, k, equals both the size of the committee represented by v and the maximum Hamming distance of v to the ballots in V.

Every vertex in G represents one candidate, so $C = \{c_i \mid v_i \in V(G)\}$, and every edge $\{v_i, v_j\}$ of G represents one voter who approves of the two candidates c_i and c_j and who disapproves of all other candidates. Formally, $V = \{b_{i,j} \mid \{v_i, v_j\} \in E(G)\}$, where $b_{i,j}$ is the bit-vector of length $\|C\|$ that has a one at positions i and j and a zero at all other positions. Obviously, the instance (C, V, k, k) of FSM-AV can be computed in time polynomial in the size of the given VERTEX COVER instance (G, k).

Before proving the correctness of the reduction, let us illustrate the construction by an example.

Example 10.5.3. Consider the graph G with six vertices and eight edges that is shown in Fig. 10.1. Since the edges $\{1, 3\}$, $\{4, 5\}$, and $\{2, 6\}$ have no vertex in

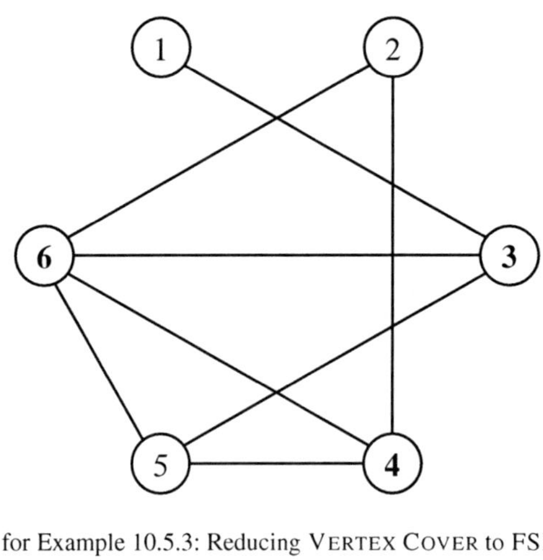

Fig. 10.1 Graph G for Example 10.5.3: Reducing VERTEX COVER to FSM-AV

[13] Note that the reduction presented here is based on (but is much simpler than) the reduction given by LeGrand (2004).

Table 10.11 Eight ballots constructed from graph G and their Hamming distance to $v = 001101$

Edge $\{v_i, v_j\}$	$\{1,3\}$	$\{2,4\}$	$\{2,6\}$	$\{3,5\}$	$\{3,6\}$	$\{4,5\}$	$\{4,6\}$	$\{5,6\}$
Ballot $b_{i,j} \in V$	101000	010100	010001	001010	001001	000110	000101	000011
$H(001101, b_{i,j})$	3	3	3	3	1	3	1	3

common, each vertex cover of G must contain at least three vertices to cover these edges. One possible vertex cover of size three for G is the set $V' = \{3, 4, 6\}$, whose elements are in boldface in Fig. 10.1. Thus $(G, 3)$ is a "yes" instance, but $(G, 2)$ is a "no" instance of VERTEX COVER.

Given the "yes" instance $(G, 3)$ of VERTEX COVER, our reduction yields the instance $(C, V, 3, 3)$ of FSM-AV, where $C = \{c_1, c_2, \ldots, c_6\}$ is the set of candidates and V consists of the eight ballots shown in Table 10.11, where each ballot is a 6-bit binary vector that corresponds to one of the edges of G. Note that the vector $v = 001101$, which represents a committee, corresponds to the vertex cover $V' = \{3, 4, 6\}$, and the last row in Table 10.11 shows that the Hamming distance between v and all the ballots in V is at most three. Thus $(C, V, 3, 3)$ is a "yes" instance of FSM-AV, as desired.

However, decrementing the parameter k to 2 gives the "no" instance $(G, 2)$ of VERTEX COVER. Our reduction then yields the FSM-AV instance $(C, V, 2, 2)$, where the election (C, V) is as described above. One can easily check that no 6-bit vector with exactly two ones has a Hamming distance of at most two to all the ballots in V. So $(C, V, 2, 2)$ is a "no" instance of FSM-AV, as desired.

Now, returning to the proof of Theorem 10.5.1, we claim that G has a vertex cover of size k if and only if there is a vector $v \in \{0, 1\}^{\|C\|}$ such that v has exactly k ones and $H(v, b_{i,j}) \leq k$ for all $b_{i,j} \in V$.

From left to right: Suppose there is a vertex cover V' of size exactly k in G. Define the vector $v \in \{0, 1\}^{\|C\|}$ to have a one at position i if $v_i \in V'$, and to have a zero at position i otherwise. By construction, there are k ones in v, and since V' is a vertex cover of G, every ballot in V has at least one of its two ones at a position that is set to one in v. Thus the Hamming distance of v to each ballot $b_{i,j}$ in V is at most k, since v can be transformed into $b_{i,j}$ by flipping at most one zero in v to one and at most $k - 1$ ones in v to zeros.

From right to left: Suppose there is a vector $v \in \{0, 1\}^{\|C\|}$ such that v has exactly k ones and $H(v, b_{i,j}) \leq k$ for all $b_{i,j} \in V$. Define the set V' to contain exactly those vertices v_i for which v has a one at position i. Clearly, $\|V'\| = k$. Since $H(v, b_{i,j}) \leq k$, each ballot $b_{i,j} \in V$ must have at least one of its two ones at a position at which also v has a one. Thus $V' \cap \{v_i, v_j\} \neq \emptyset$ for all edges $\{v_i, v_j\} \in E(G)$. That is, V' is a vertex cover of size k for G. □

If an important problem turns out to be NP-hard, this doesn't make the problem go away: We still want to solve it, at least as well as we can. So, how can we cope with NP-hardness? There are several answers to this question; let us sketch some of them below.

In practical applications, we may content ourselves with an efficient heuristic algorithm that isn't *always* correct but that does work correctly for those inputs that typically occur in practice. For example, Homan and Hemaspaandra (2009) proposed an efficient greedy heuristic for finding Dodgson winners, and they proved that under the model of voting that political scientists call impartial culture this heuristic is a "frequently self-knowingly correct algorithm" (see also the closely related work of McCabe-Dansted et al. 2008; see also the discussion in Erdélyi et al. 2008a, 2009b, 2007, which in turn discusses notions proposed by Procaccia and Rosenschein 2007; see also Erdélyi et al. 2009a). Regarding this approach to coping with NP-hardness, Sect. 10.5.3 will present some local search heuristics for minimax approval voting.

From a practical point of view, we may also be interested in solving NP-hard problems only for certain small parameters. For example, if the size of the allowed committee in FSM-AV instances is always bounded by a constant k then all $\binom{m}{k'} = \mathcal{O}(m^k)$, $k' \leq k$, possible committees where k' is the committee size can be evaluated in polynomial time, and if this constant bound on committee size is small then this naive algorithm may even be useful in practice. Similarly, if the number of voters in FSM-AV instances is bounded by a small constant, then the problem can be solved in polynomial time by an integer linear program with a constant number of constraints (see LeGrand et al. 2006, 2007). Such results fall into the area of fixed-parameter tractability and parameterized complexity (see, e.g., the excellent textbooks Downey and Fellows 1999; Flum and Grohe 2006; Niedermeier 2006 and the recent article by Buss and Islam 2008). For specific applications of fixed-parameter tractability and parameterized complexity to problems from computational social choice, we refer to the survey by Lindner and Rothe (2008).

As another approach to coping with NP-hardness, we may sometimes be satisfied with having an efficient algorithm that doesn't yield the optimal solution but rather an approximation. While approximability has proven to be a natural and useful approach to coping with the NP-hardness of many important optimization problems in computer science (see, e.g., Ausiello et al. 2003), one may wonder whether it really is sensible in the context of voting. After all, election systems are used to determine an election winner, someone society finds most acceptable among all alternatives, and if an approximation algorithm instead of a real winner outputs some candidate whose score is no worse than, say, one third of a real winner's score (i.e., the algorithm approximates the optimal solution within a factor of three), then this candidate certainly should not be considered to be a reasonable social choice.

Nonetheless, approximation can be useful in the context of voting. For example, Brelsford et al. (2008) set up a uniform framework that uses the approximability of NP-hard manipulation, control, and bribery problems as a measure of the effectiveness of specific manipulation, control, and bribery attacks. As another example, Caragiannis et al. (2009) study the approximability of the winner problem for Dodgson and Young elections, two problems known to be NP-hard (Bartholdi et al. 1989b; Hemaspaandra et al. 1997; Rothe et al. 2003), and they propose two schemes for approximating Dodgson scores. They argue that such an "approximation algorithm is equivalent to a new voting rule" (Caragiannis et al. 2009), and they prove their approximation schemes to possess certain properties that are desirable from a social-choice point of view.[14] In contrast, they prove that Young scores are NP-hard to approximate within *any* factor.

LeGrand et al. (2006, 2007) observed that there is an approximation algorithm for the minimax approval voting problem (which is defined below as the search problem corresponding to the decision problem FSM-AV), and we will present their algorithm here because it is related to the local search heuristics for this problem to be studied in Sect. 10.5.3.

First, we need some notation. The *weight* of a binary vector v, denoted by $weight(v)$, is the number of ones in v, and a k-*completion* of v is a vector v' that is obtained from v by randomly flipping $weight(v) - k$ ones to zeros if $weight(v) > k$, and by randomly flipping $k - weight(v)$ zeros to ones otherwise. Every k-completion v' of v has exactly k ones. Finally, given an election (C, V) and a vector $v \in \{0,1\}^{\|C\|}$, we denote by $MaxHD_{(C,V)}(v) = \max_{b \in V} H(v, b)$ the maximum Hamming distance of v to the ballots in V.

The minimax approval voting problem seeks to find a minimax committee of weight k so as to minimize the maximum Hamming distance to all ballots in a given approval election. Formally, it is defined as follows (note that, since this is a search and not a decision problem, we have an "Output" field instead of a "Question" field):

Name: MINIMAX-AV.
Given: A set C of candidates, a collection V of voters represented via ballots over C, and a nonnegative integer $k \leq \|C\|$.
Output: A vector $v \in \{0,1\}^{\|C\|}$ such that $weight(v) = k$ and $MaxHD_{(C,V)}(v)$ is minimum among the vectors in $\{0,1\}^{\|C\|}$ of weight k.

Theorem 10.5.2 (LeGrand et al. 2006, 2007). *There is a polynomial-time algorithm that approximates the optimal solution of* MINIMAX-AV *within a factor of three.*

Proof. The approximation algorithm for MINIMAX-AV is rather simple. Given an election (C, V) and a nonnegative integer $k \leq \|C\|$, it works as follows:

1. Choose a ballot $v \in V$ at random. (Or, to keep this algorithm in deterministic polynomial time, select the first ballot.)
2. Compute a k-completion v' of v. (Or, to keep this algorithm in deterministic polynomial time, deterministically ensure in any simple way that v' via the smallest possible number of flips ends up having exactly k ones.)
3. Output v' as a solution.

[14] In particular, both approximation schemes proposed by Caragiannis et al. (2009) satisfy the Condorcet property (Condorcet 1792, see also Fishburn 1977), which says that a Condorcet winner – a candidate that defeats every other candidate by strict majorities – should be elected whenever one exists. Also, like many other natural voting rules, one of their approximation schemes is "weakly monotonic."

Obviously, this algorithm runs in polynomial time. To estimate the approximation ratio of this algorithm, let $v \in V$ be the initial ballot chosen by the algorithm, let v' be the k-completion of v that is output by the algorithm, and let w be an optimal solution of MINIMAX-AV for the given input, i.e., $w \in \{0,1\}^{\|C\|}$ is a vector of weight k such that $MaxHD_{(C,V)}(w)$ is minimum among all such vectors.

We have to show that for each $u \in V$, $H(v',u) \leq 3 \cdot MaxHD_{(C,V)}(w)$. Since the Hamming distance satisfies the triangle inequality, we have that for each $u \in V$,

$$H(v',u) \leq H(v',v) + H(v,u).$$

Applying the triangle inequality once again gives:

$$H(v',u) \leq H(v',v) + H(v,w) + H(w,u).$$

Since w is an optimal solution, it holds that $H(v,w) \leq MaxHD_{(C,V)}(w)$ and for all $u \in V$, $H(w,u) \leq MaxHD_{(C,V)}(w)$. Furthermore, since v' is a k-completion of v, $H(v',v) \leq H(w,v) \leq MaxHD_{(C,V)}(w)$. Combining these inequalities, we obtain the desired property:

$$H(v',u) \leq 3 \cdot MaxHD_{(C,V)}(w),$$

which proves that the approximation ratio of the above algorithm is three.

10.5.3 Local Search Heuristics

The algorithm of Theorem 10.5.2 has a guaranteed approximation ratio of three even in the worst case, i.e., regardless of how unluckily we choose the initial vector and how unluckily we choose its k-completion. However, since the solution output by the algorithm depends solely on these random choices, one may wonder whether we perhaps can obtain a better result (i.e., a solution closer to the optimal solution) by picking the initial vector v and the bits to flip for computing a k-completion v' of v in a more purposeful or more clever way. In this section, we present some local search heuristics of LeGrand et al. (2006, 2007) that are motivated by improving upon the approximation algorithm of Theorem 10.5.2. Although they don't have a better guaranteed approximation ratio in the worst case, these local search heuristics seem to perform better in practice.

In computer science, local search heuristics are often applied to find a good, even though typically not optimal, solution to a problem that is hard to compute. The starting point of a local search algorithm is some solution to the problem. Then the algorithm searches in an appropriate neighborhood of the starting solution to find a modification of this solution that is (in some sense) better. The general framework of a local search heuristic for the minimax problem can be described as follows: Given a set C of candidates, a set V of voters represented via ballots over C, a nonnegative integer $k \leq \|C\|$, and a parameter r, do the following:

1. Choose a vector v with $weight(v) = k$.
2. Repeat until $MaxHD_{(C,V)}(v)$ does not change anymore for at most $m = \|C\|$ iterations:
   a. Compute the set S of vectors resulting from flipping up to r bits in v from 0 to 1, and the same number of bits from 1 to 0.
   b. Compute the set $S' \subseteq S$ of vectors that minimize $MaxHD_{(C,V)}(s)$ for all vectors $s \in S$.
   c. Randomly choose, under the uniform distribution, some element of S' to be the new v.
3. Output v as a solution.

To show that such a heuristic algorithm runs in polynomial time on a machine that can make random choices, let us have a closer look at the second step first because we will present different approaches, with different running times, for the first step later on. In the second step, v is an m-bit vector, so $MaxHD_{(C,V)}(v)$ is between 0 and m, and for each of these values the loop in the second step may be executed m times in the worst case. Since $MaxHD_{(C,V)}(v)$ can only decrease and will never increase, the second step of the heuristic may be executed m times for each value, which means the number of loop iterations is $\mathcal{O}(m^2)$. The set S can be computed in time $\mathcal{O}(m^{2r})$, since the number r of flipped bits is constant. To compute S', $MaxHD_{(C,V)}(s)$ must be computed for all $s \in S$, which can be done in time $\mathcal{O}(mn)$. Altogether, the running time of the algorithm's second step is $\mathcal{O}(nm^{2r+3})$.

The adjustable parameters in this local search heuristic are the parameter r and the starting vector v. LeGrand et al. (2006, 2007) proposed to choose $r = 1$ or $r = 2$ and proposed the following alternatives for the starting vector v:

1. A minisum solution, i.e., k candidates with the most approvals.
2. Choose a ballot v whose $weight(v)$ is closest to k so as to minimize the sum of the Hamming distances to all other ballots, and then compute a k-completion v' of v, where the bits to be flipped are chosen so as to minimize the sum of the Hamming distances to all other ballots.
3. A randomly chosen vector v with $weight(v) = k$.

Note that all these starting vectors can be computed in polynomial time (in models allowing appropriate access to randomness), so the heuristic runs in polynomial time (with access to random choices) for each of these choices. In Sect. 10.5.2, we saw that a minisum solution can be computed in deterministic polynomial time. Regarding the second choice of a starting vector, the vector v in the repeat-loop can be determined in polynomial time, since the weight of each ballot and the sum of the Hamming distances to all other ballots can be computed in polynomial time. The k-completion v' can also be computed in polynomial time, since it is the same as computing a minisum solution, either with some fixed committee members (if $weight(v) < k$) or over a reduced candidate set (if $weight(v) > k$). Finally, note that the random choice of a vector v with $weight(v)$ for the third choice of a starting vector can obviously be done in polynomial time (with access to random choices).

We will refer to these heuristics by $h_{i,j}$, where $i \in \{1, 2, 3\}$ is the choice for the starting method named via the three-item list's numbering above (so for example $i = 3$ means start with a randomly chosen vector v with $weight(v) = k$) and where j is the value chosen for the parameter r. The experimental evaluation with the data in Table 10.12 shows that these heuristics perform considerably better than the worst-case ratio of the approximation algorithm in Theorem 10.5.2 would suggest and also are significantly better than the minisum solution.

Let us explain these empirical results of LeGrand et al. (2006, 2007) in a bit more detail. Table 10.12 considers three different types of elections, named "unbiased," "biased," and "GTS 2003," where in each election the number of candidates is m, the committee size is k, and the number of ballots is n. The unbiased and biased elections consist of randomly generated ballots, each for 20 candidates and for a committee size of 10, and with the number n of ballots used being 50 or 200. In the unbiased case, the ballots are generated under the uniform distribution. In the biased case, the generation of the ballots is more complex: First, for each i, $1 \leq i \leq m$, two probabilities π_i and π_i' are randomly chosen for candidate c_i; then, candidate c_i is approved of by 40% of the ballots with probability π_i and by another 40% of the ballots with probability π_i', and the remaining 20% of the ballots are again chosen according to the uniform distribution. The GTS 2003 election, in contrast, is a real-life election performed by the Game Theory Society, with 24 candidates, a committee size of 12, and 161 ballots.

The entries in the "worst case" row of Table 10.12 represent the average approximation ratio of the worst outcome possible for the given election, i.e., a vector v with $weight(v) = k$ and such that $MaxHD_{(C,V)}$ is maximal. The entries in the minimax row are always 1.0000, since the minimax outcome is the optimal solution. All other

Table 10.12 Average approximation ratios for local search heuristics obtained by LeGrand et al. (2006, 2007)

Election with m candidates, n ballots, and committee size k	Unbiased		Biased		GTS 2003
	$m = 20$, $n = 50$, $k = 10$	$m = 20$, $n = 200$, $k = 10$	$m = 20$, $n = 50$, $k = 10$	$m = 20$, $n = 200$, $k = 10$	$m = 24$, $n = 161$, $k = 12$
Minimax	1.0000	1.0000	1.0000	1.0000	1.0000
Minisum	1.1650	1.1521	1.2119	1.1932	1.2143
Worst case	1.6746	1.4859	1.8509	1.6302	1.7143
$h_{1,1}$	1.0058	1.0320	1.0083	1.0210	1.0012
$h_{2,1}$	1.0118	1.0365	1.0112	1.0251	1.0017
$h_{3,1}$	1.0122	1.0370	1.0122	1.0262	1.0057
$h_{1,2}$	1.0004	1.0129	1.0004	1.0025	1.0000
$h_{2,2}$	1.0004	1.0164	1.0005	1.0029	1.0000
$h_{3,2}$	1.0004	1.0164	1.0005	1.0031	1.0000

entries give the average approximation ratios, always obtained from 5,000 simulations, where the four biased/unbiased elections are randomly chosen anew for each simulation but the GTS 2003 election is the same fixed election in each simulation.

To conclude, observe the following. First, Table 10.12's empirical results are good for all starting vectors (i.e., the corresponding heuristics seem to perform significantly better than what one would expect from the worst-case approximation ratio of Theorem 10.5.2). Overall in Table 10.12, the $i = 1$ cases – i.e., using a minisum solution as the starting vector – seem to be the best choice among these heuristics. Second, the higher the r the better the solution, but the running time then also increases. Trivially, a sufficiently large r always leads to an optimal solution, since we can choose r so that the neighborhood inspected becomes the whole solution space (in that case, however, the heuristic would require the same amount of time as the naive algorithm needs for computing the optimal solution, which would render the heuristic useless).

Acknowledgements Supported in part by DFG grants RO 1202/11-1 and RO 1202/12-1, NSF grants CCF-0426761, IIS-0713061, and CCF-0915792, the Alexander von Humboldt Foundation's TransCoop program, the European Science Foundation's EUROCORES program LogICCC, and two Friedrich Wilhelm Bessel Research Awards.

We are grateful to Claudia Lindner for many helpful proofreading comments and suggestions.

References

Ausiello, G., Crescenzi, P., Gambosi, G., Kann, V., Marchetti-Spaccamela, M., & Protasi, M. (2003). *Complexity and approximation* (2nd edn.). Berlin: Springer-Verlag.

Bartholdi, J., III, & Orlin, J. (1991). Single transferable vote resists strategic voting. *Social Choice and Welfare, 8*(4), 341–354.

Bartholdi, J., III, Tovey, C., & Trick, M. (1989a). The computational difficulty of manipulating an election. *Social Choice and Welfare, 6*(3), 227–241.

Bartholdi, J., III, Tovey, C., & Trick, M. (1989b). Voting schemes for which it can be difficult to tell who won the election. *Social Choice and Welfare, 6*(2), 157–165.

Bartholdi, J., III, Tovey, C., & Trick, M. (1992). How hard is it to control an election? *Mathematical and Computer Modelling, 16*(8/9), 27–40.

Brams, S. (1980). Approval voting in multicandidate elections. *Policy Studies Journal, 9*(1), 102–108.

Brams, S., & Fishburn, P. (1978). Approval voting. *American Political Science Review, 72*(3), 831–847.

Brams, S., & Fishburn, P. (1981). Approval voting, Condorcet's principle, and runoff elections. *Public Choice, 36*(1), 89–114.

Brams, S., & Fishburn, P. (1983). *Approval voting*. Boston: Birkhäuser.

Brams, S., & Fishburn, P. (2002). Voting procedures. In K. Arrow, A. Sen, & K. Suzumura (Eds.), *Handbook of social choice and welfare* (Vol. 1, pp. 173–236). Amsterdam: North-Holland.

Brams, S., & Fishburn, P. (2005). Going from theory to praxis: The mixed success of approval voting. *Social Choice and Welfare, 25*(2–3), 457–474.

Brams, S., & Sanver, R. (2006). Critical strategies under approval voting: Who gets ruled in and ruled out. *Electoral Studies, 25*(2), 287–305.

Brams, S., & Sanver, R. (2009). Voting systems that combine approval and preference. In S. Brams, W. Gehrlein, & F. Roberts (Eds.), *The mathematics of preference, choice, and order: Essays in honor of Peter C. Fishburn* (pp. 215–237). Berlin: Springer.

Brams, S., Kilgour, D., & Sanver, R. (2004). A minimax procedure for negotiating multilateral treaties. In M. Wiberg (Ed.), *Reasoned choices: Essays in honor of academy professor Hannu Nurmi* (pp. 108–139). Helsinki: The Finnish Political Science Association.

Brams, S., Kilgour, D., & Sanver, R. (2007a). A minimax procedure for electing committees. *Public Choice, 132*(3–4), 401–420.

Brams, S., Kilgour, D., & Sanver, R. (2007b). A minimax procedure for negotiating multilateral treaties. In R. Avenhaus & I. Zartman (Eds.), *Diplomacy games* (pp. 265–282). Berlin: Springer.

Brelsford, E., Faliszewski, P., Hemaspaandra, E., Schnoor, H., & Schnoor, I. (2008, July). Approximability of manipulating elections. In *Proceedings of the 23rd AAAI conference on artificial intelligence* (pp. 44–49). AAAI Press.

Buss, J., & Islam, T. (2008). The complexity of fixed-parameter problems. *SIGACT News, 39*(1), 34–46.

Caragiannis, I., Covey, J., Feldman, M., Homan, C., Kaklamanis, C., Karanikolas, N., Procaccia, A., & Rosenschein, J. (2009, January). On the approximability of Dodgson and Young elections. In *Proceedings of the 20th ACM-SIAM symposium on discrete algorithms* (pp. 1058–1067). Society for Industrial and Applied Mathematics.

Condorcet, J.-A.-N. (1792/1995). A survey of the principles underlying the draft constitution. In I. McLean & A. Urken (Eds.), *Classics of social choice* (pp. 145–150). Ann Arbor, MI: University of Michigan Press.

Conitzer, V., & Sandholm, T. (2003, August). Universal voting protocol tweaks to make manipulation hard. In *Proceedings of the 18th international joint conference on artificial intelligence* (pp. 781–788). Morgan Kaufmann.

Conitzer, V., & Sandholm, T. (2006, July). Nonexistence of voting rules that are usually hard to manipulate. In *Proceedings of the 21st national conference on artificial intelligence* (pp. 627–634). AAAI Press.

Conitzer, V., Sandholm, T., & Lang, J. (2007). When are elections with few candidates hard to manipulate? *Journal of the ACM, 54*(3), Article 14.

Dodgson, C. (1876). *A method of taking votes on more than two issues*. Pamphlet printed by the Clarendon Press, Oxford, and headed "not yet published." (Reprinted in McLean, I., & Urken, A. (1995). *Classics of social choice*. Ann Arbor, MI: University of Michigan Press; and Black, D. (1958). *The theory of committees and elections*. Cambridge: Cambridge University Press).

Downey, R., & Fellows, M. (1999). *Parameterized complexity*. Berlin: Springer-Verlag.

Duggan, J., & Schwartz, T. (2000). Strategic manipulability without resoluteness or shared beliefs: Gibbard–Satterthwaite generalized. *Social Choice and Welfare, 17*(1), 85–93.

Dutta, B., De Sinopoli, F., & Laslier, J. (2006). Approval voting: Three examples. *International Journal of Game Theory, 35*, 27–38.

Dwork, C., Kumar, R., Naor, M., & Sivakumar, D. (2001). Rank aggregation methods for the web. In *Proceedings of the 10th international world wide web conference* (pp. 613–622). ACM Press.

Elkind, E., & Lipmaa, H. (2005). Small coalitions cannot manipulate voting. In *Proceedings of the 9th international conference on financial cryptography and data security. Lecture Notes in Computer Science* (Vol. 3570, pp. 285–297). Berlin: Springer-Verlag.

Endriss, U., & Goldberg, P. (Eds.) (2008). *Proceedings of the 2nd international workshop on computational social choice*. University of Liverpool. Retrieved from http://www.csc.liv.ac.uk/~pwg/COMSOC-2008/proceedings.html.

Endriss, U., & Lang, J. (Eds.) (2006). *Proceedings of the 1st international workshop on computational social choice*. Universiteit van Amsterdam. Retrieved from http://staff.science.uva.nl/~ulle/COMSOC-2006/proceedings.html.

Ephrati, E., & Rosenschein, J. (1997). A heuristic technique for multi-agent planning. *Annals of Mathematics and Artificial Intelligence, 20*(1–4), 13–67.

Erdélyi, G., Hemaspaandra, L., Rothe, J., & Spakowski, H. (2007, August). On approximating optimal weighted lobbying, and frequency of correctness versus average-case polynomial time. In *Proceedings of the 16th international symposium on fundamentals of computation theory. Lecture Notes in Computer Science* (Vol. 4639, pp. 300–311). Berlin: Springer-Verlag.

Erdélyi, G., Hemaspaandra, L., Rothe, J., & Spakowski, H. (2008a). *Frequency of correctness versus average-case polynomial time and generalized juntas*. (Technical report TR-934). Department of Computer Science, University of Rochester, Rochester, NY.

Erdélyi, G., Nowak, M., & Rothe, J. (2008b). Sincere-strategy preference-based approval voting broadly resists control. In *Proceedings of the 33rd international symposium on mathematical foundations of computer science. Lecture Notes in Computer Science* (Vol. 5162, pp. 311–322). Berlin: Springer-Verlag.

Erdélyi, G., Nowak, M., & Rothe, J. (2008c). *Sincere-strategy preference-based approval voting fully resists constructive control and broadly resists destructive control* (Technical report). arXiv:0806.0535v4 [cs.GT]. ACM Computing Research Repository (CoRR), June 2008. Revised, September 2008.

Erdélyi, G., Hemaspaandra, L., Rothe, J., & Spakowski, H. (2009a). Frequency of correctness versus average polynomial time. *Information Processing Letters, 109*(16), 946–949.

Erdélyi, G., Hemaspaandra, L., Rothe, J., & Spakowski, H. (2009b). Generalized juntas and NP-hard sets. *Theoretical Computer Science, 410*(38–40), 3995–4000.

Erdélyi, G., Nowak, M., & Rothe, J. (2009c). Sincere-strategy preference-based approval voting fully resists constructive control and broadly resists destructive control. *Mathematical Logic Quarterly, 55*(4), 425–443.

Everaere, P., Konieczny, S., & Marquis, P. (2007). The strategy-proofness landscape of merging. *Journal of Artificial Intelligence Research, 28*, 49–105.

Faliszewski, P. (2008, May). Nonuniform bribery. In *Proceedings of the 7th international joint conference on autonomous agents and multiagent systems* (pp. 1569–1572). ACM Press.

Faliszewski, P., Hemaspaandra, E., & Hemaspaandra, L. (2006, July). The complexity of bribery in elections. In *Proceedings of the 21st national conference on artificial intelligence* (pp. 641–646). AAAI Press.

Faliszewski, P., Hemaspaandra, E., Hemaspaandra, L., & Rothe, J. (2007, July). Llull and Copeland voting broadly resist bribery and control. In *Proceedings of the 22nd AAAI conference on artificial intelligence* (pp. 724–730). AAAI Press.

Faliszewski, P., Hemaspaandra, E., Hemaspaandra, L., & Rothe, J. (2008a). Copeland voting fully resists constructive control. In *Proceedings of the 4th international conference on algorithmic aspects in information and management. Lecture Notes in Computer Science* (Vol. 5034, pp. 165–176). Berlin: Springer-Verlag.

Faliszewski, P., Hemaspaandra, E., & Schnoor, H. (2008b, May). Copeland voting: Ties matter. In *Proceedings of the 7th international joint conference on autonomous agents and multiagent systems* (pp. 983–990). ACM Press.

Faliszewski, P., Hemaspaandra, E., & Hemaspaandra, L. (2009a). How hard is bribery in elections? *Journal of Artificial Intelligence Research, 35*, 485–532.

Faliszewski, P., Hemaspaandra, E., Hemaspaandra, L., & Rothe, J. (2009b). Llull and Copeland voting computationally resist bribery and constructive control. *Journal of Artificial Intelligence Research, 35*, 275–341.

Faliszewski, P., Hemaspaandra, E., Hemaspaandra, L., & Rothe, J. (2009c). A richer understanding of the complexity of election systems. In S. Ravi & S. Shukla (Eds.), *Fundamental problems in computing: Essays in honor of Professor Daniel J. Rosenkrantz* (Chap. 14, pp. 375–406). Berlin: Springer.

Fishburn, P. (1977). Condorcet social choice functions. *SIAM Journal on Applied Mathematics, 33*(3), 469–489.

Flum, J., & Grohe, M. (2006). Parameterized complexity theory. In *EATCS texts in theoretical computer science*. Heidelberg: Springer-Verlag.

Garey, M., & Johnson, D. (1979). *Computers and intractability: A guide to the theory of NP-completeness*. New York: W. H. Freeman and Company.

Ghosh, S., Mundhe, M., Hernandez, K., & Sen, S. (1999). Voting for movies: The anatomy of recommender systems. In *Proceedings of the 3rd annual conference on autonomous agents* (pp. 434–435). ACM Press.

Gibbard, A. (1973). Manipulation of voting schemes. *Econometrica, 41*(4), 587–601.

Hemaspaandra, E., & Hemaspaandra, L. (2007). Dichotomy for voting systems. *Journal of Computer and System Sciences, 73*(1), 73–83.

Hemaspaandra, L., & Ogihara, M. (2002). The Complexity theory companion. In *EATCS texts in theoretical computer science*. Berlin: Springer-Verlag.

Hemaspaandra, E., Hemaspaandra, L., & Rothe, J. (1997). Exact analysis of Dodgson elections: Lewis Carroll's 1876 voting system is complete for parallel access to NP. *Journal of the ACM, 44*(6), 806–825.

Hemaspaandra, E., Spakowski, H., & Vogel, J. (2005). The complexity of Kemeny elections. *Theoretical Computer Science, 349*(3), 382–391.

Hemaspaandra, E., Hemaspaandra, L., & Rothe, J. (2006, August). *Hybrid elections broaden complexity-theoretic resistance to control* (Technical report). arXiv:cs/0608057v2 [cs.GT]. ACM Computing Research Repository (CoRR).

Hemaspaandra, E., Hemaspaandra, L., & Rothe, J. (2007a). Anyone but him: The complexity of precluding an alternative. *Artificial Intelligence, 171*(5–6), 255–285.

Hemaspaandra, E., Hemaspaandra, L., & Rothe, J. (2007b, January). Hybrid elections broaden complexity-theoretic resistance to control. In *Proceedings of the 20th international joint conference on artificial intelligence* (pp. 1308–1314). AAAI Press.

Hemaspaandra, E., Hemaspaandra, L., & Rothe, J. (2009). Hybrid elections broaden complexity-theoretic resistance to control. *Mathematical Logic Quarterly, 55*(4), 397–424.

Homan, C., & Hemaspaandra, L. (2006). Guarantees for the success frequency of an algorithm for finding Dodgson-election winners. In *Proceedings of the 31st international symposium on mathematical foundations of computer science. Lecture Notes in Computer Science* (Vol. 4162, pp. 528–539). Berlin: Springer-Verlag.

Homan, C., & Hemaspaandra, L. (2009). Guarantees for the success frequency of an algorithm for finding Dodgson-election winners. *Journal of Heuristics, 15*(4), 403–423.

Kilgour, D., Brams, S., & Sanver, R. (2006). How to elect a representative committee using approval balloting. In F. Pukelsheim & B. Simeone (Eds.), *Mathematics and democracy: Voting systems and collective choice* (pp. 83–95). Berlin: Springer.

LeGrand, R. (2004, November). *Analysis of the minimax procedure* (Technical report WUCSE-2004-67). Department of Computer Science and Engineering, Washington University, St. Louis, MO.

LeGrand, R., Markakis, E., & Mehta, A. (2006, December). Approval voting: Local search heuristics and approximation algorithms. In U. Endriss & J. Lang (Eds.), *Proceedings of the 1st international workshop on computational social choice* (pp. 276–289). Amsterdam: Universiteit van Amsterdam.

LeGrand, R., Markakis, E., & Mehta, A. (2007, May). Some results on approximating the minimax solution in approval voting. In *Proceedings of the 6th international joint conference on autonomous agents and multiagent systems* (pp. 1193–1195). ACM Press.

Lindner, C., & Rothe, J. (2008, October). Fixed-parameter tractability and parameterized complexity, applied to problems from computational social choice. In A. Holder (Ed.), *Mathematical programming glossary*. INFORMS Computing Society.

McCabe-Dansted, J., Pritchard, G., & Slinko, A. (2008). Approximability of Dodgson's rule. *Social Choice and Welfare, 31*(2), 311–330.

Meir, R., Procaccia, A., Rosenschein, J., & Zohar, A. (2008). The complexity of strategic behavior in multi-winner elections. *Journal of Artificial Intelligence Research, 33*, 149–178.

Niedermeier, R. (2006). *Invitation to fixed-parameter algorithms*. Oxford: Oxford University Press.

Papadimitriou, C. (1994). *Computational complexity*. Reading, MA: Addison-Wesley.

Procaccia, A., & Rosenschein, J. (2007). Junta distributions and the average-case complexity of manipulating elections. *Journal of Artificial Intelligence Research, 28*, 157–181.

Rothe, J. (2005). Complexity theory and cryptology. An introduction to cryptocomplexity. *EATCS texts in theoretical computer science*. Berlin: Springer-Verlag.

Rothe, J., Spakowski, H., & Vogel, J. (2003). Exact complexity of the winner problem for Young elections. *Theory of Computing Systems, 36*(4), 375–386.

Satterthwaite, M. (1975). Strategy-proofness and Arrow's conditions: Existence and correspondence theorems for voting procedures and social welfare functions. *Journal of Economic Theory, 10*(2), 187–217.

Zuckerman, M., Procaccia, A., & Rosenschein, J. (2008, January). Algorithms for the coalitional manipulation problem. In *Proceedings of the 19th ACM-SIAM symposium on discrete algorithms* (pp. 277–286). Society for Industrial and Applied Mathematics. Extended version appears in *Artificial Intelligence, 173*(2), 392–412 (2009).

Part V
Probabilistic Exercises

Chapter 11
On the Condorcet Efficiency of Approval Voting and Extended Scoring Rules for Three Alternatives

Mostapha Diss, Vincent Merlin, and Fabrice Valognes

11.1 Introduction

The results presented in this chapter belong to the long tradition of evaluating the voting rules on their propensity to select the Condorcet winner. This tradition dates back to de Borda (1781), who first noticed that just voting for one name and then selecting as a winner the candidate with more votes, could lead to the selection of an option that is beaten by all the other candidates in pairwise comparisons. In order to replace the rule that is now known as the Plurality rule, he suggested a new voting mechanism, which now bares his name, the *Borda Count* (BC). When m candidates are in competition, Borda suggested that each voter could give $m-1$ points to his first choice, $m-2$ points to his second choice, and so on down to one point for his next to the last alternative and zero point for the candidate he ranks last. Next, the candidate who receives the highest number of points is declared as the winner.

Unfortunately, a few years latter, Condorcet (1785) noticed the Borda Count was also plagued by that same critique Borda had raised against the Plurality rule: A candidate that would defeat each of the other alternatives in a series of pairwise elections may not be selected by the Borda Count either. Indeed, Condorcet proved in his Éssai that any scoring rule may fail to select such an alternative, which is now known as the *Condorcet winner*. By scoring rule, we mean any voting system wherein, after each voter has ranked all the alternatives in strict order from the one he likes more till his least preferred option, a decreasing number of points is awarded to each alternative according to its ranks. The candidate with the maximal number of points is then selected.

A merit of Social Choice theory has been to offer a clear mathematical framework where these pioneer works could be revisited and developed. True, all the scoring methods may fail to pick out the Condorcet winner for some preference profiles. But the sophistication of modern theory allows us to investigate in detail a connected

---

V. Merlin (✉)
CNRS and University of Caen Basse Normandie, CREM, UMR CNRS 6211, Caen, France
e-mail: vincent.merlin@unicaen.fr

J.-F. Laslier and M.R. Sanver (eds.), *Handbook on Approval Voting*, Studies in Choice and Welfare, DOI 10.1007/978-3-642-02839-7_11,

issue: What is the propensity of a given voting rule to select the Condorcet winner, given that such a candidate exits? In other words, we seek for the scoring rule which is the closest to Condorcet's ideas. This research agenda owes a lot to the contributions of Fishburn and Gehrlein. In Gehrlein and Fishburn (1978a,b), they are the first authors to derive an exact formula to estimate the *Condorcet efficiency* of scoring rules, that is their propensity to select the Condorcet winner whenever it exits. More precisely, they assume that voters have to decide among three candidates only, say a, b, and c. Next, they suppose that each voter will select independently any preference among the six possible strict preferences on $\{a, b, c\}$ with probability 1/6. They termed this assumption the *Impartial Culture* (IC) condition. At last, they consider the class of scoring rules where each voter awards one point to his first choice, $\lambda \in [0, 1]$ point to his second choice, and zero point to the last alternative. The Plurality rule is then defined by $\lambda = 0$ while it is easy to show that the Borda rule is equivalent to the case where $\lambda = 1/2$. In this framework, using the central limit theorem, the Condorcet efficiency of the Borda Count tends toward 0.9012 as the population tends to infinity and is maximal for all $\lambda \in [0, 1]$. On the other hand, the minimal value for Condorcet efficiency, 0.7573, is reached for $\lambda = 0$ (the Plurality rule) and $\lambda = 1$ (called the Antiplurality rule).

Since 1978, more results have been obtained and it is impossible to mention all of them. They propose estimations of the Condorcet efficiency with more options, for scoring run-off systems, under assumptions different from IC, with exact formulas or Monte Carlo simulations, for other Condorcet like criteria etc. For a comprehensive survey of this literature, see Gehrlein (1997). We will just detail one extra contribution. Deriving exact formulas for more than three alternatives is almost an impossible task, but van Newenhizen (1992) manages to obtain partial results under IC (and generalized versions of IC). More precisely, she managed to prove that, when each voter selects independently each of the possible strict preference among m candidates with probability $1/m!$, the Borda Count maximizes the probability that the majority outcome between any two candidates is reflected by the scoring rule outcome. It is then tempting to conjecture that, with these probabilistic assumptions and for any $m \geq 3$, the optimal scoring rule regarding the Condorcet efficiency will always be the Borda Count in the class of scoring rules.[1]

At this point, one may wonder why *Approval Voting* (AV) is concerned about the Condorcet efficiency issue. By definition, AV just asks a voter to distinguish the candidates he approves of from the ones he considers as unacceptable. The alternative with the highest degree of approbation is then selected. The preferences expressed

[1] In fact, van Newenhizen asserts that she proved that statement, but she wrongly assumed that the rankings of the different pairs were independent events. Indeed, Cervone et al. (2005) analyzed in details her arguments, and questioned the optimality of the Borda Count. More specifically, with a different probabilistic assumption called the Impartial Anonymous Assumption (see Gehrlein and Fishburn 1976), the best rule in terms of Condorcet efficiency is obtained for $\lambda \approx 0.375$, while the maximal agreement for one given pair of alternatives between the majority ranking and the scoring outcome is effectively achieved at $\lambda = 1/2$! So, the optimality of Borda Count under the IC assumption for $m > 3$ is still a conjecture, though we doubt the assertion can be falsified.

Table 11.1 A possible conflict between Approval Voting and the Condorcet criteria

Preference type	Number of voters
$\underline{a} \succ b \succ c$	4
$\underline{b} \succ a \succ c$	2
$\underline{b} \succ \underline{c} \succ a$	1
$\underline{c} \succ b \succ a$	4

by the voters are thus dichotomous, and Approval Voting will always pick out the Condorcet winner, which always exists in this context (see Ju (2010) and Xu (2010) in this volume). But some authors, like Saari and van Newenhizen (1988b) will not buy this story, and assert that the Approval Voting ballots could hide the true nature of the preferences, which may still be strict. To give an example, it is impossible for a voter who approves two candidates a and b, to further indicate whether he is truly indifferent among both candidates or whether he has a strong preference for one of them. Consider the preference profile of Table 11.1, where 11 voters rank strictly three alternatives according to their preferences. If each voter approves of just one candidate except the sole voter with preference $b \succ c \succ a$, who approves two candidates, the Condorcet winner, b, will be defeated by the Condorcet loser, c! More generally, depending on how many candidates a voter approves, Saari and van Newenhizen (1988b) prove that Approval Voting could even lead to the selection of any of the three candidates for a single profile. The 20-year-old polemic between Saari and van Newenhizen (1988a,b) and Brams et al. (1988a,b) is a good example of the difference of interpretations that can arise in this context.

Assuming that the true preferences of the voters are strict orderings, Gehrlein and Lepelley (1998) were able to study with more detail the Condorcet efficiency of Approval voting. Starting from the IC model with m alternatives, they first assume that each voter picks out his preference independently among the $m!$ strict preference with probability $1/m!$. On the top of that, they assume that each voter has the same probability q_t to approve his first t alternatives, with $\sum_1^{m-1} q_t = 1$. Thus, the IC assumption governs the selection of the strict preferences, while the selection of alternatives for Approval Voting is determined by a common probability vector $q = (q_1, \dots q_{m-1})$. With this model, Gehrlein and Lepelley (1998) are able to prove that, for any m, the Condorcet efficiency of Approval Voting is at least equal to the Plurality one, but never superior to the one of the rule where all the voters approve exactly $\lceil m/2 \rceil$ candidates. For three candidates, they are even able to derive a precise formula: the Condorcet efficiency of Approval Voting is exactly equal to that one of Plurality vote and Antiplurality vote, which is the worst scenario.

The picture is thus quite confusing. On the one hand, in a world of dichotomous preferences, Approval Voting will naturally inherit from the properties of majority voting. On the other hand, in a strict preference universe, it can fare as badly as Plurality rule, and we can conjecture that it is always dominated by the Borda Count. At this point, our conclusion is that we need a richer world to analyze the Condorcet efficiency of Approval Voting, Borda Count, and all the other systems based on points. In this chapter we will propose for the three-alternative case a framework that combines dichotomous and strict preferences. Indeed, for three alternatives, the model of Gehrlein and Lepelley (1998) really considered twelves preference types, as each voter could approve one or two alternatives. We will add to these types seven new possibilities, by allowing the voters to express indifference among two or three candidates.

In this larger framework, it is not only possible to define simply the Approval Voting, the Borda Count and the Plurality rule, but also the whole family of extended scoring rules, to which they all belong to. We will thus have a "fair ground", that do not favor a priori a specific voting rule, where we can perform again the Condorcet efficiency exercise.

The rest of the chapter is organized as follows. Section 11.2 will be devoted to the exposition of the model. We will first explain in detail the choices we made to model the preferences, and next see how the classical voting rules are defined in this context. We will conclude the section by presenting an extended version of the Impartial Culture hypothesis. Section 11.3 will be devoted to a simpler issue, the robustness issue. As in van Newenhizen (1992), we will first study the agreement between the pairwise ranking and the scoring outcome for a given pair of alternatives. In Sect. 11.4, we will derive the general formula that enables us to evaluate the value of the Condorcet efficiency for any extended scoring rule, for a large population and under an extended version of the IC assumption. The proofs, which can be somehow technical, will be presented in Appendices. Section 11.5 will conclude the chapter.

11.2 The Model

11.2.1 Preferences

Consider an election in which n voters have to choose among three candidates a, b, and c. When computing a priori the probability of some voting events, it is typically assumed that the voters have strict preferences over the set of alternatives. Thus, this leads to only six preferences types over $\{a, b, c\}$, and the Impartial Culture assumption (see Guilbaud 1952) assumes that each voter is equally likely to pick any of these preferences. To analyze Approval Voting, Gehrlein and Lepelley (1998) further assumed that each voter will approve one alternative (resp. two alternatives) with the same probability q_1 (resp. q_2). They impose $q_1 + q_2 = 1$, as approving all the alternatives has no impact on the selection of the approval winner. By doing so, they stick to the "strict ordering" paradigm. Voters are truly able to express a clear judgement on each pair of alternatives, even if they approve two of them. On the other hand, one could argue that a vote for two alternatives means that the agent is either indifferent between them, or at least would not consider his preference as a strong one. A way to model these preference types is to allow for indifference.

Indeed, in this paper we will consider five classes of preferences for the 19 different voting behaviors:

- **Class I:** Voters may have strict preferences, and just report their preferred alternative. These are voters of types 1–6 in Table 11.2.
- **Class II:** Voters may have strict preferences, but report their top two choices together with the exact information about their ranking. These are types 7–12 in Table 11.2.
- **Class III:** Voters may be indifferent about their two preferred alternatives or do not consider the difference significant enough to reveal their true strict preference. This defines three extra voting types, labeled 13–15.
- **Class IV:** Voters may just report their top choice, considering the other two as equally unacceptable (types 16–18).
- **Class V:** Voters may consider the three alternatives as equally acceptable (type 19).

The summary of the 19 possible cases is displayed in Table 11.2 where the preference order $a \succ b$ means that a is strictly preferred to b. $a \sim b \succ c$ means that individuals with this preference order are indifferent between a and b and prefer both a and b to c. A bar below a letter indicates the approved candidate(s). To paraphrase Brams and Sanver (2009), we not only ask voters to report their preferences, but we also ask them to draw a line between acceptable and unacceptable candidates. And in this volume, Sanver (2010) explores from an axiomatic point of view this "extended arrovian framework", where new rules and new conditions can be proposed.

Here, p_i denotes the probability that a voter picks the associated preference ranking on candidates a, b and c. The number of individuals with type i preference will be denoted by n_i. A voting situation $\tilde{n} = (n_1, n_2, \ldots, n_{19})$ describes the distribution of the n voters on the different preference rankings.

A population with preference types 1–12 only will lead to a situation similar to the one analyzed by Gehrlein and Lepelley (1998). On the other hand, class III and IV can be used to model dichotomous preferences. Preferences from

Table 11.2 The 19 admissible preference rankings for three candidates

Class I:	$\underline{a} \succ b \succ c$	p_1	Class II:	$\underline{a} \succ \underline{b} \succ c$	p_7
	$\underline{a} \succ c \succ b$	p_2		$\underline{a} \succ \underline{c} \succ b$	p_8
	$\underline{b} \succ a \succ c$	p_3		$\underline{b} \succ \underline{a} \succ c$	p_9
	$\underline{b} \succ c \succ a$	p_4		$\underline{b} \succ \underline{c} \succ a$	p_{10}
	$\underline{c} \succ a \succ b$	p_5		$\underline{c} \succ \underline{a} \succ b$	p_{11}
	$\underline{c} \succ b \succ a$	p_6		$\underline{c} \succ \underline{b} \succ a$	p_{12}
Class III:	$\underline{a} \sim \underline{b} \succ c$	p_{13}	Class IV:	$\underline{a} \succ b \sim c$	p_{16}
	$\underline{a} \sim \underline{c} \succ b$	p_{14}		$\underline{b} \succ a \sim c$	p_{17}
	$\underline{b} \sim \underline{c} \succ a$	p_{15}		$\underline{c} \succ a \sim b$	p_{18}
Class V:	$\underline{a} \sim \underline{b} \sim \underline{c}$	p_{19}			

class I can be envisioned as truncated preferences: Voters just report their top choice, though they are able to rank strictly all the alternatives. Thus class II, III, IV and V exclude the possibility for a voter to report a truncated preference. The reader has to be aware that in this paper, we assume naively that a voter reports her "true" approvals independently of his beliefs about the consequence of her vote. In this volume Laslier (2010) will discuss with more detail the way the voters may react to extra information and study the dynamics of the voting behaviors.

11.2.2 Voting Rules

Let $N_{x,y}(\tilde{n})$ be the number of voters that strictly prefer x to y for the voting situation $\tilde{n}$. Alternative x will defeat alternative y with respect to the majority criterion (denoted by $xM(\tilde{n})y$, or shortly xMy) whenever $N_{x,y}(\tilde{n}) > N_{y,x}(\tilde{n})$. The Condorcet winner of an election is the candidate who, when compared in turn with each of the other candidates, is preferred over the other candidate. Unfortunately, there are cyclic election examples where a Condorcet winner will not always exist (Condorcet (1785)). We may encounter a Condorcet paradox defined by aMb, bMc, cMa (or bMa, aMc, cMb). Though the concept of a Condorcet winner remains appealing whenever it exists, this fact calls for the use of other more practical voting methods in real life.

In this paper, we need to extend the classical definition of scoring rules, that fits the case of strict preferences only. Generally speaking, 'scoring rules' are voting procedures under which voters cast scores for the different candidates and the candidate with the highest total score wins the election.

When preferences are restricted to types 7–12, a scoring rule is uniquely defined by a parameter $\lambda \in [0, 1]$, such as each voter gives one point to his most preferred alternative, λ points to the second one, and zero point to the third one. This definition enables us to include in this category the Plurality rule ($\lambda = 0$), the Borda Count ($\lambda = 1/2$) and the Antiplurality rule ($\lambda = 1$). To extend these rules to preferences 1–6, we will assume that, by considering the last two alternatives as unacceptable, they have no choice but to give them zero point.[2] Only the first alternative will receive one point and the preference $\underline{a} \succ b \succ c$ is tallied by the vector of scores $(1, 0, 0)$. When a voter reports a preference type 13, 14 or 15, we will assume that he awards μ point to each of the first ranked alternatives, and zero to the last one; The preference $\underline{a} \sim \underline{b} \succ c$ is tallied by the vector $(\mu, \mu, 0)$. For preference types 16–18, one point is given to the top alternative, and γ points to the two equally unacceptable

[2] Though not considered here, we could have add an extra parameter ν to tally the preference $\underline{a} \succ b \succ c$ with the vector $(\nu, 0, 0)$. A reason to exclude this possibility is that we wanted to contain the values of the scores given by each voter in the interval $[0, 1]$. Moreover, if we interpret the preferences in Class I as truncated preferences, it seems difficult to award more than one point to the top alternative.

Table 11.3 Famous voting rules as extended scoring rules

Classes	*I*	*II*	*III*	*IV*
Tallies	$(1, 0, 0)$	$(1, \lambda, 0)$	$(\mu, \mu, 0)$	$(1, \gamma, \gamma)$
Approval Voting (AV)	$(1, 0, 0)$	$(1, 1, 0)$	$(1, 1, 0)$	$(1, 0, 0)$
Plurality (P)	$(1, 0, 0)$	$(1, 0, 0)$	$(\frac{1}{2}, \frac{1}{2}, 0)$	$(1, 0, 0)$
Antiplurality (A)	$(1, 0, 0)$	$(1, 1, 0)$	$(1, 1, 0)$	$(1, \frac{1}{2}, \frac{1}{2})$
Borda Count (BC)	$(1, 0, 0)$	$(1, \frac{1}{2}, 0)$	$(\frac{3}{4}, \frac{3}{4}, 0)$	$(1, \frac{1}{4}, \frac{1}{4})$

options; We use the vector $(1, \gamma, \gamma)$ to tally a class IV preference.[3] As class V voters are completely indifferent among the three candidates, they have no impact on the selection of the winner. Thus, to simplify the analysis, we will assume throughout this chapter that $p_{19} = n_{19} = 0$. Thus, to summarize, an extended scoring rule is characterized by a vector $\Theta = (\lambda, \mu, \gamma)$, which indicates the number of points to award to each alternative for each preference type. Following Smith (1973) and Black (1976), we will assume that the natural extension of a scoring rule defined by the parameter λ is given by $\mu = \frac{1+\lambda}{2}$ and $\gamma = \frac{\lambda}{2}$. This solution always keeps the number of points attributed by a voter to $1 + \lambda$ whenever he expresses indifference. Thus the Borda Count is defined by $\Theta_{BC} = (1/2, 3/4, 1/4)$, while the natural extensions of the Plurality and Antiplurality rules are respectively $\Theta_P = (0, 1/2, 0)$ and $\Theta_A = (1, 1, 1/2)$.

Approval Voting has also an immediate definition as an extended scoring rule in this context; It is characterized by the vector $\Theta_{AV} = (1, 1, 0)$. Table 11.3 summarizes the way we represent the major extended scoring rules in our framework.

11.2.3 Extending the Impartial Culture Hypothesis

Guilbaud (1952) was the first author to suggest a probabilistic model of voting to estimate a priori the likelihood of a Condorcet paradox. His model, now known as the Impartial Culture (IC) hypothesis, is a model of an electorate in which a randomly selected voter is equally likely to have each of the possible preferences on the candidates. The original condition includes only linear orders and assumes that $p_i = \frac{1}{m!}$, where m denotes the number of candidates. If we ignore the problem of acceptable alternatives for a moment, preferences 1–6 are just a set of strict preferences over a, b, and c, and the original IC assumptions asserts that the probability attached to the profile $\tilde{n}$ for a population of size n follows a multinomial law:

$$
\begin{aligned}
Pr(\tilde{n} = (n_1, n_2, n_3, n_4, n_5, n_6)) &= \frac{n!}{n_1!n_2!n_3!n_4!n_5!n_6!}\left(\frac{1}{m!}\right)^n \\
&= \frac{n!}{n_1!n_2!n_3!n_4!n_5!n_6!}\left(\frac{1}{6}\right)^n
\end{aligned}
$$

[3] Equivalently, we could have used $(\gamma', 0, 0)$.

When we consider the 19 possible preference types, each voter will select independently preference i with probability p_i and we obtain:

$$Pr(\tilde{n} = (n_1, \ldots, n_{19})) = \frac{n!}{\prod_{i=1}^{19} n_i!} \left(\prod_{i=1}^{19} p_i^{n_i} \right)$$

In this paper, the probability of each voting situation is calculated for voter preferences with possible ties. In this field, Gehrlein and Fishburn (1980) have already proposed an extension of the IC assumption in order to cope with indifference, called *Impartial Weak Ordering Culture* (IWOC). We follow here the same route in order to take into account the five classes of preferences we have identified. Let k_1 denote the probability that a voter's preference belongs to Class I. Likewise, k_2 denotes the probability for voters with preferences in class II, k_3 represents the probability for voters with preference type $i = 13, \ldots, 15$. k_4 denotes the probability for voters with preference type $i = 16, \ldots, 18$. Finally, k_5 denotes the probability for voters with complete indifference on candidates a, b and c. Naturally, $k_1 + k_2 + k_3 + k_4 + k_5 = 1$. The Extended Impartial Culture (EIC) hypothesis means that the preference rankings within a class are equally likely to be observed. Consequently, $p_i = \frac{k_1}{6}$ for $i = 1, \ldots, 6$; $p_i = \frac{k_2}{6}$ for $i = 7, \ldots, 12$; $p_i = \frac{k_3}{3}$ for $i = 13, \ldots, 15$; $p_i = \frac{k_4}{3}$ for $i = 16, \ldots, 18$ and $p_{19} = k_5$.

The EIC assumption is flexible enough to recover previously used assumptions. Obviously, the basic assumption of IC is obtained for the case of $k_2 = 1$ ($k_1 = k_3 = k_4 = k_5 = 0$). The IWOC model proposed by Fishburn and Gehrlein (1980) is recovered with $k_3 = k_4$, and $k_1 = 0$. The model used by Gehrlein and Lepelley (1998) can be explored by assuming that $k_3 = k_4 = k_5 = 0$, with $q_1 = k_1$ and $q_2 = k_2$. On the other hand, the dichotomous assumption states that $k_1 = k_2 = 0$. Thus, by modifying the vector $\mathbf{k} = (k_1, k_2, k_3, k_4, k_5)$, the EIC model enables us to explore different scenarios, being favorable either to the Borda Count ($k_3 = k_4 = k_5 = 0$) or the Approval Voting ($k_1 = k_2 = 0$) in terms of Condorcet efficiency. But in all the cases, these assumptions treat the three candidates in a perfectly symmetric way.

To our knowledge, the paper due to Fishburn and Gehrlein (1980) is the first work to introduce the IWOC condition in the social choice literature. In this paper, the likelihood of Condorcet paradox is re-examined when the indifference between distinct candidates is allowed for large electorates. This model has been used recently in two contributions. Gehrlein and Valognes (2001) compute the Condorcet efficiency of scoring rules for large electorates and a case of three-candidate election when the possibility of ties between the candidates is allowed. Merlin and Valognes (2004) consider two questions in the context of three-candidate and large elections. The first one deals with the probability that a scoring rule and the pairwise vote agree on pairs of candidates. The second question considers the probability that the Condorcet winner is ranked last by a scoring rule.

In the next sections, we will study two issues:

- We will derive the Condorcet efficiency of the extended scoring rules Θ for a large number of voters under the EIC assumption. We will denote this probability by $P_{CE}^{\infty}(\mathbf{k}, \Theta)$.
- But, we will first study a simpler problem, the probability that the Majority rule and a given extended scoring rule lead to the same outcome for a pair of alternatives. We denote by $P_{R}^{\infty}(\mathbf{k}, \Theta)$ the probability of the robustness of the pairwise ranking when using the extended scoring rule Θ. The Robustness issue has been introduced in the literature by Gehrlein and Fishburn (1980).

We will be able to find the optimal voting rule(s) for the maximization of $P_{R}^{\infty}(\mathbf{k}, \Theta)$, and we will explore in details the scenarios (regarding $\mathbf{k}$) which are favorable to each of the major voting rule (Borda Count, Plurality rule, Antiplurality rule, and Approval Voting). We will check whether the conclusions we obtained for $P_{R}^{\infty}(\mathbf{k}, \Theta)$ extend to $P_{CE}^{\infty}(\mathbf{k}, \Theta)$. A side exercise will be to identify the 'optimal' extensions of the scoring rules $(1, \lambda, 0)$ when we allow for indifference. We will test whether the intuition proposed by Smith and Black (keeping the total sum of weight unchanged), is right with regards to the robustness issue.

11.2.4 The Likelihood of a Condorcet Winner Under EIC

Before answering these questions, we need a preliminary result.

Theorem 11.2.1. *For the case of three-candidate election and for a large number of voters ($n \to \infty$), the EIC assumption implies that the probability that the Condorcet winner exists is given by:*

$$P_{Con}^{\infty}(EIC) = \frac{3}{4} + \frac{3}{2\pi} \arcsin\left(\frac{k_1 + k_2 + k_3 + k_4}{3k_1 + 3k_2 + 2k_3 + 2k_4}\right)$$

Proof. see Appendix A.

In the case of $k_3 = k_4 = k_5 = 0$, we recover the result of Guilbaud (1952) who showed that the probability that the Condorcet winner exists as n tends to infinity under IC condition is equal to:

$$P_{Con}^{\infty}(IC) = \frac{3}{4} + \frac{3}{2\pi} \arcsin\left(\frac{1}{3}\right)$$

The formula found by Fishburn and Gehrlein (1980) under IWOC is also recovered (with $k_1 = 0$, $k_1' = k_2$, $k_3 = k_4$, and $k_2' = 2k_3 = 2k_4$):

$$P_{Con}^{\infty}(IWOC) = \frac{3}{4} + \frac{3}{2\pi} \arcsin\left(\frac{k_1' + k_2'}{3k_1' + 2k_2'}\right)$$

11.3 The Probability Calculations: Robustness of the Ranking of a Pair

11.3.1 The Optimal Extended Scoring Rule on Pairs

The aim of this section is to derive a formula for $P_R^{\infty}(\mathbf{k}, \Theta)$ under the EIC assumption. Without loss of generality, we will estimate the probability of *aMb*, given that *aSb*. *aMb* indicates that candidate a defeats candidate b in the pairwise majority election. *aSb* means that candidate a wins election against candidate b using the scoring voting system Θ. Thus, candidate a defeats candidate b in the pairwise majority election for a voting situation $\tilde{n}$ if and only if the inequality (11.1) is satisfied. In addition, candidate a is selected against b by the extended scoring rule $\Theta = (\lambda, \mu, \gamma)$ if and only if inequality (11.2) is satisfied.

$$n_1+n_2-n_3-n_4+n_5-n_6+n_7+n_8-n_9-n_{10}+n_{11}-n_{12}+n_{14}-n_{15}+n_{16}-n_{17} > 0 \quad (11.1)$$

$$n_1+n_2-n_3-n_4+(1-\lambda)(n_7-n_9)+n_8-n_{10}+\lambda(n_{11}-n_{12})+\mu(n_{14}-n_{15})+(1-\gamma)(n_{16}-n_{17}) > 0 \quad (11.2)$$

Theorem 11.3.1. *Under the Extended Impartial Culture assumption, the probability of the robustness of the pairwise ranking when using the extended scoring rule Θ for three alternatives and n large is given by:*

$$P_R^{\infty}(\mathbf{k}, \Theta) = \frac{\pi - \arccos\left(\frac{2k_1+2k_2+\mu k_3+(1-\gamma)k_4}{\sqrt{3k_1+3k_2+k_3+k_4}\sqrt{2k_1+2\lambda^2k_2-2\lambda k_2+2k_2+\mu^2k_3+(1-\gamma)^2k_4}}\right)}{\pi}$$

Proof. See appendix B.

The first question is to know which $\Theta = (\lambda, \mu, \gamma)$ maximizes the probability that a scoring rule agrees with the majority rule on a given pair of candidates.

Corollary 11.3.1. *Consider elections using scoring systems for which there are three candidates and n voters, and for which the indifference between distinct candidates is allowed. Then for any two candidates and for large electorates ($n \to \infty$), the values of λ, γ and μ maximizing the probability that the scoring system ranks these two in the same order as they are ranked by pairwise majority are given by:*

$$\Theta^{\star} = \left(\lambda^{\star} = \frac{1}{2}, \ \mu^{\star} = \frac{4k_1+3k_2}{4(k_1+k_2)}, \ \gamma^{\star} = \frac{k_2}{4(k_1+k_2)}\right)$$

Proof. See appendix B.

The main message of Corollary 11.3.1 is that the optimal rule is a non trivial extension of the Borda Count. As soon as the probability of reporting a linear ordering and two acceptable options is non null ($k_2 \neq 0$), the only choice is to set $\lambda = 1/2$, that is to use the Borda tally for these types. Curiously, the vectors of points that we should use for indifferent voters depends on the ratio between k_1 and k_2. When $k_1 = 0$, we recover the classical extension of the Borda Count, as $\Theta^\star = \Theta_{BC} = (1/2, 3/4, 1/4)$. But a situation where $k_1 = k_2$ leads to $\Theta^\star = (1/2, 7/8, 1/8)$. $\underline{a} \sim \underline{b} \succ c$ is tallied by $(7/8, 7/8, 0)$ while $\underline{a} \succ b \sim c$ is tallied by $(1, 1/8, 1/8)$. As $k_2 \to 0$, we observe that $(\mu, \mu, 0) \to (1, 1, 0)$ and $(1, \gamma, \gamma) \to (1, 0, 0)$: The optimal rule behaves more and more like Approval Voting for voters with indifference. Indeed, for $k_2 = 0$, using $(1, \lambda, 0)$ is meaningless as preferences of types 7–12 are absent in the profile.

One can also use Theorem 11.3.1 to derive the probability of robustness for the voting rules presented in Table 11.3.

Corollary 11.3.2. *For three alternatives and a large population, we obtain, under the EIC assumption:*

$$P_R^\infty(\mathbf{k}, \Theta^\star) = \frac{\pi - \arccos\left(\dfrac{2k_1+2k_2+\frac{4k_1+3k_2}{4k_1+4k_2}(k_3+k_4)}{\sqrt{3k_1+3k_2+k_3+k_4}\sqrt{2k_1+\frac{3}{2}k_2+\left(\frac{4k_1+3k_2}{4k_1+4k_2}\right)^2(k_3+k_4)}}\right)}{\pi}$$

$$P_R^\infty(\mathbf{k}, AV) = \frac{\pi - \arccos\left(\dfrac{\sqrt{2k_1+2k_2+k_3+k_4}}{\sqrt{3k_1+3k_2+k_3+k_4}}\right)}{\pi} = \frac{\pi - \arccos\left(\dfrac{\sqrt{2-k_3-k_4}}{\sqrt{3-2k_3-2k_4}}\right)}{\pi}$$

$$P_R^\infty(\mathbf{k}, P) = \frac{\pi - \arccos\left(\dfrac{2k_1+2k_2+\frac{1}{2}k_3+k_4}{\sqrt{3k_1+3k_2+k_3+k_4}\sqrt{2k_1+2k_2+\frac{1}{4}k_3+k_4}}\right)}{\pi}$$

$$P_R^\infty(\mathbf{k}, A) = \frac{\pi - \arccos\left(\dfrac{2k_1+2k_2+k_3+\frac{1}{2}k_4}{\sqrt{3k_1+3k_2+k_3+k_4}\sqrt{2k_1+2k_2+k_3+\frac{1}{4}k_4}}\right)}{\pi}$$

$$P_R^\infty(\mathbf{k}, BC) = \frac{\pi - \arccos\left(\dfrac{2k_1+2k_2+\frac{3}{4}k_3+\frac{3}{4}k_4}{\sqrt{3k_1+3k_2+k_3+k_4}\sqrt{2k_1+\frac{3}{2}k_2+\frac{9}{16}k_3+\frac{9}{16}k_4}}\right)}{\pi}$$

11.3.2 Comparing Borda Count and Approval Voting

Theorem 11.3.1 and the Corollaries suggest that $\Theta_{BC} = (1/2, 3/4, 1/4)$, the classical Borda Count extension, is close to be the optimal rule in terms of robustness when voters can report preferences with indifference as well as indicating acceptable alternatives. However, the precedent analysis suggest that in some particular cases, especially when $k_2 \to 0$, Approval Voting may be closer to $\Theta^\star$ in term of robustness. In this section, we compare $P_R^\infty(\mathbf{k}, BC)$ and $P_R^\infty(\mathbf{k}, AV)$. We can immediately deduce from Corollary 11.3.2, that k_3 and k_4 play a symmetric role for

Table 11.4 Probabilities $P_R^{\infty}((k_2, k_{34}), BC)$ and $P_R^{\infty}((k_{34}), AV)$

$k_{34} \rightarrow$ $k_2 \downarrow$	0	0.1	0.2	0.3	0.4	0.5	0.6	0.7	0.8	0.9	1
0	0.8041	0.8074	0.8113	0.8160	0.8217	0.8289	0.8382	0.8508	0.8692	0.8994	1
0.05	0.8070	0.8105	0.8148	0.8200	0.8263	0.8343	0.8447	0.8591	0.8806	0.9190	–
0.1	0.8099	0.8138	0.8184	0.8241	0.8310	0.8399	0.8517	0.8682	0.8939	0.9467	–
0.2	0.8161	0.8207	0.8262	0.8330	0.8415	0.8525	0.8676	0.8899	0.9299	–	–
0.3	0.8228	0.8282	0.8347	0.8429	0.8533	0.8671	0.8870	0.9196	–	–	–
0.4	0.8300	0.8363	0.8442	0.8540	0.8668	0.8846	0.9122	–	–	–	–
0.5	0.8378	0.8452	0.8546	0.8666	0.8828	0.9068	–	–	–	–	–
0.6	0.8463	0.8552	0.8664	0.8813	0.9025	–	–	–	–	–	–
0.7	0.8557	0.8662	0.8801	0.8990	–	–	–	–	–	–	–
0.8	0.8662	0.8789	0.8962	–	–	–	–	–	–	–	–
0.9	0.8780	0.8939	–	–	–	–	–	–	–	–	–
1.0	0.8918	–	–	–	–	–	–	–	–	–	–
$P_R^{\infty}((k_{34}), AV)$	0.8041	0.8082	0.8128	0.8184	0.8251	0.8333	0.8437	0.8575	0.8767	0.9068	1

Table 11.5 Probabilities $\left[P_R^{\infty}((k_2, k_{34}), \Theta^{\star}) - P_R^{\infty}((k_2, k_{34}), BC)\right]$

$k_{34} \rightarrow$ $k_2 \downarrow$	0	0.1	0.2	0.3	0.4	0.5	0.6	0.7	0.8	0.9	1
0	0.0000	0.0008	0.0015	0.0020	0.0034	0.0044	0.0055	0.0067	0.0075	0.0074	0
0.1	0.0000	0.0006	0.0014	0.0020	0.0027	0.0033	0.0037	0.0037	0.0027	0.0000	–
0.2	0.0000	0.0005	0.0011	0.0015	0.0019	0.0021	0.0020	0.0012	0.0000	–	–
0.3	0.0000	0.0004	0.0008	0.0011	0.0012	0.0012	0.0005	0.0000	–	–	–
0.4	0.0000	0.0004	0.0005	0.0006	0.0006	0.0004	0.0001	–	–	–	–
0.5	0.0000	0.0003	0.0003	0.0003	0.0002	0.0000	–	–	–	–	–
0.6	0.0000	0.0001	0.0001	0.0001	0.0000	–	–	–	–	–	–
0.7	0.0000	0.0001	0.0000	0.0000	–	–	–	–	–	–	–
0.8	0.0000	0.0001	0.0000	–	–	–	–	–	–	–	–
0.9	0.0001	0.0000	–	–	–	–	–	–	–	–	–
1.0	0.0000	–	–	–	–	–	–	–	–	–	–

Borda Count and Approval Voting; Thus, we can express the formulas directly as a function of $k_{34} = k_3 + k_4$. Table 11.4 gives the two probabilities as a function of k_2 and k_{34}, and enables to compare them.

$P_R^{\infty}((k_2, k_{34}), BC)$ is always increasing with k_2; said differently, the worst scenario for the Borda rule occurs when the proportion of voters with a strict preference and only one acceptable option is high. The influence of indifferent voters is also clear: the higher the proportion of dichotomous voters is, the higher the robustness for the Borda Count is. One can also notice in Table 11.5, that the difference in terms of robustness between $\Theta^{\star}$ and Θ_{BC} is always small and inferior to 1%.

The last line of Table 11.4 gives the robustness for AV. It only depends upon k_{34}, that is, the proportion of dichotomous voters. We can immediately notice that Approval Voting will dominate the Borda Count for small values of k_2 only. At $k_2 = 0$, Approval Voting is indeed the optimal rule, but at $k_2 = 0.05$, the situation is already more favorable to BC for all the values of k_{34}. When half of the voters have strict preferences ($k_2 = 0.5$), the advantage of the Borda Count against Approval Voting is clear, ranging from 3% to 7%; It rises regularly as the proportion of dichotomous voters vanishes.

11.3.3 Optimal Extensions for Plurality and Antiplurality

A possible interpretation of Theorem 11.3.1 is to say that the best extension of the Borda Count for indifferent voters is not the one suggested by Black and Smith, but $\Theta^\star$. The same question arise for the Plurality rule and the Antiplurality rule: given λ, what is the optimal pair (μ, γ) that maximizes robustness?

Corollary 11.3.3. *Consider elections using scoring systems for which there are three candidates and n voters (n large), and for which the indifference between distinct candidates is allowed.*

Then, if $\lambda = 1$, the values of γ and μ maximizing the probability that the scoring system ranks two alternatives in the same order as they are ranked by pairwise majority are $\mu = 1$ and $\gamma = 0$. That is, the Approval Voting is the optimal extension of the Antiplurality rule, $\Theta_{A^\star} = (1, 1, 0)$.

Similarly, if $\lambda = 0$, the values maximizing $P_R^\infty(\mathbf{k}, (0, \mu, \gamma))$ are $\mu = 1$ and $\gamma = 0$. In other words, the optimal extension of the Plurality rule $\Theta_{P^\star} = (0, 1, 0)$ should behave as the AV for the classes of indifferent voters.

See the proof in Appendix B.

Corollary 11.3.4. *For all values of* $\mathbf{k}$, *under the EIC assumption*

$$P_R^\infty(\mathbf{k}, AV) = P_R^\infty(\mathbf{k}, P^\star) \geq P_R^\infty(\mathbf{k}, P)$$
$$P_R^\infty(\mathbf{k}, AV) = P_R^\infty(\mathbf{k}, A^\star) \geq P_R^\infty(\mathbf{k}, A)$$

The fact that Approval Voting does better than the classical extension of the Antiplurality is an immediate consequence of Corollary 11.3.3, as the optimal extension of the Antiplurality is Approval Voting. The fact that Approval Voting dominates the Plurality rule in terms of robustness over pairs, needs a specific proof, displayed in Appendix B.

Thus, though Approval Voting is often dominated by the Borda Count, at least it is a better option than the Plurality rule or the Antiplurality rule. By introducing voters with indifference, we have been able to differentiate it from these two rules, a result that Gehrlein and Lepelley (1998) could not obtain for the three-alternative case without dichotomous voters.

Another consequence of Corollaries 11.3.3 and 11.3.4 is that the Plurality and Antiplurality can only dominate the Borda Count in the razor thin regions where Approval Voting is dominating Borda, that is in regions where the proportion of k_2 voters is extremely low.

11.4 The Condorcet Efficiency

11.4.1 Deriving the Condorcet Efficiency for Extended Scoring Rules

By examining the robustness issue, we have already been able to raise many interesting conclusions. We are now interested in the following question: Which rule has the greatest Condorcet efficiency? In other words, which values of λ, γ and μ maximize the probability that the scoring winner will coincide with the Condorcet winner, given that the Condorcet winner exists. It is equivalent to find the solution of: aMb, aMc, aSb and aSc.

The candidate a is a Condorcet winner for the voting situation $\tilde{n}$ if and only if he beats b and c in pairwise comparisons. This is equivalent to fulfill the following inequalities:

$$n_1+n_2-n_3-n_4+n_5-n_6+n_7+n_8-n_9-n_{10}+n_{11}-n_{12}+n_{14}-n_{15}+n_{16}-n_{17}>0$$
$$n_1+n_2+n_3-n_4-n_5-n_6+n_7+n_8+n_9-n_{10}-n_{11}-n_{12}+n_{13}-n_{15}+n_{16}-n_{18}>0$$

The candidate a is chosen by the extended scoring rule Θ for the voting situation $\tilde{n}$ whenever he beats b and c. Therefore, the following inequalities should be satisfied:

$$n_1+n_2-n_3-n_4+(1-\lambda)(n_7-n_9)+n_8-n_{10}+\lambda(n_{11}-n_{12})+\mu(n_{14}-n_{15})+(1-\gamma)(n_{16}-n_{17})>0$$
$$n_1+n_2-n_5-n_6+n_7+(1-\lambda)(n_8-n_{11})+\lambda(n_9-n_{10})-n_{12}+\mu(n_{13}-n_{15})+(1-\gamma)(n_{16}-n_{18})>0$$

Theorem 11.4.1. *Consider a large population which has to choose among three alternatives. Under the EIC assumption the conditional probability that an extended scoring rule characterized by the vector Θ picks out the Condorcet winner is given by:*

$$P_{CE}^{\infty}(\mathbf{k},\Theta)=\frac{3}{P_{Con}^{\infty}(\mathbf{k})}\left(A-\frac{(k_1+k_2)}{4\pi^2}B\right)\ \textit{with,}$$

$$P_{Con}^{\infty}(\mathbf{k})=\frac{3}{4}+\frac{3}{2\pi}\arcsin\left(\frac{k_1+k_2+k_3+k_4}{3k_1+3k_2+2k_3+2k_4}\right)$$

$$A=\frac{1}{9}+\frac{1}{4\pi^2}\left[\arcsin(b)+\arcsin\left(\frac{b}{2}\right)\right]\left[\arcsin(b)-\arcsin\left(\frac{b}{2}\right)+\pi\right]$$

$$B=\int_0^1\frac{\arccos\left(\frac{4b^2+2c^2-5cb^2-2}{4-5b^2-4c^2+4cb^2}\right)dt}{\sqrt{4(3k_1+3k_2+2k_3+2k_4)^2-(3k_1+3k_2+2k_3+2k_4-tk_1-tk_2)^2}}$$

$$b = \frac{\sqrt{2}\,(k_1+k_2+k_3\mu+k_4\,(1-\gamma))}{\sqrt{(3k_1+3k_2+2k_3+2k_4)\left(k_1+k_2\,(1-\lambda+\lambda^2)+k_3\mu^2+k_4\,(1-\gamma)^2\right)}}$$

$$c = \frac{1}{2} - t\,\frac{(k_1+k_2)}{6\,k_1+6\,k_2+4\,k_3+4\,k_4}$$

Proof. See Appendix C.

The complexity of the formula obtained in Theorem 11.4.1 makes it difficult to establish precise statements, as for the robustness case. Nevertheless, we can derive from the formula exact values for the Condorcet efficiency of any Θ rule. Thus, we are able to check whether some conclusions for the robustness case can be transposed:

1. Do we still need an extremely high proportion of dichotomous voters for Approval to dominate Borda?
2. Does $\Theta^\star$ still dominate both Borda and Approval Voting?
3. Are the classical extensions of Plurality and Antiplurality rule still dominated by Approval Voting?

The next tables will help us to answer positively to these three issues. Table 11.6 displays the values of the Condorcet efficiency for Approval Voting and the Borda Count. The pattern is very similar to the one observed for the robustness issue. For $k_2 = 0$, the optimal rule is again Approval Voting (see Table 11.7 for a comparison with $\Theta^\star$). But if more than 10% of the voters belongs to Class II, the Borda Count dominates Approval Voting. When class II voters represent 50% of the population, the advantage can be rather important, from 5% to 10%. The only improvement for

Table 11.6 Probabilities $P^{\infty}_{CE}((k_2,k_{34}),BC)$ and $P^{\infty}_{CE}((k_{34}),AV)$

$k_{34} \rightarrow$ $k_2 \downarrow$	0	0.1	0.2	0.3	0.4	0.5	0.6	0.7	0.8	0.9	1
0	0.75720	0.76408	0.77184	0.78071	0.79096	0.80302	0.81756	0.83569	0.85961	0.89494	1
0.05	0.76162	0.76890	0.77714	0.78658	0.79755	0.81057	0.82641	0.84650	0.87379	0.91750	–
0.1	0.76620	0.77390	0.78265	0.79272	0.80450	0.81859	0.83594	0.85838	0.89014	0.94986	–
0.2	0.77587	0.78451	0.79442	0.80595	0.81963	0.83632	0.85755	0.88667	0.93537	–	–
0.3	0.78631	0.79606	0.80735	0.82067	0.83677	0.85699	0.88405	0.92631	–	–	–
0.4	0.79763	0.80871	0.82169	0.83728	0.85662	0.88202	0.91987	–	–	–	–
0.5	0.81002	0.82270	0.83782	0.85639	0.88041	0.91497	–	–	–	–	–
0.6	0.82369	0.83837	0.85628	0.87910	0.91110	–	–	–	–	–	–
0.7	0.83895	0.85624	0.87804	0.90795	–	–	–	–	–	–	–
0.8	0.85626	0.87716	0.90532	–	–	–	–	–	–	–	–
0.9	0.87643	0.90310	–	–	–	–	–	–	–	–	–
1.0	0.90119	–	–	–	–	–	–	–	–	–	–
P^{∞}_{CE} $((k_{34}),AV)$	0.75720	0.76623	0.77606	0.78688	0.79891	0.81253	0.82826	0.84707	0.87081	0.90438	1

Table 11.7 Probabilities $P_{CE}^{\infty}((k_2, k_{34}), \Theta^{\star})$

$k_{34} \rightarrow$ $k_2 \downarrow$	0	0.1	0.2	0.3	0.4	0.5	0.6	0.7	0.8	0.9	1
0	0.75720	0.76623	0.77606	0.78688	0.79891	0.81253	0.82826	0.84707	0.87081	0.90438	1
0.1	0.76620	0.77573	0.78617	0.79772	0.81070	0.82556	0.84309	0.86475	0.89416	0.94986	–
0.2	0.77587	0.78603	0.79726	0.80981	0.82413	0.84091	0.86145	0.88883	0.93537	–	–
0.3	0.78631	0.79728	0.80953	0.82346	0.83970	0.85945	0.88533	0.92631	–	–	–
0.4	0.79763	0.80964	0.82326	0.83908	0.85817	0.88280	0.91987	–	–	–	–
0.5	0.81002	0.82337	0.83882	0.85733	0.88089	0.91498	–	–	–	–	–
0.6	0.82369	0.83880	0.85680	0.87939	0.91110	–	–	–	–	–	–
0.7	0.83895	0.85646	0.87820	0.90795	–	–	–	–	–	–	–
0.8	0.85626	0.87723	0.90532	–	–	–	–	–	–	–	–
0.9	0.87643	0.90310	–	–	–	–	–	–	–	–	–
1.0	0.90119	–	–	–	–	–	–	–	–	–	–

Table 11.8 Probabilities $P_{CE}^{\infty}((k_3, k_4), A)$

$k_4 \rightarrow$ $k_3 \downarrow$	0	0.1	0.2	0.3	0.4	0.5	0.6	0.7	0.8	0.9	1
0	0.7572	0.7574	0.7582	0.7599	0.7629	0.7676	0.7751	0.7873	0.8077	0.8464	1
0.1	0.7662	0.7668	0.7680	0.7701	0.7737	0.7792	0.7880	0.8020	0.8261	0.8750	–
0.2	0.7761	0.7770	0.7787	0.7814	0.7858	0.7924	0.8028	0.8196	0.8495	–	–
0.3	0.7869	0.7883	0.7906	0.7942	0.7995	0.8077	0.8204	0.8417	–	–	–
0.4	0.7989	0.8010	0.8041	0.8088	0.8156	0.8261	0.8427	–	–	–	–
0.5	0.8125	0.8155	0.8198	0.8260	0.8352	0.8495	–	–	–	–	–
0.6	0.8282	0.8326	0.8386	0.8474	0.8608	–	–	–	–	–	–
0.7	0.8471	0.8534	0.8625	0.8764	–	–	–	–	–	–	–
0.8	0.8708	0.8811	0.8970	–	–	–	–	–	–	–	–
0.9	0.9044	0.9260	–	–	–	–	–	–	–	–	–
1	1	–	–	–	–	–	–	–	–	–	–

AV is that the razor thin region where it dominates BC seems a little bit larger. For example, it still dominates BC for $k_2 = 0.05$ and $k_{34} = 0.5$, while this point was already on the Borda side for the robustness issue. However, Table 11.7 indicates that $\Theta^{\star}$ is still superior to both AV and BC for all $\mathbf{k}$.

In Corollary 11.3.4, we could assess that the robustness of AV was superior to the ones of both Plurality and Antiplurality. First, one can notice that $P_{CE}^{\infty}(\mathbf{k}, A)$ and $P_{CE}^{\infty}(\mathbf{k}, P)$ can be expressed as functions of $k_{12} = k_1 + k_2$. Secondly, k_3 and k_4 plays a symmetric role: $P_{CE}^{\infty}((k_{12}, k_3, k_4), A) = P_{CE}^{\infty}((k_{12}, k_4, k_3), P)$. Table 11.8 displays the values for $P_{CE}^{\infty}((k_{12}, k_3, k_4), A)$, the Plurality case being symmetric. As it is no longer possible to derive exact results for the efficiency case, Approval is compared with the Antiplurality rule in Table 11.9. The values speak by themselves: Approval Voting is always superior to the Antiplurality rule in terms of Condorcet efficiency. A similar conclusion holds between AV and the Plurality rule.

Table 11.9 Probabilities $P^{\infty}_{CE}((k_3, k_4), AV) - P^{\infty}_{CE}((k_3, k_4), A)$

$k_4 \rightarrow$ $k_3 \downarrow$	0	0.1	0.2	0.3	0.4	0.5	0.6	0.7	0.8	0.9	1
0	0.0000	0.0088	0.0179	0.0270	0.0360	0.0449	0.0532	0.0598	0.0631	0.0580	0.0000
0.1	0.0000	0.0093	0.0189	0.0288	0.0388	0.0491	0.0591	0.0688	0.0783	0.1250	–
0.2	0.0000	0.0099	0.0202	0.0311	0.0425	0.0547	0.0680	0.0848	0.1505	–	–
0.3	0.0000	0.0106	0.0219	0.0341	0.0476	0.0631	0.0840	0.1583	–	–	–
0.4	0.0000	0.0115	0.0242	0.0383	0.0552	0.0783	0.1573	–	–	–	–
0.5	0.0000	0.0128	0.0273	0.0448	0.0692	0.1505	–	–	–	–	–
0.6	0.0000	0.0145	0.0322	0.0570	0.1392	–	–	–	–	–	–
0.7	0.0000	0.0174	0.0419	0.1236	–	–	–	–	–	–	–
0.8	0.0000	0.0233	0.1030	–	–	–	–	–	–	–	–
0.9	0.0000	0.0740	–	–	–	–	–	–	–	–	–
1	0.0000	–	–	–	–	–	–	–	–	–	–

11.5 Conclusion

The main objective of this chapter was to revisit the Condorcet efficiency issue, in a framework where voters can not only report their preferences, but also indicate the alternatives they approve of. In this way, we hoped that the rather negative conclusion obtained by Gehrlein and Lepelley (1998) on the Condorcet efficiency of Approval Voting could be circumvented. Unfortunately, our conclusion remains mitigated: we need an extremely high proportion of dichotomous voters for the Approval Voting to perform better than the Borda Count. Said differently, it is difficult to beat the Borda Count on the Condorcet efficiency ground! As a consolation, it should be noted that AV fares better that the Plurality rule and the Antiplurality rule as soon as dichotomous voters come into play.

We could have also compared Approval Voting with other methods, such as the scoring run-off rules: After the first round, the two alternatives with the highest scores go to the run-off. The candidate who is able to defeat the other one on the basis of their pairwise comparison is then declared as the winner. For three-candidate elections, it is trivial to see that the Condorcet efficiency of scoring run-off rules is equivalent to 1 minus the probability of ranking the Condorcet winner last. Tataru and Merlin (1997) derived the Condorcet efficiency for all the classical scoring rules $(1, \lambda, 0)$, and confirmed that the introduction of a run-off greatly improves the Condorcet efficiency. It is also well known (see for example Smith (1973)) that the Condorcet efficiency of the Borda run-off is one. So we may conjecture that Approval Voting would also be dominated by many scoring run-off rules. But as Sanver (2010) points out, the extended arrovian framework where voters can express preferences and indicate acceptable alternatives calls for new rules. In this context, he defines Approval Voting with run-off, as well as several other rules, which could also be evaluated on their propensity to select the Condorcet winner when it exists.

However, one has to be aware that the Impartial Culture hypothesis, as well as the extended version we used in this chapter, is a very peculiar scenario. These

Table 11.10 A list of the possible classes of preferences for four candidates

$\underline{a} \succ \underline{b} \succ \underline{c} \succ d$	$\underline{a} \succ \underline{b} \sim \underline{c} \succ d$	$\underline{a} \sim \underline{b} \sim \underline{c} \succ d$
$\underline{a} \succ \underline{b} \succ c \succ d$	$\underline{a} \succ b \sim c \succ d$	$\underline{a} \succ b \sim c \sim d$
$\underline{a} \succ b \succ c \succ d$	$\underline{a} \sim \underline{b} \succ \underline{c} \succ d$	$\underline{a} \sim \underline{b} \sim \underline{c} \sim \underline{d}$
$\underline{a} \succ \underline{b} \succ c \sim d$	$\underline{a} \sim \underline{b} \succ c \succ d$	
$\underline{a} \succ b \succ c \sim d$	$\underline{a} \sim \underline{b} \succ c \sim d$	

assumptions describe extremely symmetric societies, where no preference is more likely than another one within a class. We are far from real life elections, where the debates and exchanges of ideas will permit the emergence of some candidates at the expenses of other ones. Thus, a way to test the robustness of our results would be to derive the Condorcet efficiency with different probabilistic models. Thanks to the techniques developed by Cervone et al. (2005), Chua and Huang (2000), and Lepelley et al. (2008) an obvious extension would be to perform again the exercise under the Impartial Anonymous Culture, which can model more homogeneous voting situations. In this volume, Laslier (2010) and Lehtinen (2010) propose new voting assumptions that can also be adapted to tackle the same issue.

The reader may also wonder what happens with more than three candidates. We must admit that we have no clue at this point. Indeed, as seen in Table 11.10, the number of possible types already explodes for four candidates, as well as the number of extended scoring rules. The only hope would be to generate profiles randomly with Monte Carlo simulations, in order to test which rule(s) would fare well in terms of robustness and Condorcet efficiency. The emergence of $\Theta^{\star}$ as the optimal rule in the three alternative case let us think that many surprises await when ones want to study voting rules in a extended framework where voters can report acceptable preferences as well as weak orderings!

Acknowledgements We are indebted to Jean François Laslier and Remzi Sanver for the comments they made on the early versions of this chapter. A usual, all remaining errors are ours.

Appendix A: Proof of Theorem 11.2.1

Without loss of generality, the candidate a is a Condorcet winner for the voting situation $\tilde{n}$ if and only if she beats b and c in pairwise comparisons. This is true if the following inequalities are satisfied:

$$n_1 + n_2 - n_3 - n_4 + n_5 - n_6 + n_7 + n_8 - n_9 - n_{10} + n_{11} - n_{12} + n_{14} - n_{15} + n_{16} - n_{17} > 0$$
$$n_1 + n_2 + n_3 - n_4 - n_5 - n_6 + n_7 + n_8 + n_9 - n_{10} - n_{11} - n_{12} + n_{13} - n_{15} + n_{16} - n_{18} > 0$$

To begin, we develop two discrete variables X_1 and X_2 that are defined for each randomly selected voter's preference ranking.

$$\begin{aligned} X_1 = 1 \quad &: p_1 + p_2 + p_5 + p_7 + p_8 + p_{11} + p_{14} + p_{16} \\ -1 \quad &: p_3 + p_4 + p_6 + p_9 + p_{10} + p_{12} + p_{15} + p_{17} \\ 0 \quad &: p_{13} + p_{18} + p_{19} \end{aligned}$$

$$\begin{aligned} X_2 = 1 \quad &: p_1 + p_2 + p_3 + p_7 + p_8 + p_9 + p_{13} + p_{16} \\ -1 \quad &: p_4 + p_5 + p_6 + p_{10} + p_{11} + p_{12} + p_{15} + p_{18} \\ 0 \quad &: p_{14} + p_{17} + p_{19} \end{aligned}$$

For a given voter's preference ranking, the definitions of the p_i's and the associated rankings that they represent, indicate that a is ranked below (above) b when $X_1 > 0$ ($X_1 < 0$). Similarly, a is ranked below (above) c when $X_2 > 0$ ($X_2 < 0$). Based upon these definitions, the probability that Candidate a is a Condorcet winner for the voting situation $\tilde{n}$ is equivalent to the joint probability that $\bar{X}_i > 0$, for $i = 1, 2$.

By directly following the analysis in Gehrlein and Fishburn (1978a) as $n \to \infty$ with EIC, this probability is equivalent to the quadrivariate normal positive orthant probability $\Phi(2, \mathbf{R})$ that $\sqrt{n}\bar{X}_i > 0$ for $i = 1, 2$ with a correlation matrix $\mathbf{R}$ that is obtained from the correlations between X_1 and X_2 such that: $\mathbf{R} = (\rho)$ and $\rho = \frac{Cov(X_1, X_2)}{\sqrt{V(X_1)V(X_2)}} = \frac{k_1+k_2+k_3+k_4}{3k_1+3k_2+2k_3+2k_4}$.

This bivariate normal positive orthant probability is given in David and Mallows (1961) such that $\Phi(2, \mathbf{R}) = \frac{1}{4} + \frac{1}{2\pi} \arcsin(\rho)$. Taking into consideration the symmetry of EIC regarding candidates, the probability that a Condorcet winner exists is equal to:

$$P_{Con}^{\infty}(EIC) = 3\Phi(2, \mathbf{R}) = \frac{3}{4} + \frac{3}{2\pi} \arcsin\left(\frac{k_1 + k_2 + k_3 + k_4}{3k_1 + 3k_2 + 2k_3 + 2k_4}\right)$$

Appendix B: Proof of Theorem 11.3.1 and its Corollaries

Proof of Theorem 11.3.1

van Newenhizen (1992) and Saari and Tataru (1999) developed techniques different from those of Gehrlein and Fishburn in order to assess the probability of voting events under the IC assumptions. For them, the problem is equivalent to the computation of angles between the hyperplanes that characterize the voting events. Merlin and Valognes (2004) explained how to adapt their techniques to extended version of the IC hypothesis.

In order to apply their method to calculate the probability of this voting situation, we should modify the problem by using the scoring vector $\sqrt{k_1}(1, 0, 0)$ for $i = 1, \ldots, 6$, $\sqrt{k_2}(1, \lambda, 0)$ for $i = 7, \ldots, 12$, $\sqrt{k_3}(\mu, \mu, 0)$ for $i = 13, \ldots, 15$ and $\sqrt{k_4}(1, \gamma, \gamma)$ for $i = 16, \ldots, 18$. Taking into consideration these modifications, let W_1 and W_2 be normal vectors for hyperplanes H_1 and H_2, described by the new versions of equations (11.1) and (11.2).

$$W_1 = (\sqrt{k_1}, \sqrt{k_1}, -\sqrt{k_1}, -\sqrt{k_1}, \sqrt{k_1}, -\sqrt{k_1}, \sqrt{k_2}, \sqrt{k_2}, -\sqrt{k_2}, -\sqrt{k_2}, \sqrt{k_2}, -\sqrt{k_2}, 0, \sqrt{k_3}, -\sqrt{k_3}, \sqrt{k_4}, -\sqrt{k_4}, 0, 0)$$

$$W_2 = (\sqrt{k_1}, \sqrt{k_1}, -\sqrt{k_1}, -\sqrt{k_1}, 0, 0, (1-\lambda)\sqrt{k_2}, \sqrt{k_2}, (-1+\lambda)\sqrt{k_2}, -\sqrt{k_2}, \lambda\sqrt{k_2}, -\lambda\sqrt{k_2}, 0, \mu\sqrt{k_3}, -\mu\sqrt{k_3}, (1-\gamma)\sqrt{k_4}, (-1+\gamma)\sqrt{k_4}, 0, 0)$$

Let α_{12} be the angle between the hyperplanes H_1 and H_2.

$$\begin{aligned}\alpha_{12} &= \arccos\left(-\frac{W_1.W_2}{\|W_1\|.\|W_2\|}\right)\\ &= \pi - \arccos\left(\frac{2k_1+2k_2+\mu k_3+\gamma k_4}{\sqrt{3k_1+3k_2+k_3+k_4}\sqrt{2k_1+2\lambda^2k_2-2\lambda k_2+2k_2+\mu^2k_3+(1-\gamma)^2k_4}}\right)\end{aligned}$$

Then, the probability that a scoring rule agree with the majority rule on a given pair of candidates a and b is given by:

$$\begin{aligned}P_R^{\infty}(\mathbf{k},\Theta) &= 2.\frac{1}{2\pi}.\alpha_{12}\\ &= \frac{\pi - \arccos\left(\frac{2k_1+2k_2+\mu k_3+\gamma k_4}{\sqrt{3k_1+3k_2+k_3+k_4}\sqrt{2k_1+2\lambda^2k_2-2\lambda k_2+2k_2+\mu^2k_3+(1-\gamma)^2k_4}}\right)}{\pi}\end{aligned}$$

Proof of Corollary 11.3.1

A relative maximum of $P_R^{\infty}(\mathbf{k},\Theta)$ occurs at its critical point. To locate the critical points of $P_R^{\infty}(\mathbf{k},\Theta)$, we begin by computing the first partial derivatives of $P_R^{\infty}(\mathbf{k},\Theta)$.

$$\frac{\partial P_R^{\infty}(\mathbf{k},\Theta)}{\partial\lambda} = \frac{ak_2(1-2\lambda)}{\pi c\sqrt{bc-a^2}} \qquad \frac{\partial P_R^{\infty}(\mathbf{k},\Theta)}{\partial\mu} = \frac{k_3(c-a\mu)}{\pi c\sqrt{bc-a^2}} \qquad \frac{\partial P_R^{\infty}(\mathbf{k},\Theta)}{\partial\gamma} = \frac{k_4(a-c-a\gamma)}{\pi c\sqrt{bc-a^2}}$$

where

$$a = 2k_1 + 2k_2 + k_3\mu + k_4(1-\gamma) \qquad b = 3k_1 + 3k_2 + k_3 + k_4$$

$$c = 2k_1 + 2k_2 - 2k_2\lambda + 2k_2\lambda^2 + k_3\mu^2 + k_4(1-\gamma)^2$$

Taking first-order derivatives and setting equal to zero we get the first-order conditions:

$$\frac{\partial P_R^{\infty}(\mathbf{k},\Theta)}{\partial\lambda} = \frac{\partial P_R^{\infty}(\mathbf{k},\Theta)}{\partial\mu} = \frac{\partial P_R^{\infty}(\mathbf{k},\Theta)}{\partial\gamma} = 0.$$

Noting that, $\mu = \frac{a}{c} = 1-\gamma$, the resolution of these equalities leads to the following results:

$$\lambda^\star = \frac{1}{2} \qquad \mu^\star = \frac{4k_1 + 3k_2}{4(k_1 + k_2)} \qquad \gamma^\star = \frac{k_2}{4(k_1 + k_2)}$$

Therefore, the critical points are given in function of k_1 and k_2. Once we have found a critical point, a natural question to ask is what kind of critical point it is in the sense that it may be a relative maximum or a relative minimum.[4] To verify this context, we must calculate the leading principal minors of the Hessian matrix H at this critical point. The Hessian matrix H of the multivariate function $P_R^\infty(\mathbf{k}, \Theta)$ is the matrix of second partial derivatives. Here, H has the following form:

$$H = \begin{bmatrix} \dfrac{\partial^2 P_R^\infty(\mathbf{k}, \Theta)}{\partial^2 \lambda} & \dfrac{\partial^2 P_R^\infty(\mathbf{k}, \Theta)}{\partial \lambda \partial \mu} & \dfrac{\partial^2 P_R^\infty(\mathbf{k}, \Theta)}{\partial \lambda \partial \gamma} \\ \dfrac{\partial^2 P_R^\infty(\mathbf{k}, \Theta)}{\partial \mu \partial \lambda} & \dfrac{\partial^2 P_R^\infty(\mathbf{k}, \Theta)}{\partial^2 \mu} & \dfrac{\partial^2 P_R^\infty(\mathbf{k}, \Theta)}{\partial \mu \partial \gamma} \\ \dfrac{\partial^2 P_R^\infty(\mathbf{k}, \Theta)}{\partial \gamma \partial \lambda} & \dfrac{\partial^2 P_R^\infty(\mathbf{k}, \Theta)}{\partial \gamma \partial \mu} & \dfrac{\partial^2 P_R^\infty(\mathbf{k}, \Theta)}{\partial^2 \gamma} \end{bmatrix}$$

For a $n \times n$ Hessian matrix, the kth order leading principal minor is the determinant of the kth order principal submatrix formed by deleting the last $(n - k)$ rows and columns. A critical point is a maximum if a matrix is negative definite. This is true when the first principal minor is negative and the sign of consecutive ones alternates. The first principal minor is

$$\frac{-32k_2(k_1 + k_2)^2(4k_1^2 + 5k_1k_2 + k_2^2)^{-\frac{1}{2}}}{\pi\sqrt{4k_1 + 3k_2}\sqrt{32k_1^3 + 88k_1^2k_2 + 80k_1k_2^2 + 24k_2^3 + 9k_2^2k_3 + 24k_1k_2k_3 + 16k_1^2k_3 + 9k_2^2k_4 + 24k_1k_2k_4 + 16k_1^2k_4}} < 0$$

The second principal minor is

$$\frac{512k_2k_3(k_1 + k_2)^3(8k_1^2 + 16k_1k_2 + 8k_2^2 + 3k_2k_4 + 4k_1k_4)}{\pi^2(4k_1 + k_2)(32k_1^3 + 88k_1^2k_2 + 80k_1k_2^2 + 24k_2^2 + 9k_2^2k_3 + 24k_1k_2k_3 + 16k_1^2k_3 + 9k_2^2k_4 + 24k_1k_2k_4 + 16k_1^2k_4)^2} > 0$$

The third principal minor is

$$\frac{-65536\,\pi^{-3}k_4\,k_3\,k_2\,(k_2 + k_1)^7(4\,k_1 + 3\,k_2)^{-\frac{1}{2}}(4k_1 + k_2)^{-1}(k_2^2 + 4k_1^2 + 5k_1k_2)^{-\frac{1}{2}}}{(80\,k_1\,k_2{}^2 + 88\,k_1{}^2k_2 + 32\,k_1{}^3 + 24\,k_2{}^3 + 9k_2^2k_3 + 24k_1k_2k_3 + 16k_1^2k_3 + 9k_4k_2^2 + 24k_1k_2k_4 + 16k_4k_1^2)^{-\frac{5}{2}}} < 0$$

The principal minors satisfy the tests for a local maximum. Therefore, the point

$$\Theta^\star = \left(\lambda^\star = \frac{1}{2},\ \mu^\star = \frac{4k_1 + 3k_2}{4(k_1 + k_2)},\ \gamma^\star = \frac{k_2}{4(k_1 + k_2)}\right)$$

is a maximum.

[4] Another possibility is that it is neither a relative maximum nor a relative minimum.

Proof of Corollary 11.3.3

From Theorem 11.2.1,

$$P_R^{\infty}(\mathbf{k},(1,\mu,\gamma)) = \frac{\pi - \arccos\left(\frac{2k_1+2k_2+\mu k_3+(1-\gamma)k_4}{\sqrt{3k_1+3k_2+k_3+k_4}\sqrt{2k_1+2k_2+\mu^2 k_3+(1-\gamma)^2 k_4}}\right)}{\pi}$$

$$\frac{\partial P_R^{\infty}(\mathbf{k},(1,\mu,\gamma))}{\partial \mu} = \frac{k_3(c-a\mu)}{\pi c\sqrt{bc-a^2}}$$
$$\frac{\partial P_R^{\infty}(\mathbf{k},(1,\mu,\gamma))}{\partial \gamma} = \frac{k_4(a-c-a\gamma)}{\pi c\sqrt{bc-a^2}}$$

where

$$a = 2k_1 + 2k_2 + k_3\mu + k_4(1-\gamma)$$
$$b = 3k_1 + 3k_2 + k_3 + k_4$$
$$c = 2k_1 + 2k_2 + k_3\mu^2 + k_4(1-\gamma)^2$$

The resolution of $\frac{\partial P_R^{\infty}(\mathbf{k},(1,\mu,\gamma))}{\partial \mu} = \frac{\partial P_R^{\infty}(\mathbf{k},(1,\mu,\gamma))}{\partial \gamma} = 0$ gives first $1-\gamma = \mu = \frac{a}{c}$ and leads to the following result:

$$\mu = 1,\ \gamma = 0$$

where the Hessian matrix has the form:

$$H = \begin{bmatrix} \frac{\partial^2 P_R^{\infty}(\mathbf{k},(1,\mu,\gamma))}{\partial^2 \mu} & \frac{\partial^2 P_R^{\infty}(\mathbf{k},(1,\mu,\gamma))}{\partial \mu \partial \gamma} \\ \frac{\partial^2 P_R^{\infty}(\mathbf{k},(1,\mu,\gamma))}{\partial \gamma \partial \mu} & \frac{\partial^2 P_R^{\infty}(\mathbf{k},(1,\mu,\gamma))}{\partial^2 \gamma} \end{bmatrix}$$

The first principal minor is $-k_3(2k_1 + 2k_2 + k_4)/(2k_1 + 2k_2 + k_3 + k_4)^{\frac{3}{2}}$ $\sqrt{k_1 + k_2}\pi < 0$. The second principal minor is $2k_3k_4/\pi^2(2k_1 + 2k_2 + k_3 + k_4)^2 > 0$. Therefore, this functions satisfy the tests for a local maximum.

The same proof can be easily done for $\lambda = 0$ and the values maximizing $P_R^{\infty}(\mathbf{k},(0,\mu,\gamma))$ are $\mu = 1$ and $\gamma = 0$.

Proof of Corollary 11.3.4

It is simple to verify that $P_R^{\infty}(\mathbf{k}, AV) = P_R^{\infty}(\mathbf{k}, A^{\star}) = P_R^{\infty}(\mathbf{k}, P^{\star})$ in the sense that the probability $P_R^{\infty}(\mathbf{k}, \Theta)$ is the same for $\Theta_{AV} = (1, 1, 0)$, $\Theta_{A^{\star}} = (1, 1, 0)$ and $\Theta_{P^{\star}} = (0, 1, 0)$. The proof of $P_R^{\infty}(\mathbf{k}, AV) \geq P_R^{\infty}(\mathbf{k}, P)$ is as follows:

Let a, b and c three positive reals.

We know that:

$$\tfrac{5}{4}ab + \tfrac{5}{4}bc + 2ac \geq ab + bc + 2ac$$

$$\Leftrightarrow a^2 + \tfrac{1}{4}b^2 + c^2 + \tfrac{5}{4}ab + \tfrac{5}{4}bc + 2ac \geq a^2 + \tfrac{1}{4}b^2 + c^2 + ab + bc + 2ac$$

$$\Leftrightarrow \quad (a + b + c)(a + \tfrac{1}{4}b + c) \geq (a + \tfrac{1}{2}b + c)^2$$

$$\Leftrightarrow \quad \sqrt{a + b + c}\sqrt{a + \tfrac{1}{4}b + c} \geq a + \tfrac{1}{2}b + c$$

$$\Leftrightarrow \quad \sqrt{a + b + c} \geq \frac{a+\frac{1}{2}b+c}{\sqrt{a+\frac{1}{4}b+c}}$$

Let $a = 2k_1 + 2k_2$, $b = k_3$ and $c = k_4$. Then

$$\sqrt{2k_1 + 2k_2 + k_3 + k_4} \geq \frac{2k_1 + 2k_2 + \frac{1}{2}k_3 + k_4}{\sqrt{2k_1 + 2k_2 + \frac{1}{4}k_3 + k_4}}$$

and

$$\frac{\sqrt{2k_1 + 2k_2 + k_3 + k_4}}{\sqrt{3k_1 + 3k_2 + k_3 + k_4}} \geq \frac{2k_1 + 2k_2 + \frac{1}{2}k_3 + k_4}{\sqrt{3k_1 + 3k_2 + k_3 + k_4}\sqrt{2k_1 + 2k_2 + \frac{1}{4}k_3 + k_4}}$$

Taking into consideration the fact that the function arccos is always a decreasing function, we have

$$\frac{\pi - \arccos\left(\frac{\sqrt{2k_1+2k_2+k_3+k_4}}{\sqrt{3k_1+3k_2+k_3+k_4}}\right)}{\pi} \geq \frac{\pi - \arccos\left(\frac{2k_1+2k_2+\frac{1}{2}k_3+k_4}{\sqrt{3k_1+3k_2+k_3+k_4}\sqrt{2k_1+2k_2+\frac{1}{4}k_3+k_4}}\right)}{\pi}$$

Then:

$$P_R^{\infty}(\mathbf{k}, AV) \geq P_R^{\infty}(\mathbf{k}, P)$$

We can easily use the same principle to prove that $P_R^{\infty}(\mathbf{k}, AV) \geq P_R^{\infty}(\mathbf{k}, A)$ with $a = 2k_1 + 2k_2$, $b = k_4$ and $c = k_3$.

Appendix C: Proof of Theorem 11.4.1

The candidate a is a Condorcet winner for the voting situation $\tilde{n}$ if and only if she beats b and c in pairwise comparisons. This is equivalent to the following inequalities:

$$n_1 + n_2 - n_3 - n_4 + n_5 - n_6 + n_7 + n_8 - n_9 - n_{10} + n_{11} - n_{12} + n_{14} - n_{15} + n_{16} - n_{17} > 0$$
$$n_1 + n_2 + n_3 - n_4 - n_5 - n_6 + n_7 + n_8 + n_9 - n_{10} - n_{11} - n_{12} + n_{13} - n_{15} + n_{16} - n_{18} > 0$$

In other hand, the candidate a is chosen by the scoring rule for the voting situation $\tilde{n}$ whenever she beats b and c. Therefore, the following inequalities are satisfied:

$$n_1 + n_2 - n_3 - n_4 + (1-\lambda)(n_7 - n_9) + n_8 - n_{10} + \lambda(n_{11} - n_{12}) + \mu(n_{14} - n_{15}) + (1-\gamma)(n_{16} - n_{17}) > 0$$
$$n_1 + n_2 - n_5 - n_6 + n_7 + (1-\lambda)(n_8 - n_{11}) + \lambda(n_9 - n_{10}) - n_{12} + \mu(n_{13} - n_{15}) + (1-\gamma)(n_{16} - n_{18}) > 0$$

We begin by defining four discrete variables X_1, X_2, X_3 and X_4, which have values that are determined by the p_i probabilities that are associated with voter preference rankings:

$$X_1 = \begin{cases} 1 & : p_1 + p_2 + p_5 + p_7 + p_8 + p_{11} + p_{14} + p_{16} \\ -1 & : p_3 + p_4 + p_6 + p_9 + p_{10} + p_{12} + p_{15} + p_{17} \\ 0 & : p_{13} + p_{18} + p_{19} \end{cases}$$

$$X_2 = \begin{cases} 1 & : p_1 + p_2 + p_3 + p_7 + p_8 + p_9 + p_{13} + p_{16} \\ -1 & : p_4 + p_5 + p_6 + p_{10} + p_{11} + p_{12} + p_{15} + p_{18} \\ 0 & : p_{14} + p_{17} + p_{19} \end{cases}$$

$$X_3 = \begin{cases} 1 & : p_1 + p_2 + p_8 \\ -1 & : p_3 + p_4 + p_{10} \\ \lambda & : p_{11} \\ -\lambda & : p_{12} \\ \mu & : p_{14} \\ -\mu & : p_{15} \end{cases} \qquad X_3 = \begin{cases} 1-\gamma & : p_{16} \\ -1+\gamma & : p_{17} \\ (1-\lambda) & : p_7 \\ (\lambda-1) & : p_9 \\ 0 & : p_5 + p_6 + p_{13} \\ & \; + p_{18} + p_{19} \end{cases}$$

$$
\begin{array}{llll}
X_4 = 1 & : p_1 + p_2 + p_7 & X_4 = 1-\gamma & : p_{16} \\
\quad -1 & : p_5 + p_6 + p_{12} & \quad -1+\gamma & : p_{18} \\
\quad \lambda & : p_9 & \quad (1-\lambda) & : p_8 \\
\quad -\lambda & : p_{10} & \quad (\lambda-1) & : p_{11} \\
\quad \mu & : p_{13} & \quad 0 & : p_3 + p_4 + p_{14} \\
\quad -\mu & : p_{15} & & \quad +p_{17} + p_{19}
\end{array}
$$

$X_1 = 1(-1)$ corresponds to the situations where the candidate a is ranked over (under) b in a voter's preference structure. It is equivalent to the fact that a is the majority winner over b. Likewise, a will be ranked over (under) c in a voters's preference structure when $X_2 = 1(-1)$. The value of X_3 (X_4) represents the difference in points that the scoring rule assigns to candidates a and b (c).

Taking into account the fact that $E(X_i) = 0$ for $i = 1, 2, 3, 4$, such that $E(X_i)$ is the expectation value of the random variable X_i, the variance $V(X_i)$ is equal to $E(X_i^2)$ for $i = 1, 2, 3, 4$. Therefore, we find:

$$V(X_1) = V(X_2) = k_1 + k_2 + \frac{2}{3}k_3 + \frac{2}{3}k_4$$

$$V(X_3) = V(X_4) = \frac{2}{3}k_1 + \frac{2}{3}k_2 + \frac{2}{3}k_2\lambda^2 + \frac{2}{3}k_3\mu^2 + \frac{2}{3}k_4(1-\gamma)^2 - \frac{2}{3}k_2\lambda$$

Since $E(X_i) = 0$ for $i = 1, 2, 3, 4$, the covariance term $Cov(X_i, X_j)$ between each two random variables X_i and X_j is obtained as $E(X_i X_j)$. For our context, we obtain:

$$Cov(X_1, X_2) = \frac{1}{3}k_1 + \frac{1}{3}k_2 + \frac{1}{3}k_3 + \frac{1}{3}k_4$$

$$Cov(X_1, X_3) = Cov(X_2, X_4) = \frac{2}{3}k_1 + \frac{2}{3}k_2 + \frac{2}{3}k_3\mu + \frac{2}{3}k_4(1-\gamma)$$

$$Cov(X_2, X_3) = Cov(X_1, X_4) = \frac{1}{3}k_1 + \frac{1}{3}k_2 + \frac{1}{3}k_3\mu + \frac{1}{3}k_4(1-\gamma)$$

$$Cov(X_3, X_4) = \frac{1}{3}k_1 + \frac{1}{3}k_2 - \frac{1}{3}k_2\lambda + \frac{1}{3}\lambda^2 k_2 + \frac{1}{3}k_3\mu^2 + \frac{1}{3}k_4(1-\gamma)^2$$

Let $\bar{X}_i$ denote the sample mean value of X_i over the n individuals, for $i = 1, 2, 3, 4$. It is clear that candidate a is a Condorcet winner over b (c) when $\bar{X}_1 > 0$ ($\bar{X}_2 > 0$). Likewise, candidate a is the scoring winner over b (c) when $\bar{X}_3 > 0$ ($\bar{X}_4 > 0$). Let $CWSR(m, k, EIC)$ denote the joint probability that an alternative, say a, is both the Condorcet and extended scoring rule winner under EIC hypothesis for n voters and m alternatives. Wilks (1962) observes that in the limit ($n \to \infty$), the multivariate Central Limit Theorem states that $(\sqrt{n}\bar{X}_1, \sqrt{n}\bar{X}_2, \sqrt{n}\bar{X}_3, \sqrt{n}\bar{X}_4)$ has a four-variate normal distribution with covariances given above and $E(\bar{X}_i) = 0$ for $i = 1, 2, 3, 4$. When $n \to \infty$, the Central Limit Theorem states that the probability $CWSR(m, k, EIC)$ is given by the four-variate normal orthant probability with the correlation matrix **R**. Using variances and covariances above, the correlation matrix $\mathbf{R}$ is given such that the correlation coefficient between each two random variables X_i and X_j is given by the expression $\mathbf{R}(X_i, X_j) = \frac{Cov(X_i, X_j)}{\sqrt{V(X_i)V(X_j)}}$. Therefore, we find:

$$\mathbf{R} = \begin{bmatrix} 1 & a & b & \frac{1}{2}b \\ - & 1 & \frac{1}{2}b & b \\ - & - & 1 & \frac{1}{2} \\ - & - & - & 1 \end{bmatrix}$$

where

$$a = \frac{k_1 + k_2 + k_3 + k_4}{3k_1 + 3k_2 + 2k_3 + 2k_4}, \quad and$$

$$b = \frac{\sqrt{2}\,(k_1 + k_2 + k_3\mu + k_4\,(1-\gamma))}{\sqrt{(3k_1 + 3k_2 + 2k_3 + 2k_4)\left(k_1 + k_2\,(1 - \lambda + \lambda^2) + k_3\mu^2 + k_4\,(1-\gamma)^2\right)}}$$

In the case of $k_1 = k_2 = 0$, $k_3 > 0$ and $k_4 > 0$ the correlation matrix will have the following form:

$$\mathbf{R}^* = \begin{bmatrix} 1 & \frac{1}{2} & b & \frac{1}{2}b \\ - & 1 & \frac{1}{2}b & b \\ - & - & 1 & \frac{1}{2} \\ - & - & - & 1 \end{bmatrix}$$

The joint probability $CWSR(m, k, EIC)$ that an alternative, say a, is both the Condorcet and Scoring Rule winner under EIC for n voters ($n \to \infty$) and m alternatives is given by $\Phi(2(m-1), \mathbf{R})$, where $\mathbf{R}$ is the correlation matrix determined from the four distribution X_i ($i = 1, 2, 3, 4$) and $\Phi(t, \mathbf{R})$ is defined as a t-variate normal positive orthant probability with correlation matrix $\mathbf{R}$. In our context, the symmetry of EIC regarding candidates requires that the joint probability $CWSR(3, k, EIC)$ is equal to $3\Phi(4, \mathbf{R})$.

Recall that our aims is to find the combination $\Theta = (\lambda, \mu, \gamma)$ which guaranties the greatest Condorcet Efficiency. The Condorcet Efficiency of extended scoring voting rule Θ, $P_{CE}^{\infty}(\mathbf{k}, \Theta)$, was defined as the conditional probability that the Scoring Rule elects the Condorcet winner, given that a Condorcet winner exists. That is:

$$P_{CE}^{\infty}(\mathbf{k}, \Theta) = \frac{CWSR(3, k, EIC)}{P_{Con}^{\infty}(\mathbf{k})}$$

For the correlation matrix R^*, David and Mallows (1961) have obtained an interesting representation for $\Phi(4, \mathbf{R}^*)$ as:

$$\Phi(4, \mathbf{R}^*) = \frac{1}{9} + \frac{1}{4\pi^2}\left[\sin^{-1}(b) + \sin^{-1}(\frac{b}{2})\right]\left[\sin^{-1}(b) - \sin^{-1}(\frac{b}{2}) + \pi\right]$$

Taking in consideration the remarks above, if we consider a vector $k^* = (k_1 = 0, k_2 = 0, k_3 > 0, k_4 > 0, k_5 > 0)$, the Condorcet Efficiency of Scoring Voting Rule $P_{CE}^{\infty}(\mathbf{k}^*, \Theta)$ in the case of three candidates is given by the following formula:

$$P_{CE}^{\infty}(\mathbf{k}^*, \Theta) = \frac{1}{3} + \frac{3}{4\pi^2}\left[\sin^{-1}(b^*) + \sin^{-1}(\frac{1}{2}b^*)\right] \times \left[\sin^{-1}(b^*) - \sin^{-1}(\frac{1}{2}b^*) + \pi\right]$$

where

$$b^* = \frac{\sqrt{2}(k_3\mu + k_4(1-\gamma))}{\sqrt{(2k_3+2k_4)(k_3\mu^2+k_4(1-\gamma)^2)}} \quad \textit{and} \quad P_{Con}^{\infty}(\mathbf{k}) = 1$$

We can make an important remark: when we have $\Theta^* = (\lambda, \mu = 1, \gamma = 0)$, we get $b = 1$, which implies that $P_{CE}^{\infty}(\mathbf{k}^*, \Theta^*) = 1$. This can be interpreted such that if the society is composed by only voters whom rank the alternatives a, b and c with partial indifference, the combination that maximizes the probability that the scoring winner will equal the Condorcet winner, given that the Condorcet winner exist, is such that $\mu = 1$ and $\gamma = 0$. In other word, when voters have only weak orders on the alternatives, the Condorcet winner always exists and the Approval Voting must choose it for three candidates as $n \to \infty$ under EIC assumption. In addition, for the vector $(k_1 = 0, k_2 = 0, k_3 = 0, k_4 > 0, k_5 > 0)$ the value of γ that maximizes the Condorcet Efficiency is $\gamma = 0$. Likewise, for the vector $(k_1 = 0, k_2 = 0, k_3 > 0, k_4 = 0, k_5 > 0)$, the Condorcet winner must exist and $\mu = 1$ maximizes the Condorcet Efficiency.

Now, let us consider the impact of the value of λ (i.e. when $k_1 \neq 0$ and $k_2 \neq 0$) on the Condorcet Efficiency. Then, we must find the formula of $P_{CE}^{\infty}(\mathbf{k}, \Theta)$ as $n \to \infty$. Notice that Plackett's (1954) Reduction Procedure allows us to find $\Phi(4, \mathbf{R})$ from $\Phi(4, \mathbf{R}^*)$. Let $H(t)$ be the inverse matrix of the matrix $C(t)$ such that $C(t) = t\mathbf{R} + (1-t)\mathbf{R}^*$ for $0 \leq t \leq 1$.

$$H(t) = \frac{1}{J}\begin{bmatrix} -4(b^2-1) & 2(b^2-2c) & 4b(b^2-1) & -2b(b^2-2c) \\ - & -4(b^2-1) & -2b(b^2-2c) & 4b(b^2-1) \\ - & - & \frac{4}{3}(4-5b^2-4c^2+4cb^2) & \frac{-4}{3}(2-4b^2-2c^2+5cb^2) \\ - & - & - & \frac{4}{3}(4-5b^2-4c^2+4cb^2) \end{bmatrix}$$

where

$$J = (3b^2-2c-2)(b^2+2c-2) \quad \textit{and} \quad c = \frac{1}{2} - t\frac{k_1+k_2}{2(3k_1+3k_2+2k_3+2k_4)}$$

The Plackett's reduction procedure allows us to write $\Phi(4, \mathbf{R})$ in function of $\Phi(4, \mathbf{R}^*)$ such that $\Phi(4, \mathbf{R}) = \Phi(4, \mathbf{R}^*) + I$. Therefore, for $l \neq m$ such that $l, m \in \{[1,2,3,4] - [i,j]\}$, we find I as:

$$
\begin{aligned}
I &= \frac{c'_{ij}}{4\pi^2} \int_0^1 [\frac{1}{1-c_{ij}^2}]^{\frac{1}{2}} \arccos\left(\frac{h_{lm}}{\sqrt{h_{ll}h_{mm}}}\right) dt \\
&= \frac{c'_{12}}{4\pi^2} \int_0^1 [\frac{1}{1-c_{12}^2}]^{\frac{1}{2}} \arccos\left(\frac{h_{34}}{\sqrt{h_{33}h_{44}}}\right) dt
\end{aligned}
$$

c_{ij} and h_{ij} are the entries of the matrix $C(t)$ and $H(t)$ respectively and $c'_{ij} = \frac{\partial c_{ij}}{\partial t}$. Notice that, since the correlation matrix $\mathbf{R}^*$ is chosen to be a matrix of rank (4×4) with all but one of its correlation coefficients the same as those of $\mathbf{R}$, the term I is a single integral.

Therefore, we find:

$$
\begin{aligned}
\Phi(4, \mathbf{R}) = &\frac{1}{9} + \frac{1}{4\pi^2}[\sin^{-1}(b) + \sin^{-1}(\frac{1}{2}b)][\sin^{-1}(b) - \sin^{-1}(\frac{1}{2}b) + \pi] \\
&- \frac{k_1 + k_2}{4\pi^2} \int_0^1 [4(3k_1 + 3k_2 + 2k_3 + 2k_4)^2 - (3k_1 + 3k_2 + 2k_3 + 2k_4 - tk_1 - tk_2)^2]^{-\frac{1}{2}} \\
&\times \arccos\left(\frac{4b^2 + 2c^2 - 5cb^2 - 2}{4 - 5b^2 - 4c^2 + 4cb^2}\right) dt
\end{aligned}
$$

References

Black, D. (1976). Partial justification of the Borda count. *Public Choice, 28*, 1–15.

Brams, S., Fishburn, P., & Merril, S., III. (1988a). Rejoinder to Saari and van Newenhizen. *Public Choice, 59*(2), 149.

Brams, S., Fishburn, P., & Merril, S., III. (1988b). The responsiveness of approval voting: Comments on Saari and van Newenhizen. *Public Choice, 59*(2), 121–131.

Brams, S. J., & Sanver, R. (2009). Voting systems that combine approval and preferences. In S. J. Brams, W. V. Gerhlein, & F. S. Roberts (Eds.), *The mathematics of preference, choice and order* (pp. 215–237). Berlin: Springer.

Cervone, D. P., Gehrlein, W. V., & Zwicker, W. S. (2005). Which scoring rule maximizes Condorcet efficiency under IAC? *Theory and Decision, 58*, 145–185.

Chua, V., & Huang, C. H. (2000). Analytical representation of probabilities under the IAC condition. *Social Choice and Welfare, 17*, 143–155.

David, F. N., & Mallows, C. L. (1961). The variance of Spearman's rho in normal samples. *Biometrika, 48*, 19–28.

de Borda, J. C. (1781). *Mémoire sur les Élections au Scrutin*. Histoire de l'Académie Royale des Sciences, Paris.

de Condorcet, M. (1785). *Éssai sur l'Application de l'Analyse à la Probabilité des Décisions Rendues à la Pluralité des Voix*. Paris.

Fishburn, P. C., & Gehrlein, W. V. (1980). The paradox of voting: Effects of individual indifference and intransitivity. *Journal of Public Economics, 14*, 83–94.

Gehrlein, W. V. (1997). Condorcet's paradox and the Condorcet efficiency of voting rules. *Mathematica Japonica, 45*, 173–199.

Gehrlein, W. V., & Fishburn, P. C. (1976). Condorcet's paradox and anonymous preference profiles. *Public Choice, 26*, 1–18.

Gehrlein, W. V., & Fishburn, P. C. (1978a). Coincidence probabilities for simple majority and positional voting rules. *Social Science Research, 7*, 272–283.
Gehrlein, W. V., & Fishburn, P. C. (1978b). Probabilities of election outcomes for large electorates. *Journal of Economic Theory, 19*, 38–49.
Gehrlein, W. V., & Fishburn, P. C. (1980). Robustness of positional scoring over subsets of alternatives. *Applied Mathematics and Optimization, 6*, 241–255.
Gehrlein, W. V., & Lepelley, D. (1998). The Condorcet efficiency of approval voting and the probability of electing the Condorcet loser. *Journal of Mathematical Economics, 29*, 271–283.
Gehrlein, W. V., & Valognes, F. (2001). Condorcet efficiency: A preference for indifference. *Social Choice and Welfare, 18*, 193–205.
Guilbaud, G. T. (1952). Les théories de l'intérêt général et le problème logique de l'agrégation. *Economie Appliquée, 5*, 501–584.
Ju, B.-G. (2010). Collective choice for simple preferences. In J.-F. Laslier & R. Sanver (Eds.), *Handbook of approval voting*. Heidelberg: Springer-Verlag.
Laslier, J.-F. (2010). *In Silico* voting experiments. In J.-F. Laslier & R. Sanver (Eds.), *Handbook of approval voting*. Heidelberg: Springer-Verlag.
Lehtinen, A. (2010). Behavioral heterogeneity under approval and plurality voting. In J.-F. Laslier & R. Sanver (Eds.), *Handbook of approval voting*. Heidelberg: Springer-Verlag.
Lepelley, D., Louichi, A., & Smaoui, H. (2008). On Ehrhart polynomials and probability calculations in voting theory. *Social Choice and Welfare, 30*, 363–383.
Merlin, V., & Valognes, F. (2004). The impact of indifferent voters on the likelihood of some voting paradoxes. *Mathematical Social Sciences, 48*, 343–361.
Plackett, R. L. (1954). A reduction formula for normal multivariate integrals. *Biometrika, 41*, 351–360.
Saari, D., & Tataru, M. (1999). The likelihood of dubious election outcomes. *Economic Theory, 13*, 345–363.
Saari, D. G., & van Newenhizen, J. (1988a). Is approval voting an unmitigated evil? A response to Brams, Fishburn and Merril. *Public Choice, 59*(2), 133–147.
Saari, D. G., & van Newenhizen, J. (1988b). The problem of indeterminancy in approval, multiple and truncated voting systems. *Public Choice, 59*(2):101–120.
Sanver, R. (2010). Approval as an intrisic part of preferences. In J.-F. Laslier & R. Sanver (Eds.), *Handbook of approval voting*. Heidelberg: Springer-Verlag.
Smith, J. H. (1973). Aggregation of preferences with variable electorate. *Econometrica, 41*, 1027–1041.
Tataru, M., & Merlin, V. (1997). On the relationship of the Condorcet winner and positional voting rule. *Mathematical Social Sciences, 34*, 81–90.
van Newenhizen, J. (1992). The Borda method is the most likely to respect the Condorcet principle. *Economic Theory, 2*, 69–83.
Wilks, S. (1962). *Mathematical Statistics*. New York: Wiley.
Xu, Y. (2010). Axiomatizations of approval voting. In J.-F. Laslier & R. Sanver (Eds.), *Handbook of approval voting*. Heidelberg: Springer-Verlag.

Chapter 12
Behavioral Heterogeneity Under Approval and Plurality Voting

Aki Lehtinen

12.1 Introduction

Approval voting (AV) has been defended and criticized from many different viewpoints. In this paper, I will concentrate on two topics: preference intensities and strategic behavior. A voter is usually defined as voting sincerely under AV if he or she gives a vote to all candidates standing higher in his or her ranking than the lowest-ranking candidate for whom he or she gives a vote. There are no 'holes' in a voter's approval set.[1] Since this kind of behavior is extremely rare, it has been claimed that approval voting makes strategic voting unnecessary (Brams and Fishburn 1978). On the other hand, Niemi (1984) has argued (see also van Newenhizen and Saari 1988a,b), that even though strategic voting may be rare under AV, even sincere voting may require a considerable amount of strategic thinking under this rule. If *strategic voting* is defined by the fact that a voter gives his or her vote to a candidate who is lower in his or her ranking than some candidate for whom he or she does not vote (see, e.g., Brams and Sanver 2006), I will be studying *strategic behavior* but not *strategic voting* under AV here.

In an earlier paper Lehtinen (2008), I proposed a switch of perspective. Instead of trying to study whether strategic voting or behavior is common or easy under various voting rules, I presented a computer simulation framework for investigating the welfare consequences of strategic behavior under approval and plurality (PV) voting. The utilitarian efficiencies obtained with *Expected Utility voting behavior* (EU behavior) and with *Sincere Voting behavior* (SV behavior) are compared. Under SV behavior all voters are assumed to vote for all those candidates for which the utility exceeds the midpoint of the voter's utility scale (Merrill 1979; Brams and Fishburn 1983, p. 85; Ballester and Rey-Biel 2007). Under EU behavior voters give their votes to different candidates depending on expected-gain calculations (Merrill

[1] See, e.g., Brams and Fishburn (1978, 1983, p. 29) and Brams and Sanver (2006).

A. Lehtinen
Department of social and moral philosophy, P.O. Box 24, University of Helsinki, 00014 Helsinki, Finland
e-mail: aki.lehtinen@helsinki.fi

J.-F. Laslier and M.R. Sanver (eds.), *Handbook on Approval Voting*, Studies in Choice and Welfare, DOI 10.1007/978-3-642-02839-7_12,

1981a,b). They give a vote to a candidate under EU behavior if the expected gain from doing so is positive (Merrill 1981b; Carter 1990). The distinction between strategic and sincere behavior is thus made according to whether or not voters take their beliefs concerning the winning chances of the candidates into account. They strategize if they take such beliefs into account and they engage in sincere behavior if their actions depend only on their preferences.[2] Under PV voters *vote strategically* if they give their vote to a candidate that they do not consider the best, and sincerely otherwise.

Utilitarian efficiency is defined as the percentage of simulated elections in which the candidate that maximizes the sum of voters' utilities (the utilitarian winner) is selected (e.g., Merrill 1988). The main finding in Lehtinen (2008) was that whether or not voters engage in strategic calculations, AV yields high utilitarian efficiencies and thus often selects candidates with broad public appeal (Brams and Fishburn 1983, pp. 135, 171). AV reflects preference intensities rather well even if voters engage in strategic behavior.

It was also shown that strategic voting is beneficial under PV in the sense that utilitarian efficiencies are higher under EU than under SV behavior. I have shown elsewhere that strategic voting is beneficial in many voting rules in that it increases utilitarian efficiency compared to sincere voting (see Lehtinen 2006, 2007a,b). These results mean that from a utilitarian, and thereby welfarist point of view, strategic voting under various voting rules, and strategic behavior under AV, are beneficial. However, the traditional arguments against strategic voting are non-welfarist.[3] One important argument is that 'unequal manipulative skills may lead to destruction of our efforts to design rules with equal treatments of individuals' (Kelly 1988, p. 103). The worry is thus that if some but not all voters engage in strategic manipulation, and if the strategizers are successful in their endeavor, this would be unfair towards the other voters.

In this paper, I will study one aspect of this worry with a welfarist model that allows analyzing whether or not unequal manipulative dispositions in the voting population yield undesirable results. Only one aspect of the worry is analyzed because the model does not specify different manipulative skills but rather just different propensities to manipulate.[4] Voters are assumed to be heterogeneous in the sense that some voter *types* do not engage in strategizing at all. The robustness of approval and plurality voting with respect to behavioral heterogeneity is thus investigated. To the best of my knowledge, this paper provides the first model in which such heterogeneity is explicitly studied.[5]

---

[2] Although Brams and Fishburn (1983, p. 85) use an expected-utility terminology, their mean utility rule is classified as sincere here.

[3] See Kelly's (1988, p. 103) list of arguments and their critique by Van Hees and Dowding (2007).

[4] Different skills could be studied within the framework presented here by giving some voters better information than others. For the time being, I postpone such an analysis into the future.

[5] I am hoping that someone proves me wrong here. The need for studying heterogeneous behavior in strategic voting is often expressed in conference presentations.

Strategic voting increases utilitarian efficiency in various voting rules because it allows for expressing preference intensities (Lehtinen 2006, 2007b,a). These results depend on the counterbalancing of strategic votes: broadly accepted candidates are likely to obtain many strategic votes and lose few: the strategic votes for a candidate are counterbalanced by strategic desertions for the very same candidate but the utilitarian winner is likely to be on the receiving end of strategic votes. The logic of counterbalancing thus suggests that the beneficial effects of strategic voting may not be very robust with respect to behavioral heterogeneity. In contrast, AV differs from other commonly used voting rules in that it allows for expressing intensity information even with SV behavior (e.g., Brams and Fishburn 2005). When I began this investigation, my intuition was that AV would be fairly robust with respect to behavioral heterogeneity. After all, as voters may express preference intensities under both behavioral assumptions, one would expect AV to yield high utilitarian efficiencies whatever the behavioral assumption, and even if the voting population is behaviorally heterogeneous. However, my intuitions turned out to be completely erroneous. It is indeed AV that is sensitive to behavioral heterogeneity rather than PV! Very roughly, the explanation for poor resistance of AV to behavioral heterogeneity is that strategic behavior dramatically reduces the number of second votes and such reductions do not have a proper counterbalance.

The structure of the paper is the following. Given that the paper is heavily based on my 2008 model, I will only explain its most important features in Sect. 12.2. I refer to this paper for an explanation of the details of the signal-extraction model, an account of interpersonal comparisons in the model, a discussion of reasonable parameter values, and in general for anything about the model that is not concerned with behavioral heterogeneity. Section 12.3 describes the novel feature of the present model: the *mixed behavior* computer simulations *setups*. Simulation results are presented in Sect. 12.4. Section 12.5 presents the conclusions.

12.2 Strategic Behavior under Approval and Plurality Voting

Let X = {x,y,z} denote the set of candidates (with generic members j, k and m). The six possible types of voters and their preference orderings are presented in Table 12.1. U_k^i denotes voter i's payoff for the k-th best candidate.

Under AV, voters give a vote to any number of candidates. Let $N = 2{,}000$ denote the total number of voters, and let n_j denote the number of voters who prefer candidate j the most. Let n_j^{AV} denote the number of votes candidate j obtains

Table 12.1 Voter types and utilities

Type of voter						
t_1	t_2	t_3	t_4	t_5	t_6	U^i
x	y	z	x	y	z	U_1^i
y	z	x	z	x	y	U_2^i
z	x	y	y	z	x	U_3^i

under sincere behavior under AV, and let n^{AV} denote the total number of votes cast under AV. Let v_x^{PV}, v_y^{PV}, and v_z^{PV} denote the *vote shares* of candidates x, y and z if all voters vote sincerely under PV: $v_j^{PV} = \frac{n_j}{N}$, and let v_x^{AV}, v_y^{AV}, and v_z^{AV} denote similar vote shares under AV ($v_j^{AV} = \frac{n_j^{AV}}{n^{AV}}$). Let $p_{jk}^{i,PV}$=prob(v_j^{PV}=$\mathrm{v}_k^{PV} > \mathrm{v}_m^{PV}$) denote the probability that voter i will be decisive in creating or breaking a *first-place* tie between j and k under PV, i.e., a *pivot probability*. $p_{jk}^{i,AV}$ denotes similar probabilities under AV. The standard way of analyzing strategic behavior in models in which game-theoretical considerations are not taken into account is by way of formulating expected gains for voters.

The expected gain in utility associated with voting for candidate j under AV is (Merrill 1979)

$$E_j^i = \sum_{j \neq k} p_{jk}^{i,AV} [U^i(j) - U^i(k)]. \tag{12.1}$$

Voters give a vote to a candidate if the expected gain from doing so is larger than zero (see also Merrill 1981b; Carter 1990). Voters will always give a vote for their most preferred candidate under approval voting (Brams and Fishburn 1978).[6] The conditions for strategic voting under PV can also be deduced from these equations once $p_{jk}^{i,AV}$ are replaced with $p_{jk}^{i,PV}$, see McKelvey and Ordeshook (1972). A voter votes for the candidate who offers the highest expected gain.

12.2.1 A Signal Extraction Model for the Pivot Probabilities

Voters' beliefs are derived by combining methods of computing pivot probabilities (Hoffman 1982; Cranor 1996) with a signal-extraction model. The voters are assumed to obtain an informative but not entirely reliable signal concerning the popularity of the candidates. They compute pivot probabilities on the basis of these signals and their confidence in the quality of those signals. The idea is thus to characterize the beliefs in terms of the *reliability* of the signals and voters' *confidence* in them.

Let v_j denote a generic vote share. Voters obtain perturbed signals about vote shares:

$$\mathcal{S}_j = v_j + \rho R_i, \tag{12.2}$$

where R_i denotes a standard normal random variable, and ρ is a scaling factor that reflects the *reliability* of the signals ($\rho \in [0.005, 0.013]$).[7] The signals thus contain information concerning the real preference profile and noise. The former is modelled through the vote shares v_j. Note that these are vote shares that would come about if *everyone* engaged in sincere behavior rather than vote shares that come about when

[6] Three-way ties are ignored here.

[7] I provide arguments for why such values are reasonable in Lehtinen (2008).

some or all voters engage in strategic behavior. The vote shares are different under AV and PV because voters may give sincere second votes under AV.

Let $s^i_{\max}$ denote the predicted vote share (i.e., a signal) of the candidate who i expects to obtain the most votes, and let $s^i_{min(j,k)}$ denote the predicted vote share of j or k, whichever i predicts to receive fewer votes. I show in Lehtinen (2008) that the pivot probabilities p^i_{jk} are given by the standard normal distribution function Φ:

$$p^i_{jk} = 2\Phi(\frac{i_p - s^i_{\max}}{\sigma}), \tag{12.3}$$

where i_p is a parameter derived from the various signals which describes the closeness of the race and σ is the voter's confidence in his or her signal.[8] Very roughly, the idea is that the closer the predicted vote share (i.e., the signal) for the candidate in question is to the predicted vote share of the perceived winner, the higher the pivot probability. Voters are assumed to construct a probability distribution around their signal. $i_p = \frac{(s^i_{\max})^2 - (s^i_{\min(j,k)})^2}{2(s_{\max} - s_{\min(j,k)})}$ is the intersection point of densities for the perceived winner and the candidate in question. The distance between this intersection point and the signal for the perceived winner, $i_p - s^i_{\max}$, determine how close the race between the two candidates is perceived to be by voter i.

I refer to my 2008 paper for a detailed explanation of the technical aspects of the model. For the purposes of this paper, it is sufficient to realize that the signal-extraction framework allows modeling beliefs that range from highly accurate to highly inaccurate, and at the same time taking voters' confidence in the quality of their information into account.

12.3 Simulation and Mixed Behavior Setups

A setup is a combination of assumptions used in a set of $G = 1{,}000$ simulated elections. In each simulated election, a profile $(U^1, U^2, \ldots, U^N)$ of individual utilities is generated. Under PV, the sincere vote shares of the various candidates are computed from this utility profile by ordering the utilities for the three candidates, and by counting how many voters most prefer each candidate. Under AV the sincere vote shares are computed by counting the number of voters for whom the utility lies above the midpoint of the utility scale. The voters then obtain three signals concerning the profile (one for each candidate) according to (12.2), and formulate their pivot probabilities using (12.3). They then use (12.1) to compute the expected gains, and vote accordingly. The winner is then determined and compared to the utilitarian winner.

[8] The confidences are usually assumed to be the same for all voters.

Expected utility setups differ with respect to the reliability of voters' signals (ρ), their confidence in the signals (σ), and the degree of correlation between voter types and preference intensities (C) (see the next paragraph). In *uniform setups* voters' utilities are drawn from a uniform distribution on [0,1],[9] while in *setups with intensity correlation* voter types 3 and 5 have systematically higher and types 1 and 6 systematically lower preference intensities for their second-best candidates x and y respectively. These setups are identical to the corresponding uniform setups with respect to all parameters except voters' preference intensities. In order to generate setups with a correlation between this parameter and voter types without affecting the interpersonal comparisons or the preference orderings, the individual utilities were derived as follows.

U_1, U_2, and U_3 were first generated from the uniform distribution on [0,1] for each voter. U_1 and U_3 were then used for defining the voter's utility scale as the [U_3,U_1] interval. A voter's utility for his or her middle candidate U_2 is referred to as the *intensity*. A *standardized intensity*, $\widetilde{U}_2$ expresses what a voter's utility for his or her second-best candidate would be if the scale was the [0,1] interval. These standardized second-best utilities are referred to as *intra*personal *intensities*. The relationship between the standardized intra-personal utility and the original scale of utility is given by

$$\widetilde{U}_2 = 1 - \frac{U_1 - U_2}{U_1 - U_3}. \tag{12.4}$$

In setups with an intensity correlation, these standardized intensities were multiplied by a parameter C, $0.5 < C \leq 1$ for those who put y second (voter types 1 and 6) so that the new correlated intensities $\widetilde{U}_2^{C,1}$ and $\widetilde{U}_2^{C,6}$ were given by

$$\widetilde{U}_2^C = C\widetilde{U}_2. \tag{12.5}$$

In order to compensate for the decreases in utility for voter types 1 and 6, the intensities for voters of types 3 and 5 (i.e., for x) were given by

$$\widetilde{U}_2^C = 1 - C\widetilde{U}_2. \tag{12.6}$$

These adjustments made the average utilities for x higher and the average utilities for y lower than in the uniform setups, while keeping the overall average utility fixed.[10] In uniform setups, $C = 1$. C thus denotes the *degree of correlation* between preference intensities and voter types.

These standardized intensities were then scaled back into the original $[U_3, U_1]$ utility scale. Let U_2^* denote a voter's correlated intensity expressed in terms of the original $[U_3, U_1]$ scale. U_2^* is given by:

$$U_2^* = U_3 + \widetilde{U}_2^C (U_1 - U_3). \tag{12.7}$$

[9] The simulations were thus based on the so-called impartial anonymous culture assumption.

[10] Note that the utility for the second-best candidate in uniform setups is $1 - \widetilde{U}_2^C$ rather than $\widetilde{U}_2^C$. Since $\widetilde{U}_2^C$ is drawn from a uniform distribution on [0,1], it does not matter which one is used.

In *pure behavior setups* (PBS) all voters engage in the same kind of behavior: either EU or SV behavior. In *mixed behavior setups* (MBS) some voters engage in SV behavior and some in EU behavior. The simplest MBS is one in which voters who engage in SV behavior are randomly selected from the set of all voters.

More interesting results are likely if only some voter *types* engage in EU behavior, or if only some voter types engage in SV behavior. In *abstaining setups* all voters except those of two particular types engage in EU behavior, and these abstaining types engage in SV behavior. Let $A_R(st)$ denote a setup in which voters of types s and t engage in SV behavior, and the rest engage in EU behavior under voting rule R. Similarly, in *engaging setups* all voters except those of two particular types engage in SV behavior, and these two types engage in EU behavior. A setup in which only types s and t engage in EU behavior is denoted $E_R(st)$.

12.4 Simulation Results

12.4.1 Non-systematic Behavioral Heterogeneity

The simulations were run with 0.005, 0.009, and 0.013 for both σ and ρ. The results will be shown only for the setups in which $\rho = \sigma$.[11]

A setup in which one-half of all voters were randomly selected to engage in EU behavior, and the rest in SV behavior was tried. Figures 12.1 and 12.2 show utilitarian efficiencies under AV and PV, respectively, when the probability of any given voter to engage in EU behavior is 0.5. UE_{SV} and UE_{EU} stand for utilitarian efficiency under SV- and EU behavior, respectively. Let EA_R(random) denote such a setup. Let us say that behavioral heterogeneity is *systematic* if there are systematic differences between the different voter types with regard to behavioral dispositions, and *non-systematic* otherwise. The setups in this section thus concern non-systematic behavioral heterogeneity.

It is easy to see from these figures that strategic behavior under AV and strategic voting through EU behavior under PV yield reasonably high utilitarian efficiencies. They are higher under AV, and particularly so under SV behavior. The reason for this is rather simple. Candidate x is practically always the utilitarian winner in setups in which correlation between intensities and voter types is high (C is small), but because the voting population is generated with the impartial anonymous culture (IAC), under PV it is selected only in one-third of the simulated elections under SV behavior. Under AV, however, voters are able to express preference intensities also under SV behavior, and the utilitarian efficiencies are correspondingly higher. These setups, while they may depict real-world elections in a realistic way, are not very interesting because the results simply reflect the relationships that hold under the pure behavior setups, but in a mitigated form.

[11] The full sets of data are available from the author on request.

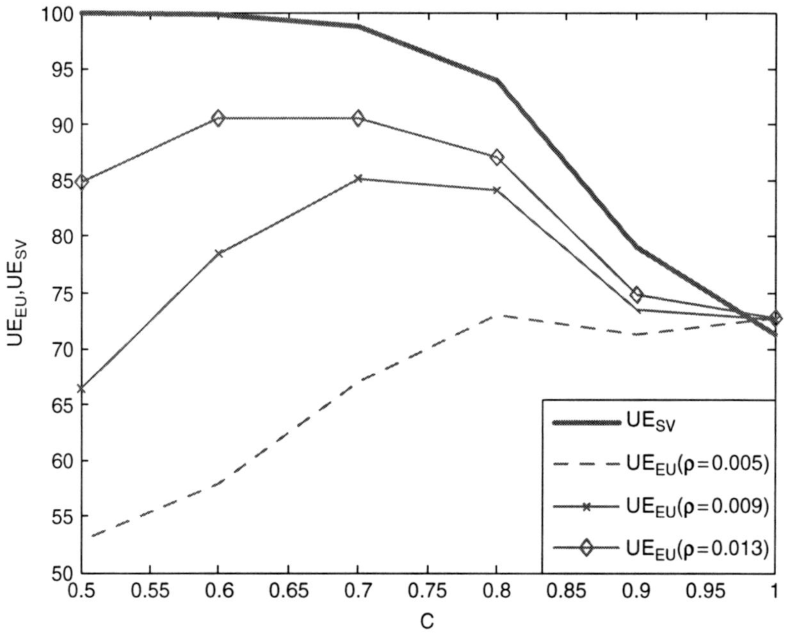

Fig. 12.1 Utilitarian efficiencies under EA_{AV}(random)

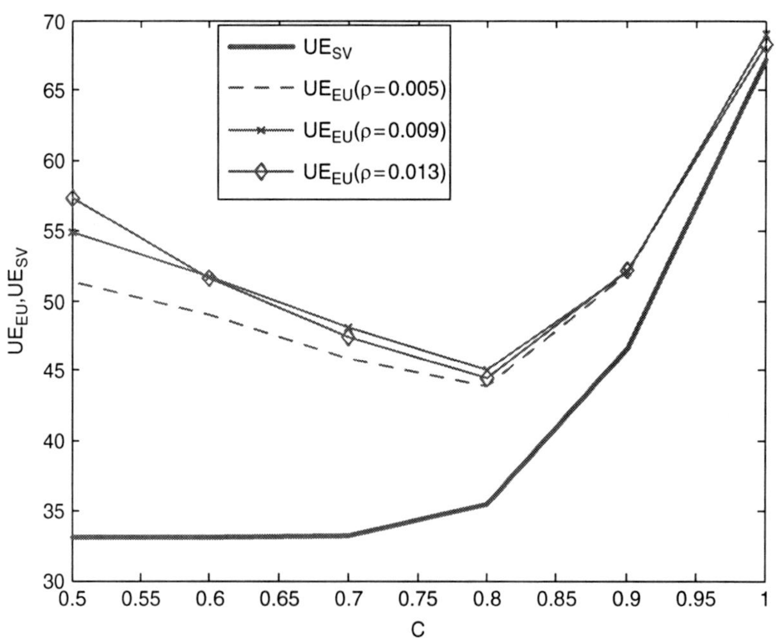

Fig. 12.2 Utilitarian efficiencies under EA_{PV}(random)

12.4.2 Systematic Behavioral Heterogeneity

12.4.2.1 Engaging Setups: Plurality Voting

The investigated setups were chosen in such a way as to provide the maximum amount of understanding on how various different heterogeneities affect the utilitarian efficiencies. In most setups only two illustrative voter types were selected to engage in SV behavior or EU behavior. The setups discussed below are not very realistic in that *all* voters within each voter type are assumed to engage either in EU or in SV behavior. It is highly likely that reality is much more complex in this respect. As the model is based on non-cooperative behavior, it is not assumed that there is a coordinating agent who could enforce one or the other behavioral assumption within a voter type.

The logic of counterbalancing suggests that the utilitarian efficiencies should be lower under most MBSs than under PBSs because these setups are constructed in such a way that the counterbalance is systematically removed. In most MBS's the utilitarian efficiencies are indeed lower than in the corresponding pure behavior setups.

Let us start by looking at PV. Figures 12.3, 12.4, and 12.5 show the results when two voter types only engage in EU behavior and the rest in SV behavior. In what

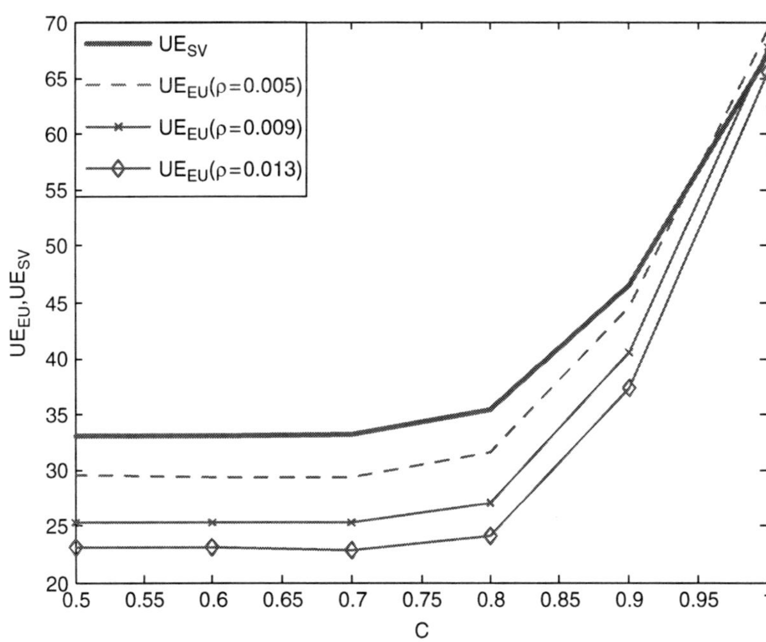

Fig. 12.3 $\mathrm{E}_{PV}(14)$

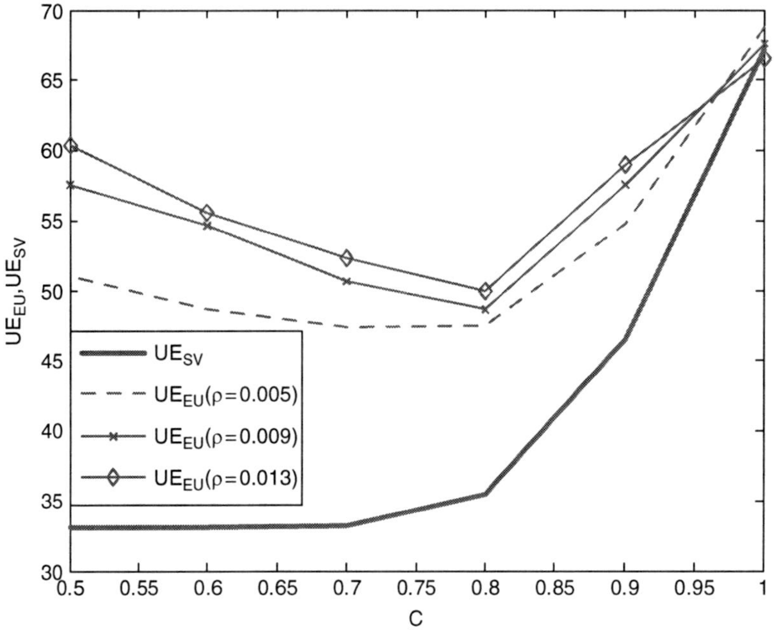

Fig. 12.4 $\mathrm{E}_{PV}(25)$

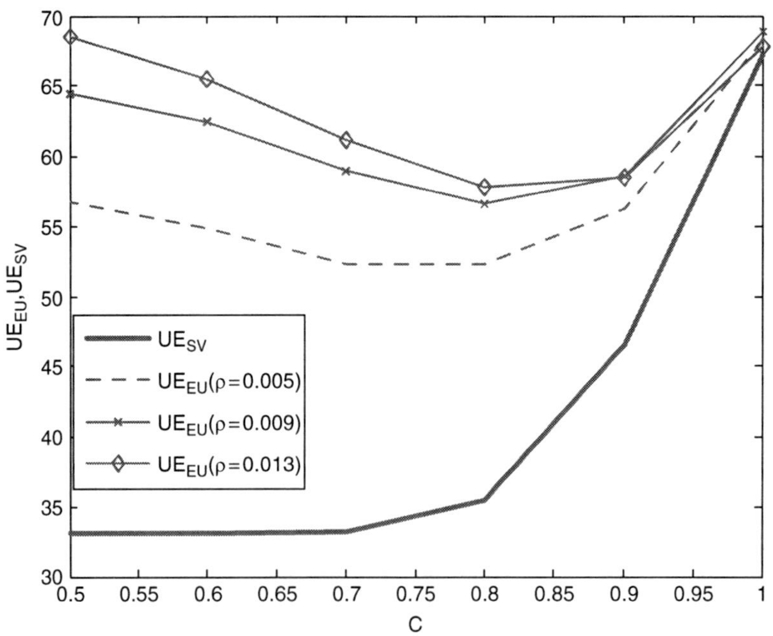

Fig. 12.5 $\mathrm{E}_{PV}(36)$

follows, the figure titles include only the name of the setup: all results concern utilitarian efficiencies.

Strategic voting becomes more welfare-increasing under $E_{PV}(36)$ than under E_{PV}(random), remains roughly the same under $E_{PV}(25)$, and it becomes welfare-diminishing under $E_{PV}(14)$. Explaining these findings is easy once the logic of counterbalancing is invoked. First, under $E_{PV}(14)$ only voters who prefer x the most engage in strategic voting. But x is usually the utilitarian winner in setups with strong correlation. Welfare-increasing strategic voting is thus theoretically possible only in those $E_{PV}(14)$ setups in which the correlation is not very high (i.e., C is close to one), and in which x is not the utilitarian winner. In all other setups strategic voting can only be harmful because it may only *decrease* the probability that the utilitarian winner wins. Second, under the $E_{PV}(36)$ setups there is a proper counterbalance: even though voters of type 6 may vote strategically for y, they do so much more seldom than voters of type 3 vote for x. Utilitarian efficiencies are higher than under the pure behavior setups because the 'wrong' kind of counterbalance is removed. Note that from the point of view of utilitarian efficiency, it is more important that there are not too many voters who vote strategically for z than those who vote strategically for y. This is because there may often be enough strategic votes for z to make it win, but y is usually the loser in any case. This also explains why utilitarian efficiencies are somewhat lower under the $E_{PV}(25)$ than the $E_{PV}(36)$ setup. Here strategic votes for z rather than for y counterbalance those for x.

Figures 12.6, 12.7, and 12.8 show the findings from the setups in which voter types who engage in strategic behavior consider the same candidate second-best.

As one might expect by now, the highest utilitarian efficiencies come from the $E_{PV}(35)$ setup where x is the only candidate to obtain strategic votes in the first place, and the worst from $E_{PV}(16)$ where y is the only candidate in this position.

Note that even though strategic voting is welfare-diminishing in some setups, the results shown thus far have been rather supportive of PV. If the main worry about strategic voting is that it benefits one particular group at the expense of everyone else, then the results show that this worry is mainly not warranted. In the $E_{PV}(14)$, the strategic voters hurt mostly themselves by their actions! They prefer x the most, but their strategic voting makes it less likely that x will be selected. It is thus clear that if they were to have perfect information about the behavioral propensities of the different voter types, they would switch to SV behavior. In a word, their strategic voting is not model-consistent because if voters knew that they are the only ones who engage in strategic behavior, they would realize that they have no incentive to act according to strategic behavior as it is specified in the model.[12] Another way to approach the issue is to note that since the signals depend on voters' preferences but not on their behavioral propensities, they give a systematically misleading picture of the winning chances of the various candidates.[13] I do not attempt to provide an

[12] Model-consistency is also known as the rational expectations hypothesis (Muth 1961).

[13] However, note that if type-1 voters vote strategically for y, and it emerges as winner, their prior beliefs are corroborated by the outcome!

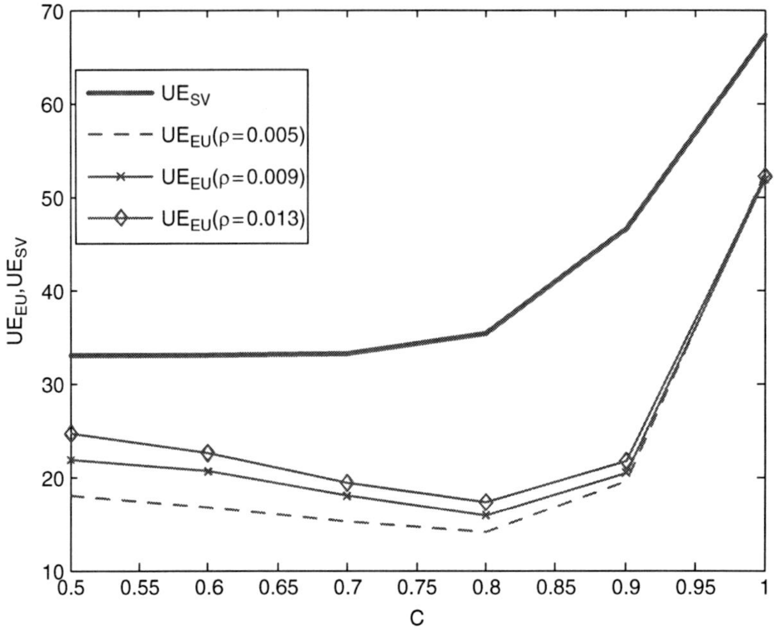

Fig. 12.6 $E_{PV}(16)$

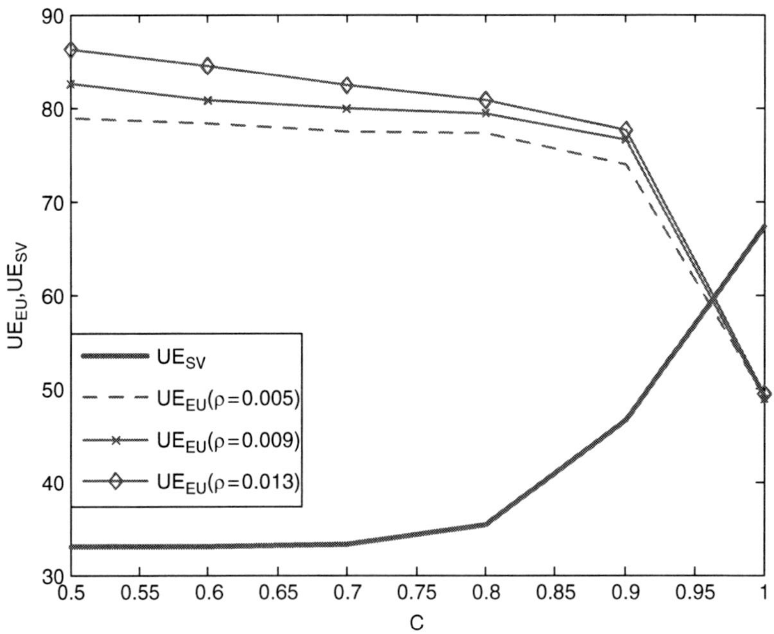

Fig. 12.7 $E_{PV}(35)$

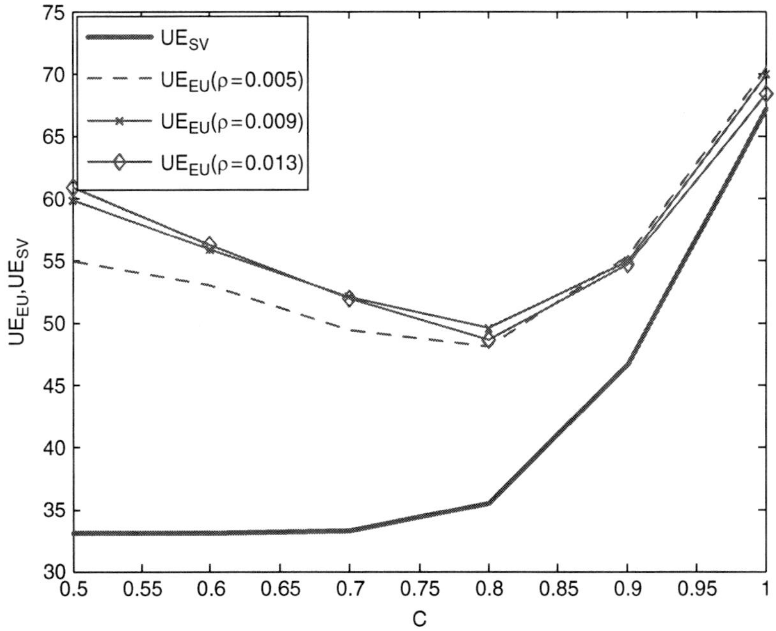

Fig. 12.8 $\mathrm{E}_{PV}(24)$

account in which the behavioral propensities are taken into account in a formal way in this paper. $\mathrm{E}_{PV}(16)$ causes more concern than $\mathrm{E}_{PV}(14)$ because it is not always the same candidate who loses the strategic votes. Nevertheless, even in this setup the outcomes are usually better under the pure behavior SV setup for the very types that engage in strategic voting, and they have an incentive to switch into sincere behavior. It is inevitable that someone must lose as a result of strategic behavior, but the results show that under an utilitarian evaluation, strategic voting is welfare-decreasing only when it harms the strategisers themselves.

12.4.2.2 Engaging Setups: Approval Voting

Let us now see what happens under AV in engaging setups. Figures 12.9, 12.10, and 12.11 show the utilitarian efficiencies under AV.

The utilitarian efficiencies are completely different from those under PV: they are highest in E(14) setups, and lowest under E(36) setups. In other words, strategic behavior under AV yields low utilitarian efficiencies precisely when strategic voting is particularly welfare-increasing under PV, and vice versa.

The key to understanding these results lies in the difference in the number of voters who give a second vote under SV behavior and under EU behavior (cf. Saari 2001). Many voters give second votes under SV behavior. Under the uniform setups

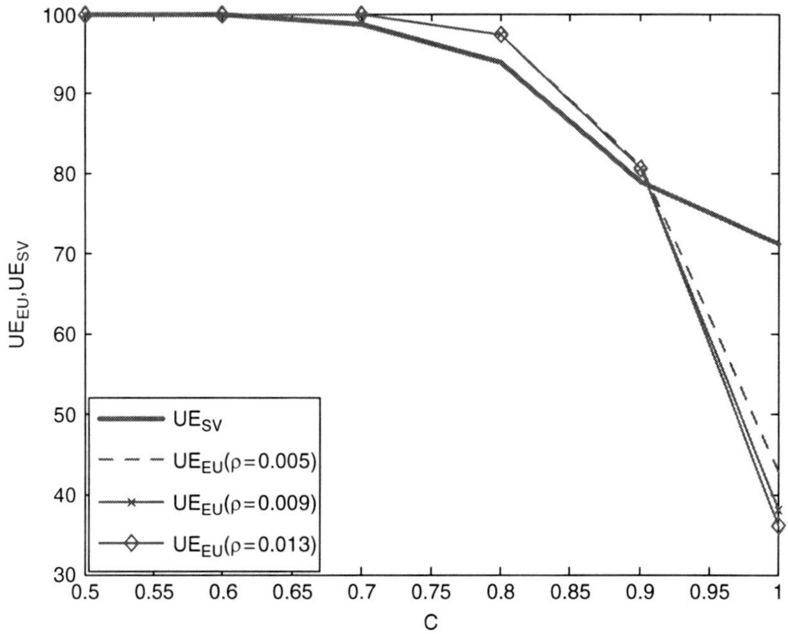

Fig. 12.9 $E_{AV}(14)$

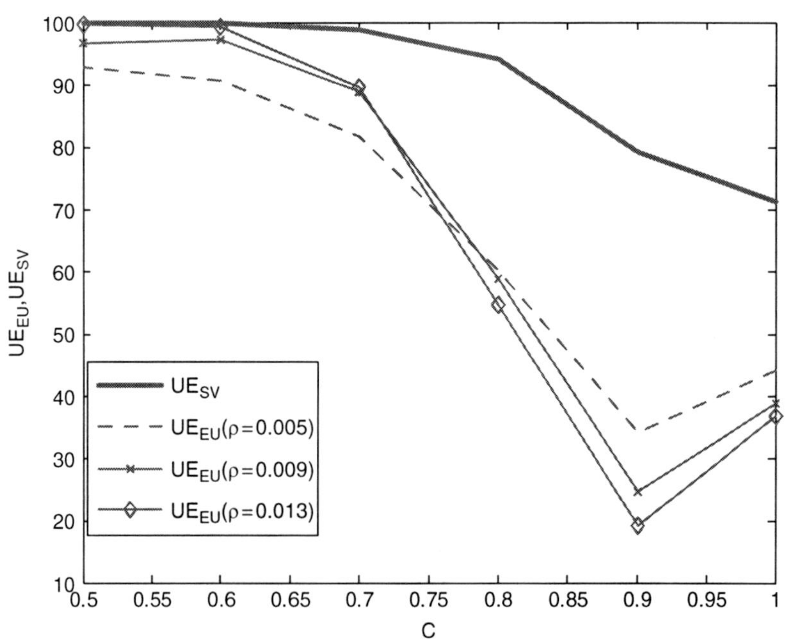

Fig. 12.10 $E_{AV}(25)$

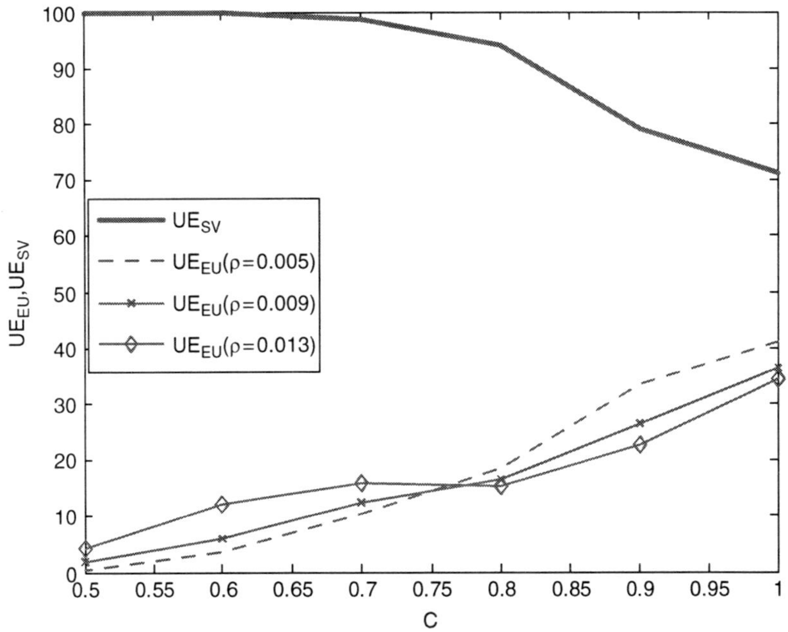

Fig. 12.11 $E_{AV}(36)$

exactly one-half of each voter type do so. Under EU behavior voters continue to give second votes, but they do so much more rarely. This reduction in the second votes is the main consequence of strategic behavior under AV. In pure behavior setups the utilitarian efficiencies are rather high because counterbalancing still ensures that the utilitarian winners obtain more second votes than the other candidates. However, in the E(36) setups, although x obtains more second votes from type-3 voters than y obtains from type-6 voters, what really matters is the dramatic reduction in second votes for x (compared to SV behavior), together with the fact that z obtains all the second votes it does under SV behavior. z is thus almost always the winner in these setups. In E(25) setups the counterbalancing is rectified by the fact that the reduced number of second votes from type-5 voters is counterbalanced by the reduced number of second votes for z from type-2 voters. It is thus more important that the reduced number of second votes for the utilitarian winner are counterbalanced by a similar reduction for the *second-best* candidate (in utilitarian terms) than the *worst*. The reason for this is that in the engaging setups there are still four voter types who give different amounts of second votes, and counterbalancing among these second votes is more important than counterbalancing among the strategically determined second votes.

Although the findings seem to support AV superficially, the setups in which strategic behavior is welfare-diminishing are in fact more worrisome than under PV. Consider, for example, $E_{AV}(36)$. This is a setup in which those who prefer z the most are the only ones to engage in strategic behavior. They give much fewer second votes to x (and y) than under sincere behavior. As a consequence, their best candidate z often wins. Unlike in the $E_{PV}(14)$ and in the $E_{PV}(16)$ setups, upon learning the behavioral differences between the voter types, they would not have an incentive to switch into SV behavior. $E_{AV}(36)$ is thus a setup in which the one group of voters is indeed able to inflict harm on others by strategising: if they acted sincerely, the results would be better for the whole electorate.

12.4.2.3 Abstaining Setups

Figures 12.12, 12.13, and 12.14 show utilitarian efficiencies under setups in which two voter types abstain from strategic behavior. As expected, $A_{PV}(35)$ exemplifies a catastrophe because the only voter types to abstain from strategic behavior are those that may vote strategically for x under EU behavior. But why are efficiencies higher under $A_{PV}(24)$ than under $A_{PV}(16)$? The reason is again that strategic votes for the *second-best* candidate are more likely to lower utilitarian efficiency than those for the worst candidate, because the worst candidate rarely wins the election anyway. Thus, under $A_{PV}(24)$ those who might vote strategically for z refrain from doing so but under $A_{PV}(16)$ such voters would have voted strategically for y. Figures 12.15, 12.16, and 12.17 show the corresponding results under AV.

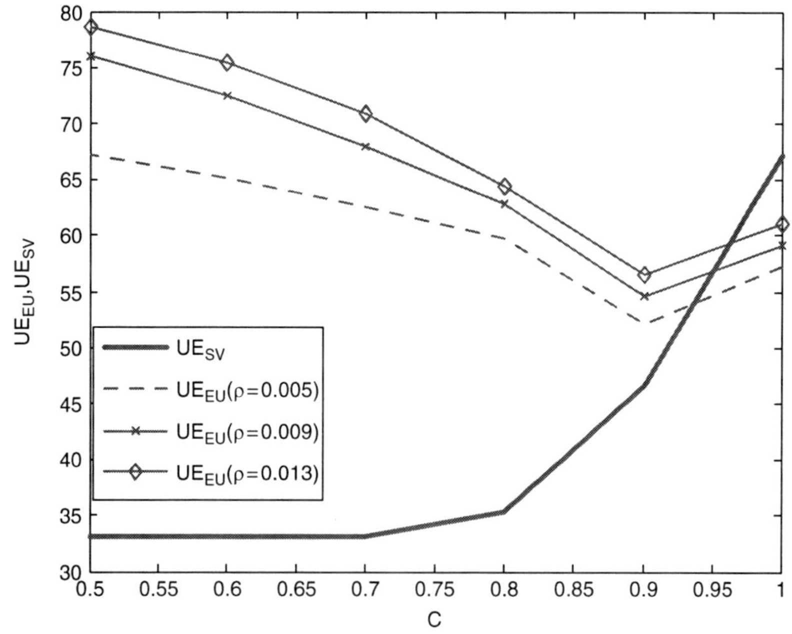

Fig. 12.12 $A_{PV}(24)$

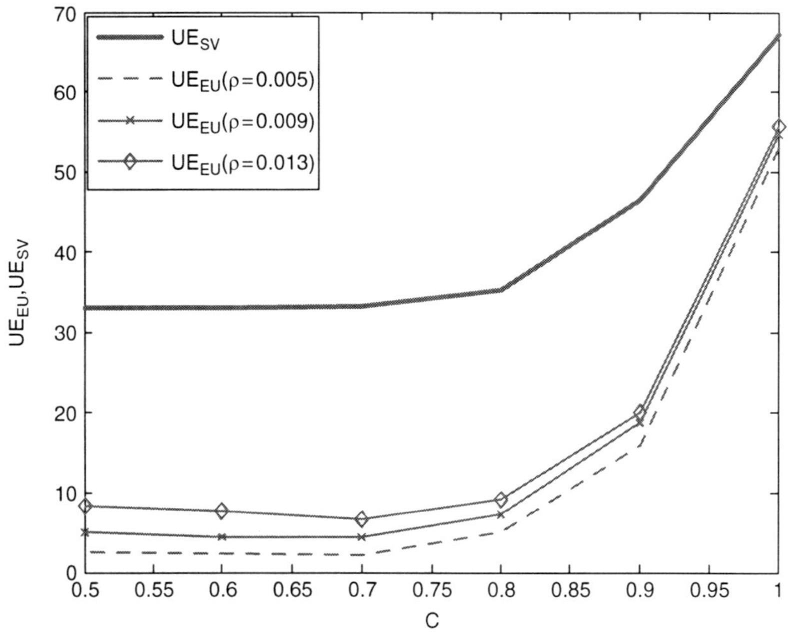

Fig. 12.13 $A_{PV}(35)$

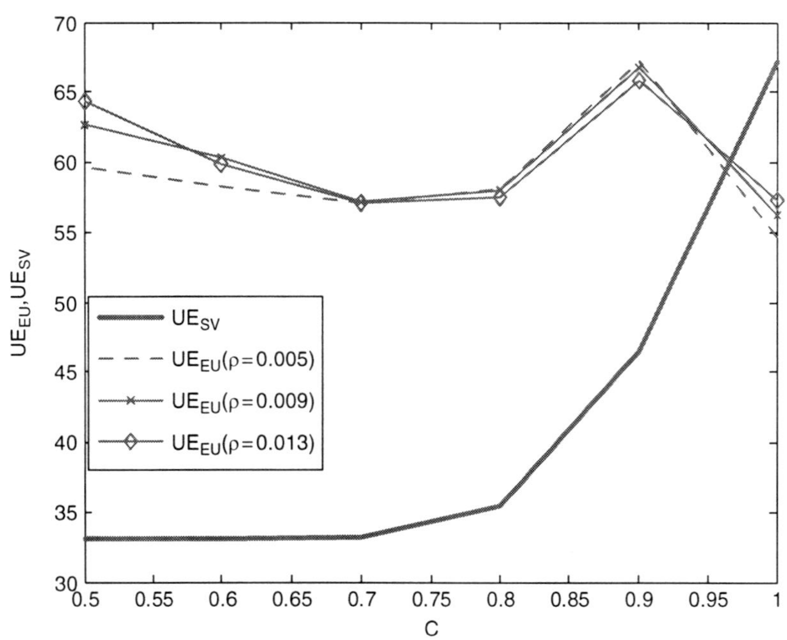

Fig. 12.14 $A_{PV}(16)$

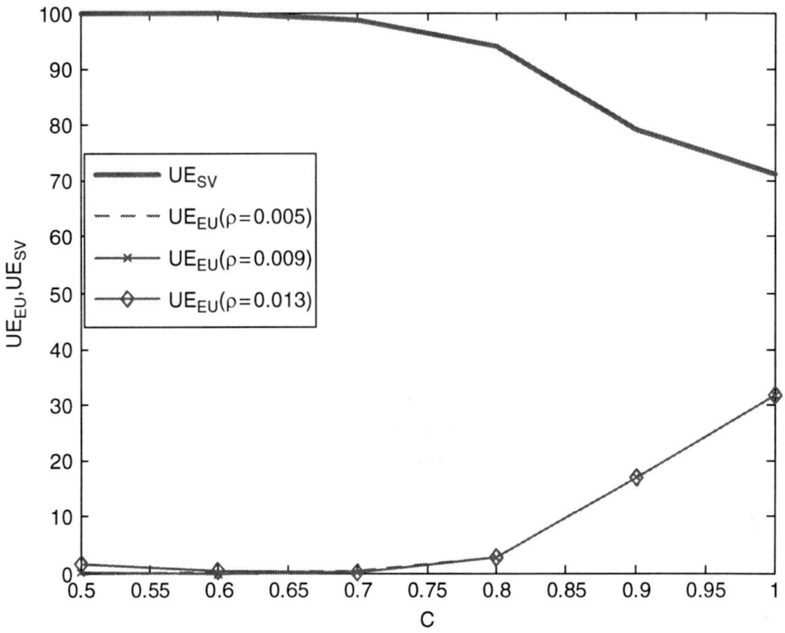

Fig. 12.15 $A_{AV}(24)$

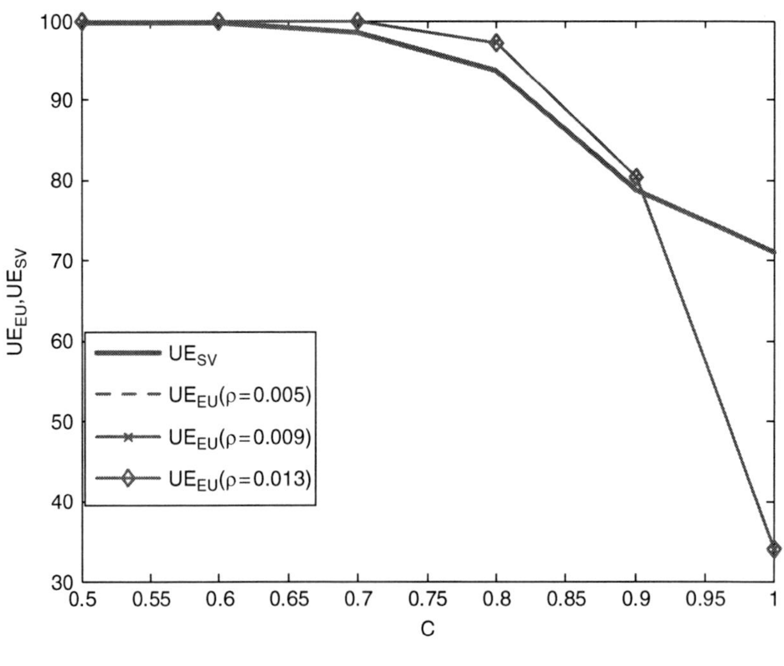

Fig. 12.16 A_{AV} (35)

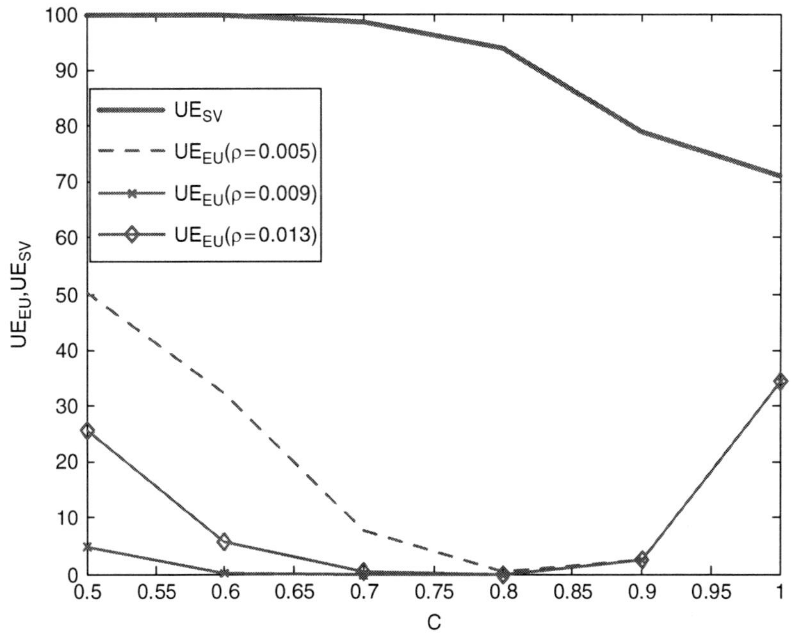

Fig. 12.17 $A_{AV}(16)$

The utilitarian efficiencies are now very low except in $A_{AV}(35)$ where those who put x second refrain from strategic behavior and give plenty of second votes for x. Note that $A_{AV}(24)$ and $A_{AV}(16)$ are setups in which those who refrain from strategic behavior consider the *same* candidate second-best. Hence, if they abstain from strategising, this will often result in the victory of their second-best candidate. The utilitarian efficiencies are low in setups where that second-best alternative is not the utilitarian winner. Furthermore, the efficiencies are lower under $A_{AV}(24)$ than under $A_{AV}(16)$ because x is able to win some elections even when those who put y second give their sincere second votes for it, but x has no chance against z because there are more of those who give their sincere second votes to z under $A_{AV}(24)$ than those who give their sincere second votes to y under $A_{AV}(16)$.

Let us now look at setups in which those who refrain from strategic behavior consider the same candidate best. Figures 12.18, 12.19, and 12.20 show utilitarian efficiencies in $A_{AV}(14)$, $A_{AV}(25)$, and $A_{AV}(36)$ setups.

It seems clear that utilitarian efficiencies remain high if at least some voter types give sincere second votes to x, but if the only types that abstain from strategic behavior put x first, then utilitarian efficiencies are understandably very low because y and z obtain a large number of sincere second votes from type-1 and type-4 voters, and the strategic second votes from the other voter types are not a sufficient counterbalance to these sincere votes.

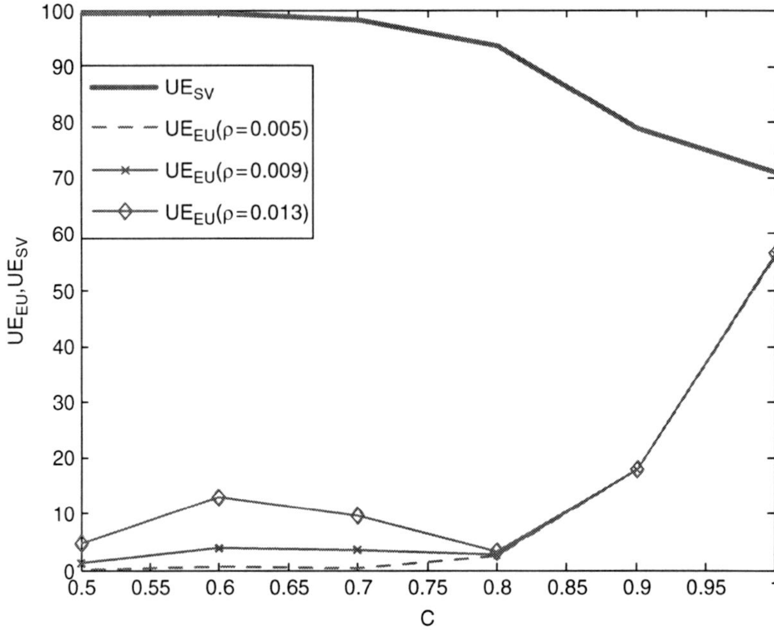

Fig. 12.18 $A_{AV}(14)$

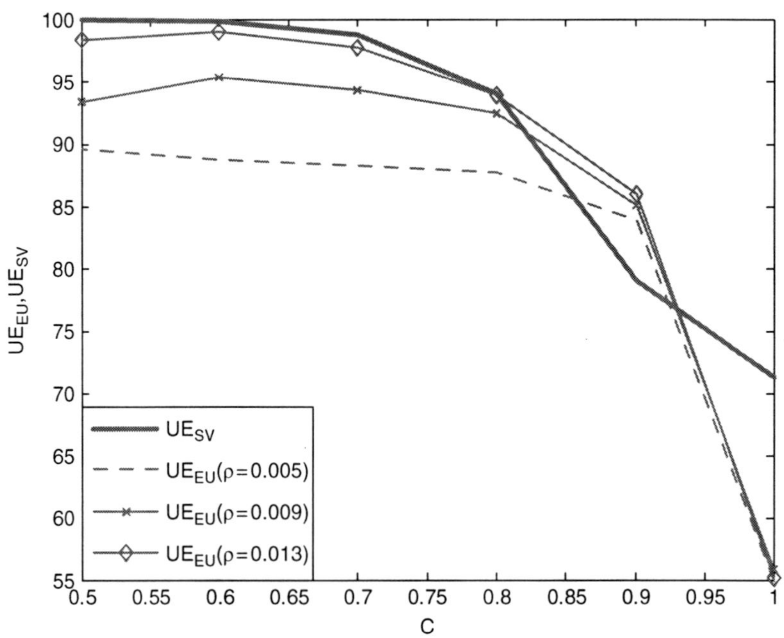

Fig. 12.19 $A_{AV}(25)$

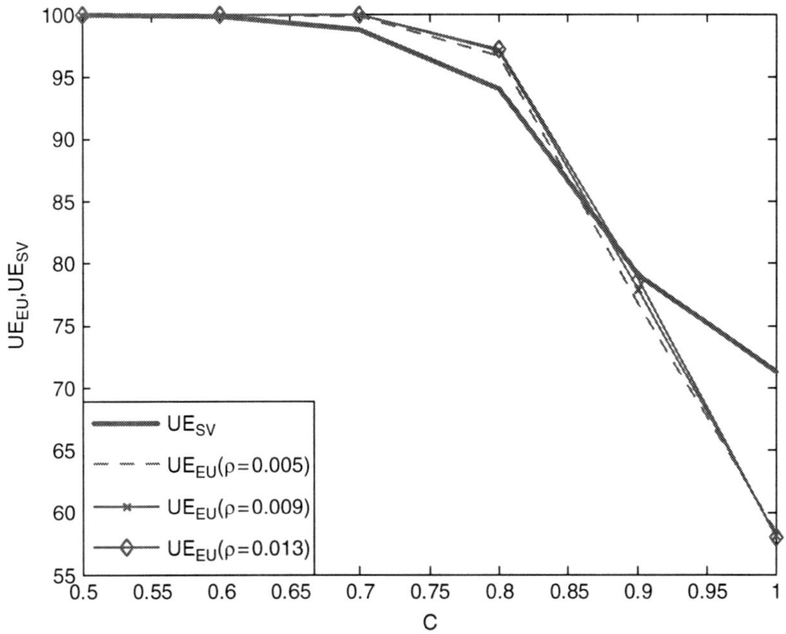

Fig. 12.20 $A_{AV}(36)$

12.4.3 A Comparison

The previous findings have provided detailed information concerning how the different combinations of behavioral assumptions matter for utilitarian efficiency. It may be somewhat difficult to derive an overall judgement concerning the two rules on the basis of them. In order to provide an explicit comparison, setups in which two randomly selected voter types engage in EU behavior were investigated. Let E_R(random) denote such a setup under voting rule R. Figures 12.21 and 12.22 show the findings from such setups.

The utilitarian efficiencies remain somewhat higher under PV than under AV. Perhaps the most important aspect of these results is that, on average, strategic voting remains welfare-increasing even in setups with the most extreme kind of behavioral heterogeneity. A simulation was run also for the case in which two randomly selected voter types abstained from strategic behavior, and the rest engaged in EU behavior. The results were highly similar to those in Figs. 12.21 and 12.22, and will thus not be shown here.

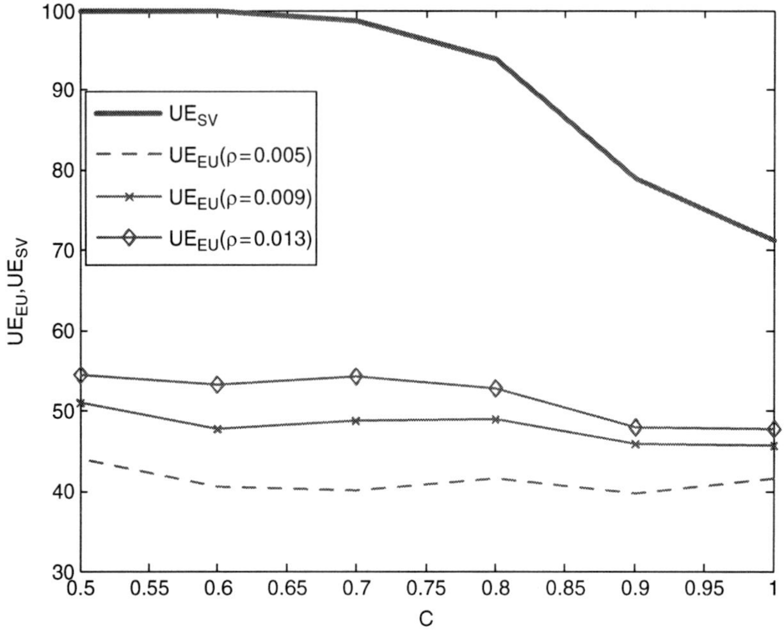

Fig. 12.21 E_{AV}(random)

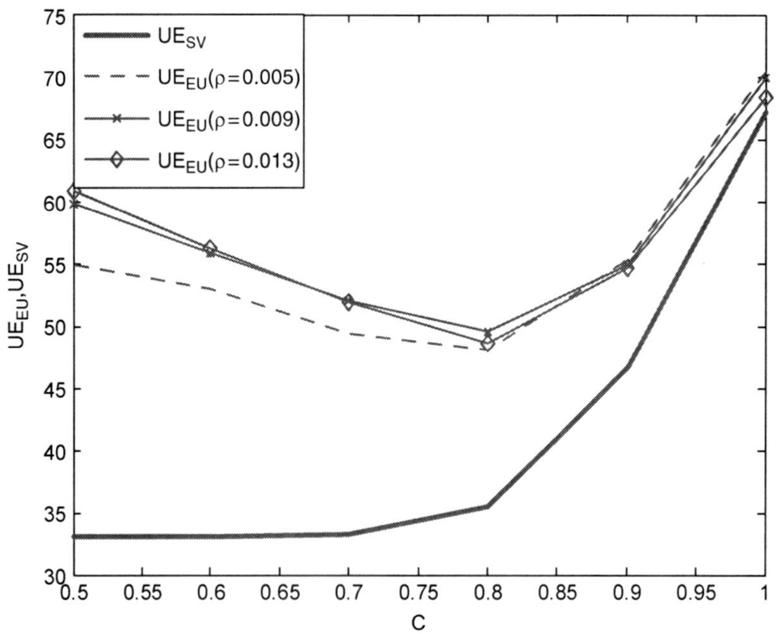

Fig. 12.22 E_{PV}(random)

12.4.3.1 The Consequences of Intensity Information in the Signals

As explained in detail in Lehtinen (2008), all the simulations discussed thus far are unrealistic for two reasons. First, it is psychologically unrealistic to assume that voters engage in strategic voting if they consider the second-best candidate almost as bad as the worst one. Second, given that the signals already contain some information on the preference intensities under AV but not under PV, the previous setups are likely to yield lower utilitarian efficiencies for PV than for AV. To rectify these weaknesses in the model, voters were also assumed to obtain some intensity information under PV, and to vote strategically only if their intensity exceeds a threshold-level τ. As in Lehtinen (2008), the threshold was assumed to be rather low: $\tau = 0.2$.

Let U denote the sum of utility for all candidates, and U(j) the sum of utility for candidate j. Let $\lambda \in [0.1]$ denote the relative share of intensity information in the signals. A *composite signal* consists of a combination of preference and intensity information, and a random term:

$$S_{i,j} = \lambda v_j + (1-\lambda)\frac{U(j)}{U} + \rho R_i, \tag{12.8}$$

where R_i and ρ have the same interpretations as before. When $\lambda = 1$, the pivot probabilities are based only on information on preference orderings under PV. The findings from simulations with full information are shown in Figs. 12.23 and 12.24.

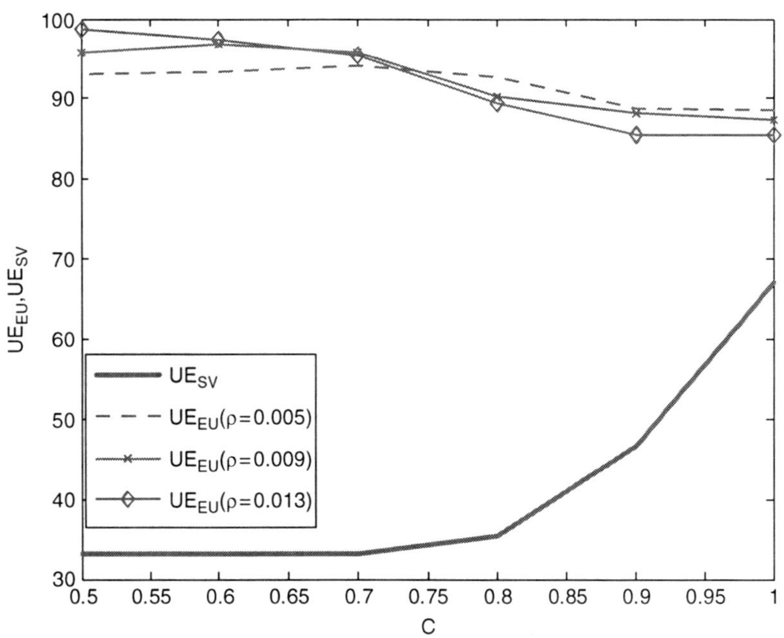

Fig. 12.23 $E_{PV}(\text{random}, \tau = 0.2, \lambda = 0)$

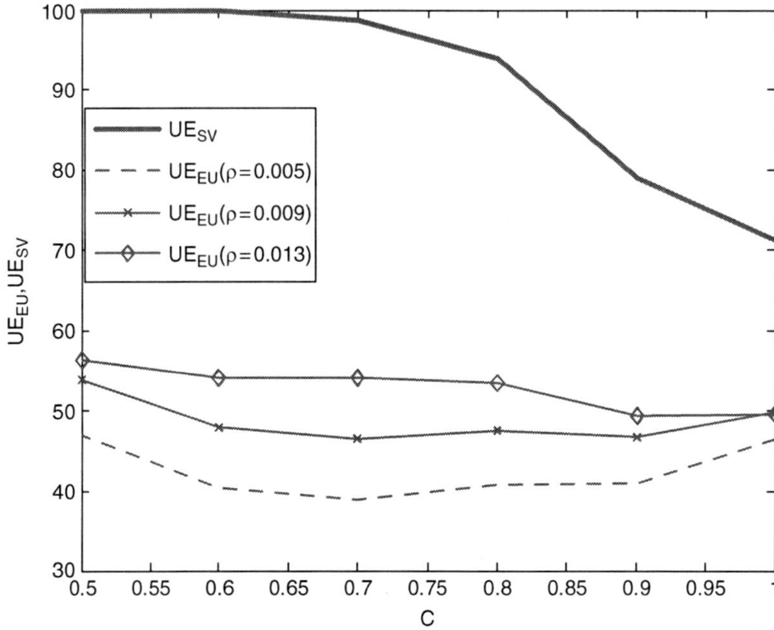

Fig. 12.24 $E_{AV}(\text{random}, \tau = 0.2, \lambda = 0)$

Under PV the utilitarian efficiencies are considerably higher in setups with full intensity information ($\lambda = 0$), but full intensity information is not all that important under AV: the utilitarian efficiencies remain relatively low.

12.5 Conclusions

As expected, utilitarian efficiencies are lower in the mixed behavior setups than in pure behavior setups. The results depend heavily on which voter types engage in strategic and sincere behavior. Strategic voting and strategic behavior continue to be welfare-increasing in many mixed behavior setups, but in some cases strategic behavior leads to a catastrophe.

The findings are somewhat surprising. AV is much more sensitive to behavioral heterogeneity than PV. The main reason is that under the standard specification of sincere behavior, many second votes are given under AV. Strategic behavior decreases the number of second votes dramatically, and if only some voter types abstain from giving second votes, the reduction in these second votes is often sufficient to change the winner. If the reduction in second votes concerns the utilitarian winner, it is often not selected. Even though there is counterbalancing among the strategically given second votes, this does not matter so much because the difference in the number of sincere and strategic second votes trumps the counterbalancing among the strategic second votes.

When strategic voting is welfare-diminishing under PV, the voter types that engage in it typically obtain a worse outcome for themselves than they would have obtained under the pure behavior SV setup. As such voters would not have an incentive to continue to vote strategically if they knew that they are the only ones to do so, it does not seem very likely that such strategic voting will be found in the real world: strategic voting is only welfare-diminishing under PV when voters who engage in it do not act in a model-consistent fashion. The worry that some particular groups would be able to benefit from strategic voting at the expense of everyone else thus really has to be formulated in a non-welfarist way: when particular groups benefit from strategic voting, they typically increase the overall welfare at the same time.

The consequences of behavioral heterogeneity are usually exactly the opposite in the two voting rules: when EU behavior is welfare-increasing in a mixed behavior setup under PV, it is welfare-decreasing under AV, and vice versa. It is then not surprising that when strategic behavior is welfare-diminishing under AV, the voter types that engage in it typically obtain a better outcome for themselves than they would have obtained under the pure behavior SV setup. This means that those voters really have an incentive to engage in strategic behavior. The worry about unequal manipulative propensities thus turned out to be an argument against AV.

The findings concerning the comparison of AV and PV can be summarized as follows. AV yields higher utilitarian efficiencies than PV when there is no behavioral heterogeneity or when heterogeneity is of the non-systematic type. PV is much more resistant to systematic heterogeneity, particularly if voters obtain perturbed information on preference intensities. An overall judgment concerning preference intensities and strategic behavior in the two rules depends on the relative magnitude between the various parameters. Given that there seems to be no particular reason why behavioral heterogeneity should be of the systematic type, it may be that the findings reported here are not so devastating for AV after all. An empirical investigation concerning behavioral heterogeneity might provide a fuller picture.

References

Ballester, M. A., & Rey-Biel, P. (2007). *Sincere voting with cardinal preferences: Approval voting* (working paper).

Brams, S. J., & Fishburn, P. C. (1978). Approval voting. *The American Political Science Review, 72*(3), 831–847.

Brams, S. J., & Fishburn, P. C. (1983). *Approval voting*. Boston: Birkhäuser.

Brams, S. J., & Fishburn, P. C. (2005). Going from theory to practice: The mixed success of approval voting. *Social Choice and Welfare, 25*(2–3), 457–474.

Brams, S. J., & Sanver, M. R. (2006). Critical strategies under approval voting: Who gets ruled in and ruled out. *Electoral Studies, 25*(2), 287–305.

Carter, C. (1990). Admissible and sincere strategies under approval voting. *Public Choice, 64*(1), 43–55.

Cranor, L. F. (1996). *Declared-strategy voting: An instrument for group decision-making*. PhD thesis, Department of Engineering and Policy, Sever Institute of Technology, University of Washington.

Hoffman, D. (1982). A model for strategic voting. *SIAM Journal on Applied Mathematics, 42*(4), 751–761.
Kelly, J. S. (1988). *Social choice theory: An introduction*. Berlin: Springer-Verlag.
Lehtinen, A. (2006). Signal extraction for simulated games with a large number of players. *Computational Statistics and Data Analysis, 50*, 2495–2507.
Lehtinen, A. (2007a). The Borda rule is also intended for dishonest men. *Public Choice, 133*(1–2), 73–90.
Lehtinen, A. (2007b). The welfare consequences of strategic voting in two commonly used parliamentary agendas. *Theory and Decision, 63*(1), 1–40.
Lehtinen, A. (2008). The welfare consequences of strategic behaviour under approval and plurality voting. *European Journal of Political Economy, 24*(3), 688–704.
McKelvey, R. D., & Ordeshook, P. C. (1972). A general theory of the calculus of voting. In J. F. Herndon & J. L. Bernd (Eds.), *Mathematical applications in political science* (pp. 32–78). Charlottesville, VA: University Press of Virginia.
Merrill, S. I. (1979). Approval voting: A 'best buy' method for multi-candidate elections? *Mathematics Magazine, 52*(2), 98–102.
Merrill, S. I. (1981a). Strategic decisions under one-stage multi-candidate voting systems. *Public Choice, 36*(1), 115–134.
Merrill, S. I. (1981b). Strategic voting in multicandidate elections under uncertainty and under risk. In M. J. Holler (Ed.), *Power, voting, and voting power* (pp. 179–187). Würzburg: Physica-Verlag.
Merrill, S. I. (1988). *Making multicandidate elections more democratic*. New Jersey: Princeton University Press.
Muth, J. (1961). Rational expectations and the theory of price movements. *Econometrica, 29*(3), 315–335.
Niemi, R. G. (1984). The problem of strategic behavior under approval voting. *The American Political Science Review, 78*(4), 952–958.
Saari, D. G. (2001). Analyzing a nail-biting election. *Social Choice and Welfare, 18*(3), 415–430.
Van Hees, M., & Dowding, K. (2007). In praise of manipulation. *British Journal of Political Science, 38*, 1–15.
van Newenhizen, J., & Saari, D. G. (1988a). Is approval voting an 'unmitigated evil'? A response to Brams, Fishburn, and Merrill. *Public Choice, 59*(2), 133–147.
van Newenhizen, J., & Saari, D. G. (1988b). The problem of indeterminacy in approval, multiple, and truncated voting systems. *Public Choice, 59*(2), 101.

Chapter 13
In Silico Voting Experiments

Jean-François Laslier

13.1 Introduction

This chapter is devoted to computation-based simulations of voting. To perform such a simulation requires two things. On the one hand, one has to specify what might be called the "economic environment," that is the number of voters, the number of alternatives, and the voter preferences (or tastes, values, utilities, opinions, ...) over the alternatives. On the other hand, one has to specify the decision process, which is itself made of two ingredients: firstly the material decision procedure, for instance the formal voting rule, and secondly the individual behavior, that is how a voter decides to place in the urn one ballot rather than another, given her preferences and any other relevant information.

The random generation of a profile of voter preferences is usually called a *culture*. For instance choosing n individual preferences uniformly and independently among the $K!$ linear orderings of K alternatives is called the impartial culture of size (n, K). Several different cultures will be studied in this chapter. Different cultures may be relevant for different research purposes and to model different real-life voting situations: juries, project assessment, committee decisions, political elections, ... The individuals will often be called "voters" and will receive female pronouns, and I shall sometime refer to alternatives as "candidates" and use male pronouns in that case.

The second ingredient is the decision process, here: *voter behavior* under a *voting rule*. For instance, under the familiar Plurality rule, a voter can vote for her preferred candidate, not taking into account what she knows about the relative chances of winning of the various candidates. This behavior called sincere voting is well defined under Plurality rule (up to indifferences). Different decision processes, sometime including strategic considerations will be studied. I will restrict attention to some practical voting rules which, up to unavoidable ties, select a winner: Plurality, Borda, Copeland and Approval voting. I will not compute choice correspondences like the Uncovered set, the Essential set or the Yolk which are set-valued in practice.

J.-F. Laslier
Laboratoire d' Económétrie, École Polytechnique, 91128 Palaiseau, France
e-mail: jean-francois.laslier@polytechnique.edu

J.-F. Laslier and M.R. Sanver (eds.), *Handbook on Approval Voting*, Studies in Choice and Welfare, DOI 10.1007/978-3-642-02839-7_13,

Section 13.2 introduces the four kinds of cultures this chapter deals with: (1) Common interest cultures, in the tradition of Rousseau and Condorcet; (2) Impartial culture, often considered by mathematicians and social choice theorists; (3) Distributive cultures, suitable for the study of the "Divide a dollar" problem and interesting for Normative Economics; and (4) Spatial Euclidean cultures, often met in Theoretical Politics. Section 13.3 introduces the voting rules and voter behavior under scrutiny. Section 13.4 contains the results and Sect. 13.5 concludes.

13.2 Cultures

13.2.1 Rousseauist Cultures

In this section I model individual preferences that may differ because of mistakes individuals make when forming their opinion about a pre-existing truth. This the typical approach of Condorcet: differences of opinions are due to differential information or to mistakes with respect to some underlying truth that collective decision-making can discover. Such a conceptual framework is called "project assessment" by Nurmi and Salonen (2008). It is the framework of the original "Condorcet Jury theorem" (Condorcet 1785) whose philosophy follows from Rousseau's ideal (Rousseau 1762) of a "general will." I therefore refer to Rousseau and call such cultures Rousseauist cultures.

To model the notion of individual mistakes I suppose that there exists an underlying true ranking of the alternatives, say

$$1 \succ 2 \succ \ldots \succ K$$

and that each voter is correct with probability $p(k,k') = p(k',k)$ when comparing k and k'. I suppose that these mistakes are independent from one voter to another. The alternative number one, the "true" best alternative will be called the *Rousseau* alternative.

I use the two-parameter formulation of Truchon and Drissi-Bakhkhat (2004) and Truchon (2008) which states that for some $\alpha \geq 0$ and $\beta \geq 0$, for all $k < k'$, the mistake probability is the following function of the rank difference $k' - k$:

$$p(k,k') = \frac{e^{\alpha+\beta(k'-k-1)}}{1+e^{\alpha+\beta(k'-k-1)}}.$$

Note that for $k < k'$:

$$1/2 \leq p(k,k') < 1.$$

I suppose that each voter gives one consistent opinion on each pair: if individual i reports that she prefers k to k', she does not also reports that she prefers k' to k. But I do not require the individual to be consistent across pairs: If i reports that she prefers k to k' and k' to k'', she may reports that she prefers k'' to k (of course this implies that she has made at least one mistake). This framework is called the *Rousseauist culture* of size (n, K) and parameters (α, β).

The information reported by each individual is therefore a tournament (complete and asymmetric binary relation) over the set of alternatives The most likely reported tournament is the true ranking of the alternatives.

For $\beta = 0$ the probability of a mistake does not depend on the ranks of the alternatives, as in Young (1988). The above model with $\beta \geq 0$ is more flexible and it seems reasonable to postulate that the probability of an error is larger when comparing two alternatives closer one to each other in the true underlying ranking. When β or α is large, $p(k, k')$ tends to one, which means that the voter's expertise is very good.

With a number n of voters, the individual preferences over pairs of alternatives define a vote matrix M of size $K \times K$, in which the entry $m_{k,k'}$ is the number of voters who prefer k to k', with $m_{k,k'} + m_{k',k} = n$. For convenience one can define on the diagonal $m_{k,k} = n/2$.

A *Condorcet winner* can be defined in this framework:[1] it is an alternative k such that $m_{k,k'} \geq n/2$ for all k'. In the simulations I will always take n odd, so that there is no need to distinguish strict from large inequalities in this definition. As usual a Condorcet winner needs not to exist but, if it exists, it is unique.

Any rule based on pairwise comparisons may be computed in this framework. For instance the *Borda rule* may be applied, even if some individual preferences are not transitive: as it is well known, the Borda score of an alternative k is the sum:

$$bs(k) = \sum_{k'=1}^{K} m_{k,k'}$$

and a *Borda winner* is and alternative with highest Borda score.[2]

13.2.2 Impartial Culture

The *Impartial culture* for n voters and K alternatives is obtained by choosing each individual preference at random uniformly among the $K!$ linear orderings of the alternatives, and independently of the preferences of the other voters. One thus obtains the uniform probability distribution over the set of profiles of linear orders. In this culture, there is a complete symmetry among alternatives: learning something on the relative ranking by some individuals of some alternatives gives no information on the other individuals or alternatives.

---

[1] Even if some individual rankings are not transitive.

[2] The reader will easily make the connection with the other, equivalent, definition of the Borda rule using sum of ranks.

This mathematically simple culture has been widely studied in the social choice literature. For instance it is known that the probability in this culture of the existence of a Condorcet winner decreases with the number of voters and the number of alternatives. See Gehrlein and Fishburn (1979) or Gehrlein (1997).

13.2.3 Distributive Cultures

Distributive cultures describe societies of complete antagonism. They are generated as follows. One unit of a divisible good (a "cake") has to be shared among n individuals. Each individuals wants her share to be as large as possible, and does not care about the other shares. The set of alternatives is here infinite, it is the n-simplex:

$$\Delta_n = \left\{ x \in \mathbb{R}^n : 0 \leq x_i, \sum_{i=1}^{n} x_i = 1 \right\}.$$

Theoretical models of redistributive politics (Lindbeck and Weibull 1987; Myerson 1993; Lizzeri 1999; Laslier 2002; Laslier and Picard 2002) use economic environments which are identical or related to this set of alternatives. There is no obvious "natural" probability distribution over this set, and I will use several such distributions, presented in the next paragraphs.

13.2.3.1 Consensual Redistributive Culture

Here, I use the projection on the simplex Δ_n of the uniform distribution on the cube $[0,1]^n$. This distribution is most easy to simulate: one chooses at random (independently and uniformly between 0 and 1) numbers y_i^k for $i = 1, \ldots, n$ and $k = 1, \ldots, K$ and then computes

$$x_i^k = \frac{y_i^k}{\sum_{j=1}^{n} y_j^k}.$$

Ties can be neglected so that this process defines a random profile of linear orders on K alternatives for n voters by setting

$$x^k \, P_i \, x^{k'} \iff x_i^k > x_i^{k'}.$$

Note that even if this culture describes a situation of complete antagonism, it is not clear wether alternatives in this culture are typically very unequal distributions or close to the equal split. A first observation is that, when the number n of individuals is large, the probability distribution on Δ_n tends to concentrate around the point of equal division $(1/n, 1/n, \ldots, 1/n)$. This can be seen in the simulations by computing the standard deviation

$$d(x) = \sqrt{\frac{1}{n}\sum_{i=1}^{n}(x_i - \frac{1}{n})^2}$$

which is also the Euclidean distance between the point x and the equal division $(1/n, 1/n, \ldots, 1/n)$. This random quantity tends, in expectation, to 0 when n is large. (See the Appendix.) Therefore this culture may be seen as consensual, or even egalitarian, because it describes a society who tends to imagine solutions to the pure redistribution problem which are close (according to the Euclidean distance) to the perfectly egalitarian one.

But inequality is usually measured not by the standard deviation but by specific indices such as the Gini index of inequality. Let u_i be the share of the i-th poorest individual and let v_i be the total share of the i-th poorest individuals:

$$u_1 \leq u_2 \leq \cdots \leq u_n$$

$$v_i = \sum_{j \leq i} u_j.$$

In the case where the shares are all identical, $u_i = 1/n$ and $v_i = i/n$. The increasing numbers v_i, for $i = 1, \ldots, n$, define the concentration of the distribution and one can measure how concentrated (or "unequal") the distribution is by the Gini index

$$gini(x) = \frac{2}{n}\sum_{i=1}^{n}\left(\frac{i}{n} - v_i\right).$$

This coefficient[3] is between 0 and 1.

If one measures inequality by the Gini index of inequality, one reaches a different conclusion: When n is large the expected value of the Gini index tends to 1/3. This is a non-degenerated value and, arguably, a relatively small one. For these reasons I chose to call this culture the *consensual* distributive culture. Figure 13.1 shows the distribution of the values taken by this index in a society of 99 individuals.

Empirical averages for the standard deviation and the Gini index are provided in Table 13.1.

13.2.3.2 Inegalitarian Distributive Cultures

In order to introduce these cultures, consider first that the scale of shares obtained by the individuals is fixed and linear: the poorest individual gets some amount t, the

[3] I use here a slightly simplified version of the Gini index. For a discrete distribution, the exact formula is $\frac{2}{n}\sum_{i=1}^{n}\left(\frac{i-1/2}{n} - \frac{v_i+v_{i-1}}{2}\right)$.

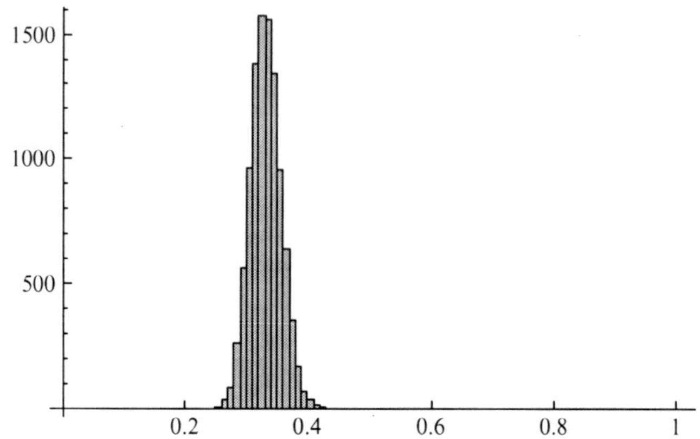

Fig. 13.1 Consensual distributive culture, 99 individuals: histogram for the Gini index

Table 13.1 Consensual distributive culture: standard deviation and Gini index depending on the number of voters

n	3	5	11	49	99	999
Standard deviation	0.16	0.10	0.05	0.01	0.006	0.00006
Gini index	0.25	0.28	0.31	0.33	0.33	0.33

second poorest gets $2t$, the third gets $3t$, and so on up to the richest individual who gets nt. Then the total amount is

$$(1 + 2 + \cdots + n)t = \frac{n(n+1)}{2}t$$

and, in order that the individual shares add to 1, one sets $t = \frac{2}{n(n+1)}$.

A feasible alternative is the assignments of these shares to individuals. There are thus $n!$ different possible alternatives. This defines a culture if these fixed shares are randomly assigned to the individuals. One picks at random K of these redistributions, independently and uniformly to define the *linear-inegalitarian distributive culture* of size (n, K). Note that in this culture, unlike the previous case the amount of inequality is the same in any alternative.

Starting from this idea, one can generalize it and define more or less egalitarian redistributive cultures by changing the linear scale $(t, 2t, \ldots, nt)$ to a non linear one; then one can measure inequality by the usual Gini index. This is what I will do now.

To generate preference profiles, consider a one-parameter family of concentration curves

$$x \mapsto x^e$$

for $e \geq 1$. In a society of n individuals with such concentration, the i poorest individuals together get

Table 13.2 Gini coefficient depending on the parameter e

e	1	1.2	1.5	2	3	4	5	6
Gini	0	0.1	0.2	0.33	0.5	0.6	0.66	0.7

$$v_i = \left(\frac{i}{n}\right)^e$$

which means that the poorest individual has $u_1 = \left(\frac{1}{n}\right)^e$, the second poorest has $u_2 = \left(\frac{2}{n}\right)^e - \left(\frac{1}{n}\right)^e$, and so on up to the richest individual who has $u_n = 1 - \left(\frac{n-1}{n}\right)^e$.

The inequality of such a redistribution depends on e (it is approximately independent of n). The Gini coefficients are given in Table 13.2.

Typical real values for the Gini index of income distributions at the national level are 0.25 in Sweden, 0.33 in France, 0.45 in the US, 0.59 in Brazil and more than 0.7 in some African countries.

I define the $(e_{\min}, e_{\max})$-*inegalitarian distributive culture* of size (n, K) as follows. For each k independently (with $1 \leq k \leq K$) a parameter e_k is picked at random uniformly on the interval $[e_{\min}, e_{\max}]$, then alternative k is chosen according to the concentration parameter e_k, by assigning randomly the specified shares to the individuals.

As mentioned above, distributive cultures are interesting models of Politics: an alternative is a political platform that offers some amount to the different voters. One problem with this approach is that it is not reasonable to imagine that actual political platforms can target individual voters one by one. But certainly they can target social groups. To this respect, remark that the distributive culture introduced in this section well describes a situation where there are not n individuals but n *groups* of individuals, the groups being of equal size. Each of the K candidates then chooses to favor more or less the various groups. For this reason I find pertinent, as a model of large politics, to consider inegalitarian distributive cultures for relatively small values of the parameter n.

13.2.4 Spatial Cultures

These cultures stem from the spatial theory of voting. In the Euclidean space $\mathbb{R}^d$ with d dimensions, each voter i has a bliss point ω_i and a utility function defined on $\mathbb{R}^d$ which is decreasing with the distance to ω_i:

$$u_i(x) = -\|x - \omega_i\|$$

An alternative is a point in $\mathbb{R}^d$. A culture is defined by the number of dimensions d and the probability distributions for n bliss points and K alternatives. The next paragraphs describe the spatial cultures that will be considered in this paper.

13.2.4.1 Uni-dimensional Spatial Culture

As it is well known, for $d = 1$, a preference profile generated by the above single-peaked utility functions always has a Condorcet winner, which is the available alternative closest to the median bliss point (see for instance Austen-Smith and Banks 1999 for details). This culture is an easy way to generate profiles with this property. In the simulations I pick bliss points at random uniformly in the interval $[0, 1]$.

13.2.4.2 Multi-dimensional Cultures

With more than one dimension, it may be the case or not that a Condorcet winner exists. From the theory (McKelvey 1986), one may expect that (as soon as there are more than three voters) if the number of candidate is large, the probability that one of them is a Condorcet winner becomes very small; but this clearly cannot be true in general and indeed depends on the probability distributions on bliss points and alternatives. For simplicity I will use only uniform distributions over bounded boxes (intervals, rectangles,...), with the same probability distributions for bliss points and for alternatives.[4]

If the box is very thin, then the profiles become similar to one-dimensional profiles. If the rectangle or the box is far from degenerated, the uniform choice of bliss points and alternatives may lead to think that, if the numbers of voters and alternatives are large, the profile can be qualitatively described by a model with a continuum of voters and alternatives uniformly distributed. This convergence question is a delicate theoretical issue. McKelvey and Tovey (2010) study the convergence problem in the case of the Yolk and Tovey (2010) provides further mathematical insights on the same problem. The simulation results presented below give some empirical clues for this question.

13.3 The Decision Process: Voting Rule and Behavior

I will be essentially interested in Plurality rule, the Borda rule, the Copeland rule and Approval Voting, when voters vote sincerely or strategically. Strategic behavior is introduced in a heuristic way as "responsive voting" without reference to equilibrium considerations.

[4] Multi-dimensional Gaussian distributions are another option, used by Merrill (1984) for simulations and by Laslier (2006) for the theory.

13.3.1 Sincere Voting

I consider three voting rules for which sincere behavior is defined naturally and without ambiguity. The Plurality rule, the Borda rule and the Copeland rule, a familiar Condorcet-consistent aggregation rule. For Approval Voting, sincere voting (in the usual definition) does not provide a well-specified behavior. Behavior under this rule must be responsive and is thus described in the next section.

For a profile of strict preferences, the *Plurality* score of an alternative k is simply the number of individuals whose best-preferred alternative is k. Plurality rule defines as winners the alternatives with largest plurality score. If preferences are not strict, the definition is naturally completed by saying that if d distinct alternatives tie at the first rank in an individual preference ordering, then this individual gives, in the Plurality count, $1/d$ point to each of them. If the individual preferences are not transitive, the Plurality rule has no straightforward extension and I shall not use it.

Some pieces of notations will be useful. Recall that $m_{k,k'}$ denotes the number of voter who (sincerely) prefer alternative k to alternative k'. The collum-sum of the matrix $M = \left(m_{k,k'}\right)$ provides the candidates' *Borda* scores.

$$bs(k) = \sum_{k'=1}^{K} m_{k,k'}.$$

Replacing the matrix M by the matrix $T = \left(t_{k,k'}\right)$, with

$$t_{k,k'} = \begin{cases} 1 & \text{if } m_{k,k'} > n/2 \\ 0 & \text{if not} \end{cases}$$

one obtains the "tournament" matrix where $t_{k,k'} = 1$ means that a majority of voters prefers k to k'. The *Copeland* score $cs(k)$ of an alternative k is the number of other alternatives k beats. It is easily computed from the tournament matrix.

$$cs(k) = \sum_{k'=1}^{K} t_{k,k'}$$

whose possible values ranges from 0 if k is a Condorcet loser to $K-1$ if k is a Condorcet winner. A Copeland winner is an alternative with maximal Copeland score. If there exists a Condorcet winner, this alternative is the unique Copeland winner.

13.3.2 Responsive Voting

Here the voters respond to an announced candidate score vector. The proposed reaction functions are derived from the theory of strategic voting: the voter holds some belief on the other voters' actions and rationally responds to this belief. The available information is essentially the same for all voters: it is a public signal about the popularity of the various candidates. In reality, such a signal is derived from the results of previous elections, from pre-electoral polls, or from any similar public information. Although different behaviors may appear as "rational" behavior within some fully specified game-theoretic models, the choices made here have the advantage of being comparable among different voting rules and cultures. In particular the public signal always takes the form of an announced ranking of the candidates. I do not need to suppose that a voter knows the other voters' preferences, or holds beliefs about them. In order to know what to do, a voter only has to figure out what the others do.

The important point is that individual rational behavior cannot be defined, except in general terms, knowing only the preferences. The answer to the question "Is it rational for me to cast this ballot" depends on what I believe the other voters decide. It follows that a simulation approach to strategic voter behavior has to take the form of what is called here "responsive voting."

13.3.2.1 Plurality Voting

Given announced scores for the various candidates, the voter votes for her preferred candidate among the two candidates with highest scores. If ties occur at the first places in the score vector, I introduce a small noise in the score vector to randomly break the ties and let the voter decide among two candidates only. Then, clearly, votes gather on two candidates only. If one of these two candidates is a Condorcet winner then this candidate wins, but it is possible that a Condorcet winner exists but votes nevertheless gather on other candidates. More exactly any candidate k except a Condorcet loser can be elected, provided that people believe that votes gather on k and some other candidate k' which is losing in front of k according to majority rule. On that point, see Cox (1997) and Myerson (2002) for the theory and Blais et al. (2008) for experiments. The theory, if not predictive, delivers a clear-cut message here, the path-dependence effect is so important that simulation work does not appear to be of interest. I will therefore not study strategic response in the case of Plurality voting.[5]

13.3.2.2 Approval Voting

I use the strategic best-response function introduced and justified in Laslier (2009). Given approval scores for the various candidates (and if there are no ties in the first places) the voter considers the top-ranked candidate k_1, the "leader." She votes for or

[5] Lehtinen (2008) studies Plurality and Approval in some three-alternative societies, with voters' out-of-equilibrium beliefs based on perturbation of sincere voting vote shares. He concludes that Approval voting has a high utilitarian efficiency and Plurality has a low utilitarian efficiency, which is improved by strategic behavior. See also Lehtinen's contribution in this volume.

against the other candidates by comparing them to k_1. In order to decide whether she votes for or against the leader k_1 himself, she compares k_1 to k_2, the second-ranked candidate (the "challenger"). (See the example below.)

This response function has a fixed point if and only if there exists a Condorcet winner. In that sense, strategic approval voting is Condorcet-consistent. If there is no Condorcet winner, the best response function is still well-defined, but the beliefs have to be specified. In the simulations I compute the first five iterations of this function, starting from a Condorcet-consistent sincere rule (I report on the Copeland rule, but using other Condorcet-consistent rules leads to the same conclusions). If there is a Condorcet winner, this candidate will not be defeated, and if there is no Condorcet winner, the procedure has no fixed points but looking at the first iterations may give a sense of what alternatives are selected by a society of strategic voters using approval voting.

13.3.2.3 Borda Voting

I use the following response function, which is inspired from the previous one and could probably receive the same strategic justification. Given Borda scores $bs(k)$ of the various candidates k, and if there are no ties, the voter considers in turn $k_1, k_2, \ldots, k_K$ the candidates ordered according to the score vector bs. First she compares the leader k_1 to his main challenger k_2. If she prefers k_1, she puts k_1 at the first place in her ballot and k_2 at the last place, thereby giving as many points as possible to k_1 and as few points as possible to k_2. If she prefers k_2 to k_1 she does just the contrary. Then she turns to k_3, the third-ranked candidate in bs. She only compares k_3 to k_1, because if there is to be a tie between two candidates involving k_3, it will most likely be a tie between k_3 and k_1. If she prefers k_3 to k_1, she gives as many points as possible to k_3, and if she prefers k_1 to k_3, she gives k_3 as few points as possible. She will thus put k_3 in her ballot in the position 2 or in the position $K-1$. Then she continues filling her Borda ballot this way until all the candidates have been compared with k_1.

Up to my knowledge, this heuristics has not been published for the study of strategic behavior under Borda rule. It amounts to suppose that the voter considers that the most likely ties are ordered by the score vector: $\{k_1, k_2\}$ is by far the most likely, followed by $\{k_1, k_3\}$, $\{k_1, k_4\}$, etc.

13.3.2.4 Example

Suppose that there are five candidates A, B, C, D, E and that the announced score vector ranks the candidates as follows:

$$s(C) > s(A) > s(B) > s(D) > s(E)$$

Consider an individual whose preference P_i is:

$$A \; P_i \; B \; P_i \; C \; P_i \; D \; P_i \; E$$

- Under Plurality rule this individual will compare the leading candidates C to his challenger A, and therefore vote "A" because she prefers A to C.
- Under Approval Voting, she will vote "$\{A, B\}$": The leader is C, thus A and B are approved because they are better than C, and D and E are not approved because they are worse than C. And C himself is not approved because he is worse than the main challenger (A).
- Under the Borda rule (with the scale $4, 3, 2, 1, 0$) she will give 4 points to A, 0 points to C, 3 points to B, 1 points to D, 2 points to E, thereby submitting the Borda-style ballot "$A \succ B \succ E \succ D \succ C$."

13.3.2.5 Discussion

The reaction functions that I use are not the only possible ones but they have the advantage of being derived from the same idea, which is prominent in the strategic voting literature: each voter considers that his vote is going to make a difference in the case of a tie between two candidates and responds individually to her subjective beliefs about the chances of the candidates by considering the likelihood of possible ties.

I suppose that all voters simultaneously respond to the same belief, defined by a score vector. The interpretation is natural here in terms of pre-election polls. This way of doing has the advantage that it makes possible to study strategic behavior out of equilibrium.

13.4 Results

I present (when possible) results for the winning alternative with sincere and responsive voting under Plurality, Borda, Copeland and Approval Voting. For responsive voting behavior, the winner may depends on the announced ranking of candidates (the score vector) to which the voters react.

For Approval Voting, I choose to look at the iterated reactions starting from the Copeland ranking of alternatives. This is because Approval Voting, to many respect is a Condorcet-like voting method and Copeland is a Condorcet-consistent rule. For Borda, I naturally chose to iterate starting from the sincere Borda ranking.

13.4.1 Results for Rousseauist Cultures

In this culture, sincere plurality voting is not well-defined[6] and I thus concentrate on the other rules. The first observation is that if the number of voters is large, because

[6] Because individual preferences need not be transitive.

Table 13.3 Rousseauist culture: probability of a Condorcet (a Condorcet–Rousseau) alternative for five voters and five alternatives

$\binom{n=5}{K=5}$	$\beta = 0$	$\beta = 0.3$	$\beta = 0.5$	$\beta = 0.7$	$\beta = 1$
$\alpha = 0$	0.31 (0.07)	0.40 (0.20)	0.52 (0.30)	0.60 (0.36)	0.69 (0.42)
$\alpha = 0.4$	0.41 (0.22)	0.63 (0.43)	0.74 (0.52)	0.81 (0.58)	0.86 (0.62)
$\alpha = 0.7$	0.54 (0.40)	0.78 (0.61)	0.86 (0.68)	0.90 (0.72)	0.93 (0.75)
$\alpha = 1$	0.70 (0.60)	0.87 (0.76)	0.92 (0.81)	0.96 (0.84)	0.98 (0.86)

Table 13.4 Rousseauist culture: probability of a Condorcet (a Condorcet–Rousseau) alternative for 5 voters and 11 alternatives

$\binom{n=11}{K=5}$	$\beta = 0$	$\beta = 0.3$	$\beta = 0.5$	$\beta = 0.7$	$\beta = 1$
$\alpha = 0$	0.32 (0.06)	0.48 (0.29)	0.61 (0.39)	0.69 (0.44)	0.74 (0.48)
$\alpha = 0.4$	0.47 (0.33)	0.78 (0.62)	0.88 (0.69)	0.91 (0.72)	0.93 (0.74)
$\alpha = 0.7$	0.69 (0.61)	0.92 (0.81)	0.95 (0.85)	0.97 (0.86)	0.98 (0.87)
$\alpha = 1$	0.87 (0.83)	0.98 (0.94)	0.99 (0.95)	0.99 (0.95)	1.0 (0.95)

they are supposed to be independent, all voting rules detect the Rousseau winner with a high probability. I thus focus attention on small size societies (or "juries") and take $I = 5$ and $I = 11$.

Such a profile may have a Condorcet winner or not, but it is important to keep in mind that even if there exists a Condorcet winner, this alternative may different from the true best alternative, the "Rousseau" one. Tables 13.3 and 13.4 show, for the case of $K = 5$ alternatives and $n = 5$ or 11 voters, the probability of the event "there exists a Condorcet winner" and, in brackets, the probability of the event "the Rousseau alternative is a Condorcet winner." These probabilities have been estimated from 1,000 draws of the above model. For instance, for $\alpha = 0.4$ and $\beta = 0.5$ (the values I will use later) and for $n = K = 5$, in $74 - 52 = 22\%$ of cases there is a Condorcet winner but this candidate is nevertheless the "wrong" one.

In order to evaluate voting rules and behaviors in such cultures, it is natural to observe the rank, according to the true ranking, of the chosen alternative. For Copeland rule and for Borda rule, this is well defined. For responsive voting behavior, this depends on the score vector to which the voters react.

For Approval Voting, I look at the iterated reactions starting from the Copeland ranking of alternatives. For Borda, I naturally iterate starting from sincere Borda ranking.

Some results are reported in Tables 13.5 and 13.6. These tables report average ranks so, in reading them, one is interested in having ranks as small as possible.

Intuitively, in such cultures, applying the Borda rule seems a better way to discover the best alternative than applying Condorcet-consistent choice rules, for the following reason.

If the randomly generated profile is very homogeneous and close to the true ranking, then all voting rules should agree. The difference between voting rules thus comes from the cases where enough mistakes have been done by the voters

Table 13.5 Rousseauist culture: average rank of the chosen alternative for five alternatives

$\alpha = 0.4,\ \beta = 0.5,\ K = 5$	$n = 5$	$n = 11$
Pr. of Condorcet	0.745	0.877
Pr. of Rousseau–Condorcet	0.518	0.693
Rule	Average rank	Average rank
Copeland	2.05	1.727
AV1	1.724	1.384
AV2	1.769	1.441
AV3	1.405	1.235
AV4	1.661	1.366
AV5	1.759	1.444
Borda	1.268	1.116
Borda 1	2.586	2.832
Borda 2	1.557	1.342
Borda 3	2.140	2.155
Borda 4	1.605	1.403
Borda 5	1.901	1.654

Table 13.6 Rousseauist culture: average rank of the chosen alternative for 15 alternatives

$\alpha = 0.4,\ \beta = 0.5,\ K = 15$	$n = 11$
Pr. of Condorcet	0.865
Pr. of Rousseau–Condorcet	0.670
Rule	Average rank
Copeland	2.35
AV1	1.41
AV2	1.47
AV3	1.24
AV4	1.40
AV5	1.46
Borda	1.10
Borda 1	14.41
Borda 2	4.90
Borda 3	1.54
Borda 4	5.35
Borda 5	2.32

and, most importantly, from mistakes made when comparing the Rousseau winner to other alternatives. Because the probability of mistake decreases with the rank difference between alternatives, this probability is the largest when comparing the Rousseau alternative to alternative number 2, the second-best one. It follows that the probability that the Rousseau alternative is detected as a Condorcet winner may be relatively low: as an extreme case (if $\alpha = 0$) one may have that voters are wrong half of the time when comparing adjacent alternatives even if they are very skilled at comparing distant ones.

With the chosen parameters ($\alpha = 0.4$, $\beta = 0.5$, n small) one obtains 0.518, 0.693 and 0.670 in Tables 13.5 and 13.6. As a consequence, the performance of the Copeland method is rather poor: the average rank of the Copeland winner is 2.05, 1.727, and 2.35.

The Borda rule also gathers information from other pairwise comparisons, so that its performance is better than the performance of Copeland: The Borda winner has average ranks of 1.268, 1.116 and 1.10 with the same parameters.

From the simulations, we also learn that the Borda rule, in this favorable framework, behaves poorly with respect to manipulation, in contrast with Approval Voting.

The lines "Copeland, AV1, AV2, ..., AV5″ depict successive strategic responses to the Copeland ranking, under Approval Voting. One can see that responsive voting is here beneficial.

The lines "Borda, Borda1, ..." depict the successive strategic responses to the sincere Borda ranking. One can see that strategic behavior is here detrimental. This phenomena is very spectacular in the line "Borda 1" of Table 13.6 (the average rank is there 14.41 out of 15 candidates) and it can also be observed with a smaller number of candidates. Here is the explanation.

Consider the case of four alternatives, with true ranking

$$1 > 2 > 3 > 4,$$

and suppose first that all the voters agree on that ranking. Then sincere Borda obviously provides the true ranking. But then the strategic response of any voter is to rank the second alternative last, and to rank at the second place the least dangerous alternative, that is the one with the lowest score, the alternative 4, the worst one. This provides the (very un-sincere!) Borda ballot:

$$1 > 4 > 3 > 2.$$

Now suppose that a fraction ε of the voters by mistake think that 2 is better than 1. These voters tend to strategically rank 1 at the very last position, giving him as few points as possible in the Borda count, precisely because 1 is ranked first and appears thus as the most dangerous challenger for 2. Moreover those voters, just like the ones who made no mistake, will also put at the second position alternative 4, casting the Borda ballot

$$2 > 4 > 3 > 1.$$

The Borda score of alternative 1 (with the Borda scale 3, 2, 1, 0) is thus $(1-\varepsilon)\cdot 3+\varepsilon\cdot 0 = 3-3\varepsilon$ and the Borda score of alternative 4 is $(1-\varepsilon)\cdot 2+\varepsilon\cdot 2 = 2$. It follows that, for $\varepsilon > 1/3$, the alternative with highest Borda count is now alternative 4, the worst one!

If there are only three alternatives, this phenomenon does not happen,[7] but if there are more alternatives, it becomes more frequent, even for small mistake

[7] Many studies (Favardin et al. 2002; Myerson 2002; Lehtinen 2007) concentrate on three-alternative cases, and thus miss this effect.

Table 13.7 Impartial culture: probability of a Condorcet winner

(1,000 or 100 draws)	$n = 3$	$n = 5$	$n = 11$	$n = 99$
$K = 3$	0.947	0.939	0.920	0.914
$K = 5$	0.840	0.826	0.746	0.755
$K = 15$	0.591	0.514	0.445	0.41
$K = 50$	0.41	0.24	0.20	

probabilities. This situation occurs often in the simulation, which explains why the first-order strategic Borda winner is very badly ranked in these cultures.

One can see how strategic thinking with the Borda rule gives rise to erratic behavior, as reflected in these simulations. Of course this curious pattern is a consequence of the extreme assumption that all the voters react strategically and simultaneously to the same information. For instance the effect will be mitigated if some fraction of the voters vote sincerely by principle. Borda himself defended his method by saying that it is "intended for honest men." This is a lucid remark, but one should stress that, in the culture studied here, all voters share the same goal. Thus it is not clear that they should be labelled as "dishonest" when trying individually to be as efficient as possible in reaching the common will, even if they end up in a collective failure to do so.

13.4.2 Results for Impartial Cultures

Table 13.7 provides the frequency of the existence of a Condorcet winner, for several profile sizes. One can see that this probability decreases when the number of candidates grows. To compare voting schemes in the impartial culture, Table 13.8 indicates (for 11 voters and for 3, 5 and 15 candidates) the probability for a given voting scheme, to elect the (sincere) Borda winner.[8] For very small values of K, there is in general a Condorcet winner and, most often, this alternative is also the Plurality and the Borda winner and indeed the winner under most voting rules. So figures for $K = 3$ are all very large. But when the number of individuals grows, things are quite different. One can notice in particular that strategic voting makes the Borda prediction totally unstable: for $K = 15$. According to the table, the probability that the strategic response to sincere Borda voting still elects the Borda winner is only 0.163.[9]

[8] Ties are broken randomly so that the Borda winner is always unique.

[9] Note that this has little to do with the probability that the Borda rule be manipulated, as usually defined. Here all the voters vote responsively. Of course, if the number of individuals is not small, the probability that a single vote makes a difference is tiny.

Table 13.8 Impartial culture with 11 voters: probability of choosing the Borda winner

$(n = 11)$	$K = 3$	$K = 5$	$K = 15$
Pr. Condorcet	0.920	0.746	0.445
Rule		(Borda)	
Plurality	0.800	0.586	0.248
Copeland	0.889	0.780	0.696
AV1	0.857	0.703	0.426
AV2	0.857	0.682	0.435
AV3	0.889	0.770	0.569
AV4	0.857	0.721	0.491
AV5	0.857	0.686	0.447
Borda	1	1	1
Borda 1	0.863	0.514	0.163
Borda 2	0.846	0.690	0.545
Borda 3	0.879	0.547	0.388
Borda 4	0.860	0.698	0.464
Borda 5	0.852	0.690	0.489

Table 13.9 Consensual distributive culture: probability of a Condorcet winner

(1,000 or 100 draws)	$n = 3$	$n = 5$	$n = 11$	$n = 99$
$K = 3$	0.751	0.790	0.752	0.761
$K = 5$	0.379	0.348	0.351	0.39
$K = 15$	0.004	0.005	0.002	0.00

13.4.3 Results for Consensual Distributive Cultures

Table 13.9 provides the frequency of the existence of a Condorcet winner, for various profile sizes. One can see that this probability tends quickly to 0 when the number of candidates grows.

A specific feature of this culture is that the Gini index of inequality tends to 1/3 when the number n of voters tends to infinity. For instance, for $n = 11$, the expected value of this index is 0.31. This is what would be obtained on average if there were no vote but a random choice. Table 13.10 provides the average values of the Gini index for the winning alternative according to different voting schemes. Note that the three value $K = 3, 5, 15$ chosen for the number of candidates give rise to preference profiles which are quite different the ones from the others since the frequency of Condorcet winners goes from 75% to 2%. In terms of inequality, one notices that Plurality voting does slightly worse than a random choice whereas the other schemes do slightly better.

The interpretation of these results must be related to the shape of the probability distribution over the set of alternatives that this culture defines. The main question that is raised when comparing voting rules in redistributive settings is to know to what extend a voting schemes tends to select more or less egalitarian alternatives. But in the consensual distributive culture, the probability distribution over the set of

Table 13.10 Consensual redistributive culture with 11 voters: average inequality of the chosen alternative

($n = 11$)	$K = 3$	$K = 5$	$K = 15$
Pr. Condorcet	0.752	0.351	0.002
Rule		Gini	
Random choice	0.31	0.31	0.31
Plurality	0.32	0.36	0.40
Copeland	0.30	0.30	0.29
AV1	0.31	0.30	0.31
AV2	0.31	0.31	0.31
AV3	0.30	0.30	0.31
AV4	0.31	0.31	0.30
AV5	0.31	0.31	0.30
Borda	0.31	0.30	0.29
Borda 1	0.31	0.31	0.31
Borda 2	0.31	0.31	0.31
Borda 3	0.31	0.31	0.31
Borda 4	0.31	0.30	0.30
Borda 5	0.31	0.31	0.31

alternatives is such that existing alternatives tend to be similar to that respect. It is therefore delicate to disentangle by simulation this effect from the effect of the different voting rules. Since there is no reason to believe that the consensual distributive culture is close to any "real" culture (despite its mathematical simplicity), I conclude that one should rather use a different approach in order to study by simulation the redistribution problem. This is why I introduced the other kind of distributive cultures, called "inegalitarian distributive cultures," which I will study now.

13.4.4 Results for Inegalitarian Distributive Cultures

For the simulations I chose the parameters

$$e_{\min} = 1.3$$
$$e_{\max} = 4.$$

With this values, the interval of possible values for the Gini index of inequalities roughly covers the actual national values of this coefficient for income distributions. But the Gini formula is not linear with the parameter e. With the uniform distribution for the parameter e on the interval $[e_{\min}, e_{\max}]$, one obtains a probability distribution for the Gini index as depicted in Fig. 13.2.

Table 13.11 provides the probability of the existence of a Condorcet winner, for various profile sizes. One can see that this probability varies a lot with the number of candidates and, perhaps more surprisingly, the number of individuals.

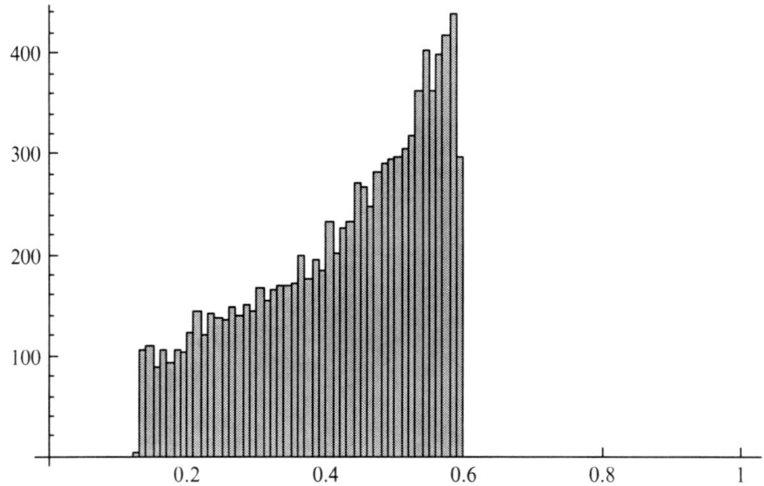

Fig. 13.2 Inegalitarian distributive culture: histogram for the Gini index

Table 13.11 Inegalitarian distributive culture: probability of a Condorcet winner

(1,000 or 100 draws)	$n = 3$	$n = 5$	$n = 11$	$n = 99$
$K = 3$	0.803	0.800	0.825	0.941
$K = 5$	0.419	0.403	0.500	0.81
$K = 15$	0.013	0.013	0.029	0.21
$K = 50$	0.00	0.00		

Voting rules and behaviors are compared with respect to the degree of inequality proposed by the winning candidate, measured by the Gini index. Table 13.12 shows such results for the case of $n = 11$ individuals (or 11 groups of individuals of equal size). One can see that Plurality rule behaves very poorly, indeed worse than the mere random choice of an alternative.

This is an interesting observation, and an argument against Plurality voting with three or more candidates. The phenomenon is spectacular when the number of voters and candidates is large. For instance with $n = 99$ voters and $K = 15$ candidates, the plurality winners has an average Gini index of 0.57, whereas a random choice yields 0.42. The intuition is that in order to be the preferred alternative of several voters, a candidate must propose to each of them more than what is proposed by the other candidates. In order to do so, the candidate should propose very small shares to the voters which are not targeted, hereby proposing a relatively unequal distribution. Following this mechanism, in this setting, Plurality rule promotes inequality.

Other voting rules, such as Copeland and Borda do not suffer this pathology and designate alternatives with smaller Gini index. Voter strategic behavior in that case is detrimental to equality, although not as detrimental as sincere Plurality.

Table 13.12 Inegalitarian redistributive culture with 11 voters: average inequality of the chosen alternative

$(n = 11)$	$K = 3$	$K = 5$	$K = 15$
Pr. Condorcet	0.825	0.500	0.029
Rule		Gini	
Random choice	0.42	0.42	0.42
Plurality	0.42	0.47	0.54
Copeland	0.36	0.32	0.26
AV1	0.36	0.35	0.35
AV2	0.36	0.35	0.37
AV3	0.36	0.34	0.36
AV4	0.36	0.34	0.36
AV5	0.36	0.35	0.35
Borda	0.36	0.32	0.26
Borda 1	0.36	0.39	0.40
Borda 2	0.36	0.35	0.36
Borda 3	0.36	0.37	0.36
Borda 4	0.36	0.35	0.36
Borda 5	0.36	0.36	0.35

13.4.5 Results for Spatial Cultures

13.4.5.1 Results for Uni-dimensional Culture

Results for the uni-dimensional case are reported in Table 13.13 for $n = 11$ voters. This table indicates for the various rules how frequent is the election of the Condorcet winner. Since a Condorcet winner always exists in this culture, the Copeland rule and the Approval Voting responses to the Copeland ranking always elect the Condorcet winner.

Plurality does not often elects the Condorcet winner. Note that, since positions are chosen at random and uniformly, up to some border effects, all candidates are equally likely to be chosen by the Plurality rule.

The Borda rule does much better under sincere voting than Plurality. The iterated strategic responses to the Borda ranking do very badly in the first iteration, as was already seen in previous sections but the situation here improves with successive iterations.

In spatial cultures, it makes sense to consider large numbers of voters. Further exploration in this direction, which are not reported here, confirm the above findings.

13.4.5.2 Results for Multi-dimensional Culture

Results for a uniform two-dimensional case are reported in Table 13.14 for $n = 11$ voters. Bliss points and alternatives are drawn from a rectangle of size 1×0.5. In

Table 13.13 Uni-dimensional culture with 11 voters: probability of choosing the Condorcet winner

$(n = 11)$	$K = 3$	$K = 5$	$K = 15$
Pr. Condorcet	1	1	1
Rule		(Condorcet)	
Plurality	0.736	0.528	0.305
Copeland	1	1	1
AV1	1	1	1
AV2	1	1	1
AV3	1	1	1
AV4	1	1	1
AV5	1	1	1
Borda	0.868	0.777	0.616
Borda 1	0.850	0.508	0.158
Borda 2	0.996	0.557	0.166
Borda 3	0.983	0.732	0.223
Borda 4	0.999	0.739	0.271
Borda 5	0.998	0.806	0.340

Table 13.14 Bi-dimensional rectangular culture with 11 voters: probability of choosing the Condorcet winner

$(n = 11)$	$K = 3$	$K = 5$	$K = 15$
Pr. Condorcet	0.989	0.966	0.833
Rule		(Copeland)	
Plurality	0.782	0.538	0.233
Copeland	1	1	1
AV1	0.989	0.966	0.833
AV2	0.989	0.966	0.833
AV3	0.989	0.998	0.935
AV4	0.989	0.967	0.860
AV5	0.989	0.966	0.841
Borda	0.874	0.538	0.624
Borda 1	0.861	0.479	0.122
Borda 2	0.988	0.500	0.133
Borda 3	0.966	0.661	0.316
Borda 4	0.990	0.674	0.393
Borda 5	0.965	0.753	0.476

this cultures, it is possible that no Condorcet winner exist, but this is a rare phenomenon if the number of alternatives is not large. I therefore take as a reference point a Condorcet-consistent voting scheme: the Copeland rule. This table indicates for the various rules how frequent is the election of the Copeland winner. Since a Condorcet winner often exists in this culture, the Copeland rule and the Approval Voting responses to the Copeland ranking usually coincide (and elect the Condorcet winner).

Plurality behaves differently Note that, since I choose positions at random and uniformly, up to some border effects, all candidates are equally likely to be chosen by Plurality rule.

The Borda rule does much better under sincere voting than Plurality. The iterated strategic responses to the Borda ranking do very bad in the first iteration, as was already seen in previous sections when the number of candidates is not very small but the situation here improves with successive iterations.

In spatial cultures, it makes sense to consider large numbers of voters. Further exploration in this direction confirm the above findings One point should nevertheless be stressed. With uniform probability distributions on boxes, drawing a large number of points tends to produce ever more symmetric patterns. Empirical distributions tend to resemble the uniform continuous distribution, which is a very specific situation (mistakenly considered as "general" by Tullock 1967). For instance, here are some results obtained in the four-dimensional culture on the box $1 \times 1 \times 1 \times 1$. With $K = 3$ alternatives and $n = 5$ voters a Condorcet winner exists most often (observed frequency: 97%). Increasing the number of alternatives makes this frequency decrease, in conformity with the "chaos" ideas of Spatial Voting theory. With $K = 50$ alternatives, the observed frequency is 46%. But picking uniformly many voters makes this frequency increase. With $K = 50$ alternatives and $n = 99$ voters, the observed frequency is 91%.

Note finally that the above study of multi-dimensional cultures is restricted to uniform choices in rectangles. This introduces a symmetry with respect to the center of the boxes, but this symmetry is not essential to the notion of spatial preference profiles. We therefore have restricted our attention to a very particular, and maybe specific case. It would be interesting to go beyond that case and to study spatial patterns justified by political questions rather than by mathematical simplicity.

13.5 Conclusion

This study confirms what has been observed theoretically and empirically: (1) Voting rules exist that improve substantially on Plurality rule. (2) Apart the voting rule itself, the behavior of voters is of primary importance to predict the outcome of an election and therefore to assess the quality of a voting rule.

One point that is emphasised by these simulations is that the way we should judge voting rules depends also on the context. What are the numbers of voters and options? What are these options? Are we dealing with a problem of shared interest or a conflict? All these questions are relevant and suggest new observations. For instance, we noted that Plurality rule tends to promotes unequal sharings in distributive problems. We noted that sincere Borda voting is very efficient is a certain kind of Jury problem.

With respect to strategic voting, it appears that the use of the Borda rule may generate substantial perverse effects, in particular if there are many alternatives on the agenda. Comparatively, Approval voting does not seem to generate such pathologies. This may be related to the fact that strategic behavior under Approval Voting is usually sincere (in fact in the model which was used, such is always the case) according to the usual definition of sincerity for Approval Voting ("If you approve A and not B then you prefer A to B"). On the contrary, strategic behavior under the Borda rule may produce very un-sincere votes.

Appendix

Standard Deviation in the Consensual Distributive Culture

The variables y_i, $i = 1, \ldots, n$ are independent and uniform on $[0, 1]$. Let $s = \sum_{i=1}^{n} y_i$ and $x_i = y_i/s$. The standard deviation $d(x)$ of x is such that

$$
\begin{aligned}
d(x)^2 &= \frac{1}{n} \sum_i \left(\frac{y_i}{s} - \frac{1}{n} \right)^2 \\
&= \frac{1}{s^2} \frac{1}{n} \sum_i \left(y_i - \frac{s}{n} \right)^2
\end{aligned}
$$

When n tends to infinity, the sum $\frac{1}{n} \sum_i \left(y_i - \frac{s}{n} \right)^2$ tends to the theoretical variance of the uniform distribution on $[0, 1]$, that is a fixed positive number. Because, s/n tends to the expected value $1/2$, one can see that $d(x)$ tends to 0 as n^{-1}.

Gini Index in the Consensual Distributive Culture

The index is:

$$
gini = \frac{2}{n} \sum_{i=1}^{n} \left(\frac{i}{n} - v_i \right)
$$

where v_i denotes the (partial ordered) sum of the i smallest values of the n variables described above. One can write

$$
gini = \frac{2}{n} \sum_{i=1}^{n} \left(\frac{i}{n} - \sum_{j=1}^{i} \frac{u_i}{s} \right).
$$

For n large, s is close to $n/2$ so that $gini$ is close to

$$
\frac{2}{n^2} \sum_{i=1}^{n} i - \frac{4}{n^2} \sum_{i=1}^{n} \sum_{j=1}^{i} u_i
$$

The first term tends to 1. For the second term, note that $\sum_{j=1}^{i} u_i \simeq i^2/(2n)$ and $\sum_{i=1}^{n} i^2 \simeq i^3/3$. Thus the second term tends to 2/3 and one can see why *gini* tends to 1/3.

References

Austen-Smith, D., & Banks, J. S. (1999). *Positive political theory 1: Collective preferences*. Ann Arbor: University of Michigan Press.

Blais, A., Laslier, J.-F., Sauger, N., & Van der Straeten, K. (2008). *Sincere, strategic, and heuristic voting under four election rules: An experimental study* (mimeo).

Condorcet (1785). *Essai sur l'application de l'analyse à la probabilité des décisions rendues à la pluralité des voix*. Paris: Imprimerie Royale.

Cox, G. W. (1997). *Making votes count: Strategic coordination in the world's electoral systems*. Cambridge: Cambridge University Press.

Favardin, P., Lepelley, D., & Serais, J. (2002). Borda rule, Copeland method and strategic manipulation. *Review of Economic Design, 7*, 213–228.

Gehrlein, W. V. (1997). Condorcet's paradox and the Condorcet efficiency of voting rules. *Mathematica Japonica, 45*, 173–199.

Gehrlein, W. V., & Fishburn, P. C. (1979). Proportions of profiles with a majority candidate. *Computers and Mathematics with Applications, 5*, 117–124.

Laslier, J.-F. (2002). How two-party competition treats minorities. *Review of Economic Design, 7*, 297–307.

Laslier, J.-F. (2006). Spatial approval voting. *Political Analysis, 14*, 160–185.

Laslier, J.-F. (2009). The leader rule: A model of strategic approval voting in a large electorate. *Journal of Theoretical Politics, 21*, 113–136.

Laslier, J.-F., & Picard, N. (2002). Distributive politics and electoral competition. *Journal of Economic Theory, 103*, 106–130.

Lehtinen, A. (2007). The Borda rule is also intended for dishonest men. *Public Choice, 133*, 73–90.

Lehtinen, A. (2008). The welfare consequences of strategic behavior under approval and plurality voting. *European Journal of Political Economy, 24*, 688–704.

Lindbeck, A., & Weibull, J. (1987). Balanced-budget redistribution as the outcome of political competition. *Public Choice, 52*, 273–297.

Lizzeri, A. (1999). Budget deficit and redistributive politics. *Review of Economic Studies, 66*, 909–928.

McKelvey, R. D. (1986). Covering, dominance, and institution-free properties of social choice. *American Journal of Political Science, 30*, 283–314.

McKelvey, R. D. & Tovey, C. A. (2010). Aproximation of the yolk by the LP yolk. *Mathematical Social Sciences, 59*, 102–109.

Merrill, S., III (1984). A comparison of efficiency of multicandidate electoral systems. *American Journal of Political Science, 28*, 23–48.

Myerson, R. B. (1993). Incentives to cultivate favored minorities under alternative electoral systems. *American Political Science Review, 87*, 856–869.

Myerson, R. B. (2002). Comparison of scoring rules in Poisson voting games. *Journal of Economic Theory, 103*, 219–251.

Nurmi, H., & Salonen, H. (2008). More Borda count variations for project assessment. *AUCO Czech Economic Review, 2*, 109–122.

Rousseau, J.-J. (1762). *Du contrat social*. Amsterdam: Marc-Michel Rey.

Tovey, C. A. (2010). A critique of distributional analysis in the spatial model. *Mathematical Social Sciences, 59*, 88–101.

Truchon, M. (2008). Borda and the maximum likelihood approach to vote aggregation. *Mathematical Social Sciences, 55*, 96–102.

Truchon, M., & Drissi-Bakhkhat, M. (2004). Maximum likelihood approach to vote aggregation with variable probabilities. *Social Choice and Welfare, 23*, 161–185.

Tullock, G. (1967). The general irrelevance of the general impossibility theorem. *The Quarterly Journal of Economics, 81*, 256–270.

Young, H. P. (1988). Condorcet's theory of voting. *American Political Science Review, 82*, 1231–1244.

Part VI
Experiments

Chapter 14
Laboratory Experiments on Approval Voting

Jean-François Laslier

14.1 Introduction

This chapter reviews the experimental work about Approval Voting which follows the now standard practice of Experimental Economics. The principle of Experimental Economics is to observe individual behavior in situations where the experimenter can control individual preferences. The classical way to induce and control preferences is to use money, that is to pay the subjects more or less, depending on what they do and, in group experiments, what the other subjects do. This methodology has slowly gained popularity among the economists and is summarized in several classical references (for instance Davis and Holt 1993; Kagel and Roth 1995).

Although there also exist experiments designed to study the behavior of political parties, economic experiments about voting essentially deal with voter behavior, and this will be my focus here. Typically, each participant is instructed of how much money he or she will personally earn, depending on the option chosen by the group through some voting procedure. The act of voting is therefore quite similar, in the lab, to what it is in the real life. By chosing the payoff scheme, the designer of the experiment builds in the lab a "small world" that is supposed to mimic some feature of the real world. For instance the experimenter can mimic a society divided in two by deciding that there are two options only, and some participants will earn a lot if one option is chosen and nothing if the other option is chosen, and conversely for the other participants. Several of these preference profiles will be mentionned in the sequel.

A key-stone result of the political theory of electoral systems is Maurice Duverger's statement that proportional representation creates conditions favorable to foster multi-party development, while a plurality system tend to favor a two-party system (Duverger 1951). This statement rests on the consideration of voter's behavior and specifically on the idea that, under the plurality rule, voters will not vote for candidates who have no chance to win the election. Voters deserting

J.-F. Laslier
Laboratoire d' Econométrie, École Polytechnique, 91128 Palaiseau, France
e-mail: jean-francois.laslier@polytechnique.edu

J.-F. Laslier and M.R. Sanver (eds.), *Handbook on Approval Voting*, Studies in Choice and Welfare, DOI 10.1007/978-3-642-02839-7_14,

non-viable candidates in turn implies that the political supply, that is the set of candidates who run for office, as well as their political platforms, is in part determined by the voting rule.

It is therefore clear that the question of voters' behavior in front of different voting rules has important implications for political science. Unfortunately, the co-determination of the political supply and of the voter behavior makes it is difficult for historical or sociological research to tackle the counterfactual question of how voters would behave if the electoral system were changed. And, obviously, the problem is not simpler dealing with a voting rule (like approval voting) which is seldom used in practice.

For these reasons, pure theoretical research, in the manner of formal economic theory, has been called to analyse the possible consequences of voters' behavior under different voting rules (Riker 1982; Cox 1997). In particular, game theorists, who are used to the intricacy of rational behavior in collective situations, have applied their methodology and worked under the hypothesis of more or less perfectly rational voters (Moulin 1983; Myerson 1993, 1995; Austen-Smith and Banks 1996; Taagepera 2007). Formal theory has contributed to our understanding of voting rules by following the lead of strategic voting as a key factor to explain actual behavior, and maybe to predict behavior in new situations.

The results obtained in this field by formal political science mainly confirm and precise Duverger's claims, but the methodology of formal analysis raises two problems: a validity question about the rational voter hypothesis, and an epistemological question about equilibrium predictions or absence of predictions.

As to the first problem, many authors in political science have questioned the assumption that voters act rationally (for instance Lazarfeld et al. 1948; Delli Caprini and Keeter 1991; Sniderman et al. 1991; Sniderman 1993; Lupia and McCubbins 1998; Kuklinski and Quirk 2000). Since this assumption is made in most formal models, the critic tends to extend to the whole methodology of formal theorizing (Green and Shapiro 1994). Empirical investigation on that point has provided mixed evidence and both the idea that voters are purely determined and the opposite idea that they are perfectly autonomous and rational contains a part of empirical truth (Blais 2000, 2002, 2004; Blais and Bodet 2006). But a detailed description and understanding of when and why some voters engage in strategic considerations when voting is still lacking.

The second problem is less often raised but not less important. The rationality postulate may be difficult or impossible to test because its consequences are not well-defined. The usual practice in micro-economics and game theory is to look for Nash equilibria. A Nash equilibrium is a situation in which each individual action is specified,[1] and such that any intelligent player who would be told what the other are doing would not wish to change his action (Myerson 1991). Unfortunately, it turns out that, in many voting games, there are a plethora of Nash equilibriua, and almost anything can happen in a Nash equilibrium. For instance, if voters are only interested in the result of the election and vote according to plurality rule, the prediction

[1] An "action" can be a random choice, in which case a probability distribution must be specified.

that a given candidate receives no vote is always part of a Nash equilibrium. The announcement to voters that no-one is voting for this candidate is self-fulfilling, for it confirms voters in their intention to desert this candidate. Consequently, the Nash equilibrium concept should be viewed as a minimal rationality requirement but not as a predictive theory. This problem points to a fundamental incompleteness of the rationality paradigm in interactive situations. It has given rise in game theory to the development of the concepts of equilibrium refinement, which aim at making more precise predictions by ruling out some Nash equilibria.

Note that if the multiplicity of equilibrium is only an artifact of the rational paradigm, then it is just a methodological problem for the users of this paradigm, and people interested in results rather than technical problems can leave it to technicians. But if the multiplicity of equilibrium describes some true feature of the voting mechanisms, then it is of some importance for political science. The idea that voting in a democracy expresses some true will of the society implies that the result of an election should relate to the fundamentals of this society, that is to individual voters' will. It is hardly compatible with the idea that the outcome of the election cannot be predicted from individual opinions.

Theorists working within the rational voter paradigm have recognized these difficulties, which all call for a finer understanding of what could be the actual behavior of possibly strategic voters. On the conceptual side they studied refinements of equilibrium for voting situations (Myerson and Weber 1993; Myerson 2002; De Sinopoli et al. 2006; McKelvey and Patty 2006; Laslier 2009). On the empirical side they designed experiments in which the voters' decisions can be carefully observed.

This paper is a survey about laboratory experiments on approval voting. It leaves aside observation of AV in some real situations (see Brams and Fishburn 2005) as well as framed field experiments (see Alós–Ferrer and Granić 2010; Baujard and Igersheim 2010). I report on three sets of experiments, which differ with respect to the preference profile they use. The next section is devoted to experiments by Baron et al. (2005) on a "Divided society"; Sect. 14.3 is devoted to the seminal experiments of Forsythe et al. (1996) on the "Split majority"; and Sect. 14.5 is devoted to experiments by Van der Straeten et al. (2010) on the "Single-peaked domain". The last section concludes.

14.2 Experiments on a Divided Society

14.2.1 The Preference Profile

As to the outcome of approval voting election, the main prediction to be tested is that this method tends to favor consensual outcomes. The mechanism at work is very simple and can be seen on the following, basic example. The society is split into two groups of voters, say A and B, of equal size. There are three alternatives, a b, and c. Alternative a favors group A and is detrimental to group B, conversely, alternative b favors group B and is detrimental to group A. The third alternative, c,

Table 14.1 Payoffs for the experiments in Baron et al. (2005)

Voter type	Number of voters	a	b	c
"A"	28	10	0	6
"B"	28	2	12	8

Table 14.2 Proposals for the first experiment in Baron et al. (2005)

	Income of group A	Income of group B
a	Increased by 10%	Increased by 2%
b	Increased by 0%	Increased by 12%
c	Increased by 6%	Increased by 8%

is equally appreciated by both groups as a rather good choice, but not as good as each group's favorite alternative; alternative c is the consensual one, and the best alternative from the point of view of the whole society.

In Table 14.1 is an example of a payoff scheme leading to such a preference profile. This scheme is used by Altman et al. (2005) in their laboratory protocol, that will be described in this section. But prior to describing this laboratory experiment, it is useful to first describe a preliminary experiment done by the same authors on the internet.

14.2.2 First Protocol, on the Internet

A first experiment made by Altman, Baron and Kroll does not use money. It takes place on the internet, the groups are abstract and the alternatives are presented as favoring more or less "people who live in your country" and "people who live in another country of the same size". More exactly, the participants were asked to assume the following.

- The economic effects are all that matter in this vote.
- You make $10,000 per year (in the currency of that time).
- The groups are similar in their standard of living.
- About half of the voters in each group actually vote.
- You are always in group A.

The three proposals are described as in the Table 14.2, with variations in the figures that maintain the qualitative comparison among options.

14.2.3 Results of the Internet Protocol

If voters can vote for only one option, then 66% vote for the proposition a, that favors their own group, 4% vote for the opposite proposal b, and 30% vote for the consensus c. If we imagine the two groups to behave in the same way, the selfish proposals a and b will receive $(66 + 4)/2 = 35\%$ each, to be compared with the 30% received by the consensus. Consequently, a selfish proposal will be chosen.

With approval voting, the results are:

$$\{a\} : 36\%,\ \{b\} : 2\%,\ \{c\} : 16\%,\ \{a,b\} : 2\%,\ \{a,c\} : 42\%,\ \{b,c\} : 1\%$$

so that the approval scores are:

$$a : 80\%,\ b : 5\%,\ c : 59\%$$

If we imagine the two groups to behave in the same way, the selfish proposals a and b will receive $(80 + 5)/2 = 42.5\%$ to be compared with the 59% received by the consensus: the consensual option is chosen.

So one can observe that, in such a situation, using approval voting rather than standard voting increases the support for the consensual option c.

14.2.4 Second Protocol, in the Laboratory

The previous protocol did not allow for any control of the participants' preferences and is thus subject to the usual criticisms of opinion surveys: participants are asked how they would react in some (very!) hypothetical situation and their answer is of no consequence. So why should we trust their answers?

In a second experiment by the same authors, participants are paid. Moreover, instead of abstract A and B, the groups are real too since the students are recruited at two American universities (the University of Pennsylvania and the St. Lawrence University) and their type is the university they belong to. The instructions given to the student are of the form: "If proposal a is chosen, each participant at SLU will get ten tokens and each participant at UPenn will get two tokens." (Two tokens are one US dollar.)

14.2.5 Results of the Laboratory Protocol

If voters can vote for only one option, then 45% vote for the proposition a, that favors their own group, 2% vote for b, and 54% vote for the consensus c. If we imagine the two groups to behave in the same way, the selfish proposals a and b will receive $(45 + 2)/2 = 23.5\%$ each, to be compared with the 54% received by the consensus. So the collective best proposal will be chosen, unlike what was observed with the previous protocol.

With approval voting, the results are:

$$\{a\} : 19\%,\ \{b\} : 0\%,\ \{c\} : 15\%,\ \{a,b\} : 10\%,\ \{a,c\} : 48\%,\ \{b,c\} : 8\%$$

so that the approval scores are:

$$a : 77\%, \; b : 18\%, \; c : 71\%$$

If we imagine the two groups to behave in the same way, the selfish proposals a and b will receive $(77 + 18)/2 = 47.5\%$ each, to be compared with the 71% received by c: the consensual option is chosen, as was observed with the previous protocol.

In that case, c is chosen under both voting rules, and this victory is large in both cases: 71% against 47.5% with approval voting, and 54% against 23.5% with standard plurality.

14.2.6 Conclusion on Divided Society Experiments

Altman et al. (2005) write that "in general results were similar" in their two experiments, and draw some general conclusions. For instance they write:

> "(...) people can take the opportunity to approve proposals that are somewhat less good for their own group but better for the whole. Approval voting can thus favor compromise among competing groups. For example, workers may fear that a trade agreement would threaten their jobs, but they may also care about increased access to goods and about benefits to other workers elsewhere. If they were offered enough options, they might approve a free trade agreement if they saw it as sufficiently beneficial for all."

But in fact the results are quite different under the two protocols, even if the differences are not explained. And the second protocol does not show clear differences between the two voting rules under consideration. Thus, far-reaching conclusions, if true, can hardly be derived from this set of experiments alone.

Other experiments on voting have concluded that the behavior of a voter could only be understood as a response to this voter's idea about what the other voters are doing. This was demonstrated in the early laboratory experiments of McKelvey and Ordeshook (1985) and Forsythe et al. (1993), and is emphasized by Béhue et al. (2009) and Blais et al. (2010). Altman, Baron and Kroll's tests are one shot and are more akin to some individual morality test than to a decision procedure, specifically the internet one.

As a morality test, these experiments show that a relatively large share of the population might be ready to approve a collectively optimal option next to their selfish preferred one, if they are given the opportunity to do so.

14.3 Experiments on a Split Majority

14.3.1 The Preference Profile

In two papers, Forsythe et al. (1993, 1996) studied experimentally three-candidate elections under various electoral rules: the standard plurality voting with single-name

Table 14.3 Payoffs for the experiment in Forsythe et al. (1996)

Voter type	Number of voters	Orange	Green	Blue
"*O*"	4	\$1.60	\$1.20	\$0.30
"*G*"	4	\$1.20	\$1.60	\$0.30
"*B*"	6	\$0.60	\$0.60	\$1.90

ballots, approval voting and the Borda rule. The design borrows from Felsenthal et al. (1988) and Rapoport et al. (1991). The experiments of Forsythe et al. used 14 individual subjects. Candidates are labelled "Orange", 'Green", and "Blue", and subjects belong to three groups labelled "*O*", "*G*", and "*B*" of, respectively 4, 4, and 6 students.[2] The payoff to a student depends on his or her group and on which candidate is elected, as in Table 14.3.

One can see that the electorate is in fact divided in two. A minority of 6 voters out of 14 would like to see the Blue candidate elected and is indifferent between the Orange and Green candidates. That is the group "B". On the other hand, a majority of $4 + 4 = 8$ voters rejects Blue and would like to see Orange or Green elected, but this majority is split into two groups of equal size: the 4 "O" voters favor the Orange candidate and the 4 "G" voters favor the Green candidate. The Blue candidate is a Condorcet loser: he would be defeated by 8 votes (the "O" and "G" voters) against 6 in any pairwise comparison.

This society is thus facing a coordination problem under plurality voting: if the "O" and "G" voters fail to coordinate their vote, the minority candidate may win. This split-majority example is the first example given by Borda (1784), at the beginning of his *mémoire*, to prove that "plurality rule does not always indicate the voters' will." (It may be worth noticing that Borda himself considers as obvious that Majority Rule here indicates "the voters' will".) This example is also a clear case for two-round majority rule.

14.3.2 The Protocol

Here I briefly describe the protocol used by Forsythe et al. (1996) to compare voting rules. More details can be found in the original article. The electorate is made of 14 subjects, in accordance with the Table 14.3. Series of eight consecutive elections take place, with scores of the candidates being announced each time. Participants therefore have the time to learn and, maybe, develop strategies. During a session, 28 subjects are present. From one series of eight elections to another, groups and voters' types are reshuffled. The table of payment is known to all participants and, of course, each participant knows his or her own type. Under the "Poll" treatment, participants are moreover asked to state their voting intention before each vote, without commitment, and the results of these polls are announced before the vote. The

[2] Felsenthal et al. used group voting: a player decides for a whole group.

studied voting rules are plurality, approval and Borda. Abstention is always allowed and ties are resolved randomly. It is interesting to quote the exact phrasing that describes these rules for the purpose of the experiment:

Plurality "If you do not abstain, you may vote for a most one candidate. To do this, place a check next to the candidate for whom you are voting."

Approval "If you do not abstain, you may cast one vote each for as many candidate as you wish. To do this, place a place a check next to each candidate for whom you are voting."

Borda "If you do not abstain, you must give two votes to one candidate and one vote to one of the other candidates. To do this, write "2" next to the candidate to whom you are giving two votes and write "1" next to the candidate to whom you are giving one vote"

These rules are so simple that, in the laboratory, one does not have to explain how ballots are counted: people naturally understand that votes are added.

14.3.3 Results

The initial experiment of Forsythe et al. (1993) on plurality voting showed a high frequency of strategic voting (individuals not voting for their favored alternative), and this is again observed here. The analysis performed by Forsythe and his coauthors is driven by the consideration of the possible "equilibria" under these voting rules. These equilibria are theoretical predictions derived from the strategic voting model of Myerson and Weber (1993). The main feature of this model is that a voter perceives the probability p_{ij} that his or her vote is decisive between two candidates i and j as being proportional to the difference between the scores of i and j.[3]

- Under plurality rule, there are three equilibria. In two of them, votes from the majority gather on one of the two majority candidate (Orange or Green), and the minority simply votes for its candidate (Blue). In a third equilibrium, the majority fails to coordinate and split its vote, leading to the election of the Condorcet loser, Blue.
- Under approval voting, there are also three equilibria. In two of them, a majority candidate is elected. For instance the Orange-voters can cast a ballot $\{O\}$, the Green-voters cast the ballot $\{G, O\}$, and the Blue-voters (the minority) cast the ballot $\{B\}$. This results in scores of 8, 4, and 6 respectively for Orange, Green and Blue. Another equilibrium exists, in which all three candidates tie with a score of 6 in expectation. This third equilibrium is mixed: it involves Orange and

[3] More recent models have more solid probabilistic foundations and obtain that this probability p_{ij} is decreasing not in proportion with, but much faster than the vote margin: see Myerson (2002), Laslier (2009).

Table 14.4 Agregate results for the experiment in Forsythe et al. (1996): Frequency of the Condorcet loser winning

	Without polls	Without polls
Plurality	0.2604	0.1979
Approval	0.1910	0.1458
Borda	0.0972	0.1111

Green voters balancing their votes in a precise way between single-approval and double-approval.

- Under the Borda rule there exists only a mixed equilibrium, with a tie among all three candidates, which involves voters from each group to balance their votes among two different ballots in some precise proportions.

The observations about plurality rule are simple, and very well in line with the theory: votes gather on two candidates, and these candidates can be any two. This process is made easier with polls, which play the same role as elections. With plurality voting, polls are useful as a coordinating device, and people do not manipulate the polls. This is clearly in line with historical remarks and the so-called "Law" of Duverger (1951) and with modern formal political theory (Cox 1997).

With approval voting, polls are also important, but not in such a direct manner. Indeed the polls tended to be invalidated when the predicted winner was a minority candidate. Pools have obviously an important impact on elections, but this impact need not be straightforward. About this experiment, Forsythe et al. write: "... it appears that the focal information conveyed by polls under plurality rule was information differentiating the two majority candidates, while polls under approval voting and Borda rule served primarily to determine the level of threat posed by the minority candidate, as perceived by the supporters of the majority candidates."

As to the result of the election, the question posed by this experiment is to avoid the election of the Condorcet loser. Table 14.4 indicates the percentage of elections which result in the election of the Condorcet loser candidate, Blue. This paradoxical phenomenon occurs quite frequently under plurality rule: more than 25% of the time without polls and nearly 20% of the time with polls.

Since the example was cooked by Borda (1784) to explain that using the Borda rule should avoid the paradox, it should come as no surprise that the Borda rule indeed works in that case. Notice that this remark holds here in a situation where voters often vote insincerely under the Borda rule,[4] whereas Borda's argument rests on the assumption of sincere voting. Indeed, the results of this experiment confirm that the Borda rule seems to be relatively immune to strategic manipulation when the number of candidates is small, and, in general, strategic behavior under the Borda rule tend to make this rule more Condorcet-consistent (see Béhue et al. 2009; Kube and Puppe 2009; Laslier 2010).

Using approval voting also seriously reduces the frequency of the paradox, compared with plurality voting Notice that, under approval voting, the existence of

[4] See the original article for measures of the extend of strategic voting in the experiment.

pre-election polls leads to a (statistically significative) reduction of the paradox frequency, from 19.10% to 14.58%.

14.3.4 Conclusion on Split Majority Experiments

The experiments of Forsythe et al. are designed to observe in the laboratory a phenomenon that has been identified in practice and in theory: the paradoxical situation in which a bad candidate is elected under plurality voting because its opponents fail to coordinate on a unique candidate. They show that using the Borda rule or approval voting can limit the prevalence on this phenomenon.

From the methodological point of view, these experiments demonstrate the importance of individual strategic voting, and therefore of voters' information about voting intentions. For controlled experiments, this requires to hold in the laboratory series of elections or to have pre-elections polls.

14.4 Experiments on a Political Domain

14.4.1 The Preference Profile

This section reports on experiments done by Blais, Laslier, Laurent, Sauger and Van der Straeten (Blais et al. 2007, 2010; Van der Straeten et al. 2010) using a preference profile that mimics the classical one-dimensional political domain. Voters, as well as candidates, are located along a single political axis, and each voter prefers the candidates which are closest to his own position. This image of the political landscape is by far the most common in the literature and, even if this image is incomplete, it indeed captures several stylized facts about politics.

The basic setting is as follows. 21 subjects vote among five alternative candidates, located at five distinct points on a left–right axis that goes from 0 to 20: an extreme left candidate, a moderate left, a centrist, a moderate right, and an extreme right (see Fig. 14.1).

Each subject is also assigned a position, on the same political axis, that determines the payoff she will earn. The incentive for a subject is that the elected candidate be as close as possible to her position. Precisely, the subjects are informed that they will be paid 20 Euros (or Canadian dollars) minus the distance between the elected candidate's position and their own assigned position. For instance, a voter whose assigned position is 11 will receive 10 euros if candidate A wins, 12 if E wins,

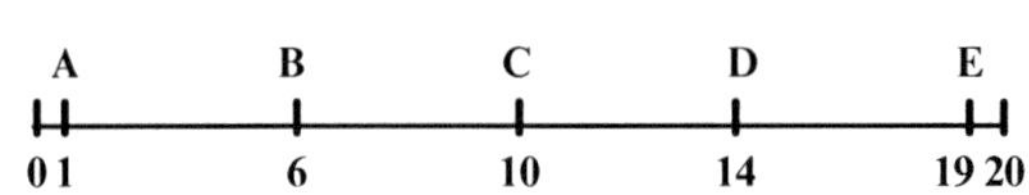

Fig. 14.1 Positions of the five candidates in Blais et al. (2007)

15 if B, 17 if D, and 19 if C. It is in the voter's interest that the elected candidate be as close as possible to her own position.

The induced preference profile, in the words of Social Choice Theory, is a standard single-peaked profile. The centrist option (candidate C) is a Condorcet winner. If every voter was to vote sincerely for the candidate that is closest to her position, candidates A and E would each receive four votes. Four voters have B as their closest candidate, four have D, and three have C; the last two voters (positions 8 and 12) are equally distant from C and B or from C and D.

14.4.2 The Protocol

These experiments were primarily designed to compare the one round plurality system (labelled 1R) and the two-round majority system (2R), which are by far the most common voting rules used for electing single candidates. (Proportional systems raise different questions.) Some sessions also included approval voting (labelled AV) and single transferable vote with the Hare system of transfers (labelled STV).

The elections were performed by series of four, the results of each election being made public each time. The set of options and the payoff scheme are identical for all elections. In each group of participants, 2 or 3 series of 4 elections are held successively. The four elections are held with the same voting rule. At the beginning of each series the participants are assigned randomly drawn positions on the 0–20 axis. There are a total of 21 positions, and each participant has a different position. The participants are informed about the distribution of positions: they know their own position, they know that each possible position is filled exactly once but they do not know by whom. Voting is anonymous. After each election, ballots are counted and the results (the five candidate scores) are publicly announced. The participants are informed from the start that only one election will be randomly drawn as the "decisive" election, the one which will actually determine the payoffs.[5]

14.4.3 Results: Electoral Outcomes

The overall results are as follows. Table 14.5 shows the aggregate results for all elections and Table 14.6 the same results restricted to the last two elections of each series of four. The extremist candidates (A and E) are never elected. In 1R elections candidate C (the centrist candidate, a Condorcet winner in our case) is elected in about half of the elections, and candidates B or D are elected in the remaining half (with B being elected more often than D). In 2R elections, the picture is similar but in AV elections and STV elections, it is very different. In AV elections, C is almost always elected, and in STV elections, C is never elected.

[5] This is customary in Experimental Economics; it has the advantage of keeping the subjects equally interested in all elections and of avoiding insurance effects.

Table 14.5 Elections won (all)

	1R	2R	AV	STV
C	49%	54%	79%	0
B or *D*	51%	45%	21%	100%
A or *E*	0	0	0	0
Total	92	92	24	16

Table 14.6 Elections won (last two)

	1R	2R	AV	STV
C	52%	50%	100%	0
B or *D*	48%	50%	0	100%
A or *E*	0	0	0	0
Total	46	46	12	8

With this kind of protocol one can see, with 1R plurality, the same path-dependence effect that was observed in the experiments by Forsythe et al. described in the previous section. Votes gather on two "viable" candidates, but there are three possibly viable candidates: the centrist one and the two moderate ones. The extreme candidates are not viable. With 2R majority voting, votes gather for the first round, on three candidates, instead of two but the outcome of the election is still unpredictable since the centrist candidate (who is winning in any runoff) may be eliminated at the first round. With STV (Hare system) if the centrist candidate is not eliminated before the two extreme candidates then the two moderate candidates receive the transferred votes from extremist voters, so that the centrist candidate is eliminated before the two moderate ones. This is why the Condorcet winner is never elected with STV.

- With 1R and 2R systems.
  - The elected candidate can be any one of the three viable, non-extreme, candidates.
  - Who is elected is not predictable: it depends on initial focalization effects.
  - The Condorcet winner is elected approximately half of the time.
- With AV: Who is elected is predictable: the electorate quickly evolves to always elect the Condorcet winner
- With STV: The Condorcet winner is always eliminated

14.4.4 Results: Individual Behavior

14.4.4.1 Number of Approbations and Sincerity

An approval voting ballot is "sincere" if and only if there do not exist two candidates *a* and *b* such that the voter strictly prefers a to b and nevertheless approves of *b* and

Table 14.7 Number of approved candidates per ballot

Number of candidates	0	1	2	3	4	5
Number of ballots	0	124	277	92	9	2
% of ballots	0	24.6	54.9	18.3	1.8	0.4

does not approves of a. Van der Straeten et al. (2010) count the number of pairs (a, b) that exhibit such as a violation of sincere voting. 6 sessions with 4 elections and 21 voters make 504 ballots and thus 5,040 pairs. Insincere comparisons are very rare in the data: 78 observed pairs, that is 1.5%.

This definition of sincere voting leaves one degree of freedom to the voter since it does not specify at which level, given her own ranking of the candidates, the voter should place her threshold of approbation. With 5 candidates most voters have 6 sincere ballots (including the "empty ballot"). Table 14.7 indicates how many ballots contain 0, 1, 2, etc. approved candidates. The empty ballot is never observed and most ballots approve 2 candidates. On average the number of approved candidate is 1.98 out of five.

14.4.4.2 Strategic Behavior

Theories of strategic behavior rest on the idea that a rational voter is responding to his or her guess about the other voters' actions. Consider that the results of the previous election delivers such information. They consider "myopic" voters, who each suppose that the others will vote this time as they did last time.[6] The behavior of strategic voters then follows the "Leader Rule" of Laslier (2009). This rules makes a prediction as to the vote of each voter, with respect to each candidate, for each election after the initial one, and these predictions are unique except in some cases of indifference.

The voter considers the result of the previous election (the number of votes that each candidate received last time, not including the individual's own approvals) and focuses in particular on the candidate who obtained the largest number of votes, say a_1. All other candidates are evaluated with respect to this one: the voter approves all candidates she prefers to a_1 and disapproves all candidates she finds worse than a_1. The main candidate is itself evaluated by comparison with the second-ranked candidate (the "main challenger"): the voter approves the main candidate if and only if she prefers this candidate to her main challenger. The voter therefore places her "Approval Threshold" around the main candidate, either just above or just below.

This theory produces 1,776 unique predictions (out of $21 * 6 * 5 * 3 = 1{,}890$ votes). As one can see in Table 14.8, this theory correctly predicts $576 + 996 = 1{,}572$ of the 1,776 predicted votes, that is 88%. Table 14.8 distinguishes these

[6] They also consider the alternative hypothesis of voters with perfect foresight, which are able to formulate exact anticipations. The findings are identical under the two hypothesis.

Table 14.8 88% of the approbations are predicted by the "strategic" model

	Approval = 1	Approval = 0	*Total*
Prediction = 1	**576**	137	713
Prediction = 0	67	**996**	1,063
Total	643	1,133	1,776

Table 14.9 82% of approbations are predicted by the "average target" model

	Approval = 1	Approval = 0	*Total*
Prediction = 1	**695**	295	990
Prediction = 0	26	**802**	828
Total	721	1,097	1,818

Table 14.10 87% of approbations are predicted by the "best two" model

	Approval = 1	Approval = 0	*Total*
Prediction = 1	**619**	119	738
Prediction = 0	121	**995**	1,116
Total	740	1,114	1,854

predictions, right or wrong, according to the observed vote. The theory tends to slightly overestimate the number of approved candidates.

Some authors (Merrill 1981; Lehtinen 2010) have proposed, as a possible behavior under approval voting that a voter would approve of the candidates which provide her a utility above the average. Table 14.9 indicates the predictions of this "Average target" theory on the last three periods of each series (so that one can compare with the previous theory). The number of unique predictions is 1,818 due to the fact that for some voters, some candidates are exactly "average" candidates. This theory predicts $695 + 802 = 1{,}497$ votes, out of 1,818, that is 82%.

Recall that more than half of the ballots are two-name ballots and that insincere ballots are very rare. It is thus natural to test the heuristics model which simply states that the voter approves precisely her two preferred candidates. Table 14.10 indicates the predictions of this theory relative to the last three periods of each series (to compare with the previous theories). There are 1,854 unique predictions.[7] On the set of possible predictions this simple theory predicts $619 + 995 = 1{,}615$ votes, that is 87% .

Of course this nice result (87% of explained votes) is due to the fact that we calibrated the theory to an observed variable (the number of approval per ballot). In other circumstances, the number of approbations per ballot might be notably different. Here are figures reported for in situ tests during real elections. In France, 2002, the oberved average was 3.15, out of 16 candidates, see Laslier and Van der Straeten (2008). Alós–Ferrer and Granić (2010) observe in one case an average of

[7] Some voters are indifferent between two candidates as their second choice. In that case the "best two" candidates are not well-defined.

2.25 parties out of 17, which is not too far from two, but the standard deviation is 1.14 so that a model which predicts a fixed number of 2 approvals can only be rejected. In another case where they observe an average of 1.86 with again a large standard deviation (0.87). Baujard and Igersheim (2010) make similar observations: an average of 2.23 out of 12, but 2-name ballots represent only about 30% of the ballots. In those cases, obviously, the votes cannot be explained correctly by a model which stipulates that voters approve exactly two candidates.

14.4.5 Conclusion on Political Domain Experiments

The experiments of Blais et al. are designed to observe in the laboratory the behavior of voters in a situation that has been identified, in practice and in theory, as politically relevant: the standard situation in which a centrist candidate is surrounded by two moderate candidates, with extremist, non-viable candidates also running for office. Then the centrist candidate, even being a Condorcet winner, can fail to be elected under one-round plurality voting and under two-round majority voting because votes can gather on the two moderate candidates. The experiments show that this phenomenon does not appear with approval voting. This is easily understandable since, with approval voting, a strategic vote for a candidate that is not a first best does not preclude the voter from also voting for the candidates she prefers.

From the methodological point of view, these experiments confirm the importance of individual strategic voting in one-round plurality or two-round majority elections. For approval voting, with five candidates, votes are easy to explain. Several alternative theories can be found that coincide and explain more than 80% of the observed votes, even if the strategic voting theory performs better than the other theories.

14.5 Conclusion

14.5.1 Consensual Outcomes

All the experiments show that approval voting, compared to other methods, tends to favor consensual outcomes. This result is obtained in different settings, which mimics idealized problems of collective choice or were designed to study other, more common voting rules.

In one case ("divided society"), the society is divided in two groups of equal size; the problem is that the social optimum is the first choice of no group. This is a problem of "parochialism": defending one's own group rather than the general interest.

The second case ("split majority") is totally different: there is a clear majority and the social optimum is the majoritarian one, that is to follow the majority's will; the problem is that the majority may split its vote among two candidates, and thus the election may fail to detect the, majoritarian, social optimum.

The third case ("single-peaked domain") is closer to actual politics and takes elements from the two previous problems: like in the second setting, the social optimum is the majoritarian outcome (a "Condorcet winner"), but the structure of the society is such that a division in two groups of equal size may appear, like in the first setting. Indeed some voting rules may create, or at least make effective, a left–right division. This is obtained in a single-peaked profile whenever only one left-wing and one right-wing candidate are perceived as viable.

Approval voting seems to be immune to these pathologies and able to help choosing consensual, socially optimal, outcomes. This is an important result, that does not conflict with the field observation, nor with the theoretical knowledge, nor with simulation work, nor with the observations made during *in situ* tests during French and German elections.

Note that in all these experiments, the set of available alternatives (options, candidates, etc.) is fixed. Note also that these laboratory experiments were done with a small number of alternatives: 3 in most cases, and 5 in the experiments by Blais et al., only 2 of them being in practice viable. In most countries, the number of effective parties is much larger, so increasing the number of viable and non-viable candidates would bring the experimental protocol closer to political realism. Moreover, the way parties adapt their platforms must also be considered. But the value-added of laboratory research, in that case, is not to be found on the side of the political supply but rather on the side of voters' behavior.

14.5.2 Voting Behavior

Theories of individual behavior under approval voting have delimited the set of ballots an individual could a priori reasonably cast. For instance, the requirements of sincerity and admissibility drastically narrow down the number of ballots for each voter; typically, this number drops from 2^K to $(K - 1)$. But this is not enough still, because a predictive theory must aim at making essentially unique predictions. This is achieved by the rationality postulate of utility maximization, at a cost: strategic behavior is not a priori, but is a response to a belief about what the others do. This point is nicely put on the stage in the laboratory when we allow repetition. Then, voters have the possibility to base their thinking on informed beliefs, as it is the case in the real world.

In the case of approval voting, voter's behavior seems to be best described by the theory of strategic voting. But even those who do not trust theories of perfect rationality should not be afraid by such a statement. In effect, unlike what happens with other voting rules, voting strategically under approval voting is usually simple and natural, and does not contradict voter's sincerity. This is what is seen in these experiments when individual votes (rather than group outcomes) are analyzed. Specific preference profiles should be invented and used to push the strategic theory at its limit, when its predictions would be counter-intuitive. In the present state of the art and as concerns approval voting, strategic voting is pretty satisfactory.

References

Alós–Ferrer, C., & Granić, D.-G. (2010). Approval voting in Germany: description of a field experiment. In J.-F. Laslier & R. Sanver (Eds.), *Handbook on approval voting*. Heildelberg: Springer.

Austen-Smith, D., & Banks, J. (1996). Information aggregation, rationality, and the Condorcet jury theorem. *American Political Science Review*, *90*, 34–45.

Baron, J., Altman, N. Y., & Kroll, S. (2005). Approval voting and parochialism. *Journal of Conflict Resolution*, *49*, 895–907.

Baujard, A., & Igersheim, H. (2010). Framed field experiments on approval voting in a political context. In J.-F. Laslier & R. Sanver (Eds.), *Handbook on approval voting*. Heildelberg: Springer.

Béhue, V., Favardin, P., & Lepelley, D. (2009). La manipulation stratégique des règles de vote: une étude expérimentale. *Recherches Economiques de Louvain 75*, 503–516.

Blais, A. (2000). *To vote or not to vote? The merits and limits of rational choice*. Pittsburgh: University of Pittsburgh Press.

Blais, A. (2002). Why is there so little strategic voting in Canadian plurality rule elections? *Political Studies*, *50*, 445–454.

Blais, A. (2004). Y a-t-il un vote stratégique en France? In B. Cautres & N. Mayer (Eds.), *Le nouveau désordre électoral: les leçons du 21 avril 2002*. Paris: Presses de la Fondation nationale des sciences politiques.

Blais, A., & Bodet, M. (2006). *Measuring the propensity to vote strategically in a single-member district plurality system*. Mimeo., university of Montréal.

Blais, A., Labbé-St-Vincent, S., Laslier, J.-F., Sauger, N., & Van der Straeten, K. (2010). Strategic vote choice in one round and two round elections: an experimental study. *Political Research Quarterly*, in press.

Blais, A., Laslier, J.-F., Laurent, A., Sauger, N., & Van der Straeten, K. (2007). One round versus two round elections: an experimental study. *French Politics*, *5*, 278–286.

Brams, S. J., & Fishburn, P. C. (2005). Going from theory to practice: the mixed success of approval voting. *Social Choice and Welfare*, *25*, 457–474. Reprinted in J. F. Laslier & R. Sanver (Eds.), *Handbook on approval voting*. Heildelberg: Springer.

Cox, G. W. (1997). *Making votes count : Strategic coordination in the world's electoral systems*. Cambridge: Cambridge University Press.

De Borda, J.-C. (1784). Mémoire sur les élections au scrutin. *Histoire de l'Académie Royale des Sciences*, *1781*, 657–665.

De Sinopoli, F., Dutta, B., & Laslier, J.-F. (2006). Approval voting: three examples. *International Journal of Game Theory*, *35*, 27–38.

Davis, D. D., & Holt, C. A. (1993). *Experimental economics*. Princeton: Princeton University Press.

Delli Caprini, M. X., & Keeter, S. (1991). Stability and change in the US public's knowledge of politics. *Public Opinion Quarterly*, *55*, 581–612.

Duverger, M. (1951). *Les Partis Politiques*. Paris: Armand Colin.

Felsenthal, D. S., Rapoport, A., & Maoz, Z. (1988). Tacit cooperation in three alternative non-cooperative voting games: a new model of sophisticated behavior under the plurality procedure. *Electoral Studies*, *7*, 143–131.

Forsythe, R., Rietz, T. A., Myerson, R., & Weber, R. J. (1993). An experiment on coordination in multicandidate elections: the importance of polls and election histories. *Social Choice and Welfare*, *10*, 223–247.

Forsythe, R., Rietz, T. A., Myerson, R., & Weber, R. J. (1996). An experimental study of voting rules and polls in three-way elections. *International Journal of Game Theory*, *25*, 355–383.
Green, D. P., & Shapiro, I. (1994). *Pathologies of rational choice theory : A critique of applications in political science*. New Haven: Yale University Press.
Kagel, J. H., & Roth, A. E. (1995). *The handbook of experimental economics*. Princeton: Princeton University Press.
Kube, S., & Puppe, C. (2009). (When and How) Do voters try to manipulate? *Public Choice*, *139*, 39–52.
Kuklinski, J. H., & Quirk, P. J. (2000). Reconsidering the rational public: cognition, heuristics, and mass opinion. In A. Lupia, M. McCubbins, & S. Popkin (Eds.), *Elements of reason: Cognition, choice, and the bounds of rationality*. Cambridge: Cambridge University Press.
Laslier, J.-F. (2009). The leader rule: model of strategic approval voting in a large electorate. *Journal of Theoretical Politics*, *21*, 113–136.
Laslier, J.-F. (2010). *In silico* voting experiments. In J. F. Laslier & R. Sanver (Eds.), *Handbook on approval voting*. Heildelberg: Springer.
Laslier, J.-F., & Van der Straeten, K. (2008). A live experiment on approval voting. *Experimental Economics*, *11*, 97–105.
Lazarfeld, P., Berelson, B. R., & Gaudet, H. (1948). *The eople's choice: How the voter makes up his mind in a presidential campaign*. New York: Columbia University Press.
Lehtinen, A. (2010). Behavioral heterogeneity under approval and plurality voting. In J. F. Laslier & R. Sanver (Eds.), *Handbook on approval voting*. Heildelberg: Springer
Lupia, A., & McCubbins, M. (1998). *The democratic dilemma: Can citizens learn what they need to know?* New York: Cambridge University Press.
McKelvey, R., & Patty, J. (2006). A theory of voting in large elections. *Games and Economic Behavior*, *57*, 155–180.
McKelvey, R. D., & Ordeshook, P. C. (1985). Rational expectations in elections: some experimental results based on a multidimensional model. *Public Choice*, *44*, 61–102.
Merrill, S. III (1981). Strategic decisions under one-stage multi-candidate voting systems. *Public Choice*, *36*, 115–134.
Moulin, H. (1983). *The strategy of social choice*. Amsterdam: North-Holland.
Myerson, R. B. (1991). *Game theory: Analysis of conflict*. Cambridge: Harvard University Press.
Myerson, R. B. (1993). Incentives to cultivate favored minorities under alternative electoral systems. *American Political Science Review*, *87*, 856–869.
Myerson, R. B. (1995). Analysis of democratic institutions: structure, conduct and performance. *Journal of Economic Perspectives*, *9*, 77–89.
Myerson, R. B. (2002). Comparison of scoring rules in Poisson voting games. *Journal of Economic Theory*, *103*, 219–251.
Myerson, R. B., & Weber, R. J. (1993). A theory of voting equilibria. *American Political Science Review*, *87*, 102–114.
Rapoport, A., Felsenthal, D. S., & Maoz, Z. (1991). Sincere versus strategic behavior in small groups. In T. Palfrey (Ed.), *Laboratory research in political economy*. Ann Arbor: University of Michigan Press.
Riker, W. H. (1982). *Liberalism against populism: A confrontation between the theory of democracy and the theory of social choice*. San Francisco: W.H. Freeman.
Sniderman, P. M. (1993). The new look in public opinion research. In A. Finifter (Ed.), *Political science: The state of the discipline II*. Washington: American Political Science Association.
Sniderman, P. M., Tetlock, P. E., & Brody, R. A. (1991). *Reasoning and choice: Explorations in political psychology*. Cambridge: Cambridge University Press.
Taagepera, R. (2007). Electoral systems. In C. Boix & S. Stokes (Eds.), *The Oxford handbook of comparative politics*. Oxford: Oxford University Press.
Van der Straeten, K., Laslier, J.-F., Blais, A. & Sauger, N. (2010). Sincere, strategic, and heuristics voting under four election rules: an experimental study. *Social Choice and Welfare*, in press.

Chapter 15
Framed Field Experiments on Approval Voting: Lessons from the 2002 and 2007 French Presidential Elections

Antoinette Baujard and Herrade Igersheim

15.1 Introduction

Competitive elections are an essential feature of representative democracies; thus, the choice of voting method is partly constitutive of the form of the democracy. Clearly, this engenders fundamental debates on the properties that acceptable voting rules should and should not exhibit. These debates take place primarily in two spheres: the public and the scientific. Let us here consider an example from France. The President of the French Republic is elected by direct universal suffrage, on the basis of a two-round plurality vote. In other words, run-off voting ensures that the elected President always obtains a majority. On each round, each voter can vote for one and only one candidate. If no candidate receives a majority of votes in the first round of voting, there is a run-off between the two highest-scoring candidates. The winner of this latter round is the winner of the election. Hence, each round is determinant for the result and considered as an important source of information on citizens' political preferences. The results of the first round of the 2002 French presidential election were a shock for a large part of the population: contrary to the predictions of the opinion polls, the candidate for the extreme Right, Jean-Marie Le Pen, and the sitting president, Jacques Chirac, were selected for the second round. This surprise has contributed to serious public debate on the mechanisms of the two-round single-name vote. This discussion focuses in particular on the tension between tactical and sincere voting, with many citizens pleading for the adoption of a voting method which would allow better expression of their true preferences.

Alongside the debates in the public sphere, voting rules have been the subject of extensive theoretical study since the works of Borda (1781) and Condorcet (1785). Theoreticians have established numerous results that illuminate the properties of different voting rules. In particular, Brams and Fishburn (1983) have shown that Approval Voting (henceforth, AV) is endowed with many favorable properties (e.g. providing strong incentives for sincere voting, and having a high probability of

A. Baujard (✉)
CREM, University of Caen Basse-Normandie, Caen, France
e-mail: antoinette.baujard@unicaen.fr

J.-F. Laslier and M.R. Sanver (eds.), *Handbook on Approval Voting*, Studies in Choice and Welfare, DOI 10.1007/978-3-642-02839-7_15,

electing the Condorcet candidate). Let us briefly recall the principle of AV: AV is a voting rule in which voters can approve of as many candidates as they want, and the winner of the election is the candidate who obtains the highest number of approvals. At first sight, the principle of AV seems quite simple to understand and to apply. Further, AV appears to have good prospects for meeting the expectations of voters, since it theoretically allows them to give their opinion about all the candidates – as opposed to a two-round voting system that restricts them to picking a single candidate.

Only field experiments – i.e. experiments carried out in the voting stations, with real voters and citizens – could determine whether AV is accepted by the electorate, confirm (or invalidate) theoretical claims made about its properties in political contexts, and, above all, show that voters' aims can be combined with scientific results in a manner that elaborates a better democracy. Although laboratory experiments are of obvious interest regarding the properties of AV and voter rationality (see, in this volume, Laslier 2010b), since they provide the only suitable protocol to control preferences, it proves hard to convince the public and policymakers of the relevance of lab findings for real elections, in which the political context determines not only strategic information and beliefs, but also the expression of voters' rationality. More generally, conventional lab experiments are often criticized, first, for providing biased and unrepresentative results, since it is mostly students who participate, and, second, for factoring out the wider political context, even though *the context itself is relevant to the performance of subjects*.

However, large-scale field experiments can hardly be conducted within a political context. Harrison and List (2004) stress that a main feature of the natural field experiment is that "the environment is one where the subjects naturally undertake these tasks and where the subjects do not know that they are in an experiment." But conducting such a field experiment would imply that the voting system was dependent on the experiment rather than on the Constitution, that different voting rules were being used for different groups of voters (due to different rules applying to control groups), and that the voters did not know in advance which official rule had been chosen: for Constitutional reasons at the very least, these traits straightforwardly rule out large-scale natural field experiments within political contexts. However, there is a bridge between lab and natural field experiments: these are *framed field experiments*, which Harrison and List (2004) describe as being undertaken "in naturally occurring settings, in which the factors that are at the heart of the theory arise endogenously, and on which the remaining controls needed to implement the experiment are then imposed. In other words, rather than impose all the controls exogenously on a convenient sample of college students, Harrison and List (2004)[1] locate a population in the field in which one of the factors of interest arises naturally and can be easily identified, and then add the necessary controls." In the context of the French presidential election, identifying the relevant population is straightforward: all official voters are good candidates for the experiment. Further, by tying

[1] The working paper here referred to is now published as Harrison and List (2008).

the framed field experiment closely to the official election, the experimental circumstances (site, date, process, etc.) should here mimic exactly the official voting circumstances.

What we could now call a French school of large-scale framed field experiments in a political context was inaugurated in 2002. During the 2002 French presidential election, Balinski, Laraki, Laslier, and van der Straeten conducted a framed field experiment (henceforth, the 2002 experiment) with 5,000 voters, in order to test AV in a large-scale election. A similar experiment – though with the design slightly modified in line with the results from tests of Evaluation Voting (henceforth, EV) conducted by Baujard and Igersheim – was used during the 2007 French presidential election, with the participation of over 5,500 voters (henceforth, the 2007 experiment). In each case, a large team of researchers and student volunteers worked on the realization of the experiment. The objectives were, first, to evaluate the feasibility of this kind of large-scale experiment in political elections; second, to check whether AV is comprehensible to, and accepted by, a large proportion of the public; third, to compare the results obtained under AV with those of the official election, in order to investigate how AV behaves statistically with real electoral preferences, and to determine the extent to which different voting rules may yield different outcomes; fourth, to facilitate an extended analysis of voters' behavior with respect to AV, and on the French political supply structure, based on the data collected.

The remainder of the paper is organized as follows: Sect. 15.2 is a presentation of the experimental design; Sect. 15.3 sets out wider lessons from AV framed field experiments; Sect. 15.4 surveys the specific lessons on AV derived from the present experimental data; Sect. 15.5 concludes.

15.2 Experimental Design

As emphasized above, (framed) field experiments on AV are necessary in order that voters' expectations concerning the capacity for democratic expression embodied in a voting rule can be integrated with the hypotheses put forward by theoreticians about its properties. In France, the presidential election is the appropriate setting in which to run such a large-scale experiment. First, it aims at selecting one winner from a list of candidates which is the same all over France, using an official voting method – the two-round vote – which is similarly uniform. Second, it has the highest rate of participation (and thus promises the most representative experimental results) of all the official ballots.

The idea of conducting a large-scale experiment on AV in the context of a presidential election may have first been raised by Mann in his PhD on AV, defended in 1995 in the École Polytechnique. The first large-scale experiment on AV was conducted on April 21st, 2002, during the first round of the French presidential election, by a team of researchers from the Laboratoire d'Économétrie of the École Polytechnique: Balinski, Laraki, Laslier and van der Straeten (Balinski et al. 2002, 2003; Laslier and van der Straeten 2004, 2008). This seminal experiment generated an original protocol, on the basis of which every large-scale political experiment subsequently conducted in France has proceeded.

Numerous experiments took place in parallel with the first round of the next French presidential election, on April 22nd, 2007: one by Balinski and Laraki on "majority judgment" (Balinski and Laraki 2007b, c), another by Farvaque, Jayet and Ragot on the single transferable vote (Farvaque et al. 2009), and a third by Baujard and Igersheim on AV and EV (Baujard and Igersheim 2007a, b, c; Baujard and Igersheim 2009).[2] All three experiments were based on the protocol introduced by the seminal 2002 experiment. The major differences between the four protocols lie, of course, in the voting rules tested and some corresponding specific improvements. Balinski and Laraki experimented regarding the "majority judgement" method in three polling stations in Orsay: on this voting rule, each voter evaluates every candidate in a common language of grades, and a candidate wins the election if she has the highest final grade based on the median of these evaluations (on this method and similar techniques, see Bassett and Persky 1999; Gehrlein and Lepelley 2003; Balinski and Laraki 2007a; Laslier 2009a). Farvaque, Jayet and Ragot tested the single transferable vote method in two voting posts in Faches-Thusmenil in Nord-Pas-de-Calais. Finally, Baujard and Igersheim conducted their experiment on AV and EV in six polling stations: three in Illkirch-Graffenstaden (Alsace), two in Louvigny (Basse-Normandie), and one in Cigné (Pays de Loire). Since what is at stake here is AV, we restrict our attention to the seminal 2002 experiment and to the 2007 Baujard-Igersheim experiment. In each case, the experimental voting stations were located near the official ones, and only those voters who were registered for the official vote could actually participate in the experiment.

Let us now present the protocol in detail. An essential preliminary step was to present the experimental materials and instructions to many different sectors of the public – administrative staff and students, individuals and groups (to favor brainstorming) – in order to test the material and ensure that it was clear and informative. The second step was a pilot experiment. Once the actual participants had been identified and contacted, the experiment proper took place; and the last step was the circulation of results and subsequent public debate.

15.2.1 The Experimental Voting Rules and Ballots

In line with the subject of this book, we restrict our attention here to experiments on approval voting. Although both the 2002 and 2007 experiments were concerned to test AV, certain other rules, such as EV, were also under consideration in the 2002 pilot experiment and in the 2007 experiment proper. This explains some important distinctions between the 2002 and 2007 ballots.

---

[2] See also Alós-Ferrer and Granić (2010) in this volume regarding the January 2008 experiment conducted by Carlos Alós-Ferrer in Messel (Germany).

Frame 1 Experimental ballot of the 2002 experiment

Expérience de vote : *Quel scrutin pour quelle démocratie ?*

Bulletin de vote

Bruno Mégret	
Corinne Lepage	
Daniel Gluckstein	
François Bayrou	
Jacques Chirac	
Jean-Marie Le Pen	
Christiane Taubira	
Jean Saint-Josse	
Noël Mamère	
Lionel Jospin	
Christine Boutin	
Robert Hue	
Jean-Pierre Chevènement	
Alain Madelin	
Arlette Laguiller	
Olivier Besancenot	

Règlement du vote par assentiment : L'électeur vote en mettant des croix dans la deuxième colonne du bulletin. Il peut mettre des croix pour autant de candidats qu'il le souhaite, mais pas plus d'une croix par candidat. Est élu le candidat qui obtient le plus de croix.

In the 2002 experiment, the experimental ballot (see Frame 1) presented the list of the sixteen candidates in the official election, and participants were requested to mark a cross next to the name of the candidates they wanted to approve. Notice that the order of the candidates on the 2002 and 2007 ballots is the official Constitutional order.

With EV, voters assess candidates by giving them a grade or score on a predefined scale – for instance, integers from 0 to 99 (as in http://rangevoting.org/), from 0 to 20 (as in European school grades), or from −2 to +2 (as in http://votedevaleur.info/). A candidate wins the election under EV if she has the highest sum of grades. Among the many ways in which EV systems can be set up, the rule tested in the Institut d'Etudes Politiques of Paris in January 2002 was a 10 points EV, which could be called EV10. Baujard and Igersheim chose to test the rule based on a three level scale, which Hillinger (2004a, b, c, 2005) calls EV3. It appeared, indeed, to be the most simple model, which avoids major problems of interpersonal comparison of grades, and prevents any problem of confusion between being indifferent to and disliking a candidate (see Baujard and Igersheim 2007a for arguments in favor of this option). Note, though, that Felsenthal (1989) has proposed that EV3 is really an extension of AV, interpreted as a combination of approval and disapproval voting. For each option, voters would be able to choose between three options: approve, disapprove, and abstain. A candidate's score is the difference between the number of approvals and the number of disapprovals, and the winner is the candidate with the highest score. Felsenthal has studied the possibilities for manipulation in the case of a low number of voters. He has shown that, under the assumptions of perfect information and voter rationality, the collective result would be the same under either EV3 and AV. He has also shown that, for each voter, the probability to be decisive is higher with EV3 than with AV[3]. Hillinger (2004a, b, c, 2005) sets out the general proprieties of EV and defends his preference for EV3 on the basis of pragmatic arguments. The problem with more fine-grained scales derives from the difficulty for voters to attribute specific meaning to the different points on the scale, and for this sense to be interpersonally comparable among different voters.

Let us now go further. Any three level system of EV is, on a theoretical basis at least, equivalent: thus, the (2,1,0) scale should provide results equivalent to (1,0,−1).[4] The latter scale is attracting the most support both from theoretical studies and from laymen. It is indeed a very attractive way to evaluate candidates: the meaning of −1 is quite easy to intuit, relative to +1. This straightforward interpretation would guarantee homogeneous interpretations of grades among voters: approved (+1), disapproved (−1), and acceptable or indifferent (0). In spite of these uncontroversial advantages, there are two main hindrances for (−1,0,1) EV3. First, there is the problem of confusion that could occur between the average grade (acceptable) and actual indifference (that is to say, an incompletely graded ballot paper). We have, in field experiments, to accept this possibility and consider how to treat ballots in which no grade is given for one or several candidates. This amounts to indifference, and thus it makes sense to attribute 0 to these candidates. A rule in which we attributed −1 to a candidate who has received no grade would be hardly accepted, since indifference would normally be graded 0. We thus have to consider the effect of attributing a grade of 0 to these candidates. We have observed, for instance, that the candidate Schivardi was very often not evaluated in 2007; in practice, this meant that no cross would appear in any box next to his name. See Frame 2 as an illustration. In a (2,1,0) EV3, all abstentions as regards this candidate would be considered as a zero; it thus does not raise his score. If we had taken a (−1,0,1) EV3, each abstention would have provided him with a relatively higher score than a candidate who was evaluated −1. This would tend to lead to higher average scores for candidates who attract more indifference – because, for instance, voters do not know who they are or what political values they represent – than for candidates

[3] See Laslier and Sanver (2010) and Núñez (2010) in this volume for a definition of "decisiveness."

[4] Let us remark, though, that, intuitively, it is not obvious that similar individual preferences would induce equivalent results with these two versions of EV3 in actual political contexts as opposed to theoretical settings.

Frame 2 Experimental ballot for the 2007 experiment, both sides

Cette expérience, qui vise à étudier les comportements des électeurs face à un mode de scrutin différent, est simultanément réalisée dans trois communes de France.

Centre d'analyse stratégique

PARTICIPEZ A UNE EXPERIENCE DE VOTE

A l'occasion des élections présidentielles françaises de 2007, notre équipe de recherche teste deux nouveaux modes de scrutin à un seul tour.

Nous vous proposons de participer à cette expérience en remplissant les **deux bulletins de vote** qui suivent.

Nous sommes à votre entière disposition pour répondre à vos questions.

Nous vous remercions par avance de respecter le secret et la sérénité du scrutin.

Merci de votre participation !

Vote par note

Bulletin de vote expérimental n° 1

Instructions :

Vous donnez une note à chacun des 12 candidats: soit 0, soit 1, soit 2 (2 étant la meilleure note et 0 la plus mauvaise).

Pour cela, mettez une croix dans la case correspondante. Si vous ne souhaitez pas noter un candidat, ne cochez aucune case de la ligne.

Attention : si plus d'une seule case par ligne est cochée, le bulletin est nul dans sa totalité.

Le candidat élu avec le mode de scrutin expérimental n° 1 est celui qui comptabilise le plus de points.

	2	1	0
Olivier Besancenot			
Marie-George Buffet			
Gérard Schivardi			
François Bayrou			
José Bové			
Dominique Voynet			
Philippe de Villiers			
Ségolène Royal			
Frédéric Nihous			
Jean-Marie Le Pen			
Arlette Laguiller			
Nicolas Sarkozy			

Vote par approbation

Bulletin de vote expérimental n° 2

Instructions :

Vous indiquez, parmi les 12 candidats, quels sont ceux que vous soutenez.

Pour cela, entourez le nom du ou des candidats que vous soutenez. Vous pouvez entourer un seul nom, plusieurs noms ou aucun nom.

Attention : entourez les noms un à un. Si plusieurs candidats sont entourés ensemble, le bulletin est nul dans sa totalité.

Le candidat élu avec le mode de scrutin expérimental n° 2 est celui qui reçoit le plus grand nombre de soutiens.

Olivier Besancenot
Marie-George Buffet
Gérard Schivardi
François Bayrou
José Bové
Dominique Voynet
Philippe de Villiers
Ségolène Royal
Frédéric Nihous
Jean-Marie Le Pen
Arlette Laguiller
Nicolas Sarkozy

who attract relatively more approvals and disapprovals. In an election, it is quite unacceptable to take the risk of favoring – or indeed electing – a candidate nobody knows. Second, the average grade is equal to the total of the scores divided by the number of ballots. With (−1,0,1) it is likely that the winning candidate would win the election with an average negative score. It seems to us to be quite unsatisfactory that the legitimacy of the winning candidate should be based on a negative score. For these two reasons, we selected the (2,1,0) EV3 rule. We would expect voters to consider grade 2 as denoting "approved or preferred candidates," grade 1 as denoting "acceptable candidates," and 0 as either "to be rejected" or "indifferent candidates." This is the only method which avoids the risk of electing a candidate who attracts mostly indifference, and which is at the same time also simple, transparent, and permissive of a wide scope of expression.

The AV and EV experimental ballots were registered on the same sheet of paper, as shown in Frame 2, in order to enable comparisons between how participants vote under both voting rules. This particularity introduced a change in the protocol: while voters were requested to indicate with a cross the grade they attributed to each candidate under EV, they were asked to circle the names of the candidates they wanted to approve of under AV – and not to mark a cross next to them, as in 2002 – in order to avoid any ambiguity between AV and EV.

15.2.2 The Pilot Experiments

Pilot experiments are necessary in fieldwork in order to guarantee the quality of the experimental protocol. Balinski, Laslier and van der Straeten conducted a pilot experiment on January 23rd, 2002, in the Institut d'Etudes Politiques. As mentioned above, this used a slightly different protocol since, in addition to AV, an EV from 0 to 10 points was also tested. See Balinski et al. (2002) for a brief presentation.

Baujard and Igersheim carried out their pilot experiment on March 20th, 2007. This took place from 11:15 to 14:00 in Caen University Restaurant A, on Campus 1 in Caen, and was made possible by the assistance of around ten colleagues and PhD students. This restaurant serves up to 2000 persons a day in a very short period of time, the clientele comprising students and any other persons working in the university. The date was chosen to be close enough to the presidential election for information about candidates to be available to the participants, but also far enough away that the organizers would be able to change the protocol in case this was indicated as necessary by the results of the pilot. The pilot took place on a Tuesday as this was supposed to be the busiest day of the week for this university restaurant. The protocol was basically the same as the official one and will be presented below. No previous information had been given out; the potential participants learned of the experiment when entering the restaurant, through posters and flyers which were personally given to each of them as they entered the hall (see the poster on Frame 3). The experimental voting station was just next to the exit. Voters did not spend more than 3 min engaged in the whole process.

The pilot experiment was a success, with 447 persons, mostly students, participating, and 300 of these agreeing to devote some more time to filling in a questionnaire. The results of the questionnaire were especially interesting.

Representative negative points were the following. First, the lack of a blank was resented; this remark led to a modification of the protocol. Second, there were

Frame 3 Poster to inform people of the experiment

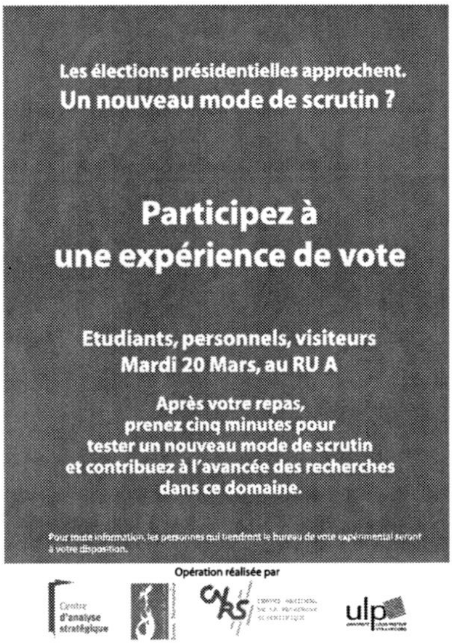

Pilote Experiment

Louvigny Experiment

complaints that there was an excessive delay before getting access to the results; this was due to the fact that the Centre d'Analyse Stratégique (CAS, Paris) – our main partner for the 2007 experiment – did not allow us to circulate the results before the end of the legislative election, which was held in late June 2007, i.e. 2 weeks after the second round of the presidential election and thus 3 months after the pilot experiment. Third, they said that they received information about the experiment too late; it seems that many other people would have liked to participate and that they would have appreciated the chance to alert friends. Fourth, a small number of people expressed strong disagreement with the voting methods tested, while also saying or otherwise indicating that they did not understand them. We assumed that those who knowingly disagreed with the voting method itself, or with the very idea of conducting such an experiment, would have rather decided not to participate.

Representative positive points were the following. First, the experimenters were congratulated for having taken the initiative to conduct an experiment on voting rules and on the presidential election. Second, appreciation was expressed for the organization and the organizational team: in particular, the location in the university restaurant, the reproduction of the official conditions (see Sect. 2.4), the simplicity of the protocol and the fact that it required very little time, the clear explanations in the documents, the fact that it was organized by researchers, and the welcoming and friendly attitude of organizers were commended. Third, participants liked the fact that we were addressing students, either because they felt that society rarely

Frame 4 Information meeting

Mardi soir, au foyer des anciens, une réunion d'information était organisée par le CNRS

Source: Ouest France, April 19th 2007, p. CAN19.
A picture illustrating the information meeting that was held in Louvigny on April 17th.

concerned itself with what students thought, or because they considered this a way for them to educate themselves in political thinking; one said that the experiment would induce students to think hard before voting. Fourth, the voting methods inspired many positive reactions: the students were glad to be engaged in "questioning the voting rule as an institution"; they appreciated the opportunity it afforded them to enhance their expressive ability through their vote: one said that the plurinominal systems "helped to go beyond his or her fear of voting by being able to approve several candidates" and another concluded that they engendered "a less silent vote," giving more relevant information for journalists to analyse political opinions. The experiment was considered as another way in which to organize a poll, but under "more satisfying conditions than the polls that are run by private institutes". Many concluded by saying they would be glad to see these voting rules being officially adopted, even though some were pessimistic that this could ever happen.

The pilot experiment led to some cosmetic modifications of the experimental design and of organizational aspects. In particular, blanks for each candidate were authorized under the evaluation rule, and some questions in the questionnaire were rephrased.

15.2.3 Information for Participants

The next step of the experimental protocol consisted in providing information about the experiment to the persons who would participate.

Three media were used for this. First, thanks to the active help of the town councils, 1 week before the election, a letter was sent to each registered voter of the six voting stations, explaining the principle of AV and requesting her participation in the experiment. In some cases, named letters were sent; in others, the letter was included in the local associations paper which is sent to each house in the town.

Second, information meetings were scheduled in each town, as illustrated by the newspaper picture in Frame 4 (see also the poster announcing the experiment and the information meeting in Louvigny, in Frame 3).

Third, general media relayed the information. In Orsay in 2002, an article in the municipal bulletin, sent 1 month in advance, announced the experiment. For the 2007 experiment, newspapers, local radio and national TV spots were of great help to make sure each voter knew about the experiment before coming to the voting stations. Nevertheless, we should acknowledge that around ten persons for our six polling stations came and regretted that they had not been informed about it: most of them were students who had been out of town or had just moved abroad. They generally claimed that the most effective way to inform people would have been to have sent the information letter along with the official voting papers; we had to answer that this "solution" was strictly forbidden.

15.2.4 The Experiments

On April 21st, 2002, Balinski, Laraki, Laslier and van der Straeten ran their field experiment during the first round of the French presidential election in six polling stations located in two towns: the single voting station of Gy les Nonains, in the region Centre, with 482 registered voters, and five voting stations out of the twelve of Orsay, in the region Ile-de-France, representing 4,237 registered voters.

On April 22nd, 2007, Baujard and Igersheim ran their experiments under similar conditions in six polling stations in three towns: the two voting stations of Louvigny in the region Basse-Normandy, representing 1948 registered voters, the single voting station of Cigné in the region Pays-de-Loire, representing 378 registered voters, and three voting stations out of the sixteen of Illkirch-Graffenstaden in the Alsace region, representing 3,211 registered voters. From Basse-Normandy to Alsace, these towns are very far from each other and thus the voters of the six stations belonged to a wide scope of political patterns, in terms of the respective electorate, social and economic class, size and rural/urban characteristics. Therefore, voters' reactions to the experiment could be expected to be different and/or more diverse.

On the day of the election, April 21st, 2002, the town councils in each locality allowed the researchers of the Laboratoire d'Econométrie to set up experimental voting posts in the immediate vicinity of the official voting stations (either in the same room or in an adjacent room), where voters – once they had made their official vote – were requested to proceed. Frame 5 shows the organisation of the building – a primary school – where the experimental and official voting posts were located in Louvigny for the 2007 experiment. The other picture gives an idea of the tags, pathways and badges. Yellow was used for communications regarding the experimental vote; this color was chosen, with the explicit agreement of the town councils, because it was the only color that was not easily associated with a political party. Tags in white, conversely, signaled official voting posts.

Frame 5 Pictures of the 2007 experiment in Louvigny. Orientating voters in the voting building

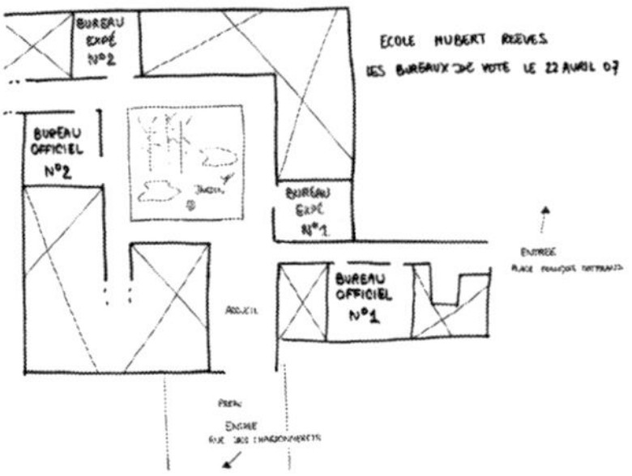

Map of the voting posts in the school building

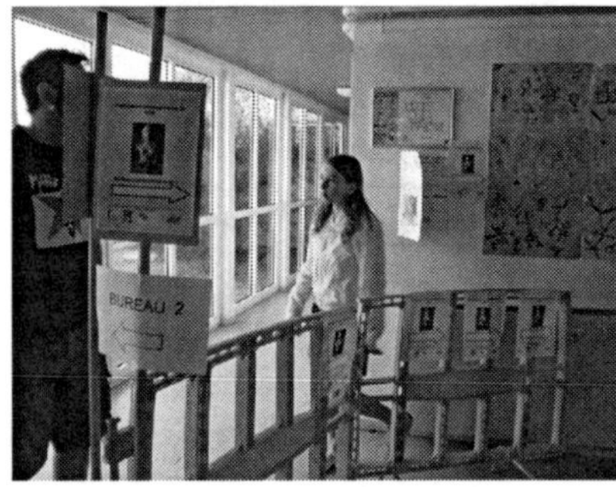

Pathways to orientate voters to experimental or official voting posts

Frame 6 Pictures of the 2007 experiment in Louvigny: The anonymous voting process

Voting booths. Once the experimental ballot is filled in, it is put in an envelope, just as in the official voting process.

See in the back the table where experimental ballots were distributed to arriving voters; on the board there are posters asking for silence.

After filling in her vote in the locked booth, this lady has registered with the experimental assessors. Each person can vote only once.

The test of AV reproduced the modus operandi of the official elections; that is, the research team respected a similar rhythm, with similar opening and closing hours, and reduced waiting time. The staff was also similar – with a president of the polling station and assessors – as was the voting equipment: envelopes, ballot papers, polling booth, ballot box, and a book to gather participants' signatures. The rules to guarantee conditions of anonymity were also similar, with the first voter of the day checking the empty box, the box then being locked and opened at closing time in front of voters; silence in the polling station was also maintained. The pictures of the vote in Louvigny in Frame 6 give an idea of the process.

15.2.5 *Questionnaires*

Many oral remarks by participants were noted by the 2002 researchers, and these have provided many insights into rules and experiments in general, as is reported

Frame 7 Pictures of the 2007 experiment in Louvigny: Answering questionnaires

in Laslier and van der Straeten (2004). This is why more systematic questioning of participants was undertaken in the 2007 experiment. After the experimental ballot, participants were invited to fill in a questionnaire about the experiment and the two voting methods. They could either answer it right away or send it afterwards by post or e-mail. Notice that the questionnaires were available on the experiment official internet site [URL: http:unicaen.fr/crem/vote]. Frame 7 shows the serious attitude of participants in answering questionnaires. Contrary to expectations – we would have thought that participants would forget to reply, or would choose not to devote more time to it – a significant number of people did send back their questionnaires, by both e-mail and post.

These questionnaires, presented in Frame 8, provide rich information about how participants reacted to the experiment and, more specifically, about the voting rules under test. The lessons drawn therefrom will be presented in the next section.

15.2.6 Participating in a Public Debate

The final, and very significant, stage of the framed field experiments was for information about the results to be fed back to the participants and the public, and for the organizers to take part in the public debate on voting methods.

The actual work of registering the results had been completed just after the day of the experiment, but the results could not be made public before the end of the official elections. In 2007, the Centre d'Analyse Stratégique (CAS) had even required the experiment team not to give out any information on the results per candidate (both for the pilot and the experiment proper) until late June, i.e. after the end of the legislative elections.

Frame 8 Questionnaire for the 2007 experiment

EXPERIMENTATION DE NOUVEAUX MODES DE SCRUTIN
Questionnaire sur l'expérience

Nous vous remercions par avance de bien vouloir nous aider à évaluer notre expérience en répondant aux questions suivantes. Répondre à ce questionnaire devrait vous prendre **moins de 5 minutes.**

		Oui	Un peu	Plutôt non	Non	Sans opinion
	Sur le vote officiel					
1	Vous avez aujourd'hui voté pour un candidat aux élections présidentielles. Parmi les informations qui ont déterminé votre choix, lesquelles ont été les plus déterminantes ? - Les programmes des candidats ? - Les informations issues des sondages ?	o o	o o	o o	o o	o o
2	Avez-vous changé d'avis sur votre choix de vote ou déterminer votre choix ces 15 derniers jours ?	o	o	o	o	o
3	Avez-vous voté au 1er tour en tenant compte de ce qui pourrait arriver au 2ème tour ?	o	o	o	o	o
4	Souhaitez-vous que le candidat pour lequel vous venez de voter soit présent au 2ème tour du scrutin ?	o	o	o	o	o
5	Estimez-vous que le raisonnement que vous suivez au moment de voter est différent depuis les dernières élections présidentielles en France (2002) ?	o	o	o	o	o
	Sur le vote expérimental					
6	Le principe du mode de scrutin par note vous semble-t-il clair ?	o	o	o	o	o
7	Le principe du vote par approbation vous semble-t-il clair ?	o	o	o	o	o
8	Pensez-vous que des chercheurs doivent étudier les modes de scrutin ?	o	o	o	o	o
9	Connaissez-vous d'autres modes de scrutin que le scrutin majoritaire à deux tours, le scrutin proportionnel et ceux de cette expérience ?	o	o	o	o	o
10	Si le vote par approbation ou le vote par note était le mode de scrutin officiel, cela influencerait-il le raisonnement que vous tenez au moment de voter ?	o	o	o	o	o

11 Pour quelles élections officielles estimez-vous que la méthode par note pourrait être utilisée ? (Cochez une ou plusieurs cases)
- O Pour les élections présidentielles
- O Pour les élections législatives
- O Autres. Précisez :......................................
- O Vous ne pensez pas que la méthode par note puisse être utilisée pour un scrutin officiel

12 Pour quelles élections officielles estimez-vous que la méthode par approbation pourrait être utilisée ? (Cochez une ou plusieurs cases)
- O Pour les élections présidentielles
- O Pour les élections législatives
- O Autres. Précisez :......................................
- O Vous ne pensez pas que la méthode par approbation puisse être utilisée pour un scrutin officiel

	Sur l'expérience en elle-même	Oui	Un peu	Plutôt non	Non	Sans opinion
13	Etes-vous satisfait d'avoir participé à cette expérience ?	o	o	o	o	o
14	Si vous aviez des questions sur l'expérience, avez-vous trouvé des interlocuteurs et les réponses que vous attendiez ?	o	o	o	o	o
15	Seriez-vous prêt à participer à nouveau à une expérience scientifique sur votre comportement de vote ?	o	o	o	o	o
16	Estimez-vous avoir été suffisamment informé sur cette expérience ?	o	o	o	o	o

17 Comment avez-vous été informé de l'existence de l'expérience ? (Cochez une ou plusieurs cases)
- O Vous n'avez pas été informé avant le jour du scrutin
- O Par bulletin municipal
- O Par courrier adressé à votre domicile
- O Par les affichages dans votre commune
- O Par la presse
- O Par le bouche à oreille
- O Autres. Précisez :......................................

18 Qu'avez-vous apprécié dans cette expérience ?
....

19 Qu'avez-vous désapprouvé dans cette expérience ?
....

20 Selon vous, quels sont les différents enjeux de cette expérience ?
....

21 Autres commentaires :
....

Merci de bien vouloir nous remettre ce questionnaire, une fois rempli par vos soins, soit :
- directement auprès des expérimentateurs présents dans les bureaux de vote le 22 avril 2007 ;
- par courrier à l'adresse suivante :
  Expérimentation de nouvelles méthodes de vote – Antoinette Baujard
  Université de Caen, CREM, Campus de Claude Bloch, 14 032 Caen Cedex
- Vous pouvez également remplir un questionnaire disponible sur le site internet de l'expérience (http://www.unicaen.fr/crem/vote) et le renvoyer par mail à: Antoinette.Baujard@unicaen.fr

Toute l'équipe de chercheurs des Universités de Caen et de Strasbourg qui travaille sur les modes de scrutin vous remercie de votre participation.

Centre d'analyse stratégique

The presentation of the different steps of the experiment in newspapers helped to create a debate not only between the organizers and the public, but also among the public more broadly. For the 2007 experiment, for instance, Frame 9 gives a short list of the different papers, TV, radio or internet programmes that discussed it. To illustrate this, the article in *Le Monde*, a national newspaper, is reproduced in Frame 11, and the pictures used in different regional newspapers in Frame 10.

The participation in various conferences organized by the Centre d'Analyse Stratégique (see Baujard and Igersheim 2007c, and the debate with sociologists and lawyers in the annexes of Baujard and Igersheim 2007a), generated some interesting discussion with specialists. The publication of the report of the first results (see Baujard and Igersheim 2007b) instigated wider public debate. As well as this, the information meetings that had been organized in late June to set out the specific results for each town had led each voter participating in the experiment to realize concretely what the new voting rule would have meant in the particular case of their community. Last but not least, the publication of the official report in December 2007 (see Baujard and Igersheim 2007a) also fed this public debate with in-depth analyses of the acceptance of the new voting rules, the voting strategies of voters and their electoral preferences. It is now available on the CAS website [URL: http://www.strategie.gouv.fr] and on the experiment website [URL: http://www.unicaen.fr/crem/vote]. A caricature published in a local newspaper of Mayenne, inspired by the results of Cigné in 2007, illustrates with humor the kind of debate that was inspired by the experiment: see Frame 12.

Frame 9 A short list of the media coverage of the 2007 experiment

National newspapers: *Le Monde* (April, 26th 2007, p.3); *Ouest France*, national edition (April 23rd 2007, p.3; June 25th, 2007, p.3).

Regional newspapers: *Ouest France*, local edition (April 19th 2007, p.7 and p.CAN16; April 23rd 2007, p.CAN17 and p.LAV16; June 26th, p.CAN18); *Les Dernières Nouvelles d'Alsace* (April 19th 2007; June 26th 2007); *Le Journal de la Haute Marne* (April 3rd 2007, p.34); *Le Courrier de la Mayenne* (April 23rd 2007; June 29th 2007; July 5th 2007), *Loopy* (June 2007, p.6–7).

Magazines: *Territoires* (n°479, June 2007, p.16); *Sciences et avenir* (July 2007, p.28); *Sciences Humaines* (n°187, October 2007, p.14).

Television broadcasts: France 3 Basse-Normandie, *Programme 19/20* (April 22nd 2007 and June 25th 2007); France 3 Alsace, *Programme 12/13* and *Programme 19/20* (April 22nd 2007; June 26th 2007); France 3 Pays de Loire, *Programme 19/20* (April 22nd 2007).

Websites: *Dépêches AFP* (April, 20th 2007 and April, 22nd 2007) taken up by several websites, among which were: LCI, TF1, France 2, Yahoo actualité, Linternaute, Orange.

Radio programmes: France Info (April, 21st 2007); France Bleu Alsace (April 20th, 2007; June 26th 2007); NRJ and Nostalgie, Edition Basse-Normandie (June 26, 2007).

Frame 10 Presentation of the 2007 experiment in regional newspapers

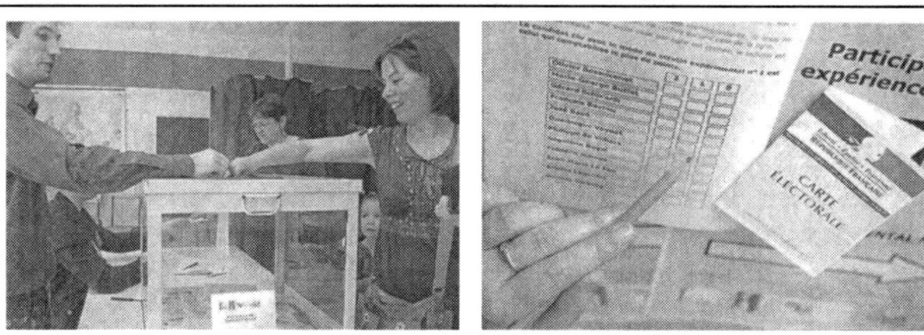

Une bonne partie des votants de Louvigny a pu voter trois fois, hier. Ils ont pris part à un test lancé par le CNRS.

Pictures to illustrate the experiment taken up in *Ouest France* and in *Les dernières nouvelles d'Alsace*, April 23rd 2007.

The 2007 experiment had benefited from wide media coverage thanks to the active help of the Centre d'Analyse Stratégique, a governmental institution. Yet, curiously, no effort has been made to circulate the actual results, nor to mark the publication of the final report where these results were analyzed.

15.3 General Lessons from AV Experiments

The most important objectives of these experiments were to evaluate the feasibility of undertaking this type of large-scale experiment in political elections, and to investigate whether AV is comprehensible to and accepted by the public. The lessons drawn from the experiments regarding these two objectives are set out below.

Frame 11 Presentation of the 2007 experiment in a national newspaper

Page trois Election présidentielle 3

Une équipe de chercheurs a permis aux électeurs de six bureaux de vote d'expérimenter d'autres modes de scrutin : donner une note aux candidats, en choisir plusieurs...

Une expérience : noter au lieu de voter

« Nous avons suivi la geste électorale de A à Z »

Le Monde, April 26th, 2007, p.3, with a short presentation on the first page.

15.3.1 Participation

The experimental design was such that only voters who turned out for the official vote could take part. Therefore, the participation rate in the experiment is defined as the ratio of the number of participants over the number of voters who turned out.

Table 15.1 presents the participation rates for the 2002 experiment in Gy and Orsay: the rates were unexpectedly high, above 75% in Orsay, and above 90% in Gy.

Frame 12 Presentation of the 2007 experiment in a national newspaper

Le Courrier de la Mayenne, July 5th, 2007, p.10

A caricature to illustrate the extensive presentation of the results of the 2007 experiment. Specific attention to the Cigné results was made. Recognize Nicolas Sarkozy – the incumbent French president – in front of the mirror saying "Mirror, tell me who is the best president", François Bayrou – a centrist candidate – in the mirror answers "it's not you". The tag "CNRS (National Center for Scientific Research)" on the mirror is supposed to designate the organizer of the experiment, even though other institutions, such as CAS, participated in financing and organizing it. It is generally the case that the public straightforwardly associates any research with CNRS, so journalists did not agree to provide a more faithful list of the institutions concerned.

Table 15.1 Participation rates and votes cast in the 2002 experiment

	Gy 1 polling station	Orsay 5 polling stations	All
Official vote			
Registered voters	482	4,237	4,719
Votes cast	395	2,951	3,346
Experimental vote			
Participants	365	2,232	2,597
Participation rate (%)	92.4	75.6	77.6
Spoiled	1	9	10
Votes cast	364	2,223	2,587
Votes cast (%)	99.7	99.6	99.6

Table 15.2 Participation data and votes cast in the 2007 experiment

	Cigné 1 polling station	Louvigny 2 polling stations	Illkirch 3 polling stations	All
Official vote				
Registered electors	378	1,948	3,211	5,537
Votes cast	318	1,760	2,526	4,604
Experimental Vote				
Participants	233	1,063	1,540	2,836
Participation rate (%)	73.3	60.4	61	61.6
Spoiled	1	12	10	23
Votes cast	232	1,051	1,530	2,813
Votes cast (%)	99.6	98.9	99.4	99.2

Table 15.2 presents the participation rates for the 2007 experiment in Cigné, Louvigny, and Illkirch-Graffenstaden:[5] they are still high but not as markedly so as in the previous experiment – around 60% on average over the 6 polling stations – and, as expected, the highest rate was for Cigné, which, like Gy, is a small village. Several points explain why the rate of participation was higher in 2002. First, the smaller the village, the higher are the participation rates. One can, indeed, easily imagine that in small communities people know each other much better. They meet and speak at the polling station, and thus everybody can observe who votes and who does not and, accordingly, who has taken part in the voting experiment and who has not. In 2006, Gerber, Green and Larimer conducted a large-scale field experiment in order to determine if and how social pressure encourages citizens to take part in a ballot. According to them, "higher turnout was observed among those who received mailings promising to publicize their turnout to their household or their neighbors. These findings demonstrate the profound importance of social pressure as an inducement to political participation" Gerber et al. (2008, p. 33). Clearly, the high participation rates in Gy in 2002 and in Cigné in 2007 are a consequence of the same kind of social pressure.

Second, one must stress that Orsay is an unusual city, close to Paris, whose population, economy and national reputation are based on famous universities and scientific research. In addition to the University of Paris 11, which is one of the biggest universities in France, and important research centers such as the CNRS, a very high number of famous French engineering schools (e.g. École Polytechnique, Supelec, Sup Optique) are located in Orsay and nearby. Thus, a statistically significant proportion of residents of Orsay are researchers or students. All these elements make the acceptance of a scientific experiment easier, no matter what the experiment is. Conversely, Illkirch and Louvigny, despite being located near two big cities with significant academic traditions (Strasbourg and Caen), are more representative of the national reception of science.

Third, the official participation rate was much higher in 2007 (83.8% in 2007 against 71.6% in 2002); and for those voters who were not used to voting and were not particularly interested in politics, participation in the official vote might have been effort enough. One can easily imagine that a voting experiment would not arouse their interest: indeed, the experimental assessors overheard comments from people who had just left the official polling stations such as: "I already did vote for the 'real' ballot; that is enough" or "even to take part in the official voting is pretty good!"

Fourth, the 2007 official election was also characterized by a very high rate of proxies (the experiment occurred during the school vacations for Illkirch), which complicated the participation of away voters in the experiment. Even though experimental voting by proxy was authorized in the 2007 experiment, few voters who had a proxy for somebody else decided to use it in the experimental vote. In Louvigny,

[5] Here, unlike Baujard and Igersheim (2007a, b, 2009), we regard the blank ballots as significant, in order that the 2002 data and the 2007 data may be compared.

Table 15.3 Answers to questionnaire qu. 13: "Are you satisfied with having taken part in this experiment?"

	Nb. Occurences	%
Yes	1,041	88.7
A little	87	7.4
Mostly not	6	0.5
No	6	0.5
No opinion	34	2.9
Total	1,174	100

Table 15.4 Answers to questionnaire qu. 8: "Do you think that researchers must study voting methods?"

	Nb. Occurences	%
Yes	855	69.1
A little	156	12.6
Mostly not	34	2.7
No	137	11.1
No opinion	55	4.4
Total	1,237	100

for instance, there were fewer than fifteen experimental proxy votes, while the city council of Louvigny granted around eighty proxies.

Finally, the 2007 experimental ballot design was different, longer, and hence more demanding, than the 2002 one, since it also tested EV; this could have reduced the rate of participation.

Putting the different rates of participation on one side, however, the most difficult question is, rather: why was the participation rate so high? Good, repeated and early information, and the support of communities, provide part of the explanation, but do not explain everything. Both the 2002 and 2007 research teams noted that voters were sincerely interested in the subject and the aim of the experiment. Most participants had brought with them the information letter they had received at home. Some of them had already filled in the experimental ballot enclosed with it. Many participants had a discussion with a member of the research team or with another participant about the voting methods under test. The answers to the 2007 questionnaire confirm these observations. Indeed, 96.10% of those who filled in the questionnaire responded positively to the question "Are you satisfied with having participated in this experiment?"(see Table 15.3).[6] 81.7% thought that researchers should continue to study alternative voting methods (see Table 15.4) and 92% responded positively to the question of whether they would be willing to take part in such an experiment again (see Table 15.5). Furthermore, almost 50% of the 626 participants who answered the open question "What did you appreciate in this experiment?" mentioned the very fact that this kind of experiment had been instigated at all. Conversely, less than 25% of the 150 participants who answered the open question "What did you dislike in this experiment?" wrote that they disapproved of it (see Tables 15.6 and 15.7).

[6] That is, 96.10% of respondents answered "yes" or "a little" to question 13 of the 2007 questionnaire.

Table 15.5 Answers to questionnaire qu. 15: "Would you be ready to take part again in a scientific experiment on your voting behavior?"

	Nb. Occurences	%
Yes	1,020	87.4
A little	54	4.6
Mostly not	21	1.8
No	56	4.8
No opinion	16	1.4
Total	1,167	100

Table 15.6 Answers to questionnaire qu. 18: "What did you appreciate in this experiment?"

Items (several items per answer)	Nb. Occurrences
Initiative of such an experiment, its consequences	305
Extended expressive possibilities of both voting methods	247
Concrete procedure of the experiment	75
Simplicity of both voting methods	48
Number of positive answers (for 1,275 questionnaires)	626

Table 15.7 Answers to questionnaire qu. 19 question: "What did you dislike in this experiment?"

Items (1 item per answer)	Nb. Occurences
Experimented voting methods	75
Lack of anterior information	37
Initiative of the experiment, its organization	34
Disappointment regarding the small scale of the experiment	4
Number of negative answers (for 1,275 questionnaires)	150

Thus, our first major general lesson is that such a large-scale experiment on voting is feasible and very well accepted by voters (on this issue, see also Laslier 2009b). Further, the answers to the 2007 questionnaire show that voters consider research on voting methods to be very useful and have expectations from its results.

15.3.2 Positive Public Response to AV

In traditional experimental economics, specific questionnaires are used to test participants' understanding before the experiment takes place; for example, in experiments on voting methods, prior questions are posed about how to fill in a ballot and how to compute the election outcome. In large-scale experiments in the field, it is very difficult to respect such protocols. One can argue a posteriori, through inspection of the experimental ballots, that the participants have understood what was asked of them, since overall the experimental ballots have been properly filled in. But it is impossible to demonstrate that they have perfectly understood the rules of AV: one can only make plausible conjectures based on the facts that, first, AV's rules (like EV's) are very simple, and, second, comprehension has been confirmed individually by some voters who are known to have explained it perfectly to organizers as well as to journalists.

With this caveat in mind, one can nevertheless plausibly claim that a satisfying level of understanding and acceptance of the principle of AV is indicated in both experiments (2002 and 2007). This second major lesson is derived from the analysis of three kinds of data: (1) the expression rate, (2) the questionnaires, and (3) the number of approvals.

First, out of 2,597 approval ballots cast in the 2002 experiment, only ten were spoiled; and out of 2,836 ballots cast in the 2007 experiment, only 23 were spoiled. Thus, as one can observe in Tables 15.1 and 15.2, the votes cast in percentage terms, which is equal to the number of non-spoiled ballots over the number of participants, is systematically higher than 99%. This suggests that almost all participants understood AV and engaged in the experimental vote according to its principle.

Second, the answers to the 2007 questionnaire corroborate this observation. Indeed, 83.5% of respondents answered positively to the question "Does the principle of Approval Voting seem clear to you?" – note that this is slightly less than for EV (89.2 %) – see Tables 15.8 and 15.9. As well as this, 75.1% of participants opined that AV could be used for official elections (presidential, legislative, and other), as against 87.9% for EV – see Tables 15.10 and 15.11. As regards the open questions ("What did you appreciate in this experiment?", "What did you dislike in this experiment?", see Tables 15.6 and 15.7), we have already stressed that we received 626 positive remarks against only 150 negative ones. Further, 295 voters sing the praises of AV and EV (247 note that AV and EV enable greater voter expression and 48 emphasize their simplicity) against only 75 voters who dislike them. Consequently, one sees voters taking a strong stance in favor of these new voting methods – with the small reservation that in 2007 EV seemed to arouse a little bit more excitement than AV. This can be explained in two different ways: on the one hand, it could have been caused by the slight change in the protocol. In 2007, voters were asked to circle the names of the candidates they wanted to approve of, and not to mark a cross next to them as in 2002: checking a name with a cross is perhaps a

Table 15.8 Answers to questionnaire qu. 7: "Does the principle of Approval Voting seem clear to you?"

	Nb. Occurences	%
Yes	826	66.9
A little	205	16.6
Mostly not	80	6.5
No	98	7.9
No opinion	25	2.0
Total	1,234	100

Table 15.9 Answers to questionnaire qu. 6: "Does the principle of Evaluation Voting seem clear to you?"

	Nb. Occurences	%
Yes	987	78.6
A little	133	10.6
Mostly not	41	3.3
No	76	6.1
No opinion	19	1.5
Total	1,256	100

Table 15.10 Answers to questionnaire qu. 12: "For which official election do you think that Approval Voting could be used?"

	Nb. Occurences	%
Presidential election	503	32.4
Legislative election	567	36.5
Other election (council, ...)	97	6.2
No election	387	24.9
Total	1,554	100

Table 15.11 Answers to questionnaire qu. 11: "For which official election do you think that Evaluation Voting could be used?"

	Nb. Occurences	%
Presidential election	720	40.2
Legislative election	723	40.4
Other election (council, . . .)	130	7.3
No election	216	12.1
Total	1,789	100

Table 15.12 Number of approved candidates in the 2002 experiment

Approvals	0	1	2	3	4	5	6	7	8	9	10	>10
Ballots	36	287	569	783	492	258	94	40	16	6	1	5
% of ballots	1.39	11.09	21.99	30.27	19.02	9.97	3.63	1.55	0.62	0.23	0.04	0.19

Average number: 3.15 candidates out of 16, over 2,587 non-spoiled ballots

Table 15.13 Number of approved candidates in the 2007 experiment

Approvals	0	1	2	3	4	5	6	7	8	9	10	>10
Ballots	120	736	905	673	264	75	23	13	1	1	1	1
% of ballots	4.27	26.16	32.17	23.92	9.38	2.67	0.82	0.46	0.04	0.04	0.04	0.04

Average number: 2.23 candidates out of 12, over 2,813 non-spoiled ballots

more intuitive procedure in an electoral context. On the other hand, the principle of EV itself – i.e. giving a grade to each candidate – is very well-known (from school, sporting events, etc.), and thus perhaps more easy to understand and to adopt.

Third, a last clue as to how voters understood and accepted the principle of AV is given by the number of approvals conceded by each participant (the number of candidates approved of). The 2002 and 2007 distributions are presented in Tables 15.12 and 15.13 and in Figs. 15.1 and 15.2 respectively. In 2002, on average in the two towns, each voter approved of 3.15 candidates out of 16, the distribution around this value being rather smooth (in particular, one-name approval ballots are few). In 2007, each voter approved of 2.23 candidates out of 12, which is still high, but lower than in 2002. This difference is a consequence of the fact that there is a larger number of zero-name and one-name approval ballots in 2007. The first point affords of various explanations, but principally it should be recalled that this experiment was

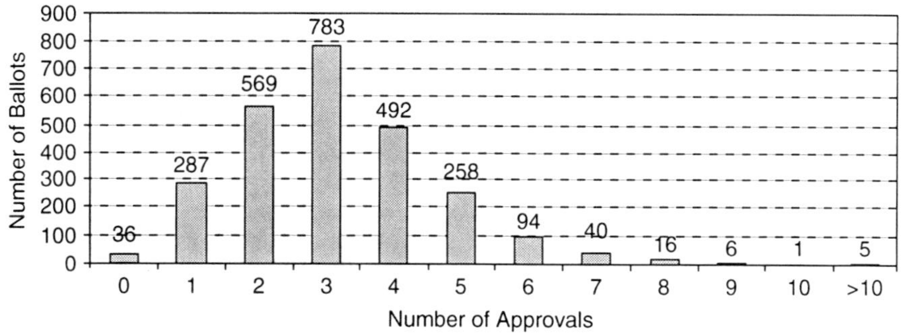

Fig. 15.1 Number of approved candidates in the 2002 experiment

conducted in parallel with a test of EV. For the evaluation rule, far fewer blanks were counted (45 against 120 for AV), and a significant proportion of these were blank for the approval voting test as well: 10 ballots are blank for both AV and EV – the exact number of spoiled ballots as in 2002! Besides this, a significant percentage of approval blanks corresponds to ballots attributing few grade 1s and no grade 2s, which reveals little enthusiasm towards the candidates: in this regard, we should learn something about the trigger under which a voter does not concede an approval. The second point, regarding the significantly higher number of one-name approval ballots in 2007, is clarified when it is noted that many of these ballots were supporting Nicolas Sarkozy, the incumbent president, and furthermore the only candidate of the traditional Right-wing in 2007. Facts about the French political supply structure in 2007, rather than a substantial desire to retain the official plurality vote, thus seem to explain this characteristic.[7] Hence, both the specifics of the 2007 protocol and the particularities of French political supply in 2007 influence the difference between the two average numbers of approvals. But the same conclusion remains: whatever the circumstances, voters do actually make use of the possibility of broader expression generated by AV; this thus strongly suggests that they understand and accept its principle.

15.4 Specific Lessons Regarding AV

The two lessons from Sect. 15.3 essentially concerned the global features of both experiments. We now review the research that has led to interesting findings based on deeper analyses of the data collected. First, to learn more about AV in an experimental context, and in view of the fact that many theoreticians claim that different voting rules may yield different outcomes (Cox 1997; Cox and Katz 2002), we

[7] On this point, see the next section and Baujard et al. (2009b).

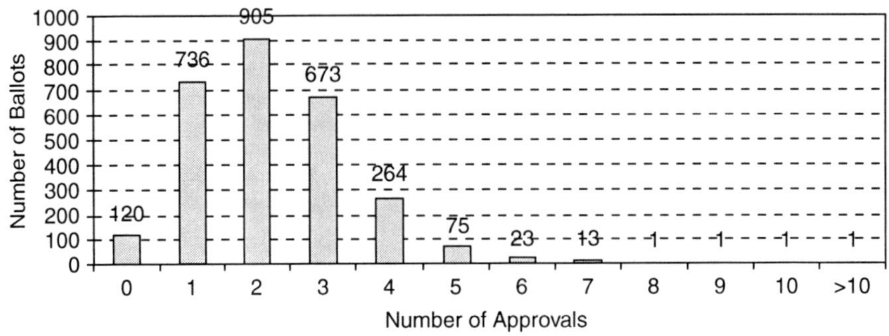

Fig. 15.2 Number of approved candidates in the 2007 experiment

examine whether AV is able to modify the overall ranking of candidates compared with the official election system. Second, the elaboration of an original behavioral model enables us to link approval voting to single-name balloting. Finally, an extended analysis of the French political supply structure, based on AV experimental data, leads to an attempt to define what a consensual candidate is.

15.4.1 Comparing the Outcomes of AV with the Official First-Round Vote

The results of the 2002 and 2007 experiments are given in Tables 15.14 and 15.15. Candidates are ordered by scores in the French official election. These tables provide the scores, i.e. the proportion of voters who approved of or voted for each candidate, and the rankings for every town, in 2002 and in 2007, in which the experiments were conducted. But one cannot compare directly the approval and the official columns in a town since the hypothesis of a participation bias in the experiment cannot be excluded (see Laslier and van der Straeten 2004; Baujard and Igersheim 2007a). Indeed, Laslier and van der Straeten show that in 2002 only a small proportion of voters for Jean-Marie Le Pen, the candidate of the extreme Right-wing, agreed to participate in the experiment. Applying the same method to the 2007 data, the analysis of the participation bias reveals that voters of Nicolas Sarkozy, the only candidate of the traditional Right-wing, were over-represented and these of François Bayrou, the centrist candidate, were slightly under-represented in the data from the AV experiment. Conversely, the results of AV and the official election can be compared at the national level (the columns headed "France"): here both participation and sampling bias have been corrected.[8]

[8] Notably via the behavioral model described in Sect. 4.2 of this article.

Let us comment on these results. A first conclusion is obvious: rankings under AV and official voting are very different, not only in detail but in their major features. In the 2002 experiment, Laslier and van der Straeten (2004) observed that "the ranking of candidates is modified in favor of Lionel Jospin, François Bayrou, Jean-Pierre Chevènement, Noël Mamère and Alain Madelin .... The candidates that AV seems to put at a disadvantage without unambiguity are J.-M. Le Pen, A. Laguiller, O. Besancenot, Jean Saint-Josse et Robert Hue." Some candidates, such as the centrist candidate François Bayrou, seem to benefit from AV, unlike others who are disadvantaged by it. This tendency is clearly corroborated by the 2007 experiment; most notably, in 2007, the winner under AV for France (François Bayrou) is different than the winner of the official ballot (Nicolas Sarkozy).

Second, some political parties receive numerous approvals, and so find greater representation under AV than they do in the official vote – in which they are, indeed, almost nonexistent. Both in 2002 and 2007, Green candidates (Noël Mamère, Corinne Lepage, Dominique Voynet) greatly benefit from AV. Conversely, some others lose from this new method of voting: notably the extreme Right candidates, and Jean-Marie Le Pen in particular. The specific feature of 2007 is that "little" candidates of the Extreme left (Olivier Besancenot especially, but José Bové and Arlette Laguiller too) gain from AV, while the opposite conclusion is valid in 2002 (Robert Hue, Arlette Laguiller).

Third, no candidate attracted an absolute majority of approvals in 2002, nor in 2007 at the national level (in his best showing, that of 2007, François Bayrou came

Table 15.14 AV's results, Extrapolations to France – 2002

	Gy				Orsay				France			
	Approval		Official		Approval		Official		Approval		Official	
Chirac	38.2	1	19.6	1	36.2	2	18.8	2	36.7	1	19.9	1
Le Pen	32.7	2	19.6	1	11.7	12	8.7	4	25.1	4	16.9	2
Jospin	23.9	3	11.1	4	43.2	1	20.7	1	32.9	2	16.2	3
Bayrou	23.4	4	6.7	6	35.2	3	10.3	3	27.1	3	6.8	4
Laguiller	17.6	9	13.0	3	15.1	10	3.7	8	16.8	9	5.7	5
Chevènement	18.4	7	4.7	8	32.3	4	8.6	5	22.4	6	5.3	6
Mamère	18.4	7	4.7	8	30.6	5	8.3	6	24.3	5	5.2	7
Besancenot	17.0	10	2.8	11	17.7	9	3.1	10	17.6	8	4.2	8
Saint-Josse	20.3	6	9.6	5	5.8	15	0.7	15	13.5	11	4.2	9
Madelin	21.2	5	5.2	7	21.3	6	4.9	7	20.4	7	3.9	10
Hue	10.2	12	3.1	10	11.7	11	2.6	15	11.3	14	3.4	11
Mégret	17.0	10	2.8	11	6.1	14	1.1	14	13.8	10	2.3	12
Taubira	9.1	14	0.5	16	20.6	7	3.6	9	12.6	13	2.3	13
Lepage	9.9	13	2.8	11	19.3	8	2.8	11	13.4	12	1.9	14
Boutin	5.8	16	0.8	15	8.1	13	1.4	13	6.7	15	1.2	15
Gluckstein	7.1	15	1.8	14	3.8	16	0.7	16	5.5	16	0.4	16
Total	290.1		100		318.6		100		297.1		100	

N.B.: For France, the official results of the second round were: Jacques Chirac (82.2%) and Jean-Marie Le Pen (17.8%).

Table 15.15 AV's results, Extrapolations to France - 2007

	Cigné				Louvigny				Illkirch				France			
	Approval		Official		Approval		Official		Approval		Official		Approval		Official	
Sarkozy	37.2	3	29.6	1	37.9	3	28.5	2	51.1	1	38.6	1	35.9	3	31.2	1
Royal	49.8	1	26.3	2	51.3	1	30.7	1	37.7	3	18.3	3	41.6	2	25.9	2
Bayrou	40.5	2	20.8	3	49.8	2	23	3	51	2	23.2	2	42.8	1	18.8	3
Le Pen	7	10	4.6	5	7.2	10	4.1	5	15.2	6	10.4	4	13.9	7	10.4	4
Besancenot	26.1	4	4.2	6	28.1	4	5	4	20.3	4	3.4	5	27.9	4	4.1	5
De Villiers	12.6	7	5.8	4	8	9	1.7	7	9.1	9	1.2	7	11.1	9	2.2	6
Buffet	9.3	8	2	7	10.1	7	1.3	8	5.2	10	0.4	10	9.8	10	1.9	7
Voynet	14.9	6	0.3	12	18.3	5	2.2	6	16.3	5	2.3	6	16.6	5	1.6	8
Laguiller	7.9	9	2	7	9.6	8	1.2	9	9.3	7	0.8	9	11.4	8	1.3	9
Bové	19.1	5	2	7	13.3	6	1.1	10	9.2	8	1	8	15.2	6	1.3	10
Nihous	6.1	11	2	7	4.5	11	1.1	10	2.3	11	0.2	12	4.4	11	1.2	11
Schivardi	3.7	12	0.7	11	1.3	12	0.2	12	1.1	12	0.2	11	1.9	12	0.3	12
Total	234		100		239.4		100		228		100		232.5		100	

N.B.: For France, the official results of the second round were: Nicolas Sarkozy (53.1%) and Ségolène Royal (46.9%).

near 43% for French simulated results out of all ballots, but note that such a score is possible because, in 2007, Bayrou was approved of by 51% of voters in Illkirch, and by 49.8% in Louvigny).

Another major lesson can thus be drawn from the three comments above: the analysis of AV's results does lead to different conclusions compared to those of the official ballot, especially because AV carries different information on voters' political preferences. In particular, with AV, we learn that voters may be interested in candidates who are ignored by the official ballot. Further, since AV allows the voters to give their opinion on each candidate, one can expect the winner to be different than the one designated by the two-round vote, as in 2007. Finally, it must be emphasized that EV's results in 2007 confirm in every respect all the preceding observations.

15.4.2 Examining the Conversion of Approval into Single-Name Balloting

So far, we have examined the raw results and compared the ranking of the candidates under AV with that of single name balloting, on the basis of the result of the official first-round vote. But AV's results cannot be reduced to scores: AV ballots provide much richer information than this, since the number of voters who approve of various groups of candidates can be computed. An agreement matrix, such as the one in Table 15.16, gives the number of voters who simultaneously approve of two candidates. For an election with 16 candidates, as in 2002, the agreement matrix for AV has 256 values: each candidate cross-referenced with all the others. The matrix is thus symmetric and the diagonal is equal to the score of each candidate. The agreement matrix for Gy les Nonains (see Laslier and van der Straeten 2004: 15) can be read as follows: among the 139 voters for Jacques Chirac, the outgoing president from the traditional Right-wing, 51 approve of Jean-Marie Le Pen, 15 approve of Lionel Jospin, the candidate of the traditional Left-wing, and so on (see also Table 15.17 for the 2007 experiment; this table can be read in the same manner as the previous one).

But an important piece of information is missing: namely, the priority that voters afford to a particular candidate in the official election process. In other words, does the intersection between the voters for Jacques Chirac and for Jean-Marie Le Pen contain those who voted for Chirac in the official first round, who also approve of Le Pen? Or, on the contrary, are they those who voted for Le Pen in the official first round, who also approve of Chirac?

In order to induce from the AV data some clues as to how voters convert their approval vote into first-round voting (or, more precisely, into single-name balloting, since the model does not distinguish between both voting methods), one can resort to formalization and set up a behavioral model (for further explanation regarding this model, see Laslier 2004; Laslier and van der Straeten 2004, 2008; Baujard and Igersheim 2007a). In this model, a parameter μ_c, called the "lever," is assigned to each candidate c: μ_c estimates the propensity of candidate c to be not only one of the approved candidates under AV, but also the chosen candidate in the official vote. More precisely, if a voter approves the set B of candidates, she votes for c with a null probability if $c \notin B$ and with a probability equal to $\frac{1}{\sum_{d \in B} \mu_d} \mu_c$ if $c \in B$. Hence, the higher μ_c, the greater the propensity of candidate c to convert approvals into votes in a single-name ballot.

Table 15.18 presents the levers computed with Gy's data (normalized to 1 for Jacques Chirac) in the 2002 experiment. One can observe that both Jacques Chirac and Jean-Marie Le Pen have the highest levers, showing that these two candidates are more capable than others of turning approvals into official votes. Of course, the values of the levers also partially reflect the phenomenon of tactical voting – or its absence – which can "artificially" increase the official score of a major candidate. Here, one can note that the lever of Lionel Jospin, the traditional Left-wing candidate and a priori the principal challenger to Jacques Chirac, is particularly weak compared to Chirac's and Le Pen's. This is not surprising, since Jospin's famous defeat during the 2002 French presidential election is generally explained by a lack of strategic voting in his favor. But, above all, the values of the levers depend on the type of support a candidate arouses among the voters. Hence, from Table 15.16, one notes that the 119 voters who approve Jean-Marie Le Pen give 2 approvals to other candidates on average. Conversely, the 62 voters who approve Olivier Besancenot give 4.4 approvals to other candidates on average. According to the behavioral model, then, the participants who approve Jean-Marie Le Pen are more able to convert their approval into an official vote for this candidate than those who approve Olivier Besancenot. In other words, the votes for the former are much more concentrated than those for the latter.

As for 2007, Table 15.19 shows us that Nicolas Sarkozy, the candidate of the traditional Right-wing, has a much higher lever than any other candidate. This

Table 15.16 Agreement Matrix for Gy les Nonains – 2002

	Chirac	Le Pen	Jospin	Bayrou	Laguiller	Chevènement	Mamère	Besancenot	Saint-Josse	Madelin	Hue	Mégret	Taubira	Lepage	Boutin	Gluckstein
Chirac	139	51	15	47	10	28	11	10	36	48	3	31	5	9	6	3
Le Pen	51	119	10	22	18	17	9	13	21	22	5	44	3	5	4	4
Jospin	15	10	87	14	21	17	40	24	11	5	26	0	23	9	5	4
Bayrou	47	22	14	85	10	25	13	9	10	33	3	13	8	14	7	2
Laguiller	10	18	21	10	64	13	19	24	10	3	18	6	11	11	7	12
Chevènement	28	17	17	25	13	67	10	11	10	19	2	7	8	11	4	3
Mamère	11	9	40	13	19	10	67	32	7	9	15	3	15	10	4	12
Besancenot	10	13	24	9	24	11	32	62	10	8	16	9	16	13	3	15
Saint-Josse	36	21	11	10	10	10	7	10	74	18	5	13	5	6	6	4
Madelin	48	22	5	33	3	19	9	8	18	77	2	15	4	10	6	3
Hue	3	5	26	3	18	2	15	16	5	2	27	0	5	4	3	7
Mégret	31	44	0	13	6	7	3	9	13	15	0	62	1	2	4	4
Taubira	5	3	23	8	11	8	15	16	5	4	5	1	33	7	4	3
Lepage	9	5	9	14	11	11	10	13	6	10	4	2	7	36	5	4
Boutin	6	4	5	7	7	4	4	3	6	6	3	4	4	5	21	1
Gluckstein	3	4	4	2	12	3	12	15	4	3	7	4	3	4	1	26

Table 15.17 Agreement Matrix for the 6 polling stations – 2007

	Sarkozy	Royal	Bayrou	Le Pen	Besancenot	De Villiers	Buffet	Voynet	Laguiller	Bové	Nihous	Schivardi
Sarkozy	1216	231	588	221	116	169	23	99	56	52	53	14
Royal	231	1176	577	40	386	0	150	303	147	194	27	11
Bayrou	588	577	1340	101	271	94	65	223	87	120	43	14
Le Pen	221	40	101	312	45	95	6	19	24	17	16	5
Besancenot	116	386	271	45	637	25	112	161	148	171	28	19
De Villiers	169	0	94	95	25	242	36	14	17	13	16	9
Buffet	23	150	65	6	112	36	198	92	68	60	9	8
Voynet	99	303	223	19	161	14	92	456	75	117	14	7
Laguiller	56	147	87	24	148	17	68	75	250	69	16	7
Bové	52	194	120	17	171	13	60	117	69	309	14	15
Nihous	53	27	43	16	28	16	9	14	16	14	91	9
Schivardi	14	11	14	5	19	9	8	7	7	15	9	38

suggests strongly that the participants who approved him in the experiment also voted for him in the official election. Indeed, the 1,216 voters who approved him gave approval to 1.33 other candidates on average: this is the lowest average for the 2007 experiment (see Table 15.17). One can note further that Jean-Marie Le Pen and, to a lesser extent, François Bayrou and Ségolène Royal, the candidate of the traditional Left-wing, also have high levers, even if not comparable to Nicolas Sarkozy's. Among the candidates of the Left-wing (traditional and alternative) – Ségolène Royal, José Bové, Marie-Georges Buffet, Olivier Besancenot, Dominique Voynet, Gérard Schivardi, Arlette Laguiller–Ségolène Royal has the highest lever, showing that there is substantial strategic voting in her favor. Indeed, the voters who approve one or many candidates of the Left wing, Ségolène Royal excepted, give support to 2.33–3.11 other candidates on average.

Now, these levers make it possible to extrapolate the results of AV if the experiments had been conducted over the whole country.[9] This indicates that the elaboration of a behavioral model, which explains the conversion of approval voting into single-name balloting and vice versa, is crucial if we are to be fully aware of the mechanisms of AV. Another way to address this issue is to examine how voters transform their approvals under AV into grades under EV. In other words, we can develop a behavioral model that explains the conversion of AV into EV. The data Baujard and Igersheim collected during the 2007 French presidential election permits the development of such a model, which should improve the understanding of AV.

For first results and remarks on this issue, one can start by computing the correlation coefficients between AV and EV for every candidate. Let us recall that the linear correlation coefficient, or the Pearson's coefficient, is obtained by dividing the covariance of two variables by the product of their standard deviations. It ranges between -1 (decreasing linear relation) and $+1$ (increasing linear relation). Hence, the closer the correlation coefficient of a candidate is to 1, the more voters give her an approval and a grade 2. Conversely, the closer it is to -1, the more voters give her no approval and a grade 0. A correlation coefficient of 0 corresponds to no linear relation between variables. Recall that a strong correlation is considered to exist between two variables whenever the correlation coefficient exceeds 70%.

In our case, there seem to be positive linear correlations between the AV and EV scores for all candidates. Table 15.20 shows that the candidates who obtain a

Table 15.18 Levers in 2002

Chirac	1	Saint-Josse	0.9
Le Pen	1.2	Madelin	0.4
Jospin	0.7	Hue	0.5
Bayrou	0.5	Mégret	0.3
Laguiller	0.4	Taubira	0.1
Chevènement	0.4	Lepage	0.5
Mamère	0.4	Boutin	0.2
Besancenot	0.2	Gluckstein	0.2

[9] The results of the extrapolation are given in Sect. 4.1.

Table 15.19 Levers in 2007

Sarkozy	1	Buffet	10^{-7}
Royal	0.00388	Voynet	$9{,}57\times10^{-7}$
Bayrou	0.00465	Laguiller	10^{-7}
Le Pen	0.56	Bové	10^{-7}
Besancenot	5.77×10^{-7}	Nihous	0.000013
De Villiers	0.000454	Schivardi	0.0000366

Table 15.20 Correlation coefficient per candidate between AV and EV – 2007

Sarkozy	80.7	Buffet	52.0
Royal	77.0	Voynet	60.8
Bayrou	77.2	Laguiller	51.2
Le Pen	73.5	Bové	58.9
Besancenot	67.6	Nihous	53.3
De Villiers	62.4	Schivardi	39.9

Table 15.21 Frequencies of conversion from AV to EV – 2007

	No Approval into...			Approval into...		
	Grade 0 (%)	Grade 1 (%)	Grade 2 (%)	Grade 0 (%)	Grade 1 (%)	Grade 2 (%)
Sarkozy	73	21	5	2	18	80
Royal	67	29	04	2	26	72
Bayrou	53	42	6	3	29	68
Le Pen	90	9	2	5	41	54
Besancenot	71	26	3	3	43	54
De Villiers	85	13	2	8	42	50
Buffet	77	21	2	6	53	42
Voynet	67	30	2	4	47	49
Laguiller	74	23	4	5	49	47
Bové	77	21	2	5	50	45
Nihous	90	9	1	9	46	45
Schivardi	92	8	0	16	46	38

significantly high correlation coefficient between AV and EV are Nicolas Sarkozy, Ségolène Royal, François Bayrou and Jean-Marie Le Pen. Other candidates, Olivier Besancenot excepted, obtain weaker correlation coefficients. This corroborates our previous remark: Nicolas Sarkozy and Jean-Marie Le Pen benefit from high concentration in their voters. With regards to Ségolène Royal, the fact that she seems to benefit from strategic voting should explain the difference between her rather high coefficient correlation in comparison with those of the other candidates of the Left-wing. François Bayrou, the centrist candidate, is able to receive approvals and maximum grades from the voters of both political sides, Left and Right.

Further, one can consider Table 15.21, which exhibits the frequencies of conversion from an approval or no approval into grades 0, 1 or 2. For instance, 80% of voters who give an approval to Nicolas Sarkozy give him a grade 2 and 73% of voters who do not give him an approval give him a grade 0. Further, the columns "No Approval into Grade 1" and "Approval into Grade 1" make it possible to analyse precisely how voters vote in AV: do they give approval both to candidates they like *and* those they merely tolerate, or only to the candidates they really prefer? Two kinds of behavior can be characterized: let us define "bonus" behaviour as consisting in giving an approval to the most preferred candidates only, whereas "malus" behaviour consists in giving approval to all the candidates a voter likes or is indifferent to: thus, on the malus strategy, only the least-preferred candidates are punished. For the 2007 data, one can count 2095 malus ballots and 5231 bonus ballots. At first sight, then, bonus behavior seems to be the most common.

15.4.3 Analyzing the French Political Context

A third strand of analysis can be dedicated to examining the French political supply structure. Indeed, we have stressed that AV's results are interesting since not only do they consist of scores, but also of the numbers of voters who approve of various groupings of candidates. Hence, if two candidates are supported by the same set of voters, one could argue that they correspond to the same political supply. AV data thus enables us to go beyond simple observations of the political sympathies claimed by the candidates themselves and to define more precisely the position of each candidate as perceived by voters. For further insights in this regard, see Laslier (2010a).

Laslier (2004, 2006) has proposed a method through which to analyse the political context, and so to build "a kind of *official photograph* of voters' preferences" (Laslier, 2006: 160). To do so, Laslier brings together three main ingredients: first, the spatial theory of voting according to which the utility function of a voter v is a decreasing function of the usual distance between the position of a candidate c, y_c, and her ideal point,[10] x_v: $\|y_c - x_v\|$, where y_c and x_v belong to the Euclidian space $\mathbb{R}^k$ and $\|.\|$ is the usual distance. But a further element is added: a candidate is evaluated by a voter not only through her political tendency (or position, deduced from correlations between the candidates' approvals), but also through her "valence" (which is a characteristic of each candidate and depends on the total number of approvals she receives). In other words, even if two candidates are located at the same point, they won't necessarily receive the same number of approvals: the one with the higher valence will get more. The second ingredient is the random utility model according to which the decision of a voter to give an approval to one or another candidate is a random variable whose probability increases (1) with the valence of the candidate, and (2) with increasing proximity of the candidate to her ideal point. The third ingredient is a principal component analysis. Finally, based on these three elements, Laslier offers a picture of the political space which is "purely endogenous ... without reference to an a priori specified set of issues" (Laslier 2006: 163). Further, the definition of the distance between two candidates thus

[10] Note that the distribution of the voters' ideal points is assumed to be normal and widely dispersed: it thus opens a promising line of research which would aim to replace this hypothesis with a more realistic one.

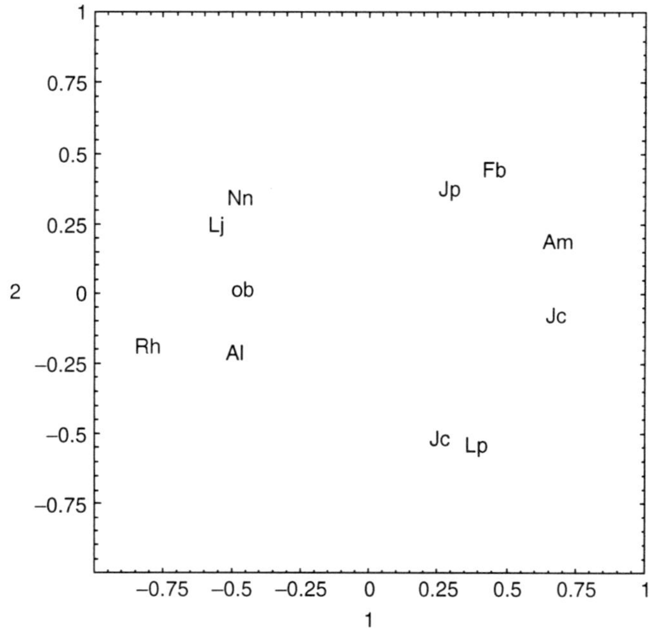

Fig. 15.3 Gy, main plane – 2002

obtained is specific to this model. Note that some other definitions of distance could be considered, such as the difference in the number of approvals. Laslier stresses that "in practice, different distances often provide the same qualitative findings and lead to essentially the same interpretations" (Laslier 2006: 169).

Let us briefly present the political space of Gy in 2002 with this original method (Laslier 2004: 179–180; Laslier 2006: 174–175). Figure 15.3 shows that the first axis can be interpreted as the Left–Right axis. Indeed, Arlette Laguiller (Al – alternative Left-wing), Olivier Besancenot (Ob – alternative Left-wing), Lionel Jospin (Jp – traditional Left-wing) and Noël Mamère (Nm – Green Party) are located on the left side of the picture, while Jean-Marie Le Pen (Lp – nationalist Right-wing), Jean Saint-Josse (Js – a candidate of the Right-wing who created his own political party "Hunting, Fishing, Nature and Traditions"), Jacques Chirac (Jc – outgoing president, traditional Right-wing), François Bayrou (Fb – centrist candidate), Alain Madelin (Am – liberal Right-wing) and Jean-Pierre Chevènement (Jp – centrist candidate) are on the right. But one could argue further that the political space does not seem to be unidimensional, since Jean-Marie Le Pen, the so-called extreme Right candidate, is not on the extreme right, and François Bayrou, the so-called Centrist candidate, is not at the center of the picture. Figure 15.4 shows indeed that each candidate of the Right lies on a different plane, and thus represents a distinct political programme, as determined by the second and the third axes.

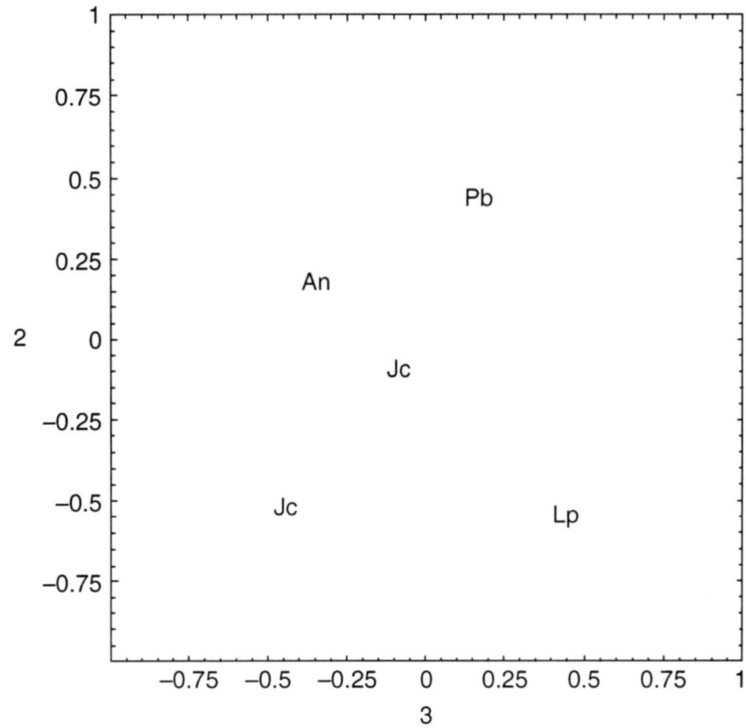

Fig. 15.4 The Right in Gy – 2002

A second way to analyse the French political supply structure using the AV data is to consider more closely the agreement matrix as in Baujard and Igersheim (2007a) and Baujard et al. (2009a, b). Indeed, by computing differently the AV data from the 2007 experiment (see Table 15.22), we derive a "second" matrix of agreement which gives the proportion of voters who support a given candidate (by column) and simultaneously another one (by row). Here the diagonal is always equal to 100% but, contrary to Table 15.17, this matrix is not symmetric: the proportion of voters for Jean-Marie Le Pen (312 approvals) who also supported Nicolas Sarkozy is 71%, while the proportion of voters for Nicolas Sarkozy (1,216 approvals) who also supported Jean-Marie Le Pen is only 18%; the difference makes sense since we do not consider the same set of voters in computing the percentage. This property of asymmetry suggests different lessons depending on whether rows or columns are at stake. In the columns, we read the propensity of voters of a candidate to support other candidates – in other words, the dilution of the support of this candidate. In the rows, we read the propensity of voters for other candidates to support the former, that is, a candidate's ability to attract voters for other candidates.

In order to compare candidates, and thus to point out some characteristics of the French political supply in 2007, the information contained in Table 15.22 is

Table 15.22 Agreement Matrix in % – 2007

	Schivardi	Laguiller	Besancenot	Bové	Buffet	Royal	Voynet	Bayrou	Nihous	Sarkozy	De Villiers	Le Pen
Schivardi	100	3	3	5	4	1	2	1	10	1	2	2
Laguiller	18	100	23	22	34	13	16	6	18	5	7	8
Besancenot	50	59	100	55	57	33	35	20	31	10	10	14
Bové	39	28	27	100	30	17	26	9	15	4	5	5
Buffet	21	27	18	19	100	13	20	5	10	2	1	2
Royal	29	59	61	63	76	100	66	43	30	19	14	13
Voynet	18	30	25	38	46	26	100	17	15	8	6	6
Bayrou	37	35	43	39	33	49	49	100	47	48	39	32
Nihous	24	6	4	5	5	2	3	3	100	4	7	5
Sarkozy	37	22	18	17	12	20	22	44	58	100	70	71
De Villiers	24	7	4	4	2	3	3	7	18	14	100	30
Le Pen	13	10	7	6	3	3	4	8	18	18	39	100

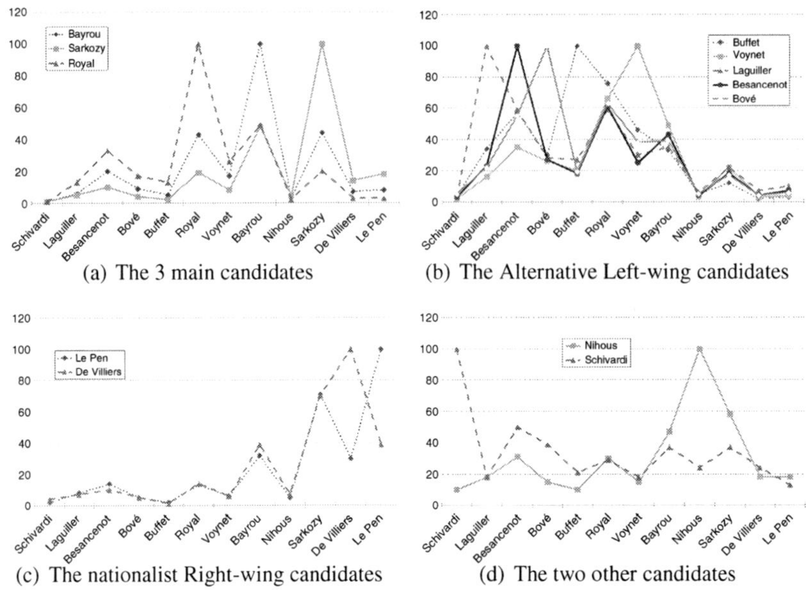

(a) The 3 main candidates (b) The Alternative Left-wing candidates

(c) The nationalist Right-wing candidates (d) The two other candidates

Fig. 15.5 Representation of columns of 2007 agreement matrix

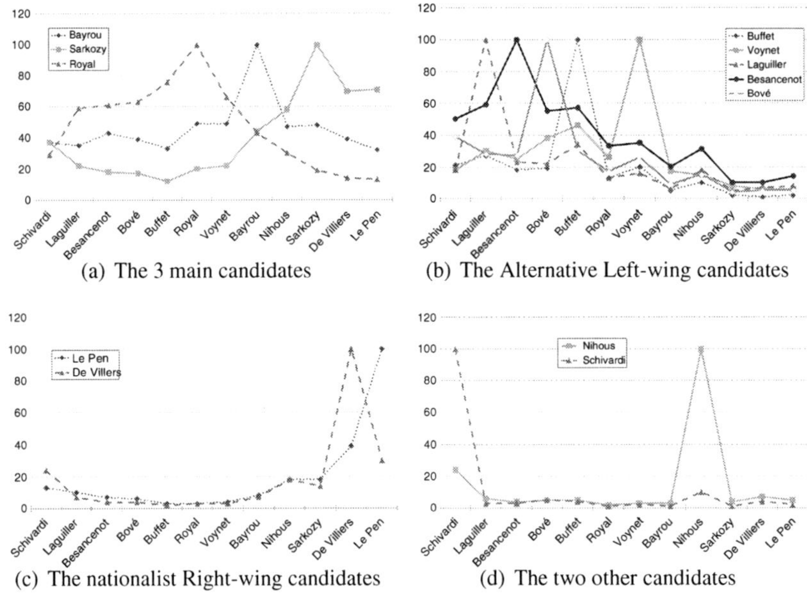

(a) The 3 main candidates (b) The Alternative Left-wing candidates

(c) The nationalist Right-wing candidates (d) The two other candidates

Fig. 15.6 Representation of rows of 2007 agreement matrix

represented in graphs 15.5 and 15.6. Note that both for Table 15.22 and graphs 15.5 and 15.6, the candidates are ordered to make reading and understanding easier, that is, to obtain nice curves, which are as far as possible single-peaked. Baujard et al. (2009b: 18) emphasize that: "this has eventually generated an order which is close to the standardly used ideological axis from Left-wing to Right-wing". Finally, five different political supplies stem from the observation of the curves – which correspond, visually, to alike or different trends: "1- Nicolas Sarkozy matches Right-wing voters; 2- Ségolène Royal matches voters of the traditional Left-wing; 3- François Bayrou matches voters refusing bipolarization; 4- Philippe de Villiers and Jean-Marie Le Pen match voters of the nationalist right-wing; 5- Dominique Voynet, Olivier Besancenot, José Bové, Marie-Georges Buffet and Arlette Laguiller match voters of the governmental and alternative Left-wing. Moreover, the two last candidates, Frédéric Nihous and Gérard Schivardi, are special cases who attract very few supporters; for this reason, we claim they do not really represent a homogeneous political supply" (Baujard et al. 2009b: 20).

Further, still based on the 2007 AV data, Baujard et al. (2009a), in a work in progress, elaborate a way to highlight the consensual candidates in the 2007 French presidential election. For this purpose, they propose an original definition of consensus based on attractiveness, dilution and symmetry of approvals.

15.5 Concluding Remarks

This paper has presented the two main framed field experiments on AV: the first one conducted by Balinski, Laraki, Laslier and van der Straeten during the 2002 French presidential election, and the second one by Baujard and Igersheim during the 2007 French presidential election.

Several lessons can be drawn from the experiments: (1) Such experiments are feasible, and very well accepted by voters; (2) The principle of AV is easily understood and accepted by the public; (3) Within the observed political context, compared to the official first-round vote, AV modifies the overall ranking of candidates: in 2002, under AV, Jacques Chirac and Lionel Jospin, rather than Jean-Marie Le Pen, would have reached the second round, and in 2007, the winner of the French presidential election would have been François Bayrou and not Nicolas Sarkozy; (4) Further, the behavioral model which links approval voting to single-name balloting suggests a new notion of "lever," which allows us to compute the probability of an approved candidate also to be chosen by a voter in the official vote. From the comparison of the levers of all candidates one can point out the high support of one candidate and/or the presence of significant levels of tactical voting in favor of another, and so on. Furthermore, since AV enables voters to give their opinion on every candidate, AV data gives in essence a very good representation of how voters perceive the political supply.

Research on the data from the AV experiments is far from being exhausted. We have indicated above three new research leads: first, to pursue the study of "how voters vote" under AV by examining the conversion from AV to EV; second, to consider a more realistic distribution of the voters' ideal points into Laslier's graphical model (Laslier 2004, 2006); third, to develop an original definition of a consensual candidate based on the experimental data on AV. These, and the many other fruitful leads which we can expect will emerge, should allow us to make progress along the path of fairness and democracy.

References

Alós-Ferrer, C., & Granić, Đ. G. (2010). Approval voting in Germany: description of a field experiment. In J.-F. Laslier & R. Sanver (Eds.), *Handbook on approval voting* (pp. 397–411). Heidelberg: Springer.

Balinski, M., & Laraki, R. (2007a). A theory of measuring, electing and ranking. *Proceeding of the National Academy of Sciences USA*, *104*(21), 8720–8725.

Balinski, M., & Laraki, R. (2007b). Election by majority judgement: experimental evidence. Cahiers du Laboratoire d'Econométrie de l'Ecole Polytechnique, no. 28, 2007.

Balinski, M., & Laraki, R. (2007c). Le jugement majoritaire : l'expérience d'Orsay. *Commentaire*, *30*(118), 413–420.

Balinski, M., Laraki, R., Laslier, J.-F., & van der Straeten, K. (2003). Le vote par assentiment: une expérience. Cahiers du Laboratoire d'Econométrie de l'Ecole Polytechnique, no. 2003-13.

Balinski, M., Laslier, J.-F., & van der Straeten, K. (2002). Compte-rendu de l'expérience de vote du 23 janvier 2002 à l'IEP. Mimeo, mars 2002.

Bassett, G. W. Jr., & Persky, J. (1999). Robust voting. *Public Choice*, *99*(3–4), 299–310.

Baujard, A., & Igersheim, H. (2007a). Expérimentation du vote par note et du vote par approbation lors des élections présidentielles françaises du 22 avril 2007. Rapport au Centre d'Analyse Stratégique, décembre 2007. 289 p.

Baujard, A., & Igersheim, H. (2007b). Expérimentation du vote par note et du vote par approbation lors des élections présidentielles françaises du 22 avril 2007. Premiers résultats. Rapport d'étape au Centre d'Analyse Stratégique, juin 2007.

Baujard, A., & Igersheim, H. (2007c) L'expérience du vote par note. Expérimenter des modes de scrutins différents. In *Organiser l'expression citoyenne. Pratiques électorales, déroulement des scrutins, technologies du vote. Un dimanche au bureau de vote. Actes du colloque du 5 avril 2007*, no. 10. Rapports et documents du Centre d'Analyse Stratégique, La Documentation Française, avril 2007, pp. 48–53. 60–62.

Baujard, A., & Igersheim, H. (2009). Expérimentation du vote par note et du vote par approbation le 22 avril 2007. Premiers résultats. *Revue Economique*, *60*(1), 189–202.

Baujard, A., Igersheim, H., & Senné, T. (2009a). Measuring consensus. An analysis based on experimental data. Mimeo CREM, Université de Caen.

Baujard, A., Igersheim, H., & Senné, T. (2009b). The political supply in the 2007 French presidential elections: An analysis based on experimental data. Mimeo CREM, Université de Caen; *Annales d'Economie et Statistiques* (Submited for publication).

Borda, J. (1781). *Mémoires sur les Élections au Scrutin*. Paris: Histoire de l'Académie Royale des Sciences.

Brams, S., & Fishburn, P. (1983). *Approval voting*. Boston: Birkhaüser.

Condorcet, J. (1785). *Essais sur l'Application de l'Analyse à la Probabilité des Décisions Rendues à la Pluralité des Voix*. Paris: Imprimerie Royale.

Cox, G. (1997). *Making Votes Count: Strategic coordination in the World's electoral system*. Cambridge: Cambridge University Press.

Cox, G., & Katz, J. (2002). *Elbridge gerry's salamander: The electoral consequences of the reapportionment revolution*. Cambridge: Cambridge University Press.

Farvaque, E., Jayet, H., & Ragot, L. (2009). Quel mode de scrutin pour quel vainqueur ? Une expérience sur le vote préférentiel transférable. *Revue d'Économie Politique*, *120*(2), 221–246
Felsenthal, D. S. (1989). On combining approval with disapproval voting. *Behavioral Science*, *34*, 53–60.
Gehrlein, W. V., & Lepelley, D. (2003). On some limitations of the median voting rule. *Public Choice*, *117*(1), 177–190.
Gerber, A. S., Green, D. P., & Larimer, C. W. (2008). Social pressure and voter turnout: evidence from a large-scale field experiment. *American Political Science Review*, *102*(1) , 33–48.
Harrison, G. W., & List, J. A. (2004). Field experiments. *Journal of Economic Literature*, *42*(4), 1009–1055.
Harrison, G. W., & List, J. A. (2008). Naturally occurring markets and exogenous laboratory experiments: a case study of the winner's curse. *Economic Journal*, *118*, 822–843.
Hillinger, C. (2004a). On the possibility of democracy and rational collective choice. Discussion Paper, no. 2004-21, University of Munich.
Hillinger, C. (2004b). Utilitarian collective choice and voting. Discussion Paper, no. 2004-25, University of Munich.
Hillinger, C. (2004c). Voting and the cardinal aggregation of judgments. Discussion Paper, no. 2004-09, University of Munich.
Hillinger, C. (2005). The case for utilitarian voting. *Homo Oeconomicus*, *23*, 295–321.
Laslier, J.-F. (2004). *Le vote et la règle majoritaire, analyse mathématique de la politique*. CNRS-édition.
Laslier, J.-F. (2006). Spatial approval voting. *Political analysis*, *14*, 160–185.
Laslier, J.-F. (2009a). A note on comparing median evaluations in single peaked-domains. Cahier no. 2009-23, Laboratoire d'économétrie, École Polytechnique, June 2009.
Laslier, J.-F. (2009b). Lessons from in situ experiments during french elections. In B. Grofman, A. Laurent & B. Dolez (Eds.), *In situ and laboratory experiments on electoral law reform: french presidential elections*, Heidelberg: Springer (in press).
Laslier, J.-F. (2010a). Describing the society through approval data. In J.-F. Laslier & R. Sanver, (Eds.), *Handbook on approval voting* (pp. 455–468). Heidelberg: Springer.
Laslier, J.-F. (2010b). Laboratory experiments on approval voting. In J.-F. Laslier & R. Sanver (Eds.), *Handbook on approval voting* (pp. 339–356). Heidelberg: Springer.
Laslier, J.-F., & Sanver, R. (2010). The basic approval voting game. In J.-F. Laslier & R. Sanver (Eds.) *Handbook on approval voting* (pp. 431–451). Heidelberg: Springer.
Laslier, J.-F., & van der Straeten, K. (2004). Election présidentielle : une expérience pour un autre mode de scrutin. *Revue Française de Science Politique*, *54*, 99–130.
Laslier, J.-F., and van der Straeten, K. (2008). Approval voting in the french 2002 presidential election: a live experiment. *Experimental Economics*, *11*, 97–195.
Núñez, M. (2010). Approval voting in a large electorate. In J.-F. Laslier & R. Sanver (Eds.) *Handbook on approval voting* (pp. 165–197). Heidelberg: Springer.

Chapter 16
Approval Voting in Germany: Description of a Field Experiment

Carlos Alós–Ferrer and Đura-Georg Granić

16.1 Introduction

The 2008 state elections in the German state of Hesse were expected to be extremely close. However, nobody expected that forming a new government would reveal itself to be impossible and, after long months of unsuccessful attempts, new elections would have to be called for almost exactly 1 year later.

On the original election day, January 21st 2008, we carried out a field experiment on approval voting in the German town of Messel, with the explicit permission and friendly support of the Hessian Ministry for the Interior and for Sport, the head election organizer (Mr. Wolfgang Hannappel), the mayor of the Messel district (Mr. Udo Henke), and the election commissioner (Mr. Dieter Lehr). This collaboration allowed us to install separate voting booths in each of the three different voting areas in the Messel district. Voters had been previously contacted per post and asked to take part in a secondary hypothetical vote after casting their official vote. In this second vote, Approval Voting was offered as an alternative voting system.

Our motivation was twofold. First, we were inspired by the experiment of Laslier and Van der Straeten (2004, 2008)[1] in the French Presidential Elections of 2002 and wanted to conduct an analogous study in Germany. We believe that conducting such field experiments in different countries is crucial to establish the practical applicability of the method.[2] Second, the particularities of the German electoral system allowed us to conduct two simultaneous experiments with the same voters, one where Approval Voting was used to select a candidate under a winner-take-all procedure (as in previous experiments elsewhere), and one where votes were cast

[1] See also Laslier (2006).

[2] The bottom-line motivation, as in Laslier and Van der Straeten (2004, 2008) or Brams and Fishburn (2005), is to show that Approval Voting could readily be incorporated into the political process. The desirability of such a development is founded on the method's sound theoretical properties, as shown by Brams and Fishburn (1978) and made explicit by the characterization results of Fishburn (1978a, 1978b) and Alós-Ferrer (2006).

C. Alós–Ferrer (✉)
Department of Economics, University of Konstanz, Box 150, D-78457 Konstanz, Germany
e-mail: Carlos.Alos-Ferrer@uni-konstanz.de

J.-F. Laslier and M.R. Sanver (eds.), *Handbook on Approval Voting*, Studies in Choice and Welfare, DOI 10.1007/978-3-642-02839-7_16,

for political parties rather than candidates, with an ensuing proportional system to determine representation in the (state) parliament.

Indeed, most German state elections are idiosyncratic in that voters are asked to cast two different votes. The first, for the district election ("Wahlkreisstimme") is given to a named candidate, and the results are determined by the winner-takes-all procedure with simple majority. Half the seats in the state parliament are allocated through this method (direct seats). The second vote, for the state election ("Landesstimme"), determines the percentage of the total seats (not the remaining ones) to be allocated to each different party which reaches at least 5% of the votes.[3] Hence, although approval voting is typically considered for candidate elections only, it was natural, in our setting, to ask voters to provide approval ballots both for district candidates and for state parties.[4]

Before the election, we sent a letter to the 3,017 citizens of Messel who were eligible to vote.[5] This letter explained the experiment's objective and the way it would be carried out. Additionally, the letter was published in the local city hall bulletin. To the best of our knowledge, ours is the first field experiment to try out this method in Germany.

As mentioned above, the main purpose of the investigation was to contribute to the empirical testing of Approval Voting. We were particularly interested in examining any differences between the outcomes of the hypothetical vote and the voting system currently in use. The results brought additional insights for political and economic theory as well as understanding the (rather delicate) political situation in Hesse.

We asked the voters in Messel to fill out two different voting forms: one for the district election and another for the state election. Thus we had two different sources of data. For the electoral district vote, there was a relatively small number of candidates (8 in total) to choose from. For the state election, there was a relatively large number of parties (17 in total).

16.2 The Official Election

Messel is part of the hessian electoral district 51 (Darmstadt-Dieburg). There were eight candidates in this district for the district elections, each representing one of the major parties: the CDU (conservative), the SPD (socialist), the Greens, the FDP (liberal), the Republicans (extreme right), the Left (extreme left), the Free Voters (mostly concerned with local issues), and the NPD (extreme right). The candidate

[3] There are minor complications if a party manages to capture a larger number of direct seats than the total percentage would allow it to have, or if a party which does not reach the 5% barrier obtains some direct seats. These difficulties are essentially dealt with by increasing the number of seats in parliament.

[4] This raises a number of interesting theoretical considerations. See Alós-Ferrer and Granić (2010) for a discussion.

[5] To preserve voter anonymity beyond all doubt, we provided the election officials with the letters and they were the ones to attach address labels and actually send them.

for the SPD Party won the direct seat. In the state election, there were 17 different parties from which to choose: the CDU, the SPD, the Greens, the FDP, the Republicans, the Animal Protection Party, the Civil Liberties Party, the PSG (communist), the Popular Vote Party, the Grey Party (oriented towards senior citizen issues), the Left, the Violet Party (oriented towards spiritual issues), the Family Party, the Free Voters, the NPD, the 'Hessian Pirates' (an organization of computer hackers) and the UB Party ('Independent Citizen Politics').

The SPD and the CDU received the greatest percentage of the vote statewide, with roughly equal numbers of votes for each party. However, none of the traditional coalitions (CDU + FDP and SPD + the Greens) could reach an absolute majority. Only five parties received more than 5% of the vote, enabling them to sit in the state parliament. These were the CDU, the SPD, the FDP, the Greens and the Left. This pattern was also found in Messel, with the exception that the Left Party received only 4.9% of the vote and thereby just missed out the State Parliament barrier. This difference is statistically meaningless.

The 3,017 registered voters were divided among three voting stations: Messel I, Messel II and 'Grube Messel', with 1,326, 1,401 and 290 eligible voters respectively. On the election day, 1,909 voters took part personally in the official election (Messel I: 847, Messel II: 902, Grube Messel: 160). Additionally, 282 voters cast an absentee vote through the post. Thus a total of 2,191 voters voted in the election. This represented 72.6% of the eligible voting population, which is relatively high in comparison to other electorates. The participation figures were similar to those for previous elections in Messel. This supports the notion that the (announced) experiment had no negative effects on voter participation.

16.3 The Experiment

Only people who voted at the voting stations took part in our experiment. Thus the absentee voters are not included in the data for this experiment. Of the 1,909 voters, 967 (50.65%) took part in the study (Messel I: 461, 54.43% of voters; Messel II: 407, 45.12% of voters; Grube Messel: 99, 66.88% of voters). There were 6 invalid votes in total (4 in Messel I, 2 in Messel II).[6] Our sample is thus composed of a total of 961 voters (Messel I: 457; Messel II: 405; Grube Messel: 99).

The results of both the district and state elections differ from those of the official election. These differences are especially pronounced in the official election. Here we will present a descriptive summary of our results, structured in four different sections[7]: the District vote; the State Vote; a hypothetical 'Messel-State Parliament'; and further miscellaneous observations (for example, Coalition results).

[6] It is of course quite hard to cast an invalid vote under approval voting. These six voters wrote comments on the ballot instead of using it for voting. One of them actually stapled a long declaration on the political situation in Messel to the ballot.

[7] A more detailed analysis is presented in Alós-Ferrer and Granić (2010) where, among other topics, we tackle the spatial representation of the Messel electorate's preferences using the spatial method described in Laslier (2006).

Remark 16.3.1. Although the results in Messel in previous elections were representative of the results in the whole state of Hesse, it is of course not statistically possible to use our sample for making predictions about future political outcomes in the whole of Hesse. Our discussion is for this reason to be understood purely as informative. We limit our comparisons with the results of the official election in Messel itself.

Our tables are set out as follows:

- **Candidate/Party:** Name of the Candidate or Party.
- **Votes:** percentage of the voters who voted for (approved of) the Candidate/Party. Because every voter could vote for more than one Candidate/Party, the percentage does not add up to 100% but rather to 186% for the district election and 225% for the state election.
- **Vote share:** amount of votes (approvals) for a Candidate/Party, divided by the total number of votes (this represents a renormalization of the votes, so that they sum to 100%).
- **Z-Rank:** The candidates and parties are arranged according to the size of their share of the votes in the hypothetical election (for example, the candidate with the most votes receives Z-Rank '1').
- **Official Vote:** Share of the votes in the official vote in Messel (excluding absentee votes).
- **O-Rank:** The candidates and parties are arranged according to the size of their share of the votes in the official election (for example, the candidate with the most votes receives O-Rank '1').

16.3.1 District Election

The following table summarizes the results of the district election, where voters had to elect a single candidate:

Candidate	Votes (%)	Z-Share (%)	Z-Rank	Official Election (%)	O-Rank
Hofmann, SPD	58.0	31.2	1	45.9	1
Milde, CDU	41.8	22.5	2	37.9	2
Harth, the Greens	31.4	16.9	3	4.5	4
Dr. Krug, FDP	30.3	16.3	4	6.0	3
Deistler, the Left	10.4	5.6	5	3.4	5
Herrmann, the Free Voters	7.4	4.0	6	0.8	7
Bauer, REP	3.9	2.1	7	1.1	6
Zeuner, NPD	2.8	1.5	8	0.3	8
Total	186.0	100.0		100.0	

The salient features of the table are described below:

1. The winner, according to the Approval Voting method, would have been Mrs Hofmann (SPD), just as in the official election. The results of the Approval Voting method differ from those of the official election in that according to the Approval Voting method, Mrs Hofmann would have achieved an absolute majority. 58% of voters gave her their approval in the hypothetical ballot. This information is lost in the official election results.
2. The Approval Voting method alters the ranking of the different candidates. In the official election, the greens are ranked fourth and the FDP's candidate is ranked third. Their positions are actually swapped through the Approval Voting method.
3. In the official election, the candidate for the extreme-right Republicans is in sixth place, whilst the candidate for the Free Voter Party is in seventh place. This ranking is reversed in the results of the Approval Voting method. In this particular case, the difference between the two election methods is especially large. Whilst in the Approval Voting results the candidate for the Free Voter Party can count on a not insignificant percentage support of 7.4%, in the official election he received only 0.8% of the total vote. The Republican candidate received a comparatively small percentage of the vote in the Approval Voting results, with 3.9% of the vote.
4. The voters voted for, on average, 1.86 candidates (standard-deviation: 0.874). This value is robust, as shown by the similar averages in the three different voting stations (Messel I: 1.89, Messel II: 1.83, Grube Messel: 1.84).

16.3.2 State Election

The following table summarizes the results for the State Election, where voters were asked to select a party list:

Party	Votes (%)	Z-Share (%)	Z-Rank	Official Election (%)	O-Rank
SPD	53.8	23.9	1	38.9	1
CDU	44.6	19.8	2	36.0	2
The greens	36.1	16.0	3	7.0	4
FDP	32.6	14.5	4	9.0	3
The left	12.3	5.5	5	4.9	5
Animal protection party	9.6	4.3	6	0.8	7
The family party	9.6	4.3	6	0.2	12
The free voters	7.1	3.1	8	0.5	9
Rebublican party	3.3	1.5	9	1.0	6
The popular vote	2.9	1.3	10	0.2	13

(Continued on next page)

Party	Votes (%)	Z-Share (%)	Z-Rank	Official Election (%)	O-Rank
NPD	2.8	1.2	11	0.8	7
The hessian pirates	2.8	1.2	11	0.3	10
The grey party	2.5	1.1	13	0.2	13
UB	2.1	0.9	14	0.1	15
The violet party	1.0	0.5	15	0.3	11
PSG	0.9	0.4	16	0.1	15
Civil liberties party	0.9	0.4	16	0.1	15
Total	225.0	100.0		100.0	

Several interesting conclusions can be drawn from this table:

1. With the Approval Voting System, the notion of the 'two big parties' seems less appropriate to describe the political situation. There were in fact 4 parties which received an approval rate above 30%: the CDU, the SPD, the Greens and the FDP. On this basis, the results of a state election (see 'Messel-Parliament' below) would have produced four major factions, each with a similar number of seats in Parliament. One could even infer on this basis, that the official vote's splitting of voter preferences into two political sides is an artificial product of the voting system. Parties such as the Greens and the FDP would have gained a great advantage through the Approval Voting system.
2. In Messel, according to the hypothetical election, the SPD Party received an absolute majority of the vote. Of course, under Approval voting it can be the case that more than one party receives an absolute majority. However, the SPD was the only such party in Messel.
3. Some of the parties that are categorized as 'small' become much larger with the Approval Voting system. There were three parties which in the official election received only a small percentage of the vote, and whose size grew to significantly more than 5% of the vote with the Approval Voting method. The following parties showed such an increase: the Animal Protection Party (9.6%), the Family Party (9.6%) and the Free Voters (7.1%). If we assume, with a leap of faith, that these figures are representative of the state of Hesse, then one can argue that these parties should have seats in the state Parliament (see the hypothetical 'Messel-Parliament' below).
4. The positions of the political minorities is distorted by the official voting method. After the four major parties and the Left come the Republicans (ranked sixth) and the NPD (ranked seventh). According to the Approval Voting system, however, it is the Animal Protection Party, the Family Party and the Free Voters who have the largest share of the votes after the four major parties and the Left (see above).
5. The voters voted on average for 2.25 parties (standard deviation: 1.141). This value remained robust, as the average value for the three voting stations was comparable (Messel I: 2.31, Messel II: 2.20, Grube Messel: 2.20).

It is interesting to note that the Approval Voting system altered the notion of a two-party dominant system, showing instead four parties with a significant proportion of the vote. Additionally, certain so-called 'small parties' were seen to be significantly larger in the Approval Voting system than in the official system. The reasons for these differences can be clarified as follows. Because voters each voted on average for 2.25 parties, every party receives on average 2.25 times the number of votes that they would have received in the official election. This means that the number of votes for an 'average' party in the Approval Voting system, on the basis of their votes in the official election, is obtained by multiplying the number of votes in the official election by 2.25. For the CDU and the SPD however, this factor was only 1.24 and 1.38 respectively. In contrast, the Greens and the FDP received a factor of 5.14 and 3.62 respectively. The factors for the NPD and the Republicans were higher than average, with 3.5 and 3.3 respectively. But the factors for the Animal Protection Party, the Family Party and the Free Voter Party were enormously high, with 12.0, 48.0 and 14.2 respectively.

We draw the conclusion that the current official voting method presents a distorted view of voter opinions. This method forces the voter to decide for only one party. For many voters in this situation, the notion of 'making your vote count' would be very important. According to this argument, the small parties the voters would actually prefer to vote for, are ignored, because they are small and have no chance of winning, either at the local or state level. Instead, voters believe they should give their vote to one of the larger parties, whose positions the voter generally agrees with, although they are not as appealing to the voter as those of the preferred smaller party. In this way, the 'small' parties remain small, even when a significant proportion of the voting population sympathizes with them. For example, the CDU and SPD could possibly be chosen because they are large parties, although the preferences of the voters for the FDP and the Greens are just as marked. In this way, the large parties remain large, only because they are already large, and are seen as such.

A similar argument holds for the minority parties. Because of the drive to 'make your vote count' ('only vote for someone who has a chance of winning'), many 'small' parties are deprived of votes, because they are small at this point in time (although according to the real preferences of the voters, they should not really be so small). Because the Approval Voting system allows the voter with these preferences to choose the small party *and* the larger one, the notion of 'making your vote count' is no longer relevant, and it no longer influences the voting behaviour of the voter.

The argument for 'making your vote count' does not apply so well for protest voters and for voters who strongly disagree with all of the large parties. The official voting method here leads to parties on the far ends of the political spectrum being overvalued. With the Approval Voting system, these parties receive votes only from confirmed followers of the party, whilst other small parties (with less extreme political positions, such as the Animal Protection Party, the Family Party and the Free Voters; see the results above), freed from the constrictions of 'making your vote count', find comparably broader support.

16.3.3 The Messel Parliament

In order to illustrate the way in which the application of the Approval Voting Method would change the composition of the state parliament, we constructed a hypothetical parliament, using the results of the approval vote in Messel. This is based on the assumption that the results for Messel are generalizable across the whole state. This hypothetical 'Messel-Parliament' is of course only intended to function as an illustrative picture of the potential effects of the Approval Voting method.

To enable a proper comparison, we first constructed a 'Messel-Parliament' on the basis of the official election results. That is, that we calculated the distribution of seats in the state Parliament of Hesse, by projecting the official election results onto the whole state.

The parliamentary election results, according to the official elections in Messel, look quite similar to the actual parliamentary election results for the entire state of Hesse. The only major difference is that the Left Party, with 4.9% in Messel, fell just short of the 5% lower limit. The actual statewide results showed, however, that the Left Party, with 5% of the vote, managed to get a place in the Hessian State Parliament. In order not to have this statistically irrelevant difference blur the overall picture, we have calculated the Messel-Parliament based on the assumption that 4.9% of the vote is good enough to gain a place in the Hessian Parliament.

The following table, and Fig. 16.1, show the hypothetical 'Messel-Parliament' according to the official election of the district of Messel. The distribution of seats is, as in the results of the official 2008 election, calculated according to the largest remainder method, known in Germany as the Hare-Niemeyer method. In this case, all parties that fell under the 5% lower limit (in our case 4.9%) were eliminated. The vote percentages were then recalculated on the basis of the total number of votes of the remaining parties. These results were multiplied by 110/100 (because there are 110 seats in the Hessian Parliament) and then rounded off. In accordance with the Hare-Niemeyer Rule, the remaining seats were allocated to the Parties with the largest remainders (counting from the first decimal place).

Party	Votes	% (converted)	Mandates
SPD	726	40.65	45
CDU	671	37.57	41
FDP	168	9.41	10
The Greens	130	7.28	8
The Left	91	5.10	6
Total	1,786	100.00	110

The picture is qualitatively similar to the actual distribution of seats in the actual hessian Parliament. The official election results for the whole state of Hesse show the following patterns: (1) the CDU and the SPD are the two major parties;

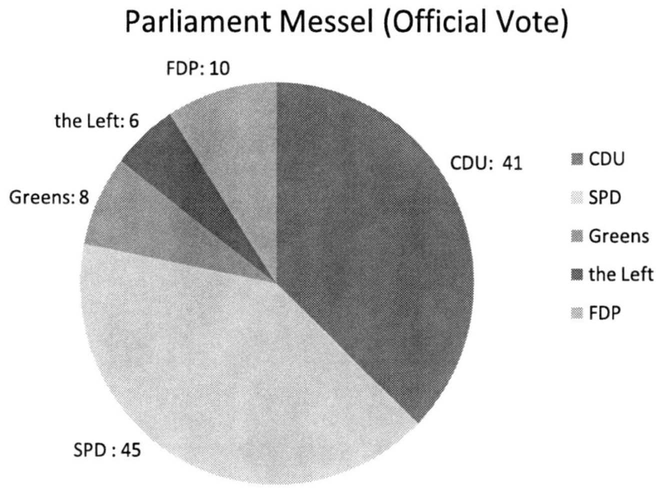

Fig. 16.1 The 'Messel Parliament', based upon the results of the official voting method

(2) neither of the "standard" coalitions, CDU + FDP and SPD + the Greens, could reach an absolute majority; and (3) the so-called 'big coalition' (CDU + SPD) would have had an absolute majority.

In order to create a hypothetical 'Messel-Parliament' from the Approval Voting data, we used a normalized voting-share. The number of approvals for the party, divided by the sum of all approvals for all parties (not through the number of voters) become the critical quantity to determine the number of elected members of any given party. To ensure a comparable distribution of our Messel-Parliament' we followed the official method of seat distribution as closely as possible. To do this, we used the Hare-Niemeyer method and maintained a 5% lower limit for election to the hessian Parliament. In this case we must decide if the 5% lower limit will be determined by votes or from vote-share. We decided that our criteria for election to the hessian Parliament would be determined by percentage of votes. This means that every party who received approvals from at least 5% of the voters, receives at least one seat in the hypothetical parliament. This procedure seemed both the simplest and most representative.[8]

The following table, and Fig. 16.2, show the hypothetical 'Messel-Parliament' as determined by the Approval Voting method. First, all parties which fell under the 5% lower limit were eliminated. Subsequently, the vote-share for the remaining parties was calculated. These results were multiplied by the factor of 110/100 (because there are 110 seats in the hessian Parliament) and rounded off. The remaining seats were divided amongst the parties with the largest remainders.

[8] See Alós-Ferrer and Granić (2010) for details.

Party	Votes	% (Votes)	Z-Share (%)	Mandates
SPD	517	53.8	26.16	29
CDU	429	44.6	21.71	24
The greens	347	36.1	17.56	19
FDP	313	32.6	15.84	17
The left	118	12.3	5.97	7
Animal protection party	92	9.6	4.66	5
Family	92	9.6	4.66	5
The free voters	68	7.1	3.44	4
Total	1,976	205.7	100.00	110

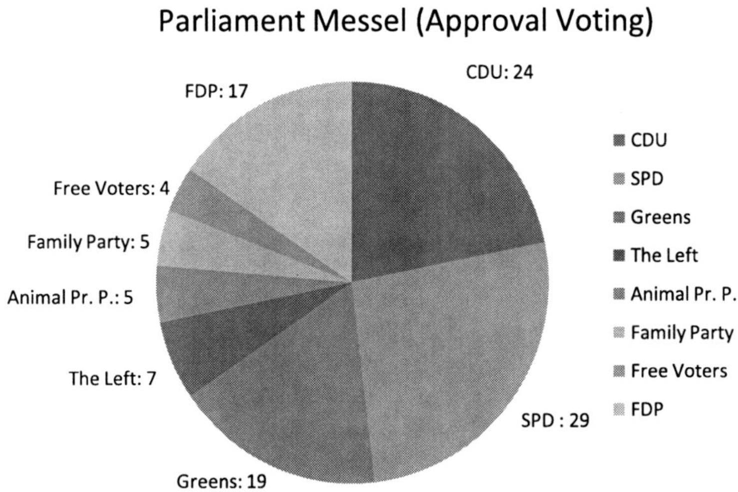

Fig. 16.2 The 'Messel Parliament', based upon the results of the Approval Voting data

The Messel Parliament based upon the results of the hypothetical voting method produced results that were quite different from those of the official state Elections: (1) according to the Messel Parliament, the majority of the seats would have been given to four different parties, the CDU, the SPD, the FDP and the Greens; (2) three small parties, which did not get any seats in the actual election, would have been elected to parliament in the hypothetical election: the Animal Protection Party, the Family Party and the Free Voters; and (3) the big coalition (the CDU and the SPD) would not have reached an absolute majority.

This hypothetical Messel Parliament allows some interesting possibilities for government formation. For example, the three new small parties in the Parliament (the Animal Protection Party, the Family Party and the Free Voters) together, could enable an SPD and Green Government, without the support of the Left or the FDP.[9] On the other hand, these three small parties would not enable the formation of a CDU and FDP government. Even a wide coalition of all parties excluding the CDU and the SPD would be theoretically possible.

Coalition	Votes (%)
SPD + Greens	27.68
CDU + FDP	25.18
SPD + FDP	9.47
Big Coalition (CDU + SPD)	9.16
FDP + Greens	6.14
SPD + Green + The Left	5.10
'Jamaica' (CDU + FDP + Greens)	4.99
'Traffic light' (SPD + Greens + FDP)	4.79

16.3.4 Further Remarks

16.3.4.1 Coalitions

An advantage of the Approval Voting Method is that the popular support for a given coalition can be assessed through the different votes, without the necessity of an additional questioning of the electorate. The data set from our study allows us to calculate how many voters voted for a given coalition. The following table shows the number of voters who voted for the different potentially interesting coalitions.

From these results it can be seen that, in our data sample, there is only a small amount of voter support for the coalition groups that actually exist.

Other specific questions can also be easily answered. For example, 517 voters voted for the SPD whilst 118 voted for the Left. Out of these only 91 voters voted both for the SPD and the Left. That means that only 17.6% of SPD voters would also vote for the Left if they had the option of doing so. In contrast, 77.1% of voters who voted for the Left would also vote for the SPD.

16.3.4.2 Number of Votes

Previous studies (such as Laslier and Van Straeten's in Orsay, France) have reported that for the Approval Voting Method, the voters choose on average three candidates to vote for. This observation seems not to generalize to our results. In our Study,

[9] We would like to remark that the impossibility of an SPD + Green government without the support of neither the FDP nor the Left was the essence of the long government formation crisis in Hesse which lasted the whole year 2008 and eventually resulted in new elections in 2009.

the voters chose on average 1.86 candidates (from 8 possible candidates) and 2.25 parties (from 17 possible parties). Because these average values were quite stable across the three different voting Stations in the district of Messel, we infer that the smaller number of candidates or parties voted for is the result of some as yet not identified psychological or cultural factor (one possible, purely economic explanation is that our participants had to provide two sets of data rather than only one, thereby opportunity costs of participating in the experiment were higher). The fact remains that the German voters in Messel 2008 chose significantly fewer from the potential options as the French voters in Orsay 2002.

16.3.4.3 Visibility of the Small Parties

Our study brought into focus another characteristic of the official voting system. All of the so-called 'small' parties face the problem of having low visibility. A significant number of voters speaking to us on polling day said that they never 'look down' far on the list of parties on the ballot paper. In other words, many voters took part in the official election without having read through the whole list of parties that they might vote for. When these voters came to the hypothetical vote, and had the possibility to vote for more than one party, the voters read the list all the way through. Some voters thought that our ballot paper was not serious, because they did not believe that parties such as 'the Hessian Pirates' or 'the Violets' were real parties. These were the same voters who only minutes before had given the official ballot paper in, with exactly the same names listed upon it.

16.4 Afterword

16.4.1 Repetition of the Election in Hessen

On January 18th, 2009, the citizens of Hesse were called to vote once again, after the previous elections held 1 year before did not enable state politicians to form a new government.

In the new elections, many of the small parties declined to participate. In total, only ten parties participated: the CDU, the SPD, the Greens, the FDP, the Republican Party, the Civil Liberties Party, the Left, the Free Voters, the NPD, and the 'Hessian Pirates'.

At the state level, the results were a disaster for the socialist party, whose previous main candidate, Andrea Ypsilanti, had infuriated supporters by attempting to form a coalition government with the radical-left "The Left".[10] As commented above, our

[10] Shortly after the election, Miss Ypsilanti took responsibility for the disastrous outcome and retired from her position as Chairman of the SPD in Hesse.

data shows that (at least in Messel), although most Left-supporters approved of the SPD, few of the SPD-supporters approved of the Left, and hence the problems faced by Miss Ypsilanti are hardly surprising. The following table shows the broad results in Hesse and in Messel, where the △ 2008 columns denote the change in percentage compared to the 2008 election:[11]

Party	% in Hesse	△ 2008 (%)	% in Messel	△ 2008 (%)
CDU	37.2	+0.4	36.4	+0.2
SPD	23.7	−13.0	23.5	−15.3
FDP	16.2	+6.8	16.4	+7.4
The greens	13.7	+6.2	15.0	+7.9
The left	5.4	+0.3	4.9	+0.3
The free voters	1.6	+0.7	1.6	+1.1
NPD	0.9	0.0	1.0	+0.1
Rebublican party	0.6	−0.4	0.4	−0.5
The hessian pirates	0.5	+0.2	0.6	+0.3
Civil liberties party	0.2	+0.2	0.1	+0.1
Total	100.0	+1.4	100.0	+1.6

Statewide, as well as in Messel, the CDU received the greatest share of votes. With an absolute majority of seats in parliament, together with FDP, the CDU formed the new state government. The picture of the vote in Messel reflects, apart from minor differences, the outcome we observe at state level quite well. The SPD voters punished their party for the attempt, against the promise in pre-election period not to do so, to form a minority government with the backing of the Left. Whilst the share of the CDU and the Left nearly stayed constant, the FDP, the Greens and the Free Voters roughly doubled their share of votes (from the participating parties, the latter three exhibit the largest multiplying factor in our experiment, see above).

Although one should be careful with the interpretation, a notable fact is that the considerable loss of votes suffered by the German Socialist Party in terms of share, both for Hesse and Messel, approximately equals the gain received by the FDP and the Greens. Excluding the possibility of fuzzy preference reversals among the whole population of Hesse, the most plausible and nearest interpretation is that a large fraction voters turned their back on the SPD and, instead, voted for the FDP and the Greens. In the context of our experiment, this development seems very natural. Not only did our experiment show that the Liberals and the Greens share a much higher acceptance rate amongst the population than the official vote suggests, they

[11] With only ten parties participating in the election in 2009, the changes of percentages do not add up to 0%, but, with 1.4% in Hesse and 1.6% in Messel, represent the total share of votes from 2008 cast for the seven parties that declined to participate.

are also the parties our participants simultaneously approved of with SPD the most (see coalition table).

16.4.2 A Second Experiment in Germany

On September 27th, 2009, we conducted a similar experiment during the nationwide German Federal elections (see Alós-Ferrer and Granić 2010 for details). This time, we selected six voting stations in the city of Konstanz, in the southern German state of Baden-Württemberg. Of the 2,879 voters who showed up at the voting stations, 1,431 (49.7%) took part in our study. The overall conclusions with regard to feasibility of the field experiment and voter acceptance were similar to our study in Hessen.

Remarkably, however, in our Konstanz study the results of the approval vote showed major differences from those of the official vote. As in Konstanz itself and most of Germany, the conservative party (CDU) received the simple majority both for the district election ("Erststimme"), where again a single candidate is selected, and the party-list election ("Zweitstimme"), where a party is elected. In contrast, the Green Party would have won both elections under approval voting (at least in the subset represented by the six selected voting stations).[12] The Green Party was approved by a 58.1% of the participants, and it was the only party to receive an absolute majority of approvals. The normalized approval vote share of the Green Party was 22.7%, coming before the CDU (16.2%) and the SPD (18.5%). In the official vote (restricted to our six voting stations), a vote share of 20.1% resulted in the Green Party coming in third, after CDU (28.6%) and SPD (21.9%). The situation was quite similar for the candidate vote.

The main political observations were also similar to the ones from the Hesse study, with four big parties arising rather than two, and some surprises among allegedly small parties. As an anecdote, the "Pirate Party" was approved of by 20.8% of the voters (normalized approval share of 8.1%, coming even before The Left), even though the official vote resulted in a vote share of only 3.7%. Other small parties also experienced large boosts, as e.g. the Animal Protection Party.

16.4.3 Final Words

Based on the observations above and our overall experience, we would like to argue that our field experiment has shown that, first, Approval Voting is a practicable method which can be easily implemented in practice, and, second, that the data

[12] Konstanz, a University city, is of course not representative for Germany. What we find interesting is the difference in results between the approval voting method and the official one.

generated by this method provide a better picture of the political preferences of the electorate than currently used methods.

References

Alós-Ferrer, C. (2006). A simple characterization of approval voting. *Social Choice and Welfare*, *27*(3), 621–625.
Alós-Ferrer, C., & Granić, Đ.-G. (2010). *Two field experiments on approval voting in Germany*. mimeo.
Brams, S. J., & Fishburn, P. C. (1978). Approval voting. *American Political Science Review*, *72*, 831–847.
Brams, S. J., & Fishburn, P. C. (2005). Going from theory to practice: the mixed success of approval voting. *Social Choice and Welfare*, *25*, 457–474.
Fishburn, P. C. (1978a). Axioms for approval voting: direct proof. *Journal of Economic Theory*, *19*, 180–185.
Fishburn, P. C. (1978b). Symmetric and consistent aggregation with dichotomous voting. In J.-J. Laffont (Ed.), *Aggregation and revelation of preferences*. Amsterdam: North Holland.
Laslier, J.-F (2006) Spatial approval voting. *Political Analysis*, *14*(2), 160–185.
Laslier, J.-F., & Van der Straeten, K. (2004). Une expérience de vote par assentiment lors de l'élection présidentielle de 2002. *Revue Française de Science Politique*, *54*, 99–130.
Laslier, J.-F., & Van der Straeten, K. (2008). A live experiment on approval voting. *Experimental Economics*, *11*, 97–105.

Part VII
Electoral Competition

Chapter 17
Classical Electoral Competition Under Approval Voting

Jean-François Laslier and François Maniquet

17.1 Introduction

In large societies, collective decisions cannot be taken directly but have to be delegated to professional decision makers. In a democracy, these delegates are typically elected through a competitive mechanism. The simplest expression of such a mechanism is the now standard *Downsian* model of Politics (Downs 1951) in which a relatively small number of candidates face a relatively large number of voters, the candidates are purely office-motivated and the voters policy-motivated. For the purpose of winning the election, each candidate freely and independently proposes a policy from a fixed and common set of possible policies. Voters are only interested in policies and not in candidates *per se*. They trust that the elected candidate will implement the policy she is proposing.

The usual case in the literature considers only two candidates under plurality voting. Then voters only face a binary choice, so that each voter simply votes for the candidate whose policy she prefers. In that case, competition for office drives the candidates to propose popular policies. In particular, if there exists a policy preferred to any other by a majority of voters – a Condorcet alternative – then both candidates propose this same policy. This statement is even an if and only if statement since, as soon as no Condorcet alternative exists, the two-party Downsian game has no pure-strategy equilibrium. Formal political science has studied this question in great details, and the literature on two-party competition under plurality rule is very large; see for instance the books of Ordeshook (1992), Roemer (2001), Mueller (2003), or Austen-Smith and Banks (2005). Following Cox (1984, 1985) and Weber (1995), this chapter is devoted to the study of multiparty electoral competition under approval voting in the Downsian political context where collective choice is delegated to office-motivated candidates. To recall, approval voting is the electoral rule under which voters are given the right to approve of as many candidates as they

J.-F. Laslier (✉)
Laboratoire d' Econométrie, École Polytechnique, 91128 Palaiseau, France
e-mail: jean-francois.laslier@polytechnique.edu

J.-F. Laslier and M.R. Sanver (eds.), *Handbook on Approval Voting*, Studies in Choice and Welfare, DOI 10.1007/978-3-642-02839-7_17,

wish, and each approval gives one point to the approved candidate. The winner of the election is the candidate having received the largest number of approvals.

Rational behavior of the voters rests on their beliefs about two things. On one hand, they have to wonder which candidate is most likely to win the election and which candidates can challenge this front-runner. On the other hand, they have to make up their mind as to the policies that each candidate would implement, if elected. Rational behavior of the candidates choosing platform campaigns, in turn, rests on their knowledge about the choices of the voters and of the other candidates. We study electoral competition in a framework where the candidates choose rationally (and simultaneously) their political platforms, and the voters react to these platforms. With more than two candidates, not only the voting rule matters but the behavior of voters is no longer as straightforward as it is with two candidates. Studying approval voting, we make the assumption that voters follow the Leader Rule, a behavioral rule which has a rational foundation (Laslier 2009) and satisfies the criterion of sincerity of Brams and Fishburn (1983) and admissibility of Dellis (2010).

We prove that when voters follow that rule, the outcome of the electoral competition among candidates converges towards the Condorcet winner policy in the following sense: if a Condorcet winner policy exists, then there exists an equilibrium that supports it; and if, moreover, the set of policies is one-dimensional and voters' preferences are single-peaked, then this equilibrium is the only one. The prediction of the model is thus that the approval voting electoral rule drives office-motivated candidates to policy moderation and makes more than two parties viable. This result should be contrasted with what happens with Plurality rule: in that case, only two parties can be viable.

In Sect. 17.2, we present the model and we recall the definition of the Leader Rule. In Sect. 17.3, we present the results. In Sect. 17.4, we discuss some possible extensions.

17.2 The Model

17.2.1 Candidates, Voters, and Preferences

There is a set X of possible policies. We consider two models below. In the first one, X is a finite set with no particular structure. In the second one, X is the real line.

In both models we make the following assumptions on voters and candidates. There is a set $\{1,\ldots,N\}$ of N voters. Voters have preferences over X. There is a set $\mathcal{C}$ of K candidates. Each candidate $c \in \mathcal{C}$ has to choose a policy

$$x_c \in X.$$

We assume that policy x_c is implemented if candidate c is elected. Consequently, a voter prefers candidate c to candidate c' if and only if she prefers x_c to $x_{c'}$, and we can equally well speak in terms of preference over candidates or preference over policies.

The way voters vote among the proposed policies x_c, $c \in \mathcal{C}$ is described below. Let us begin by describing the objectives of the candidates. As a result of the election, a fraction of the voters, which we denote by s_c, approve of policy x_c, $c \in \mathcal{C}$. The number of approvals of c is thus Ns_c. This number is called the *score* of c. The winning candidate is the one with highest score. If several candidates obtain the same, highest, score, the winner is decided by a fair lottery. We assume that the objective of a candidate is to maximize the probability of winning the election.

Let $x, y \in X$. Voters may prefer x to y, y to x, or be indifferent. We assume that the profile of voters' preferences over X is fixed. Given this preference profile of the population $\{1, \ldots, N\}$, we can compute $g(x, y) \in [0, 1]$, the fraction of the voters who strictly prefer x to y and $i(x, y) \in [0, 1]$, the fraction of the voters who are indifferent between x and y. Note that $g(x, y) - g(y, x)$ measures the relative plurality in favor of policy x against policy y. By definition,

$$g(x, y) + g(y, x) + i(x, y) = 1.$$

We suppose that the number of voters is large.

17.2.2 Individual Voting Behavior

Let us now describe how voters choose their vote. Here, we follow the behavioral rule developed in Laslier (2009) and we adapt it to the current model. A rational voter responds to the number of approval votes granted by the other voters to the various candidates (their scores). Let us assume that s_c represents the fraction of voters approving of c when we do not take account of a given voter's vote. First, the voter deduces from the scores s_c, $c \in \mathcal{C}$ a strict ranking $c_1, c_2, \ldots, c_K$ of the candidates. Candidate c_1 is the leader, according to that voter. This ranking needs to be compatible with the scores in the following sense.[1]

Definition 17.2.1. The ranking $c_1, c_2, \ldots, c_K$ of the candidates is *compatible with a score vector* $s = (s_c)_{c \in \mathcal{C}}$ if for all $k, k' \in \{1, \ldots, K\}$,

$$s_{c_k} > s_{c_{k'}} \Rightarrow k < k'.$$

If the score vector is such that the K candidates have distinct scores then there is a unique compatible ranking. That is the case analyzed in Laslier (2009). Otherwise, the candidates with identical scores can be ranked in any way, providing multiple compatible rankings.

[1] It should be clear that these rankings do not reflect voters' preferences, only voters' beliefs about which candidate will arrive first, which other candidate will arrive second, etc.

Recall that each voter has fixed preferences over X. For any list of policy positions $x = (x_1, \ldots, x_K)$ there is an induced preference relation of this voter over candidates. When the voter has strict preferences over the candidates, the Leader Rule stipulates that she approves of all the candidates she strictly prefers to c_1 and of no candidate she finds strictly worse than c_1, and she votes for c_1 if and only if she prefers c_1 to c_2. When the voter is likely to be indifferent between several candidates, the rule can be generalized as follows.

Assumption 17.2.1 (Leader Rule). *Given a strict ranking $c_1, c_2, \ldots, c_K$ of the candidates, a voter behaves as follows:*

- *If she is indifferent among all candidates, then she approves each of them with probability $1/2$, independently.*

  In all other cases, for any candidate d:

- *If the voter is not indifferent between d and c_1, she approves of d if and only if she prefers d to c_1.*
- *If she is indifferent between d and c_1 (for instance in the case $d = c_1$), she approves of d if and only if she prefers d to c_i, where c_i is the first candidate, according to the ranking $c_1, c_2, \ldots, c_K$ such that she is not indifferent between c_i and c_1.*

If the score vector is such that the K candidates have distinct scores, then this postulated behavior defines a unique ballot for any voter, except in the case where the voter is indifferent between all candidates. If the score vector contains ties, several rankings of the candidates are compatible. That may lead to different responses for some voters. For instance, let us assume that $s_1 = s_2 > s_3$ and the preferences of the voter are: Candidate 1 is preferred to candidate 3, preferred to candidate 2. If the strict ranking of the candidates compatible with scores is 1, 2, 3, then the voter approves only of 1. If the ranking is 2, 1, 3, then she approves of 1 and 2.

Let us briefly present the rationale for the Leader Rule. Assume that the scores represent how voters plan to vote, but for each voter and for each candidate she plans to approve, there is a small chance ϵ that the vote is not recorded, or that she forgets to cast that vote, etc. Then, the actual number of approvals for a candidate $c \in K$ becomes a random variable of mean $(1-\epsilon)Ns_c$. As a consequence, in spite of the fact that the expected scores of two candidates differ, there is always a positive probability that they tie, so that the vote of this voter is pivotal. Reasoning on these pair-wise ties and neglecting three-way ties when N tends to infinity a voter votes for a candidate if and only if the most likely serious tie event involving that candidate is one where the former is strictly preferred to the latter (a tie is serious if the voter is not indifferent between the two candidates). Laslier (2009) proves that it gives the above voting behavioral rule, and Nunez (2010b) presents this rule and other related models for large electorates.

17.2.3 Electorate Voting

To define the electorate response to a score vector s, suppose that s has exactly M compatible rankings. We make the assumption that a proportion $1/M$ of the population of voters adopts each of these rankings, independently of the types. For instance, in the above example $s_1 = s_2 > s_3$ among all voters sharing the same preferences, fifty per cent will behave according to the ranking 1, 2, 3, and fifty per cent according to 2, 1, 3. This assumption only makes sense in sufficiently large populations. This is precisely our definition of a large population.

Assumption 17.2.2 (Uniform tie-breaking). *Given a score vector s, each voter chooses a ranking of the candidates compatible with s, and responds to this ranking. The choice of the ranking is uniform among the rankings compatible with s, and it is independent of the voter's preferences and of the other voters' choices.*

The above assumptions form a simple and natural extension of the Leader Rule defined by Laslier (2009) to handle the possibility of ties, although it is only justified by some kind of law of insufficient reason, as is often the case for uniform rules.[2] Notice that, in any case, since each ballot is defined by the Leader Rule applied to the appropriate ranking, all ballots are sincere and admissible.

17.2.4 Equilibrium

We define an electoral competition game as one with $K + N$ players, the K candidates and the N voters. In the first stage of the game, each candidate c chooses a policy $x_c \in X$. In the second stage of the game, voters vote, using approval voting. Each vote has a given probability of not being recorded, as explained above. Depending on which votes are recorded, candidates receive numbers of approvals. The candidate with the largest number of approvals is the winner of the election. Ties are broken by a fair lottery. Voters derive utility from the (lottery over) policies that were chosen by the winning candidates. Candidates derive utility from the probability of being elected.

If we restrict our attention to the second stage of the game, then we are back to the game studied by Laslier (2009) except that candidates' score vector can now contain ties. The Leader Rule tells us how voters react to the expected vector of scores. Now, the expected vector of scores itself is completely determined by the voters' expected votes. The equilibrium notion we consider, which we call uniform consistency, is the fixed point of that relation. It is the set of pure Nash equilibrium outcomes of the second stage of the electoral game under Assumptions 17.2.1 and 17.2.2.

---

[2] The uniformity assumption is required for our proofs. Notice however that the assumption is trivially met each time there are no ties.

Definition 17.2.2. A score vector $s = (s_c)_{c \in C}$ is *uniformly consistent* with policy positions $x = (x_c)_{c \in C}$ if s is the score vector that is obtained when voters react, according to Assumptions 17.2.1 and 17.2.2, to s itself.

We are interested in the pure sub-game perfect equilibria of the electoral competition game. Using the above definition, such an equilibrium is a pair (x, s) of positions and scores such that s is uniformly consistent with x and for no candidate c there exists a unilateral deviation $x' = (x'_c, x_{-c})$ and a score vector s' uniformly consistent with x' such that the probability of c winning the election is higher in s' than in s.

For the sake of completeness, we prove the following result, which consists of adapting Laslier's result to the current framework. That results concerns cases where candidates choose policies in such a way that no voter is indifferent between any two policies. In those cases, if a strict Condorcet winner policy exists, then there is a unique score vector uniformly consistent with it. That vector is easily built by using the g function.

Recall that a Condorcet winner policy is one that is preferred to any other by a majority of voters.

Definition 17.2.3. A list of policy positions $x = (x_c)_{c \in C}$ admits a Condorcet winner policy x_ℓ if and only if for all $k \neq \ell$,

$$g(x_\ell, x_k) \geq 1/2.$$

It admits a strict Condorcet winner if the above inequality is strict for all $k \neq \ell$.

A policy x_ℓ may be a Condorcet winner in a list of policy positions $x = (x_1, x_2, \ldots, x_K)$ even if there exists a policy $y \in X$ that is preferred to x_ℓ by a majority of voters, provided y is not in the list of policy positions. We will come back to that key issue in the next section.

Proposition 17.2.1. *Assume that the list of policy positions $x = (x_c)_{c \in C}$ is such that no voter is indifferent between any pair of policies. If x admits a strict Condorcet winner policy x_ℓ, then there exists a unique score vector s that is uniformly consistent with x. This score vector is defined by: For all $k \neq \ell$, $s_k = g(x_k, x_\ell) < 1/2$, and*

$$s_\ell = \min_{k \neq \ell} g(x_\ell, x_k).$$

Proof. (1) The score vector s is consistent with x: Let $\mathcal{L}^{(2)}$ denote the set of candidates obtaining the second largest score. By construction, for all $k, k' \in \mathcal{L}^{(2)}$: $g(x_k, x_\ell) = g(x_{k'}, x_\ell)$. For all $k \in \mathcal{L}^{(2)}$, a fraction $\frac{1}{|\mathcal{L}^{(2)}|}$ of voters has ranking $\ell, k, \ldots$. Among those voters, a fraction $g(x_\ell, x_k)$ vote for ℓ, a fraction $g(x_k, x_\ell)$ vote for k, and for all $h \neq \ell, k$, a fraction $g(x_h, x_\ell)$ vote for h. Consequently,

$$s_l = \frac{1}{|\mathcal{L}^{(2)}|} \sum_{k \in \mathcal{L}} g(x_\ell, x_k) = g(x_\ell, x_{k'}) \ \forall k' \in \mathcal{L}^{(2)},$$

and for all $k \neq \ell : s_k = g(x_k, x_\ell)$. Finally, as for all $k \in \mathcal{L}^{(2)}$, $k' \notin \mathcal{L}^{(2)} \cup \{\ell\}$: $s(k) > s(k')$, by construction, $g(x_k, x_\ell) > g(x_{k'}, x_\ell)$. This implies $g(x_\ell, x_k) < g(x_\ell, x_{k'})$ so that

$$s_\ell = \min_{k \neq \ell} g(x_\ell, x_k).$$

(2) Uniqueness: Let $\mathcal{L}^{(1)}$ denote the set of candidates obtaining the largest score. Assume $\ell \notin \mathcal{L}^{(1)}$. Then,

$$s_l = \frac{1}{|\mathcal{L}^{(1)}|} \sum_{k \in \mathcal{L}^{(1)}} g(x_\ell, x_k) > \frac{1}{2}.$$

First case: $\mathcal{L}^{(1)}$ contains more than one candidate. Then, for all $k \in \mathcal{L}^{(1)}$,

$$\begin{aligned} s_k &= \frac{1}{|\mathcal{L}^{(1)}|} \sum_{\binom{k' \in \mathcal{L}^{(1)}}{k' \neq k}} g(x_k, x_{k'}) + \frac{1}{|\mathcal{L}^{(1)}|} \frac{1}{|\mathcal{L}^{(1)}| - 1} \sum_{\binom{k' \in \mathcal{L}^{(1)}}{k' \neq k}} g(x_k, x_{k'}) \\ &= \frac{1}{|\mathcal{L}^{(1)}| - 1} \sum_{\binom{k' \in \mathcal{L}^{(1)}}{k' \neq k}} g(x_k, x_{k'}). \end{aligned}$$

Summing up these scores, and recalling that $g(x_k, x_{k'}) + g(x_{k'}, x_k) = 1$ (no voter is indifferent between x_k and $x_{k'}$), we get

$$\begin{aligned} \sum_{k \in \mathcal{L}^{(1)}} s_k &= \frac{1}{|\mathcal{L}^{(1)}| - 1} \sum_{k, k' \in \mathcal{L}^{(1)}} g(x_k, x_{k'}) + g(x_{k'}, x_k) \\ &= \frac{1}{|\mathcal{L}^{(1)}| - 1} \frac{(|\mathcal{L}^{(1)}|)(|\mathcal{L}^{(1)}| - 1)}{2} = \frac{|\mathcal{L}^{(1)}|}{2}, \end{aligned}$$

so that for all $k \in \mathcal{L}^{(1)} : s_k = \frac{1}{2}$. To summarize, we have $s_k < s_l$, a contradiction. Second case: $\mathcal{L}$ contains one candidate, say 1. We must have $s_1 > \frac{1}{2}$ and for all k in the set of candidates ranked second, $s_k = g(x_k, x_1) < \frac{1}{2}$, which is inconsistent with $s_l > \frac{1}{2}$. □

There are two important directions in which the above result does not extend. First, even if the profile has a Condorcet winner, if the Condorcet winner is not strict, then it is possible that no uniformly consistent scores exist.

Example 17.2.1. Consider a set of three candidates $\{1, 2, 3\}$ such that the pair-wise comparisons among candidates are: $g(x_1, x_2) = .5, g(x_1, x_3) = .6, g(x_2, x_3) = .1$. No uniformly consistent scores exist for this profile. To see that one can check the impossibility for each ordering, strict or not, of the candidates according to s. For instance if 1 is alone at the first place in s, 2 at the second place, and 3 at the third, then the scores should be $s_1 = g(x_1, x_2) = .5$ and $s_2 = g(x_2, x_1) = .5$

also, a contradiction. If 1 and 2 tie at the first place and 3 comes third, then $s_1 = g(x_1, x_2) = .5$, $s_2 = g(x_2, x_1) = .5$, and $s_3 = (g(x_3, x_1) + g(x_3, x_2))/2 = .65 > .5$, a contradiction. The reader will easily complete this proof.

Second, if (a non-negligible fractions of) voters have indifferences, then there may be several score vectors uniformly consistent with the policy positions and even a strict Condorcet winner may fail to be ranked first in such a score vector.

Example 17.2.2. Consider a set of three candidates $\{1, 2, 3\}$ and their policy positions $x = (x_1, x_2, x_3)$ inducing the preferences described in the following table (which reads: 4 voters are indifferent between 1 and 2 and strictly prefer any of these two to 3, etc.):

4	3	2
1,2	3	3
3	1	2
	2	1

Observe that 3 is a strict Condorcet winner. Nevertheless consider the following score vector: $s_1 = 7$, $s_2 = 6$, $s_3 = 5$, in which candidate 3 is ranked last. One can easily check that, applying the Leader Rule, all voters vote for two candidates, and that this score vector is consistent with x.

These examples show that there is no pure strategy Nash equilibria under Assumptions 17.2.1 and 17.2.2 in the above games. Of course, if the profile of candidates has no Condorcet winner, it is all-the-more possible that no uniformly consistent score exists. The fact that uniformly consistent scores may fail to exist is a difficulty for the study of electoral competition under AV in the general case, under the uniform tie-breaking assumption. The results (in the next section) will thus be limited to some observations, in the case of existence of a Condorcet winner.

The result conveyed in Example 17.2.2 above involves a preference profile that is incompatible with the assumption that individual preferences are single-peaked. Recall that individual preferences are single-peaked if it is possible to order the alternatives in such a way that each agent has a unique preferred alternative and her satisfaction decreases as the selected alternative moves further away from her preferred one. In Example 17.2.2, the only orderings of alternatives that are compatible with the existence of preferences 312 and 321 (last two columns) are $1 < 3 < 2$ and $2 < 3 < 1$, which excludes preference (12)3. The last result in the next section shows that Example 17.2.2 cannot hold if preferences are single-peaked.

17.3 Results

We are now equipped to prove our two results. Both results hold even if agents show indifferences among some pairs of policy positions. They confirm the close relationship between approval voting and the Condorcet winner. Proposition 17.3.1

states that a Condorcet winner policy, if it exists, can always result from electoral competition under approval voting, and Proposition 17.3.2 states that, in single-peaked domains, this is the only possible outcome. The section is completed by showing that these results do not hold if approval voting is replaced with plurality voting.

17.3.1 Condorcet-Consistency

In the previous section, we defined a Condorcet winner by reference to a list $x = (x_1, \ldots, x_K)$ of policy positions. If we look at the entire set X of possible policies, we can define a Condorcet winner policy as one that is preferred to any other policy in X by a majority of voters. Let us note that there is no logical relation between the existence of a Condorcet winner in X and the existence of a Condorcet winner relative to a list of K policy positions in X.

The first result states that if a strict Condorcet winner exists in X, then, independently of the structure of X, all candidates choosing that policy position is an equilibrium. In view of Examples 17.2.1 and 17.2.2 above, it may be possible that some list of policy positions leads to no uniformly consistent score vector. What this proposition proves is that, still, the score vector resulting from all candidates proposing the Condorcet winner is uniformly consistent and no candidate can gain by proposing another policy.

Proposition 17.3.1. *If $x^C \in X$ is a Condorcet winner policy then the strategy profile in which all candidates choose x^C is an equilibrium of the electoral competition game. If x^C is a strict Condorcet winner then the equilibrium is strict.*

Proof. Let $x^C \in X$ be a Condorcet winner policy. Let $x = (x_1, \ldots, x_K)$ be defined by: For all $k \in \{1, \ldots, K\} : x_k = x^C$. Then, by the Leader Rule, each candidate is elected with probability $1/K$. Suppose candidate 1 (for instance) deviates to $x_1 \neq x^C$. There are now two different policy positions to choose from. Independently of how they are ranked, we have $s_1 = g(x_1, x^C) \leq .5$ and for all $k \in \{2, \ldots, K\}$: $s_k = g(x^C, x_1) \geq .5$. Candidate 1's probability of being elected is now either $1/K$ or 0. In any case, the deviation is not profitable. If x^C is a strict Condorcet winner policy, then $s_1 = g(x_1, x^C) < .5$ and for all $k \in \{2, \ldots, K\} : s_k = g(x^C, x_1) > .5$. Consequently, 1's probability of being elected decreases to 0. □

17.3.2 Median Convergence

We consider in this section the standard, one-dimensional, single-peaked model. The set of possible policies is the real line $X = \mathbb{R}$. Each voter $j \in N$ has a preferred policy p_j. Moreover, for two policies $x, y \in X$ on the left of p_j (resp., on the right of p_j), x is strictly preferred to y if and only if x is closer to p_j than

y: $y < x < p_j$ (resp., $p_j < x < y$). Let $x^m \in X$ be the median of the voters' preferred policies – as many voters have their preferred policy at the left as at the right of x^m. We suppose that this point exists and is unique. Then, as is well-known, this policy-moderated, centrist outcome, x^m is a strict Condorcet winner: for any $y \neq x^m$, $g(x^m, y) > .5$, a strict majority of the population strictly prefers x^m to y. The previous result applies and all candidate policy positions being concentrated at the median point is a strict equilibrium of the electoral competition. The following proposition, our main result, also proves it is the only equilibrium. That shows that, under approval voting, electoral competition drives candidates to propose the Condorcet policy platform.

Proposition 17.3.2. *In the single-peaked model: (i) The strategy profile in which all candidates choose the median policy position is a strict equilibrium of the electoral competition game. (ii) It is the only equilibrium.*

Proof. Point (i) follows from Proposition 17.3.1. (ii) We first note three facts related to the single-peaked profile structure. Let x, y, z be such that $x < y < z$.

- *Fact 1:* no voter is indifferent between the three positions.
- *Fact 2:* the voters (if any) who are indifferent between x and z strictly prefer y to both x and z.
- *Fact 3:* $g(y, x) \geq g(z, x)$ and $g(y, z) \geq g(x, z)$.

Next we observe that, from the definition of the Leader Rule (Assumption 17.2.1), if two candidates k, k', propose the same policy $x_k = x_{k'}$, they obtain the same number of votes and any other candidate l obtains the same number of votes as l would obtain if there was only one candidate at position x_k. Now, let $x = (x_1, \ldots, x_K)$ be some list of policy positions chosen by the candidates. For the ease of reading, and when no confusion in the course of the proof can arise, we can neglect the possibility of several candidates located at the same position and we will speak of "a set of candidates" rather than "a set of different policy positions chosen by candidates."

There is at least one candidate, say 1, with a probability of winning the election less than or equal to $1/K$. We will prove that deviating to $x'_1 = x^m$ is profitable.[3] Let $s = (s_1, \ldots, s_K)$ be a score vector associated to $x' = (x'_1 = x^m, x_2, \ldots, x_K)$. Notice that s_1 is equal to or larger than some average of $g(x^m, x_k)$, for $k \in \{2, ..., K\}$, and because x^m is a strict Condorcet winner, $s_1 > .5$. Let $\mathcal{L}^{(1)}$ denote the set of candidates obtaining the largest score.

If all the candidates except candidate 1 are located at x^m then the probability of winning goes from 0 to $1/K$ when candidate 1 deviates to x^m. We can thus suppose that some candidates are not located at x^m and we will prove that, when deviating to x^m, $1 \in \mathcal{L}^{(1)}$ and $\mathcal{L}^{(1)}$ contains at most $K - 1$ candidates, which makes the deviation profitable for candidate 1. For a contradiction, assume $1 \notin \mathcal{L}^{(1)}$. We distinguish three cases, depending on the number of candidates in $\mathcal{L}^{(1)}$.

[3] We will thus prove that the equilibrium is in dominant strategy.

Case 1 $\mathcal{L}^{(1)}$ contains exactly one candidate, say 2. Let $\mathcal{L}^{(2)}$ denote the set of candidates obtaining the second largest score. Assume that $\mathcal{L}^{(2)}$ contains more than one candidate. By Fact 1, all the scores of 2 and the candidates of $\mathcal{L}^{(2)}$ are determined by the preferences over these candidates. To simplify, let $\mathcal{L}^{(2)} = \{k, k'\}$ (the argument extends if there are more than two candidates). If $x_k < x_2 < x_{k'}$, then

$$\begin{aligned} s_2 &= \frac{g(2,k) + g(2,k')}{2} + i(2,k) + i(2,k'), \\ s_k &= g(k,2) + i(2,k), \\ s_{k'} &= g(k',2) + i(2,k'). \end{aligned}$$

We compute $s_k + s_{k'} = g(k,2) + g(k',2) + i(2,k) + i(2,k') < 1$, so that $s_k = s_{k'} < .5$, in contradiction to the fact that $s_1 > .5$. If $x_2 \notin (x_k, x_{k'})$, then (assuming, w.l.o.g., $x_k \in (x_2, x_{k'})$)

$$\begin{aligned} s_2 &= \frac{g(2,k) + g(2,k')}{2} + i(2,k), \\ s_k &= g(k,2) + i(2,k), \\ s_{k'} &= g(k',2). \end{aligned}$$

Given that $g(k,2) \geq g(k',2)$, this implies $g(k,2) = g(k',2)$ and $i(2,k) = 0$. Therefore, $s_2 > s_k$ implies $s_k < .5$, in contradiction to the fact that $s_1 > .5$. That proves that $\mathcal{L}^{(2)}$ contains one and only one candidate. We cannot have $\mathcal{L}^{(2)} = \{1\}$, as this would imply $s_2 < s_1$, a contradiction. Let $\mathcal{L}^{(2)} = \{k\}$. Therefore, we must have $s_2 > s_k > \ldots \geq s_1 \geq \ldots$. That implies $s_1 \geq g(1,2) > .5$. To have $s_k > s_1$, it must be the case that $s_2 = g(2,k) + i(2,k)$ and $s_k = g(k,2) + i(k,2)$. By Fact 2, $x'_1 \notin (x_2, x_k)$. Then, either $x'_1 < x_2 < x_3$ or $x_3 < x_2 < x'_1$. As a result, $g(1,k) > g(k,2) + i(2,k)$, a contradiction.

Case 2 $\mathcal{L}^{(1)}$ contains two candidates, say 2 and 3, with $x_2 \leq x_3$. We have $s_1 \geq \frac{g(x'_1,x_2)+g(x'_1,x_3)}{2} > .5$. Note that all voters who are indifferent between x_2 and x_3 vote exactly in the same way, as they all strictly prefer any policy in (x_2, x_3) to either x_2 or x_3, and they prefer x_2 or x_3 to any position out of (x_2, x_3). Consequently, either

$$\begin{aligned} s_2 &= g(x_2, x_3) + i(x_2, x_3) \\ s_3 &= g(x_3, x_2) + i(x_2, x_3), \end{aligned}$$

or

$$\begin{aligned} s_2 &= g(x_2, x_3) \\ s_3 &= g(x_3, x_2). \end{aligned}$$

In either case, $g(x_2, x_3) = g(x_3, x_2)$ which implies that $x_2 < x_1' = x^m < x_3$ and $s_1 = g(x_1', x_3) + i(x_2, x_3)$. By Fact 3, $s_1 \geq s_2 = s_3$, contradicting the assumption on the score vector.

Case 3 $\mathcal{L}^{(1)}$ contains three or more candidates, say 2, 3 and 4, with $x_2 \leq x_3 \leq x_4$ (a similar argument goes true if $\mathcal{L}^{(1)}$ contains more than three agents). By Assumption 17.2.2, s_2, s_3, s_4 are determined as the average between the scores that are compatible to any of the six possible strict rankings of 1, 2, 3. The scores, for each ranking, are as follows.

	s_2	s_3	s_4
234	$g(2,3) + i(2,3)$	$g(3,2) + i(2,3)$	$g(4,2)$
243	$g(2,4)$	$g(3,2)$	$g(4,2)$
324	$g(2,3) + i(2,3)$	$g(3,2) + i(2,3)$	$g(4,3)$
342	$g(2,3)$	$g(3,4) + i(3,4)$	$g(4,3) + i(3,4)$
423	$g(2,4)$	$g(3,4)$	$g(4,2)$
432	$g(2,4)$	$g(3,4) + i(3,4)$	$g(4,3) + i(3,4)$

Using $g(2,3) \leq g(2,4)$ and $g(4,3) \leq g(4,2)$, we obtain

$$s_2 \leq g(2,4) + \frac{i(2,3)}{3}$$
$$s_3 = \frac{g(3,2) + g(3,4)}{2} + \frac{1}{3}(i(2,3) + i(3,4))$$
$$s_4 \leq g(4,2) + \frac{i(3,4)}{3}$$

Given that $g(2,4) + g(4,2) \leq 1$ and $g(3,2) + g(3,4) \geq 1$, we can only have $s_2 = s_3 = s_4$ if $i(2,3) = i(2,4) = i(3,4) = 0$ and $g(3,2) + g(3,4) = 1$. Consequently, $s_2 = s_3 = s_4 = .5$ whereas $s_1 > .5$, a contradiction.

That proves that $1 \in \mathcal{L}^{(1)}$. We want to prove that all candidates in $\mathcal{L}^{(1)}$ are located at x^m. Assume two different locations are represented in $\mathcal{L}^{(1)}$: $\mathcal{L}^{(1)} = \{1, k\}$ with $x_k \neq x^m$. Then because x^m is a strict Condorcet winner, $s_1 - s_k = g(1,k) - g(k,1) > 0$, a contradiction. Moreover, we know from Case 3 above that a three candidate tie is possible only if all the scores are 0.5, which is impossible if candidate 1 is one of them. It follows that all winning candidates are at x^m. That completes the proof. □

The proofs of Propositions 17.3.1 and 17.3.2 make reasonably clear that Assumption 17.2.2 is imposed for convenience and could be weakened. What is crucial is that each voter transforms a tie in expected vote share into a strict ranking of the candidates. Without that assumption, the behavioral rule should take account of the utility voters derive from lotteries of candidates (and policy positions), whereas we are able to make the whole equilibrium notion depend only on the ordinal preferences over candidates (and policy positions). What is not crucial in the assumption, though, is that the way rankings are drawn from ties is independent of the preferences of the voters. We could easily weaken that assumption, provided the correlation is not too high. For instance, Proposition 17.3.1 relies on the impossibility to increase the probability of winning by deviating from the strategy profile at which all candidates announce the Condorcet policy. If Assumption 17.2.2 is replaced with another assumption that does not guarantee that each candidate gets a probability $1/K$ of being elected at that unanimous profile, then it is no longer an equilibrium.

17.3.3 Comparison with Plurality Voting

The above result should be contrasted with what happens under other voting rules. Consider Plurality rule and take $K > 2$ (at least 3 candidates). The models of rational voting which are similar to the one used here, such as those of Myerson and Weber (1993), Myerson (2002), or Laslier (2009), provide, as can be easily seen, the following behavior.

Rational behavior for the voter, under the plurality rule is basically to vote for the one she prefers among the two first-ranked candidates. Consider the simple case of the single-peaked model on the real line. Suppose that all candidates are at the median, each one receiving $1/K$ of the votes and having thus the probability $1/K$ of being elected. Then suppose that one candidate, say $k = 1$, moves slightly away from this position to some new position, say $x_1' = x^m + \varepsilon$ on the right of x^m. This produces a situation in which the electorate is essentially split in two: the left-wing prefers $x_2 = x_3 = ... = x_4 = x^m$ and the right-wing prefers $x_1 = x_1' = x^m + \varepsilon$.

This potentially gives to the mover almost $1/2$ of the votes while the remaining $(K - 1)$ candidates have to share the remaining votes. Under most reasonable assumption as to voters' behavior, the strength of the split-majority phenomena will be such that candidate 1 will be elected with probability 1. Therefore the situation in which all candidates propose the median is not an equilibrium, except if voters' beliefs are such that all votes gather on two candidates only. An equilibrium is obtained when only two candidates are located at the median and receive half of the votes while the other candidates receive none. This point (only two parties can survive under Plurality voting), which has been emphasized by Cox (1997) after Duverger (1954) is not valid for approval voting. Further studies on this subject also endogenize the number of candidates running for office: see Dellis and Oak (2003) and Dellis (2010).

17.4 Extensions

We need to discuss two extensions of the above model. First, we have assumed that candidates maximize their probability of winning the election. Alternatively, we could have assumed that they try to maximize their victory margin (or minimize their defeat margin), that is, for all $c \in \mathcal{C}$, candidate k maximizes

$$\frac{s(c)}{\max_{c' \in \mathcal{C} \setminus \{c\}} s(c')}.$$

This assumption is more difficult to justify in the case of approval voting than in the standard two-party plurality case because it is not clear whether one should consider the *absolute* or *relative* number of approval votes. Anyway, we conjecture that our three results remain true under this assumption.[4]

The second extension is about the source of uncertainty facing voters. We have assumed that each vote of each voter had a fixed probability of not being recorded. We might have assumed, instead, that each voter had a fixed probability of not going to vote. Under this assumption, there is some correlation between the probability that votes are not recorded. Indeed, if a voter planning to vote for c and c' does not vote, none of her two votes are recorded. Unfortunately, our results do not extend to that case. Nunez (2010a) has shown that Laslier (2009)'s result does not hold under this alternative assumption. The same kind of example as the one developed in Nunez (2010a) applies in the model we have studied in this chapter.

In conclusion, we have found a new kind of elections in which approval voting leads to electing a Condorcet winner. Compared to Laslier (2009), our results show that if policy positions are endogenous and follow from candidate competition, then strategic voting based on vote uncertainty leads to the election of the Condorcet winner when it exists. A consequence of electoral competition is that voters may be indifferent between pairs of candidates, a case which was excluded from Laslier's analysis. We showed that indifferences could prevent the general result from holding. Nonetheless, when voters have single-peaked preferences, in spite of possible indifferences, electoral competition leads all candidates to propose the median policy platform.

References

Austen-Smith, D., & Banks, J. S. (2005). *Positive political theory II: Strategy and structure*. Ann Arbor, MI: University of Michigan Press.

Brams, S. J., & Fishburn, P. C. (1983). *Approval Voting* Boston: Birkhaüser, second edition: Springer 2007.

Cox, G. W. (1984). Strategic electorate choice in multi-member districts: Approval Voting in practice? *American Journal of Political Science, 28*, 722–738.

Cox, G. W. (1985). Electoral competition under Approval Voting. *American Journal of Political Science, 29*, 112–118.

Cox, G. W. (1997) *Making Votes Count*. Cambridge: Cambridge University Press.

[4] In a previous version of the paper, we prove that conjecture in the case where preferences are so heterogeneous among voters that the number of voters being indifferent between two policies becomes negligible. The proof is available upon request.

Dellis, A. (2010). Policy moderation and endogenous candidacy in Approval Voting elections. In J. F. Laslier & R. Sanver (Eds.), *Handbook of Approval Voting*. Heidelberg: Springer.

Dellis, A., & Oak, M. (2006). Approval voting with endogenous candidates. *Games and Economic Behavior, 54*, 47–76.

Downs, A. (1951). *An Economic theory of democracy*. New York: Harper.

Duverger, M. (1954). *Les partis politiques*. Paris: Armand Colin.

Laslier, J.-F. (2009) The leader rule : A model of strategic approval voting in a large electorate. *Journal of Theoretical Politics, 21*, 113–136.

Mueller, D. C. (2003). *Public Choice III*. Cambridge: Cambridge University Press.

Myerson, R. B. (2002). Comparison of scoring rules in Poisson voting games. *Journal of Economic Theory, 103*, 219–251.

Myerson, R. B., & Weber, R. J. (1993). A theory of voting equilibrium. *American Political Science Review, 87*, 102–114.

Nunez, M. (2010a). Approval Voting and the Poisson-Myerson Environment. *Journal of Theoretical Politics, 22*, 64–84.

Nunez, M. (2010b). Approval Voting in large electorates. In J.-F. Laslier & R. Sanver (Eds.), *Handbook of Approval Voting*. Heidelberg: Springer.

Ordeshook, P. C. (1992). *A political theory primer*. London: Routledge.

Roemer, J. (2001). *Political competition: Theory and applications*. Cambridge: Harvard University Press.

Weber, R. J. (1995). Approval Voting. *The Journal of Economic Perspectives, 9*, 39–49.

Chapter 18
Policy Moderation and Endogenous Candidacy in Approval Voting Elections

Arnaud Dellis

18.1 Introduction

Approval Voting is a voting procedure in which a voter can vote for as many candidates as she wishes, and the candidate who receives the most votes wins the election. Since the seminal contribution of Brams and Fishburn (1978), Approval Voting has received considerable attention. Of course, Approval Voting has been the subject of numerous scholarly papers, as this handbook testifies. But Approval Voting has also received considerable attention outside academics. This is best exemplified by the fact that there are now several professional associations that elect their officers by means of Approval Voting.[1] Also, the use of Approval Voting in political elections has been advocated relentlessly and, at one point, was even hailed as *the* electoral reform of the twentieth century (Brams 1980, p. 105).

Proponents of Approval Voting argue that it has appealing advantages over other voting procedures. In particular, they assert that Approval Voting has the advantage to give voters more options compared to Plurality Voting.[2] Indeed, under Approval Voting voters are free to vote for more than one candidate. Proponents of Approval Voting claim that this feature would benefit mainly the centrist candidates. This is because, they argue, centrists are considered acceptable candidates by many voters and would therefore receive many 'second' votes (in addition to the 'first' votes they already receive under Plurality Voting). This claim was noted by Cox (1985, p. 112) when he wrote: "Advocates of approval voting have argued that centrist candidates are favored in multicandidate elections held under approval voting, whereas

[1]For details on the use of Approval Voting by professional and academic associations, see Brams (2007).

[2]Plurality Voting is a voting procedure in which a voter can vote for one (and only one) candidate, and the candidate who receives the most votes wins the election. Several countries (e.g., the U.S., Canada) elect their policy-makers by means of Plurality Voting.

A. Dellis
Université Laval and CIRPEE, 1025 Ave des Sciences Humaines, local 2174, Québec, QC, Canada G1V 0A6
e-mail: arnaud.dellis@ecn.ulaval.ca

J.-F. Laslier and M.R. Sanver (eds.), *Handbook on Approval Voting*, Studies in Choice and Welfare, DOI 10.1007/978-3-642-02839-7_18,

extremist candidates are sometimes favored under the plurality rule." By improving the electoral prospects of the centrist candidates, Approval Voting would lead to the adoption of moderate policies as compared to Plurality Voting. (A policy is said to be moderate compared to another policy if it is preferred by the median voter.) Obviously, such a policy moderation can have important economic implications by, for example, limiting policy changes each time a new government assumes office.

Several contributions provide a theoretical underpinning for this claim. These contributions adopt the Downsian approach to electoral competition. In this approach a given number of candidates compete for office by each choosing an electoral platform. The policy-making process has three stages. At the first stage, each candidate announces his platform. A platform is a policy the candidate is committed to implementing if he is elected. At the second stage, an election is held to select one of the candidates to become the policymaker. At the third stage, the elected candidate implements his announced platform. Contributions in this literature are of two types.

First, there are those contributions where voting is assumed to be sincere, i.e., voters base their voting decisions solely on their preferences for the different candidates (e.g., Cox 1985, 1987, 1990). The argument relies here on the elimination of the *squeezing effect* of Plurality Voting. This argument is most easily understood through the following example.[3] Consider an election with three candidates, say, a left candidate, a centrist candidate and a right candidate. Call leftist (centrist and rightist, resp.) every voter who prefers the left (centrist and right, resp.) candidate to the other two candidates. If the election is held under Plurality Voting, then every leftist will vote for the left candidate, every centrist will vote for the centrist candidate and every rightist will vote for the right candidate. The left and right candidates thus capture all the leftist and rightist votes and leave the centrist candidate with the sole centrist votes. It follows that as long as the platforms of the left and right candidates are not too divergent, the centrist candidate does not receive enough votes to win the election. The centrist candidate is said to be squeezed. This will not be the case however if the election is held under Approval Voting. Indeed, every leftist and every rightist would then be free to add on their ballot a second vote for the centrist candidate, which would improve the electoral prospects of the centrist candidate. Hence the policy moderation.[4,5]

[3] It is worth pointing out that while the example captures the key features of the argument, it is much simpler than the actual argument (as presented, for example, in Cox 1990). This simplification is made for ease of exposition.

[4] This argument was best stated by Brams and Straffin (1982, pp. 194–195) when they wrote: "[W]ith approval voting, a third candidate can always position himself in the middle between the left and right candidates – however small the distance that separates them – and win by getting many second-place votes from most of their supporters."

[5] Cox (1990) considers a candidate-positioning game and establishes the existence of a unique convergent equilibrium under Approval Voting. In this equilibrium, every candidate adopts the median voter's ideal policy. The degree of policy moderation is thus maximal under Approval Voting. By contrast, under Plurality Voting multicandidate equilibria either do not exist or are non-convergent.

Second, there are those contributions where voting is assumed to be strategic, i.e., individual voting decisions are best responses to others' voting decisions (e.g., Myerson and Weber 1993). Their argument relies on the elimination of the *wasting-the-vote effect* of Plurality Voting. To make this argument clear, consider again an election with a left, a centrist and a right candidates. Under Plurality Voting, the centrist candidate need not be elected even if a majority of voters are centrists. This can happen because of self-fulfilling prophecies. Specifically, if voters come to believe that the centrist candidate is unlikely to win the election and that the race is actually between the left and right candidates, then voters will fear wasting their only vote by casting it for the centrist candidate. Voters will thus desert the centrist candidate and vote instead for the left or the right candidate (whichever they prefer), thereby confirming the belief that the centrist candidate is unlikely to win the election. This is the so-called wasting-the-vote effect.[6] By contrast, Approval Voting is not subject to the wasting-the-vote effect since voters can then vote for as many candidates as they wish. In consequence, if the election was held under Approval Voting, then every centrist would cast a vote for the centrist candidate, which would improve the electoral prospects of this candidate. Hence the policy moderation.[7]

However, in many elections, especially political elections, candidacy is endogenous, not exogenous as was assumed in this literature. If one is serious about studying the implications of holding political elections under Approval Voting, one must then take candidacy decisions into account.[8] Doing so is justified on both theoretical and empirical grounds. Specifically, on the theoretical side, Dutta et al. (2001) show that any non-dictatorial voting procedure (that satisfies a mild unanimity axiom) is subject to strategic candidacy decisions. On the empirical side, world elections provide ample evidence of the differential consequences of alternative electoral systems on, among other things, the party system (e.g., see Lijphart 1994; Gallagher and Mitchell 2005).

In this chapter, I endogenize candidacy and revisit the claim that Approval Voting would yield policy moderation compared to Plurality Voting. Specifically, I adopt the citizen-candidate approach to electoral competition (Osborne and Slivinski 1996; Besley and Coate 1997). In this approach, a given number of potential candidates must decide whether to run for office. The policy-making process has

[6] As an illustration of the wasting-the-vote effect, Burden and Jones (2009) estimate that in the 2000 U.S. presidential election 81% of Nader's supporters voted for either Bush or Gore, not for Nader.

[7] Myerson and Weber (1993) establish that in all equilibria of their candidate-positioning game, Plurality Voting puts little restriction on the platform of the elected candidate. By contrast, under Approval Voting the platform of the elected candidate coincides with the median voter's ideal policy. The extent of policy moderation is thus maximal under Approval Voting.

[8] For a long time scholars have recognized the importance of studying candidacy decisions under Approval Voting, as Fishburn and Brams (1981, p. 426) noted: "Several people have also expressed concern about how Approval Voting would affect who enters an election and how it would influence candidates' strategies. Although we do not address this concern, it surely deserves examination."

three stages. At the first stage, each potential candidate announces whether or not he stands for election. Standing for election entails a sunk cost. A central premise of the citizen-candidate approach is that potential candidates have policy preferences and cannot commit to campaign promises. At the second stage, an election is held to select one among the self-declared candidates to become the policy-maker. At the third stage, the elected candidate chooses and implements policy.

The present analysis shows that when candidacy is endogenous, the claim that Approval Voting would lead to the adoption of more moderate policies compared to Plurality Voting needs qualification. Specifically, Approval Voting is shown to always yield policy moderation compared to Plurality Voting if two conditions are satisfied.

The first condition applies to the candidacy behavior. It requires that in equilibrium a potential candidate enters the race only if he (correctly) anticipates that he will be elected with a positive probability. This condition follows because spoiling candidacies trigger a greater multiplicity of candidates' positions. (A spoiling candidate is a candidate who stands for election not to win but because he (correctly) anticipates that his presence in the race will result in a more preferred election outcome.) This greater multiplicity of candidates' positions helps support self-fulfilling prophecies that can deter incumbent candidates from defecting and can also deter other potential candidates from entering the race. These self-fulfilling prophecies render the elimination of the squeezing and wasting-the-vote effects irrelevant, which prevents Approval Voting from always leading to more moderate policies as compared to Plurality Voting.

The second condition applies to the voting behavior. It requires that one of the following holds true.

- Either the voting behavior in Approval Voting elections is relatively sincere. Relative Sincerity is a refinement of the notion of sincere voting. Specifically, a voter is said to vote relatively sincerely if, given others' voting decisions, she votes for every candidate whose outright election she strictly prefers to the election outcome and she does not vote for any of the candidates to whom she strictly prefers the election outcome. This condition ensures that if a centrist candidate were to enter the race against a left candidate and a right candidate, every leftist and/or every rightist would then cast a vote for this centrist candidate. A centrist candidate would thus be elected outright. Anticipating this, a centrist will want to stand for election. Hence the policy moderation.
- Alternatively, voting is strategic and voters' preferences for a candidate depend exclusively on the distance between the candidate's platform and the voter's ideal policy. When voting is strategic, Plurality Voting is subject to the wasting-the-vote effect. Plurality Voting can then always deter a centrist from entering the race against a left candidate and a right candidate, and can thus result in the adoption of extreme policies. By contrast, Approval Voting is not subject to the wasting-the-vote effect and, therefore, cannot always deter a centrist from entering the race. The condition on voters' preferences ensures that this feature of

Approval Voting is not offset by its inability to deter multiple similar candidacies (i.e., multiple candidates standing at the same position).[9]

The present analysis also shows that if either of these conditions is not satisfied, then Approval Voting may result in more extreme policies as compared to Plurality Voting! Notice that, except for the condition on voters' preferences, all conditions are imposed on agents' behavior. There is therefore no a priori reason to believe that these conditions would be satisfied if political elections were held under Approval Voting. Whether they would be satisfied is an empirical question. Laboratory experiments could throw light on this question.

The remainder of the chapter is organized as follows. The next section outlines the model. The following three sections study the extent of policy moderation under different assumptions on the voting behavior. The last section concludes and discusses possible avenues for future research.

18.2 Model

I consider a community that must elect a representative to select and implement a policy (e.g., a tax rate, the location of a public facility). The set of policy alternatives X is a non-empty interval in the real line, $X \subset \mathbb{R}$, with typical element x.

Let $\mathcal{N}$ be the set of citizens, with typical element ℓ. Each citizen ℓ has preferences on X that can be represented by a continuous, bounded utility function $u_\ell : X \rightarrow \mathbb{R}$. Citizen ℓ's utility from policy x is given by $u_\ell(x) \equiv u(x - x_\ell)$ where u is strictly concave and $x_\ell \in X$ is citizen ℓ's ideal policy. Without loss of generality, I normalize $u_\ell(x_\ell) = 0$. Let m denote the median citizen's ideal policy. Throughout the paper, I abuse notation and use m to refer to the median citizen as well.

Let $\mathcal{P} \subseteq \mathcal{N}$ be the non-empty, finite set of potential candidates, with typical element i. Potential candidates are those citizens who can stand for election.

The policy-making process has three stages. At the first stage, each potential candidate decides whether to stand for election at a utility cost $\delta > 0$. Decisions are made simultaneously and non-cooperatively. Candidates cannot commit to campaign promises.[10] At the second stage, there is an election to select one among the self-declared candidates to become the policy-maker. The election winner is the candidate who receives the most votes. Ties are broken randomly. At the third stage, the elected candidate chooses and implements policy. In case nobody stands for

[9] The literature has been using different terminologies to designate multiple similar candidacies. For example, Tideman (1987) refers to multiple similar candidates as clones, whereas Myerson (2002) speaks of duplicate candidates.

[10] This assumption – a central tenet of the citizen-candidate approach – has received empirical justification from Lee et al. (2004). I argue in Sect. 18.4 that relaxing this assumption would actually strengthen the main conclusion of the analysis.

election, a default policy $x_0 \in X$ is implemented.[11] These stages are now analyzed in reverse order.

Policy selection stage. As it is the last stage of the game and candidates cannot commit to campaign promises, the elected candidate implements his ideal policy.

Election stage. Given a non-empty set of candidates $\mathcal{C} \subseteq \mathcal{P}$, let $\alpha_\ell(\mathcal{C}) = \left(\alpha_\ell^i\right)_{i \in \mathcal{C}}$ be citizen ℓ's (pure) voting strategy, where $\alpha_\ell^i = 1$ if citizen ℓ votes for candidate i and $\alpha_\ell^i = 0$ otherwise. Following a standard practice in the voting literature, I rule out weakly dominated voting strategies.[12] The profile of voting strategies is denoted by $\alpha(\mathcal{C})$. I sometimes write $\alpha(\mathcal{C}) = (\alpha_\ell(\mathcal{C}), \alpha_{-\ell}(\mathcal{C}))$, where $\alpha_{-\ell}(\mathcal{C})$ denotes the voting profile of all citizens other than citizen ℓ.

Candidacy stage. Let $e_i \in \{0, 1\}$ be a (pure) candidacy strategy for potential candidate i, where $e_i = 1$ if potential candidate i enters the race and $e_i = 0$ otherwise. The candidacy profile is denoted by $e = (e_i)_{i \in \mathcal{P}}$. For any candidacy profile e, let $\mathcal{C}(e) \equiv \{i \in \mathcal{P} : e_i = 1\}$ be the set of candidates. Citizen ℓ's expected utility is given by

$$U_\ell(\mathcal{C}(e), \alpha) = \begin{cases} \sum\limits_{i \in \mathcal{P}} p_i(\mathcal{C}(e), \alpha)\ u_\ell(x_i) - \delta e_\ell & \text{if } \mathcal{C}(e) \neq \emptyset \\ u_\ell(x_0) & \text{if } \mathcal{C}(e) = \emptyset \end{cases}$$

where $p_i(\mathcal{C}(e), \alpha)$ is the probability that potential candidate i is elected the policy-maker. I assume that when making their candidacy decision, all potential candidates anticipate the same α.

An *equilibrium* is a pair (e^*, α^*) such that: (1) for any non-empty set of candidates $\mathcal{C}$, $\alpha^*(\mathcal{C})$ is a profile of permissible voting strategies;[13] and (2) the candidacy strategy of every potential candidate i, e_i^*, is a best response to the candidacy strategies of all other potential candidates, given the voting function α^*.

[11] I assume $-u_m(x_0) > \delta$, i.e., the default policy is sufficiently distant from the median citizen's ideal policy m that a potential candidate at m prefers to run for office and implement m than letting the default policy x_0 be implemented. This assumption is sufficient for an equilibrium in pure strategies to exist.

[12] Under Plurality Voting, a voting strategy is weakly undominated for a citizen if and only if she votes for a candidate other than the candidate(s) she likes the least. Under Approval Voting, a voting strategy is weakly undominated for a citizen if and only if she votes for (all) the candidate(s) she likes the most and does not vote for (any of) the candidate(s) she likes the least. For a formal characterization of the set of weakly undominated voting strategies in Plurality and Approval Voting elections see, for example, Brams and Fishburn (1978).

[13] Which voting strategies are permissible depends on the voting procedure and on the assumption made on the voting behavior. In consequence, a different concept of permissible voting strategy will be defined in each of the following three sections.

Throughout the analysis, I shall distinguish the serious equilibria from the spoiler equilibria. An equilibrium is said to be *serious* if in equilibrium, every candidate is elected with a positive probability. By contrast, an equilibrium is said to be *spoiler* if in equilibrium, some potential candidates enter the race even though they (correctly) anticipate that they will be elected with probability zero.

It remains to define the notion of policy moderation. To this end, I introduce several additional concepts. First, two equilibria are said to be *equivalent* if they yield the same lottery over policy outcomes. Second, an equilibrium E is said to be *moderate* compared to another equilibrium $\widetilde{E}$ if (1) the median citizen prefers any policy that is implemented in E but not in $\widetilde{E}$ to all policies that are implemented in $\widetilde{E}$ and (2) the median citizen also prefers all policies that are implemented in E to any of the policies that are implemented in $\widetilde{E}$ but not in E. Finally, an equilibrium E is said to be *extreme* compared to another equilibrium $\widetilde{E}$ if $\widetilde{E}$ is moderate compared to E.[14] I am now ready to define the notion of policy moderation.

Definition 18.2.1. Approval Voting is said to yield *policy moderation* compared to Plurality Voting if the following two conditions are satisfied:

1. for every equilibrium E under Approval Voting, either there exists an equivalent equilibrium under Plurality Voting or equilibrium E is moderate compared to every Plurality Voting equilibrium; and
2. for every equilibrium $\widetilde{E}$ under Plurality Voting, either there exists an equivalent equilibrium under Approval Voting or equilibrium $\widetilde{E}$ is extreme compared to every Approval Voting equilibrium.||

18.3 Sincere Voting

In this section, I assume that citizens vote sincerely.[15] I adopt the definition of sincere voting that was proposed by Brams and Fishburn (1978).

Definition 18.3.1. Given a non-empty set of candidates $\mathcal{C}$, citizen ℓ's voting strategy $\alpha_\ell(\mathcal{C})$ is *sincere* if for every pair of candidates i and j in $\mathcal{C}$ with $u_\ell(x_i) > u_\ell(x_j)$, we have

$$\begin{cases} \alpha_\ell^j = 1 \Rightarrow \alpha_\ell^i = 1 \\ \alpha_\ell^i = 0 \Rightarrow \alpha_\ell^j = 0. \| \end{cases}$$

---

[14] For a formal definition of these concepts, see Dellis (2009a).

[15] Weber (1995, p. 43) provides the following motivation for considering sincere voting behavior: "One use of voting systems is to select from among a number of alternatives in settings where the voters have little access to information concerning either the preferences of the other voters or the intended voting behavior of the others. In these settings a voter can be presumed to vote sincerely, since the lack of information about other voters means there is no basis for voting in some clever strategic way."

Thus, under Plurality Voting a voting strategy is sincere for a citizen if she votes for (one of) her most-preferred candidate(s). Under Approval Voting a voting strategy is sincere for a citizen if whenever she casts a vote for a particular candidate i she also casts a vote for every candidate she strictly prefers to candidate i. Key to note is that under Approval Voting, there are as many sincere voting strategies as there are candidates. Indeed, a voting strategy is sincere for a citizen if she votes only for her most preferred candidate, or if she votes for her most-preferred candidate and her second most-preferred candidate, and so on.

A voting strategy is here permissible if it is both sincere and weakly undominated.[16]

The following proposition establishes that when candidacy is endogenous and voting is sincere, Approval Voting can lead to more extreme policies as compared to Plurality Voting.

Proposition 18.3.1 (Dellis 2009b). *Suppose that voting is sincere. Then, Approval Voting does not yield policy moderation compared to Plurality Voting. Furthermore, some equilibria under Approval Voting are extreme compared to all equilibria under Plurality Voting.* ‖

This result follows because Approval Voting admits a greater multiplicity of voting profiles compared to Plurality Voting.[17] The intuition underlying this result is most easily understood if one assumes that preferences are symmetric, i.e., $u_\ell(x) = u(|x - x_\ell|)$ for every citizen $\ell \in \mathcal{N}$ and every policy $x \in X$. I partition the equilibrium set into four subsets: (1) the set of one-position serious equilibria, in which only one candidate stands for election and is elected outright; (2) the set of two-position serious equilibria, in which candidates are standing at exactly two positions and all candidates are tying for the first place; (3) the set of multiposition serious equilibria, in which there are candidates standing at three or more positions and all candidates are tying for the first place; and (4) the set of spoiler equilibria, in which some of the candidates are elected with probability zero. I now discuss each of these four subsets in turn.

[16] With finitely many citizens, a single vote can be pivotal. It follows that, given others' voting strategies, a citizen's best-response set need not contain a sincere voting strategy. To avoid this issue, I assume in this section that there is a continuum of citizens whose ideal policies are distributed over the set of policy alternatives X according to some strictly increasing and continuous c.d.f. F. I further assume that the density of F is symmetric about the median citizen's ideal policy m.

[17] Interestingly, the multiplicity of permissible voting profiles under Approval Voting has been the object of a heated debate. On the one hand, Donald Saari views the multiplicity of permissible voting profiles as a vice. He argues that it makes the election outcome under Approval Voting highly indeterminate. (See, for example, Saari and Van Newenhizen 1988; Saari 2001) On the other hand, Steven Brams views the multiplicity of permissible voting profiles as a virtue. He argues that it makes Approval Voting responsive to voters' preferences. (See, for example, Brams et al. 1988; Brams and Sanver 2006).

18.3.1 One-Position Serious Equilibria

Since in these equilibria only one potential candidate enters the race, the set of one-position serious equilibria is equivalent whether the election is held under Plurality Voting or under Approval Voting. Furthermore, the one-position serious equilibria are the most moderate equilibria. This is because the sole candidate must be standing at a position which is close enough to the median citizen's ideal policy m so that no other potential candidate with an ideal policy closer to m wants to enter the race.[18]

18.3.2 Two-Position Serious Equilibria

Let x_L and x_R denote the two candidates' positions. Without loss of generality, suppose $x_L < x_R$. Notice that for all candidates to tie for the first place, it must be that x_L and x_R are symmetric around the median citizen's ideal policy m. In this way, the median citizen is indifferent between x_L and x_R and the electorate is split equally between those who strictly prefer x_L to x_R (i.e., those citizens whose ideal policy is on the left of m) and those who strictly prefer x_R to x_L (i.e., those citizens whose ideal policy is on the right of m). It must also be that no potential candidate i with ideal policy $x_i \in (x_L, x_R)$ wants to enter the race. I now argue that Approval Voting is better than Plurality Voting at deterring such potential candidates from entering the race.

Let us first consider the case where the election is held under Plurality Voting. Then, only one candidate stands at x_L and only one candidate stands at x_R. This is because several candidates standing at the same position would split their votes, thereby helping the election of the candidate(s) standing at the other position. Suppose now that a potential candidate i with $x_i \in (x_L, x_R)$ were to enter the race. Then, all the leftists would vote for the candidate standing at x_L, all the centrists would vote for candidate i and all the rightists would vote for the candidate standing at x_R.[19]

[18] Notice that with only two candidates in the race, the median citizen is decisive. It follows that if a potential candidate with an ideal policy that lies closer to m were to enter the race, he would be elected outright since he would be preferred by the median citizen. Such a potential candidate is thus deterred from entering the race only if the candidacy cost exceeds the utility gain from getting his own ideal policy implemented instead of the candidate's ideal policy. It must then be that the candidate's ideal policy is not too far away from his own ideal policy and, thus, from the median citizen's ideal policy.

[19] A leftist is defined as a citizen who prefers x_L to both x_i and x_R. Likewise, a centrist is defined as a citizen who prefers x_i to both x_L and x_R. Finally, a rightist is defined as a citizen who prefers x_R to both x_L and x_i.

Let us now consider the case where the election is held under Approval Voting. Suppose that the same potential candidate i were to enter the race. Then, the following voting profile would be permissible: all the leftists would vote for (and only for) the candidate(s) standing at x_L; all the centrists would cast a vote for candidate i and for all the candidate(s) standing at x_L or x_R (whichever they prefer); and, finally, all the rightists would vote for (and only for) the candidate(s) standing at x_R. Such a voting profile would correspond for example to a situation where following candidate i's entry in the race, citizens (correctly) anticipate that candidate i has no chance of winning the election and that the race is still between the candidate(s) standing at x_L and x_R. (Such anticipation on the part of voters could result, for instance, from focal manipulation by political leaders.) In view of this, leftists and rightists may not bother casting a 'second' vote for candidate i.

Key to note is that candidate i's vote total is the same whether the election is held under Approval Voting or under Plurality Voting. By contrast, the vote totals of the candidates standing at x_L and x_R are bigger under Approval Voting than under Plurality Voting. Indeed, under Approval Voting a candidate at x_L receives a vote from every leftist, as under Plurality Voting, but he also receives a vote from some of the centrists. Likewise, a candidate at x_R receives a vote from every rightist, as under Plurality Voting, but he also receives a vote from some of the centrists. Candidate i's vote share is thus smaller under Approval Voting than under Plurality Voting. It follows that potential candidate i can be deterred from entering the race under Approval Voting whenever he is deterred from entering the race under Plurality Voting. However, the converse is not true. The set of two-position serious equilibria under Approval Voting is thus a superset of the set of two-position serious equilibria under Plurality Voting. Moreover, those equilibria that are specific to Approval Voting are extreme compared to all equilibria under Plurality Voting. This is because the more polarized x_L and x_R are, the more difficult it is to deter a centrist from entering the race.

18.3.3 Multiposition Serious Equilibria and Spoiler Equilibria

There is no multiposition serious equilibrium and no spoiler equilibrium under Plurality Voting. This is because the votes of a defecting candidate are transferred to his closest neighbor(s). In consequence, the defection of a candidate either leaves the election outcome unchanged or results in the election of the defecting candidate's closest neighbor. With spoiler candidates in the race and/or multiple candidate positions, either the closest neighbor of the leftmost candidate lies on the left of the expected winning policy or the closest neighbor of the rightmost candidate lies on the right of the expected winning policy. Given the concavity of the utility function, the leftmost candidate and/or the rightmost candidate would then be better off defecting since he would get an electoral outcome he (weakly) prefers while saving on the candidacy cost. Hence, neither multiposition serious equilibria nor spoiler equilibria exist under Plurality Voting.[20]

By contrast, multiposition serious equilibria and spoiler equilibria can exist under Approval Voting. This is because the votes of a defecting candidate need no longer be transferred to his closest neighbor. Specifically, a candidate can be deterred from defecting if he (correctly) anticipates that his votes will not be transferred at all (i.e., voters would remove the name of the defecting candidate from their ballot and leave the rest of their ballot unchanged) or that his votes will be transferred to a candidate who is standing too far away from his ideal policy (i.e., voters would remove the name of the defecting candidate from their ballot and add the names of less-preferred candidates). Such beliefs on the part of candidates are made possible by the multiplicity of permissible voting profiles.

To sum up, the greater multiplicity of permissible voting profiles under Approval Voting compared to Plurality Voting implies that Approval Voting is better than Plurality Voting at deterring incumbent candidates from defecting and new candidates from entering the race. In consequence, the equilibrium set under Approval Voting is a superset of the equilibrium set under Plurality Voting. Some of the equilibria specific to Approval Voting are extreme compared to all Plurality Voting equilibria. It follows that when candidacy is endogenous and voting is sincere, Approval Voting does not yield policy moderation and can even result in more extreme policies as compared to Plurality Voting.

Naturally, one may object that this conclusion relies on the assumption that voting is sincere and that there is no a priori reason to believe that people would vote sincerely if political elections were held under Approval Voting. Would this conclusion be robust to strategic voting? I investigate this question in the next section.

18.4 Strategic Voting

I now assume that citizens vote strategically.[21] I define a permissible voting strategy as follows.

[20] Notice that this argument holds true because candidates are purely policy-motivated, i.e., are solely concerned by the policy that will be implemented. If instead, candidates are also office-motivated – i.e., cared about the spoils from office –, then serious contenders may not want to defect. Multiposition serious equilibria and spoiler equilibria can then exist under Plurality Voting (e.g., see Proposition 3 in Osborne and Slivinski 1996).

[21] Weber (1995, p. 45) provides the following motivation for considering strategic voting behavior: "In many political settings, voters have access to substantial information, typically gleaned from public opinion polls, concerning the expressed preferences and voting intentions of others. This information can affect each voter's perception of the relative chances of the various candidates being in contention for victory, which in turn can affect how voters cast their ballots." Cox (1997) provides empirical evidence of strategic voting behavior in political elections under Plurality Voting. Van der Straeten et al. (2010) provide evidence of strategic voting behavior in small experimental elections under Plurality and Approval Voting.

Definition 18.4.1. Given a non-empty set of candidates $\mathcal{C}$, a voting strategy $\alpha_\ell^*(\mathcal{C})$ for citizen $\ell \in \mathcal{N}$ is *permissible* if:

1. $U_\ell(\mathcal{C}; \alpha_\ell^*(\mathcal{C}), \alpha_{-\ell}^*(\mathcal{C})) \geq U_\ell(\mathcal{C}; \alpha_\ell(\mathcal{C}), \alpha_{-\ell}^*(\mathcal{C}))$ for every voting strategy $\alpha_\ell(\mathcal{C})$; and
2. $\alpha_\ell^*(\mathcal{C})$ is weakly undominated.||

Thus, a voting strategy is permissible for a citizen if it is weakly undominated and is a best response to the voting strategies of all other citizens.[22] Notice that with only two candidates in the race, a citizen is either indifferent or has a unique weakly undominated voting strategy, namely, voting for the candidate she prefers. With only two candidates, voting is thus sincere.

The following proposition establishes that when candidacy is endogenous and voting is strategic, Approval Voting always yields policy moderation compared to Plurality Voting provided that preferences are symmetric and one restricts attention to serious equilibria. Otherwise, Approval Voting does not, in general, yield policy moderation compared to Plurality Voting.[23]

Proposition 18.4.1 (Dellis 2009a). *Suppose that voting is strategic. Then, Approval Voting always yields policy moderation compared to Plurality Voting if and only if one restricts attention to serious equilibria and preferences are symmetric, i.e.,* $u_\ell(x) = u(|x - x_\ell|)$ *for every citizen* $\ell \in \mathcal{N}$ *and every policy* $x \in X$.||

The sufficiency follows because Plurality Voting is then better than Approval Voting at deterring centrists from entering the race. The necessity of symmetric preferences follows because, contrary to Plurality Voting, Approval Voting cannot deter multiple similar candidacies. Finally, the necessity of restricting attention to serious equilibria follows because the presence of spoiler candidates helps support

[22] Throughout this section, I assume that there are finitely many citizens. This is because otherwise a vote is never pivotal, in which case any weakly undominated voting strategy is permissible. I further assume that strictly less than a third of citizens have ideal policy m and that the number of citizens with an ideal policy on the left of m is equal to the number of citizens with an ideal policy on the right of m. These assumptions are made to rule out situations where too few votes are cast or where citizens with ideal policy m can elect whichever candidate they wish without the support of other citizens.

[23] Note that if candidates were able to commit to campaign promises, then Plurality Voting would always yield policy moderation compared to Approval Voting. This result has been established in both the Downsian and the citizen-candidate frameworks. Specifically, in the Downsian framework, Feddersen et al. (1990) show that when potential candidates are purely office-motivated and able to commit to campaign promises, candidacy is endogenous and voting is strategic, then in any Plurality Voting equilibrium all candidates stand at the median citizen's ideal policy. The extent of policy moderation is thus maximal under Plurality Voting, and Approval Voting cannot improve upon it. In the citizen-candidate framework, Dellis and Oak (2007) show that when potential candidates are purely policy-motivated and able to commit to campaign promises, candidacy is endogenous and voting is strategic, then all Plurality Voting equilibria are one-position serious equilibria. At the same time, two-position serious equilibria can exist under Approval Voting. As the one-position serious equilibria are the most moderate equilibria, it thus follows that Plurality Voting yields policy moderation compared to Approval Voting.

self-fulfilling prophecies that offset the elimination of the wasting-the-vote effect by Approval Voting. To make the intuition clear, I partition the equilibrium set into the same four subsets as in the previous section.

18.4.1 One-Position Serious Equilibria

The argument is the same as in the previous section. This is because in a one-position serious equilibrium, only one candidate stands for election and only deviations by one potential candidate matter (in which case voting is anyway sincere in the voting equilibrium). In consequence, the set of one-position serious equilibria is equivalent under Plurality and Approval Voting and the one-position serious equilibria are the most moderate equilibria.

18.4.2 Two-Position Serious Equilibria

Let x_L and x_R denote the two candidates' positions. Without loss of generality, suppose $x_L < x_R$. Notice that under Plurality Voting, only one candidate stands at x_L and only one candidate stands at x_R. This is because several candidates standing at the same position would split their votes, thereby helping the election of the candidate(s) at the other position. (By contrast, Approval Voting can accommodate multiple similar candidacies since there is no vote-splitting under Approval Voting.) Notice as well that since there are here exactly two candidates under Plurality Voting and exactly two candidate positions under Approval Voting, voting is then sincere in equilibrium. This implies that for all candidates to tie for the first place, it must be that x_L and x_R lie on either side of the median citizen's ideal policy m and that the median citizen is indifferent between these two positions. Finally, it must also be that no potential candidate i with ideal policy $x_i \in (x_L, x_R)$ wants to enter the race. I now argue that Plurality Voting is here better than Approval Voting at deterring such potential candidates from entering the race.[24] I further argue that the relevance of this feature for policy moderation depends on whether preferences are symmetric.

[24] Interestingly, when voting is assumed to be sincere (as in the previous section), it is instead Approval Voting that is better at deterring entry by centrists. This turn-around might seem paradoxical given that voting is anyway sincere in a two-position serious equilibrium. However, one must note that following candidate entry at a third position, voting is still sincere under Approval Voting (Brams and Fishburn 1978; Theorem 3), whereas it need no longer be sincere under Plurality Voting. The characterization of the two-position serious equilibria under Approval Voting is thus similar whether voting is assumed to be sincere or strategic. This is not the case however under Plurality Voting.

Let us first suppose that the election is held under Plurality Voting. The wasting-the-vote effect implies that Plurality Voting can always deter new candidate entry. Specifically, no other potential candidate wants to enter the race if he (correctly) anticipates that voters would believe he has no chance of winning the election and would therefore continue casting the same ballot; his candidacy would thus leave the election outcome unchanged.

Let us now suppose that the election is held under Approval Voting. I first consider the case where preferences are symmetric. Then, $u_L(x_R) = u_R(x_L)$ that is, the utility loss that a potential candidate at x_L incurs following the adoption of x_R instead of x_L is equal to the utility loss that a potential candidate at x_R incurs following the adoption of x_L instead of x_R. This implies that in equilibrium, an equal number of candidates are standing at x_L and x_R. Each policy is thus implemented with probability $\frac{1}{2}$ (as when the election is held under Plurality Voting). Now, if a potential candidate i with ideal policy $x_i \in (x_L, x_R)$ was to enter the race, he would necessarily receive a vote from every centrist (by weak undominance). At the same time, a candidate at x_L (x_R, resp.) would receive at best a vote from every citizen with ideal policy on the left (right, resp.) of m (by weak undominance again). It follows that potential candidate i can be deterred from entering the race only if x_L and x_R are not too polarized so that the centrists are not majoritarian or potential candidate i would find it too costly to stand for election. Thus, when preferences are symmetric, the set of two-position serious equilibrium outcomes under Approval Voting is a subset of the set of two-position serious equilibrium outcomes under Plurality Voting. Moreover, those equilibria that are specific to Plurality Voting are extreme compared to all equilibria under Approval Voting. This is because the more polarized x_L and x_R are, the more difficult it is for Approval Voting to deter centrists from entering the race.

I now consider the case where preferences are asymmetric. Without loss of generality, let $u_L(x_R) < u_R(x_L)$ that is, the utility loss that a potential candidate at x_L incurs following the adoption of x_R instead of x_L is greater than the utility loss that a potential candidate at x_R incurs following the adoption of x_L instead of x_R. Thus, a potential candidate at x_L has stronger incentives to stand for election compared to a potential candidate at x_R. This implies that in a two-position serious equilibrium (if one exists), more candidates are standing at x_L than at x_R, and x_R is thus implemented with a probability smaller than $\frac{1}{2}$. Now, if there are relatively many candidates standing at x_L (which can happen if $u_L(x_R)$ is significantly more negative than $u_R(x_L)$), then the probability that x_R is implemented, denoted p_R, will be significantly smaller than $\frac{1}{2}$. Potential candidates at x_R can then be deterred from entering the race. Indeed, the increase in p_R that results from their candidacy could then be too small for the utility gain to exceed the candidacy cost. In this case, there would be no two-position serious equilibrium $\{x_L, x_R\}$ under Approval Voting. By contrast, vote-splitting implies that multiple similar candidacies are deterred under Plurality Voting. This implies that in a two-position serious equilibrium under Plurality Voting (if one exists), $p_R = \frac{1}{2}$. This probability can be sufficient for a potential candidate at x_R to be willing to stand for election against a candidate at

x_L. Thus, there may exist a two-position serious equilibrium $\{x_L, x_R\}$ under Plurality Voting, whereas no such equilibrium may exist under Approval Voting (e.g., see Example 3 in Dellis and Oak 2006). The set of two-position serious equilibrium outcomes under Approval Voting is thus a subset of the set of two-position serious equilibrium outcomes under Plurality Voting, as when preferences are symmetric. However, this subset need no longer be moderate. This is because the less polarized x_L and x_R are, the larger $u_R(x_L)$ is and, therefore, the more likely it is that the presence of two or more candidates at x_L is sufficient for deterring any potential candidate at x_R from standing for office. Only the most extreme $\{x_L, x_R\}$ may then be supported as two-position serious equilibrium outcomes under both Plurality and Approval Voting. In other words, the ability of Approval Voting to accommodate multiple similar candidacies can offset its inability to deter centrists from entering the race. This can prevent Approval Voting from yielding policy moderation compared to Plurality Voting.

18.4.3 Multiposition Serious Equilibria

There is no multiposition serious equilibrium under Plurality or Approval Voting. To see this, let us first suppose that the election is held under Plurality Voting. If a multiposition serious equilibrium was to exist, then every citizen would be voting for her most-preferred candidate; otherwise, a citizen would be better off deviating and voting for another (weakly) more-preferred candidate since this candidate would then be elected outright. Also, if a multiposition serious equilibrium was to exist, then there would necessarily be two or more candidates' positions on one side, say on the left, of the expected winning position $\overline{x}$. Since utility functions are concave, the leftmost citizen would then be better off deviating and voting for her second most-preferred candidate (whose ideal policy would thus be on the left of $\overline{x}$) since this candidate would then be elected outright. No multiposition serious equilibrium can thus exist under Plurality Voting.

Let us now suppose that the election is held under Approval Voting. If a multiposition serious equilibrium was to exist, then every citizen would be voting for all the candidates she strictly prefers to the winning lottery; otherwise, a citizen would be better off adding a vote for such a candidate who would then be elected outright. Moreover, no citizen would be voting for a candidate to whom she strictly prefers the winning lottery; otherwise, she would be better off removing those candidates from her ballot since they would then be no longer in the winning set. It is then easy to check that, given the concavity of utility functions, a centrist candidate would then be receiving more votes than a leftmost or rightmost candidate. This would contradict that the equilibrium is serious.

18.4.4 Spoiler Equilibria

Finally, spoiler equilibria can exist under both Plurality and Approval Voting.[25] This is because the presence of spoiler candidates triggers a greater multiplicity of candidates' positions, which helps support self-fulfilling prophecies that deter both incumbent candidates from defecting and other potential candidates from entering the race. The presence of spoiler candidates thus offsets the elimination of the wasting-the-vote effect by Approval Voting. In consequence, Approval Voting need no longer lead to more moderate policies compared to Plurality Voting, even if preferences are symmetric.

To sum up, the fact that Approval Voting is not subject to the wasting-the-vote effect implies that Approval Voting is worse than Plurality Voting at deterring new candidate entry. It follows that the set of serious equilibrium outcomes under Approval Voting is a subset of the set of serious equilibrium outcomes under Plurality Voting. When preferences are symmetric, the greater inability of Approval Voting at deterring centrists from entering the race implies that the subset is moderate. However, when preferences are asymmetric, the inability of Approval Voting at deterring multiple similar candidacies can offset the greater inability of Approval Voting at deterring centrist candidacies, and the subset need therefore be no longer moderate. Finally, the presence of spoiler candidates renders the elimination of the wasting-the-vote effect irrelevant. In consequence, the (full) set of equilibrium outcomes under Approval Voting need not be a subset of the (full) set of equilibrium outcomes under Plurality Voting, and Approval Voting need not yield policy moderation compared to Plurality Voting.

18.5 Relative Sincerity

When the election is held under Approval Voting and multiple candidates stand for office, the notions of sincere and strategic voting put rather few restrictions on the voting behavior. This is especially clear when voting is sincere. Indeed, a citizen can then cast a vote only for the candidate(s) she prefers the most, or she can cast a vote for every candidate except the one(s) she likes the least, or she can draw the line anywhere in-between. One may thus wonder whether the above analysis effectively captures the way people would actually vote if political elections were held under Approval Voting. To address this concern, Dellis and Oak (2006) have proposed an intuitively plausible refinement of the voting behavior – called *Relative*

---

[25] Recall from the previous section that no spoiler equilibrium exists under Plurality Voting when voting is assumed to be sincere. This is because the leftmost or rightmost candidate has an incentive to defect since his votes are then transferred to his closest neighbor. By contrast, when voting is assumed to be strategic, the same candidate may not want to defect. This happens if he (correctly) anticipates that his votes would not be transferred to his closest neighbor. Hence the existence of spoiler equilibria when voting is assumed to be strategic.

Sincerity – that puts a restriction on where citizens draw the line between the candidates for whom they cast a vote and the candidates for whom they do not cast a vote. Specifically, a voting strategy is said to be relatively sincere for a citizen if, given the voting strategies of the other citizens, she votes for all the candidates she strictly prefers to the winning lottery and does not vote for any of the candidates to whom she strictly prefers the winning lottery.[26] Formally,

Definition 18.5.1. Let the election be held under Approval Voting. Given a non-empty set of candidates $\mathcal{C}$, citizen ℓ's voting strategy $\alpha_\ell(\mathcal{C}) = \left(\alpha_\ell^i\right)_{i\in\mathcal{C}}$ is *sincere relative* to the voting strategies of other citizens $\alpha_{-\ell}(\mathcal{C})$ if for every candidate $i \in \mathcal{C}$, we have

$$\begin{cases} u_\ell(x_i) > U_\ell(\mathcal{C}, \alpha(\mathcal{C})) \Rightarrow \alpha_\ell^i = 1 \\ u_\ell(x_i) < U_\ell(\mathcal{C}, \alpha(\mathcal{C})) \Rightarrow \alpha_\ell^i = 0. \| \end{cases}$$

A voting strategy is here permissible if it is both weakly undominated and relatively sincere.

The following proposition establishes that when candidacy is endogenous and voting is relatively sincere, Approval Voting always yields policy moderation compared to Plurality Voting provided that one restricts attention to serious equilibria. Otherwise, Approval Voting does not, in general, yield policy moderation compared to Plurality Voting.

Proposition 18.5.1 (Dellis and Oak 2006; Dellis 2009b). *Suppose that voting is relatively sincere under Approval Voting and is either sincere or strategic under Plurality Voting. Then, Approval Voting always yields policy moderation compared to Plurality Voting if and only if one restricts attention to serious equilibria.* ||

The sufficiency follows because the elimination of the wasting-the-vote effect and of the squeezing effect implies that Approval Voting is then unable to deter centrists from entering the race. The necessity follows because the presence of spoiler candidates helps support self-fulfilling prophecies that render the elimination of the wasting-the-vote effect and of the squeezing effect irrelevant. To make the intuition clear, I partition the equilibrium set into the same four subsets as in the previous sections.

18.5.1 One-Position Serious Equilibria

The relative sincerity refinement has no bite when less than three candidates are standing for election. In consequence, the argument is similar to the one presented in the previous sections. This implies that the set of one-position serious equilibria is equivalent under Plurality and Approval Voting and that the one-position serious equilibria are the most moderate equilibria.

[26] The relative sincerity refinement has received theoretical justification from Laslier (2009).

18.5.2 Two-Position Serious Equilibria

No two-position serious equilibrium exists under Approval Voting. To see this, assume on the contrary that such an equilibrium exists. Let x_L and x_R denote the two candidates' positions. Without loss of generality, suppose $x_L < x_R$. As in the previous sections, for all candidates to tie for the first place it must be that x_L and x_R lie on either side of the median citizen's ideal policy m and that the median citizen is indifferent between these two positions. Denote the expected winning policy by $\overline{x}$, and suppose that a potential candidate i with ideal policy $x_i \in (x_L, \overline{x}]$ was to enter the race. (A similar argument holds for $x_i \in [\overline{x}, x_R)$.) Then, all the citizens with ideal policy $x \leq x_i$ would prefer x_i to $\overline{x}$. Given the concavity of the utility functions, all these citizens would then prefer candidate i to the lottery over x_L and x_R, and would therefore cast a vote for candidate i (by relative sincerity).[27] Moreover, candidate i would be the most-preferred candidate of all citizens with ideal policy $x \in [x_i, \widetilde{x}]$, where $\widetilde{x}$ is the position at which a citizen is indifferent between x_i and x_R. All these citizens would then be casting a vote for candidate i (by weak undominance). Clearly, $\widetilde{x} > m$, which implies that candidate i would thus receive a vote from a majority of citizens. At the same time, weak undominance implies that neither citizen with ideal policy on the right (left, resp.) of m would cast a vote for a candidate at x_L (x_R, resp.) since x_L (x_R, resp.) is the candidates' position these citizens like the least. A candidate at x_L or x_R would thus receive a vote only from a minority of citizens. It follows that if potential candidate i was to enter the race, he would necessarily be elected outright. As x_L and x_R must be sufficiently polarized so that candidates at these positions are willing to stand for election, there will thus be a potential candidate i with $x_i \in (x_L, x_R)$ who wants to enter the race. Hence, there cannot exist a two-position serious equilibrium under Approval Voting when voting is relatively sincere. Recall however that two-position serious equilibria can exist under Plurality Voting.

Key to note is that this result does not depend on whether preferences are symmetric. This is because when the voting behavior is relatively sincere, leftists and/or rightists cast votes for a centrist candidate (if one enters), which implies that a centrist is always willing to enter the race. This renders the issue of multiple similar candidacies, and thus the symmetry/asymmetry of preferences, irrelevant.

18.5.3 Multi-Position Serious Equilibria

No multiposition serious equilibrium exists whether the election is held under Approval or Plurality Voting. The argument is similar to the one presented in the previous section.

---

[27] This argument implicitly assumes that following the entry of potential candidate i, citizens anticipate that the candidates at x_L and x_R would still be tying for the first place. Notice however that the same argument would hold if citizens were to anticipate that a candidate at x_R would then be elected with probability one. If instead citizens were to anticipate that a candidate at x_L would be elected with probability one, then all citizens with ideal policy $x \geq x_i$ would cast a vote for candidate i, and the rest of the argument would be similar to the argument presented in the text.

18.5.4 Spoiler Equilibria

Spoiler equilibria can exist under both Plurality and Approval Voting (e.g., see Example 4 in Dellis and Oak 2006). This is because the presence of spoiler candidates triggers a greater multiplicity of candidates' positions, which can then help support self-fulfilling prophecies that deter both incumbent candidates from defecting and other potential candidates from entering the race. This deprives the relative sincerity refinement of all its bite.

To sum up, when candidacy is endogenous and voting is relatively sincere, all serious equilibria under Approval Voting are one-position serious equilibria. Since these are the most moderate equilibria, the extent of policy moderation is then maximal under Approval Voting. This is no longer true however when one considers spoiler equilibria. Interestingly, the intuition underlying Proposition 18.5.1 corresponds to the argument that was put forward by the proponents of Approval Voting to justify that centrist candidates would benefit from having elections held under Approval Voting instead of Plurality Voting. The present analysis thus highlights two key assumptions underlying this argument, namely, the relative sincerity of the voting behavior and the absence of spoiler candidates.

18.6 Conclusion

In this chapter, I have revisited the claim that centrist candidates would benefit from having elections held under Approval Voting instead of Plurality Voting, and that Approval Voting would thus result in more moderate policies as compared to Plurality Voting. Underlying this claim is the fact that Approval Voting is subject to neither the squeezing effect nor the wasting-the-vote effect that characterize Plurality Voting. Contrary to previous contributions, I have studied this claim in a setting where candidacy is endogenous. I have shown that in such a setting, the claim must be qualified.

I conclude this chapter by discussing what I consider to be the most important limitations of the analysis and what are the promising avenues for future research. First, all the results have been derived under the assumption that the policy space is unidimensional. It would be of interest to extend the analysis to multidimensional policy spaces. Second, candidates were assumed to be purely policy-motivated. While this assumption offers an interesting contrast with the canonical Downsian model – where candidates are assumed to be purely office-motivated –, allowing for a mix of office- and policy-motivation is important. Third, information was assumed to be complete and perfect. Introducing information imperfections would be of great interest. However, as exemplified by several recent contributions (e.g., Casamatta and Sand-Zantman 2005; Grosser and Palfrey 2009), introducing information imperfections in the citizen-candidate model is not a trivial task. Fourth, only equilibria in pure strategies have been considered. Doing so was motivated by the multiplicity of equilibria in pure strategies. This said, it would be interesting to examine whether taking mixed-strategy equilibria into account would leave the above conclusions unchanged. Finally, only elections to a single office have been considered. However, many political elections serve to elect a legislature. Considering multi-seat elections is definitely a promising avenue for future research.

Acknowledgements I am grateful to Patrick Gonzalez for providing useful comments on an earlier version.

References

Besley, T., & Coate, S. (1997). An economic model of representative democracy. *Quarterly Journal of Economics, 112*, 85–114.
Brams, S. (1980). Approval Voting in multicandidate elections. *Policy Studies Journal, 9*, 102–108.
Brams, S. (2007). *Mathematics and democracy*. Princeton, NJ: Princeton University Press.
Brams, S., & Fishburn, P. (1978). Approval Voting. *American Political Science Review, 72*, 831–847.
Brams, S., Fishburn, P., & Merrill, S. (1988). The responsiveness of Approval Voting: Comments on Saari and Van Newenhizen. *Public Choice, 59*, 121–131.
Brams, S., & Sanver, R. (2006). Critical strategies under Approval Voting: Who gets ruled in and ruled out. *Electoral Studies, 25*, 287–305.
Brams, S., & Straffin, P., Jr. (1982). The entry problem in a political race. In P. Ordeshook & K. Shepsle (Eds.), *Political equilibrium* (pp. 181–195). Boston: Kluwer-Nijhoff.
Burden, B., & Jones, P. (2009). Strategic voting in the USA. In B. Grofman, A. Blais, & S. Bowler (Eds.), *Duverger's Law of Plurality Voting* (pp. 47–64). New York: Springer.
Casamatta, G., & Sand-Zantman, W. (2005). Citizen candidacy with asymmetric information. *Topics in Theoretical Economics, 5*(1), article 3.
Cox, G. (1985). Electoral equilibrium under Approval Voting. *American Journal of Political Science, 29*, 112–118.
Cox, G. (1987). Electoral equilibrium under alternative voting institutions. *American Journal of Political Science, 31*, 82–108.
Cox, G. (1990). Centripetal and centrifugal incentives in electoral systems. *American Journal of Political Science, 34*, 903–935.
Cox, G. (1997). *Making votes count*. Cambridge: Cambridge University Press.
Dellis, A. (2009a). Would letting people vote for multiple candidates yield policy moderation? *Journal of Economic Theory, 144*, 772–801.
Dellis, A. (2009b). *Splitting, squeezing and diluting: Policy moderation when candidacy is endogenous and voting is sincere* (Mimeo).
Dellis, A., & Oak, M. (2006). Approval Voting with Endogenous Candidates. *Games and Economic Behavior, 54*, 47–76.
Dellis, A., & Oak, M. (2007). Policy convergence under Approval and Plurality Voting: The role of policy commitment. *Social Choice and Welfare, 29*, 229–245.
Dutta, B., Jackson, M., & Le Breton, M. (2001). Strategic candidacy and voting procedures. *Econometrica, 69*, 1013–1037.
Feddersen, T., Sened, I., & Wright, S. (1990). Rational voting and candidate entry under plurality rule. *American Journal of Political Science, 34*, 1005–1016.
Fishburn, P., & Brams, S. (1981). Efficacy, power and equity under Approval Voting. *Public Choice, 37*, 425–434.
Gallagher, M., & Mitchell, P. (Eds.). (2005). *The politics of electoral systems*. Oxford: Oxford University Press.

Grosser, J., & Palfrey, T. (2009). A citizen-candidate model with private information. In E. Aragones, C. Bevia, H. Llavador, & N. Schofield (Eds.), *The political economy of democracy* (pp. 15–29). Fundacion BBVA.

Laslier, J.-F. (2009). The leader rule. A model of strategic Approval Voting in a large electorate. *Journal of Theoretical Politics, 21*, 113–136.

Lee, D., Moretti, E., & Butler, M. (2004). Do voters affect or elect policies? Evidence from the U.S. House. *Quarterly Journal of Economics, 119*, 807–859.

Lijphart, A. (1994). *Electoral systems and party systems*. Oxford: Oxford University Press.

Myerson, R. (2002). Comparison of scoring rules in Poisson voting games. *Journal of Economic Theory, 103*, 219–251.

Myerson, R., & Weber, R. (1993). A theory of voting equilibria. *American Political Science Review, 87*, 102–114.

Osborne, M., & Slivinski, A. (1996). A model of political competition with citizen-candidates. *Quarterly Journal of Economics, 111*, 65–96.

Saari, D. (2001). *Decisions and elections. Explaining the unexpected*. Cambridge: Cambridge University Press.

Saari, D., & Van Newenhizen, J. (1988). The problem of indeterminacy in approval, multiple, and truncated voting systems. *Public Choice, 59*, 101–120.

Tideman, N. (1987). Independence of clones as a criterion for voting rules. *Social Choice and Welfare, 4*, 185–206.

Van der Straeten, K., Laslier, J-F., Sauger, N. & Blais, A. (2010). Sincere, strategic, and heuristic voting under four election rules: An experimental study. *Social Choice and Welfare*, in press.

Weber, R. (1995). Approval Voting. *Journal of Economic Perspectives, 9*, 39–49.

Part VIII
Meaning for Individual and Society

Chapter 19
Describing Society Through Approval Data

Jean-François Laslier

19.1 Introduction

An election is not only a mean to choose among options, candidates or representatives, but it also serves as a privileged occasion for voters to express publicly their opinions and to know the opinions of the others. These two goals – choice and expression – may be contradictory but they co-exist in the minds of the voters. The usual rationale for voting is to elect someone or to have a decision taken, that is collective choice, but voters also sometime declare that their rationale for voting is to "express themselves" or to "contribute to the political debate". The choice and expression rationales also co-exist in practice since, for instance, on top of the identity of the elected candidate, the number of votes gathered by a candidate matters for deciding of her political fate, even if she is not elected.

In democratic countries with more than two parties, even when proportional rule is not used, a losing candidate who nevertheless gets many votes gains some political power in the sense that herself, her party, or the ideas she defends have better access to the media and maybe better chances for the next elections.

The same is true in democratic countries with only two main parties. For instance losing an election by a large margin may "kill" a candidate or at least foster the party to change strategy. This is known to narrow down the set of viable political options in countries using Plurality rule (Duverger 1951; Cox 1997).

In non-democratic countries, dictators take care of being elected with quasi-unanimity. Their interest in doing so is another proof that an election must be considered as is an important public signal and not only as a choice device.

This aspect should be taken into account when evaluating voting rules. The outcome of an election provides an image of the current political opinions across the electorate and this image, like any image, can be more or less distorted or reliable. Moreover the style of this image widely depends on the voting rule. After a plurality election, one only knows the scores of the various candidates, which maybe

J.-F. Laslier
Laboratoire d'Econométrie, École Polytechnique, 91128 Palaiseau, France
e-mail: jean-francois.laslier@polytechnique.edu

J.-F. Laslier and M.R. Sanver (eds.), *Handbook on Approval Voting*, Studies in Choice and Welfare, DOI 10.1007/978-3-642-02839-7_19,

enough for evaluating their "political strength," but nothing more. After a two-round election, one knows a little more (the runoff scores). But other systems provide an information which is a priori much richer. Under preferential systems, voters rank-order candidates, leaving in principle the possibility of describing the whole preference profile of the society. Under Approval Voting, the information provided by each voter is not as rich as a complete ranking of the candidates, but is much richer than a single candidate name. For instance with ten candidates, there are ten single-name ballots, 1,024 approval ballots, and more than three million preference ballots.

Consequently, the use of Approval Voting (or other voting rules which use ballots that are not single-name) raises an original question: What should be published after an election as its official "outcome"? For instance, in an approval election, the fact that all voters who approved the candidate A also approved the candidate B can hardly be considered, from a legal point of view, as different from the fact that such number of voters approved of A. Then, should not this fact be published just like A's score? I consider here that any information contained in the set of submitted ballot is bound to be published, provided of course it respects the legal constraint that defines a ballot as valid, and in particular anonymity. This point may raise a logical problem since, if the number of candidates is large, then the number of possible ballots becomes so large that the use of a particular one can be used as a quasi-signature. Brams et al. (2007) analysed an election held at the council of the Game Theory society in 2003. There were 12 members to be elected out of 24 candidates, with approval balloting. That makes $2^{24} = 16.777.216$ different possible ballots. It was observed that only 2 out of the 161 voters voted identically. Notice, however that, if the numbers of ballots is that large, and with many voters, it becomes impossible in practice to display the exhaustive list of how many voters casted each precise ballot, so that only summary of this information has to be published.

In this chapter, I will present means to summarize approval profiles in useful ways. This might certainly be used for political elections, although some may wish to argue on that point, but it obviously should be used in many other circumstances where voting takes place. For instance, committees often organize straw votes before the final one in order that the members of the committee learn about the other members' opinions and vote intentions. In a straw vote, voting is really used for the goal of sharing information. In that case, the possibility of sharing a richer information is in general an improvement. As will be seen in this chapter, starting from an approval profile, there are ways to publish rich and understandable information about voters' votes. I recommend that Approval Voting be used together with these methods, in particular during the process of decision-making for a committee.

19.2 Approval Scores

The main outcome of an election is obviously a collective decision. In the basic case considered here, it is simply the name of the elected candidate. A secondary outcome is the vector of the scores of the various candidates. With single-name balloting, there is no ambiguity as to what the score of a candidate is.[1] But with Approval Voting, such is not the case. Two kinds of "relative scores" can be computed: The percentage of voters who approved the candidate or the percentage of approbations which were given to the candidate. Let n_c be the number of voters who have approved candidate c, let C denote the set of candidates, and let n be the number of voters. The total number of approbations is then:

$$N = \sum_{c \in C} n_c.$$

The ratio

$$s(c) = \frac{n_c}{n}$$

can be called the *relative score* of the candidate c. This ratio is of course always between 0 and 1. Its meaning is clear: It is the fraction of the population who approved c. A ratio larger than 0.5 means that a majority of voters approve of c. But these relative scores do not sum to 1 in general. More exactly, notice that the ratios n_c/N, which obviously sum to 1, are proportional to the relative scores:

$$s(c) = \frac{N}{n} \cdot \frac{n_c}{N}$$

with a coefficient of proportionality (N/n) which is easily interpreted: It is simply the average number of approved candidates per voter. Therefore, one should have in mind that, independently of how the candidates' approvals are correlated,

- The percentage of voters who approve a candidate equals the percentage of approbations that are given to this candidate, multiplied by the average number of approval per ballot.
- The sum of the relative scores is the average number of approvals per ballot.

When presented with the result of an AV election, people often raise the question "What should we consider as the real score of a candidate: The fraction of the voters who approve her or the fraction of approvals she receives?" The two concepts are equivalent for a given election in a given district where the average number of approval per ballot is given; but they must be distinguished if the number of approval per ballot is not fixed, for instance when comparing results in different districts. The question thus requires to be answered.

I submit that one should use $s(c)$, the fraction of the population who approves c, rather than the ratio n_c/N, despite the (maybe disturbing) fact that these relative scores do not sum to 1. The argument in favor of this choice is the following.

If one has in mind that when a voter approves a candidate c, he does so for intrinsic reasons, independently of the presence of other candidates, then introducing new candidates will not change the voters' approval or disapproval of c. It should then

[1] The only difficulty neglected here is with the counting of spoiled or empty ballots.

be desirable that the aggregate score of c does not change either.[2] Such is the case with the relative score $s(c)$, but such is not the case with the ratio n_c/N. When new candidates are introduced and receive some approvals, the total N can only increase and thus the ratio n_c/N mechanically decreases.

Moreover this argument still has bite if voters vote strategically. Consider for instance the "leader rule" of Laslier (2009), which describes a voter's strategic behavior. Let c^* be the anticipated winner of the election. When all voters vote strategically, the relative score of a candidate $c \neq c^*$ is the fraction of the electorate who prefers c to c^*. Introducing a new candidate will only alter the votes for the other candidates if this new candidate is bound to win the election.[3] Introducing a candidate who does not challenge the leader (such a spoiler candidate could be called "irrelevant" because she is not bound to change the outcome) will not change the strategic behavior of the voters with respect to any candidate. Again it is thus reasonable to wish that the aggregate view reflected by the candidate scores be unchanged too. This will be the case for $s(c)$ but not for n_c/N.

For these reasons, it seems preferable to use $s(c)$ rather than n_c/N. The cost to pay is that the relative scores do not sum to 1 but to the average number of approval per ballot, a number which is typically around 2 or 3; see the empirical papers Baujard and Igersheim (2010), Alos-Ferrer and Granic (2010) in part IV of this book. An important consequence is that if one wishes to compare these approval scores with the result of a single-name balloting election, one will notice that scores are globally larger with Approval Voting. This mechanical effect can be observed in the figures reported in the above papers.

But the most important point is that approval scores are probably more meaningful than plurality scores. Plurality scores, be it in countries using first-past the post or in countries using plurality with a run-off, are distorted by the fact that small candidates receive only the votes of those voters who have expressive motives while the main candidates receive votes from both expressive and instrumental voters. This phenomenon vanishes with AV since these two goals are no longer contradictory for the voter.

In practice, in view of the figures reported from field experiments in France and Germany, one can have the feeling that politicians from the main parties fool themselves – and the public – when they praise themselves for receiving the support of "a majority" of the population. A voting rule that produces more meaningful figures gives a quite different image. Indeed it is a pity to notice that the most popular politicians are approved by a minority only, *even when* voters have the possibility to approve of several candidates!

On the other hand, the low plurality scores obtained by small candidates in real elections should not be trusted as a quantification of the popular support of these candidates. The plurality score of a small candidate depends on the size of her

[2] In the next chapter Sanver (2010) elaborates on that view of intrinsic approval.

[3] Or to arrive in the second position in which case it will alter the score of c^* but not the other scores.

potential supporters and on the propension of these persons to vote expressively. Notice that this implies that the plurality scores can also hardly be compared among small candidates. It is quite obvious that, for these candidates, approval scores are more trustable and, as a matter of fact, rankings of small candidates vary quite a lot when one compares plurality to approval rankings.

19.3 Endogenous Electoral Proximity

The data collected in an Approval Voting election, that is a set of Approval Voting ballots contains and makes available more information that the mere counting of each candidate's score. One can also learn about the structure of the electorate, something which is not possible using single-name balloting.

Consider for instance a group of seven voters who are to discuss about four candidates labelled a, b, c, and d. Suppose that the approval votes are as given in Table 19.1.

Looking at the scores only, one does not learn a lot about this electorate: Candidates a and c tie at the first place with four approvals and candidates b and d are just behind, with three approvals. But the Table indicates more:

- All the voters who approved b also approved a.
- No voters simultaneously approved a (or b) and d.
- Voters who approved c also approved a (two of them) or d (two of them).

From these observations, it appears that the electorate is split in two disjoint subgroups: The group of a-voters and the group of d-voters, and that c is the only candidate to gather votes from both groups. Compared to the others, c appears as a consensus candidate, even if she does not obtain more approvals than a. If one wants a picture of this vote, one should think of a and b being close one to the other, far apart from d; and c between a and d.

Notice that these observations, even if they require more information than only the scores, can be obtained from the approval ballots without breaking voters anonymity. This kind of information about the structure of the electorate and/or the positioning of parties is normally obtained through opinion polls.[4]

Table 19.1 An approval-voting profile with seven voters and four candidates

	1	2	3	4	5	6	7	Score
a	×	×	×	×				4
b	×	×	×					3
c			×	×	×	×		4
d					×	×	×	3

[4] This relates to the huge literature on spatial voting. See Stokes (1963), Poole and Rosenthal (1985, 1991), Rabinowitz and MacDonald (1989), Enelow and Hinich (1990), van Schuur (1993, 2003),

Using opinion surveys, it is possible to ask voters their opinions for different candidates and to study correlations among candidates. The statistical tools and the collected data can then be more powerful than an approval profile. But for committee decision-making, it may be useful to have a simple routine, such that an Approval Voting straw vote, rather than designing a full-vote survey. Moreover, and this is true in general, the information provided by voters through an opinion poll should not be considered as identical to the information provided by an official and decisive vote. The participation rates to opinion surveys are much lower than the participation rate to real elections, and the participation bias may be difficult to handle.

The structure of the electorate (the "electoral demand") and the structure of the set of candidates (the "electoral supply") are usually described by means of issue analysis. Researchers or journalists have identified some key issues in the political debate and try to assign positions to voters and politicians with respect to this issues. Clearly this cannot be done if one restricts attention to votes only. Looking at the above school-case example, one can be tempted to imagine that there is one (and only one) important issue which makes it possible to describe the system of voters and candidates as uni-dimensional, with c in the middle, a and b on one side, and d on the other side. Of course the preference profile is mute as to what is really this issue, this point being a matter of exogenous interpretation. But the fact that the approval data is mute about what the issues are is also an advantage, for scholars may be wrong when deciding about the set of relevant issues. The possibility of purely endogenous analysis and description of voters and candidates is valuable because it may confirm or infirm other studies which are based on some pre-conceptions, and also because it leaves open the possibility of discovering new facts.

19.4 The Canonical Representation

An approval profile with n voters and K candidates is a matrix A of size $K \times n$ of 0s and 1s. For voter $i \in \{1, \ldots, n\}$ and candidate $x \in \{1, \ldots, K\}$, one writes $A_i^x = 1$ if i approves x and $A_i^x = 0$ if not. The number of votes for candidate x is the sum $\sum_i A_i^x$.

The matrix A can be seen as providing K points $A^1, \ldots, A^K$ in the space IR^n with n dimensions. This is called the *canonical representation of the candidates*. Likewise, it can be seen as providing n points $A_1, \ldots, A_n$ in the space IR^K with K dimensions. This is called the *canonical representation of the voters*. The two representations are said to be dual, one of each other. Clearly any of the two canonical representations conveys all information contained in the data set. The problem is

Heckman and Snyder (1997), Chiche et al. (2000), Lewis and King (2000), Londregan (2000), Bailey (2001), Dow (2001), Groseclose (2001), Grunberg and Schweisguth (2002), Poole (2005), among others.

that this high-dimensional representation is too complex to be usable, therefore one must seek for appropriate ways to summarize it.

From a mathematical point of view, it is natural to consider the usual Euclidean distance on these sets, that is by definition:

$$||A^x - A^y|| = \sqrt{\sum_{i=1}^{n} \left(A_i^x - A_i^y\right)^2}.$$

This formula is simpler than it looks. If voter i approves or disapproves both x and y, $A_i^x = A_i^y$, otherwise $A_i^x - A_i^y$ is $+1$ or -1 so the sum $\sum_{i=1}^{n} \left(A_i^x - A_i^y\right)^2$ is the number of voters for whom x and y differs in terms of approbation.

Definition 19.4.1. *The* discrepancy *between candidates x and y is the number of voters who disagree on x and y: They approve one of x or y but not both. It is denoted by $d(x, y)$. The agreement between x and y is the number of voters who approve both x and y. It is denoted by $a(x, y)$.*

Notice that, knowing the number of approvals for each candidate, which can be denoted $a(x) = a(x, x)$, the same information is in fact contained in the discrepancies and in the agreements. To see that point, notice that the number of voters who approve x but not y is $a(x) - a(x, y)$, and thus the discrepancy can be expressed as:

$$d(x, y) = a(x) + a(y) - 2a(x, y).$$

It is therefore a matter of convention to use discrepancies or agreements. Of course, the canonical representation contains strictly more information; for instance, it contains the number of voters who approved of triples of candidates.

With our notation:

$$a(x, y) = \sum_{i=1}^{n} A_i^x \cdot A_i^y$$

and for the discrepancy:

$$\sum_{i=1}^{n} \left(A_i^x - A_i^y\right)^2 = d(x, y)$$

and hence

$$d(x, y) = ||A^x - A^y||^2.$$

For the example of Table 19.1, the Table 19.2 gives these discrepancies and the Table 19.3 gives the agreements.

Table 19.2 Squared distances matrix

	a	b	c	d
a	0	1	4	7
b	1	0	5	6
c	4	5	0	3
d	7	6	3	0

Table 19.3 Association matrix

	a	b	c	d
a	4	3	2	0
b	3	3	1	0
c	2	1	4	2
d	0	0	2	3

In the dual analysis the natural definition, from the mathematical point of view, is:

Definition 19.4.2. *The* discrepancy *between two voters i and j is the number of candidates about whom they disagree: x is approved by one voter but not by both.*

Comparing the two definitions, one can see that the discrepancy between voters is a less attractive concept, from the political point of view, than the discrepancy between candidates. Counting voters is a basic principle of democracy: If two citizens share the same opinion, this must really be taken into account. Counting candidates is quite different and does not refer to any fundamental principle. It is tempting to consider that the opinion of the voters is the fundamental object of study, and something which exist independently of how many candidates are presented at a given election. By contrast the number of candidates about which two voters differ does not seem to be so meaningful.[5]

Holding this position implies that the analysis one can perform using these concepts is better suited for describing the structure of the set of candidates than the set of voters. In that case, no structure is imposed with respect to the voters, but candidates are described as different points A^x which can be "close one the other" or "aligned" etc. The next section will present possible tools designed to describe in a synthetic way the structure of the set of candidates that emerges from the vote profile.

19.5 A Connection with Spatial Voting

The multidimensional scaling techniques can be used to summarize the canonical representation. The simplest (from the mathematical point of view) technique is to find the first two or three principal components of the canonical representation and

[5] Of course it is important to understand the "political supply", that is how do parties come to exist, candidates run for office, etc. But from a normative point of view, the number of parties clearly cannot be considered as fundamental.

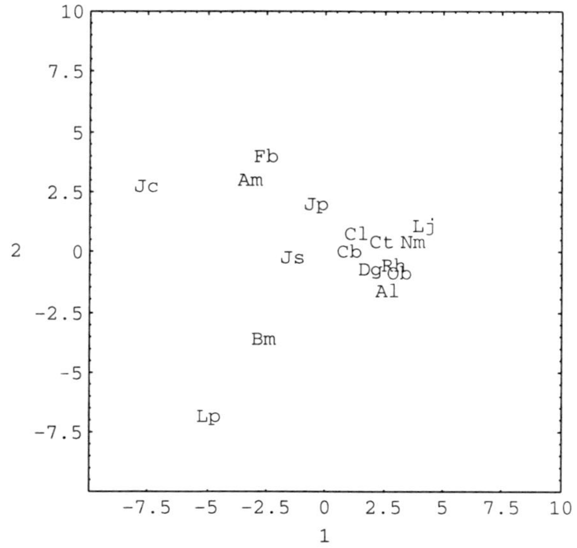

Fig. 19.1 Gy-les-Nonains, main plane for the canonical analysis

to project the K points corresponding to the K candidates from IR^n onto IR^2 or IR^3. Laslier and Van der Straeten (2002) have done such an analysis of the data collected in 2002 in Gy-les-Nonains;[6] see Fig. 19.1.

One can note that, in this picture, the candidates which are less distant are also the ones with the smallest scores. The reason is as follows. Approval scores in this example are smaller than $1/2$; two candidates who are approved by a fraction x, or less, of the electorate are thus distinguished by a fraction of the electorate smaller than $2x$, and a fraction larger than $1 - 2x$ of the voters grade them identically. For that reason, the proximity between candidates like Christine Boutin (Cb) and Daniel Gluckstein (Dg) does not mean that the same voters have approved of them. And likewise, the largest distances are found between candidates which are often approved because there is (unfortunately!) no pair of candidates who are close one to the other because they both have very large approval scores.

It follows that the analysis using directly the projection of the canonical representation in low dimension, although mathematically attractive, does not produces a meaningful picture of the political space.

If one wishes not to break the symmetry between approbation and non-approbation, one can think of different methods and distances that are more or less intuitive.[7] The following method does break the symmetry, and makes a bridge

[6] See Balinski et al. (2002), Perrineau and Ysmal (2003), Laslier and Van der Straeten (2004), Baujard and Igersheim (2010), Laslier (2010).

[7] See Laslier (1996) and Laslier (2003). This last paper studies the same data as Brams and Fishburn (2001) and Saari (2001).

between the multidimensional analysis of approval data and more standard ideas taken from the Spatial Voting literature.

Recall that the agreement $a(x, y)$ between x and y is the number of voters who approve both x and y. In Laslier (2006) a distance is introduced, which uses the logarithms of the agreement numbers.

Definition 19.5.1. *Let $a(x)$ and $a(x, y)$, the number of votes and the agreement for the candidates $x, y \in X$ be given. A logarithmic representation of the candidates is a set of locations $p_c \in IR^k$ such that the squared distances $\|p_x - p_y\|^2$ are proportional to $\rho - \ln \frac{a(x,y)}{a(x)\,a(y)}$ for an additive constant ρ called the* contrast *parameter:*

$$\|p_x - p_y\|^2 \propto \rho - \ln \frac{a(x, y)}{a(x)\,a(y)}.$$

This distance can be justified on the basis of the following hypothesis:

- Voters and candidates are located in some Euclidean space and the utility of a voter located at point x_i for a candidate located at point y_c is

$$u_i(x) = -\alpha \|p_x - p_i\|^2 + \Gamma_x + \varepsilon_{i,x}$$

  $\|p_x - p_i\|$ being the usual Euclidean distance between the location of the candidate and the ideal point of the voter, and under the hypothesis that the disturbance is distributed according to the exponential law (cumulative function $F(\varepsilon) = 1 - e^{-\varepsilon}$).
- The probability that voter i approves of candidate x is:

$$\Pr\left[A_i^x = 1\right] = \gamma_c \exp\left(-\alpha \|p_x - p_i\|^2\right), \tag{19.1}$$

  where γ_c is some positive parameter called the *valence* of candidate x, with $\Gamma_x = \log \gamma_x$, and α is a *policy salience* parameter.Under plurality voting and with two candidates x and y, the usual utility-maximization behavior would then lead to express the probability of voting for c against c' as a function[8] of the difference of unperturbed utilities:

$$\delta = -\alpha \|p_x - p_i\|^2 + \Gamma_x + \alpha \|p_y - p_i\|^2 - \Gamma_y.$$

  Such an approach is often adopted, with various statistical forms for the disturbance term (Bailey 2001). Under Approval Voting, some behavioral assumption must be made. Equation (19.1) is obtained if one assumes that voter i approves of candidate x when the utility $u_i(x)$ is larger than some fixed threshold, that we can take to be 0. In that case:

---

[8] Precisely: If ε and ε' are independent and exponential, the difference $\delta = \varepsilon - \varepsilon'$ has density $e^{-|\delta|}$ and its cumulative function is $e^{\delta}/2$ for $\delta \leq 0$ and $1 - e^{-\delta}/2$ for $\delta \geq 0$.

$$\begin{aligned}\Pr\left[A_i^x = 1\right] &= \Pr\left[\varepsilon_{i,x} > \alpha \left\| p_x - p_i \right\|^2 - \Gamma_x\right] \\ &= \exp\left(-\alpha \left\| p_x - p_i \right\|^2 + \Gamma_x\right) \\ &= \gamma_x \exp\left(-\alpha \left\| p_x - p_i \right\|^2\right).\end{aligned}$$

- Voters are distributed according to a normal density with concentration λ, with λ very small. At the limit, a large electorate is spread all over the space.

The pictures presented in Baujard and Igersheim (2010) in this book have been drawn with this method. The interested reader should consult the original articles to get familiar with the technical details. Notice that one implicitly supposes here that the voters are evenly dispersed all over the political space. It would be interesting to analyse the set of voters in the same spirit as we analyze the set of candidates, and maybe to make both analysis jointly.

19.6 What are the Results of an AV Election?

We can now come back to the question raised at the beginning of the chapter: What should be published as the "results" of an Approval Voting election?

With respect to the candidate scores, there is no serious problem: Listing the number of approvals obtained by each candidate is simple and easy to read. Moreover, for reasons given in Sect. 19.2, I suggest that, if ratios have to be used, then the $s(\cdot)$ ratios be. Their meaning is clear: the percentage of voters who approve of a candidate.

A variety of more sophisticated measures and descriptions can be produced, as shown in, or can be derived from, the previous sections. We have proposed several such methods and learned statisticians can have many other ideas.[9] But this variety raises a problem. Certainly, no single method appears to be obviously "natural" or obviously superior to the others. These measures and representations must not be considered as raw data. Even if qualitative results may not be too sensitive to the method used, there may occasionally exist important discrepancies, and since these methods are quite sophisticated, only specialists can be supposed to understand for what reason such and such method give such and such result.

What is legally defined as the outcome of the election must be raw data. I therefore suggest that the published outcome of the election be the score vector and the agreement matrix, which records for every pair of candidate (x, y), the number of voters who approved both x and y. This choice has several advantages:

1. This is raw data, which does not require an explanation.
2. This is objective data, which does not involve any analysis.

[9] See for instance Falmagne and Regenwetter (1996), Regenwetter (1997), Regenwetter and Tsetlin (2004).

3. This is small data: A table with as many rows and columns as candidates.
4. This is additive data: The result in two districts is the sum of the results in each.
5. This is valuable data: As seen in this chapter, many things can be derived from it.

Of course, one could suggest to publish more, such as the number of voters who approve of any *triple* of candidates, or even the number of voters who approve of any *subset* of candidates, that is the whole counting of all possible approval ballots. These proposals share the features mentioned above, except the third one: With ten candidates, the binary agreement table contains about 50 numbers and can be published easily, for instance in the newspapers. With triples one gets about 300 numbers which are uneasy to display. The whole data requires 1,024 numbers, with a complex structure which makes this data almost impossible to display in a readable way. If the number of candidates is small, then the whole ballot counting can be published.

One may object that this kind of data may only be useful for scholars, and not for ordinary voters. This objection is not valid for two reasons.

First point, for committees, votes often takes place in order to help the discussion and acquire information about the committee members' preferences, before being a mean to take a decision. In such a case, committee members may find useful to learn more detailed pieces of information than the simple "strength" of the alternative choices. Here, Approval Voting looks like a nice and simple tool for group decision aid, if one follows the above suggestion to publish more than the candidate scores.

Second point, which is also valid for large-scale elections, in order to appreciate the number of voters who approve both candidates x and y, one does not need any particular academic knowledge. I can thus see no reason to believe that the voters would not be able to learn from such data. This kind of information might foster and enrich the political debates. Anyone who believes in the positive value of the voter's information (this sounds like a nice definition of democracy) may therefore agree with this proposal.

References

Alós-Ferrer, C., & Granić, D.-G. (2010). Approval Voting in Germany: Description of a field experiment. In J.-F. Laslier & R. Sanver (Eds.), *Handbook on Approval Voting*. Heildelberg: Springer-Verlag.

Bailey, M. (2001) Ideal point estimation with a small number of votes: a random-effects approach. *Political Analysis, 9*, 192–210.

Balinski, M., Laraki, R., Laslier, J.-F., & Van der Straeten, K. (2002). Expérience électorale du vote par assentiment. *Pour la science*, p. 13.

Baujard, A., & Igersheim, H. (2010). Framed-field experiments on approval voting in the political context. In J.-F. Laslier & R. Sanver (Eds.), *Handbook of Approval Voting*. Heidelberg: Springer-Verlag.

Brams, S. J., & Fishburn, P. C. (2001). A nail-biting election. *Social Choice and Welfare, 18*, 409–414.

Brams, S. J., Kilgour, M., & Sanver, R. (2007) A minimax procedure for electing committees. *Public Choice, 132*, 401–420.

Chiche, J., Le Roux, B., Perrineau, P., & Rouanet, H. (2000). L'espace politique des électeurs français à la fin des années 1990. *Revue française de science politique, 50*, 463–487.

Cox, G. W. (1997). *Making votes count: Strategic coordination in the world's electoral systems*. Cambridge, MA: Cambridge University Press.

Dow, J. (2001). A comparative spatial analysis of majoritarian and proportional elections. *Electoral Studies, 20*, 109–125.

Duverger, M. (1951). *Les partis politiques*. Paris: Armand Colin.

Enelow, J. M., & Hinich, M. (Eds.). (1990). *Advances in the spatial theory of voting*. Cambridge: Cambridge University Press.

Falmagne, J.-C., & Regenwetter, M. (1996). Random utility models for approval voting. *Journal of Mathematical Psychology, 40*, 152–159.

Groseclose, T. (2001). A model of candidate location when one candidate has a valence advantage. *American Journal of Political Science, 45*, 862–886.

Grunberg, G., & Schweisguth, E. (2002). La tripartition de l'espace politique. In P. Perrineau & C. Ysmal (Eds.), *Le vote de tous les refus* (pp. 341–362). Paris: Presses de Sciences Po.

Heckman, J. J., & Snyder, J. M. (1997). Linear probability models of the demand for attributes with an empirical application to estimating the preferences of legislators. *The RAND Journal of Economics, 28*, S142–S189.

Laslier, J.-F. (1996). Multivariate analysis of comparison matrices. *Multicriteria Decision Analysis, 5*, 112–126.

Laslier, J.-F. (2003). Analyzing a preference and approval profile. *Social Choice and Welfare, 20*, 229–242.

Laslier, J.-F. (2006). Spatial approval voting. *Political Analysis, 14*, 160–185.

Laslier, J.-F. (2010). Lessons From *In Situ* experiments during French elections. working paper, Ecole polytechnique.

Laslier, J.-F., & Van der Straeten, K. (2002). Analyse d'un scrutin d'assentiment. *Quadrature, 46*, 5–12.

Laslier, J.-F., & Van der Straeten, K. (2004). Election présidentielle: une expérience pour un autre mode de scrutin. *Revue française de science politique, 54*, 99–130.

Lewis, J., & King, G. (2000). No evidence on directional vs. proximity voting. *Political Analysis, 8*, 21–33.

Londregan, J. B. (2000). Estimating legislator's preferred points. *Political Analysis, 8*, 35–56.

Perrineau, P., & Ysmal, C. (Eds.). (2003). *Le vote de tous les refus: Les élections présidentielle et législatives de 2002*, Paris: Presses de Sciences Po.

Poole, K. T. (2005). *Spatial models of parliamentary voting*. Cambridge: Cambridge University Press.

Poole, K. T., & Rosenthal, H. (1985). A spatial model for legislative roll call analysis. *American Journal of Political Science, 29*, 357–384.

Poole, K. T., & Rosenthal, H. (1991). Patterns of congressional voting. *American Journal of Political Science, 35*, 228–278.

Rabinowitz, G., & MacDonald, S. (1989). A directional theory of issue voting. *American Political Science Review, 83*, 93–121.

Regenwetter, M. (1997). Probabilistic preferences and topset voting. *Mathematical Social Sciences, 34*, 91–105.

Regenwetter, M., & Tsetlin, I. (2004). Approval voting and positional voting methods: Inference, relationship, examples. *Social Choice and Welfare, 22*, 539–566.

Saari, D. (2001). Analyzing a nail-biting election. *Social Choice and Welfare, 18*, 415–430.

Sanver, R. (2010). Approval as an intrinsic part of preference. In J.-F. Laslier & R. Sanver (Eds.), *Handbook of Approval Voting*. Heidelberg: Springer-Verlag.

Stokes, D. (1963). Spatial models of party competition. *American Political Science Review, 57*, 368–377.

van Schuur, W. (1993). Non parametric unidimensional unfolding for multicategory data. *Political Analysis, 4*, 41–74.

van Schuur, W. (2003). Mokken scale analysis: A nonparametric version of Guttman scaling for survey research. *Political Analysis, 11*, 139–163.

Chapter 20
Approval as an Intrinsic Part of Preference

M. Remzi Sanver

20.1 The Model

The collective decision making problem can be conceived as the aggregation of a vector of utility functions whose informational content depends on the assumptions made about the cardinality and interpersonal comparability of individual preferences. To be more explicit, we consider a non-empty set N of individuals and a non-empty set A of alternatives. Letting $U(A)$ be the set of real-valued "utility functions" defined over A, we model the problem through an aggregation function $f : U(A)^N \to 2^A \setminus \{\emptyset\}$. The assumptions about the cardinality and interpersonal comparability of individual preferences are formalized by partitioning $U(A)^N$ into information sets, while requiring f to be invariant at any two vector of utility functions which belong to the same information set. At one extreme, one can assume the existence of an absolute scale over which the utilities of individuals are measured and compared. This assumption partitions $U(A)^N$ into singleton information sets, hence imposing no invariance over f. At the other extreme, one can rule out any kind of cardinal information and interpersonal comparability, in which case an information set consists of the elements of $U(A)^N$ which are ordinally equivalent, i.e., induce the same ordering of alternatives for every individual.[1] When cardinality and interpersonal comparability are ruled out, the problem can be modeled through an aggregation function $f : W(A)^N \to 2^A \setminus \{\emptyset\}$ where $W(A)$ is the set of

[1]Given any ordered list $\phi = (\phi_i)_{i \in N}$ of functions from the reals to the reals and any $u \in U(A)^N$, we define $\phi \circ u \in U(A)^N$ as $(\phi \circ u)_i(x) = \phi_i(u_i(x))$ $\forall x \in A$, $\forall i \in N$. When an absolute scale exists, $u, v \in U(A)^N$ are in the same information set iff $v = \phi \circ u$ for some ϕ where each ϕ_i is the identity function. When cardinality and interpersonal comparability are ruled out, $u, v \in U(A)^N$ are in the same information set iff $v = \phi \circ u$ for some ϕ where each ϕ_i is monotonically increasing. As Sen (1986), Bossert and Weymark (2004) eloquently survey, there is a variety of cases between the two extremes.

M.R. Sanver
Department of Economics, Istanbul Bilgi University, Dolapdere Campus,
Kurtulus Deresi, Cad No 47, Istanbul 34440, Turkey
e-mail: sanver@bilgi.edu.tr

J.-F. Laslier and M.R. Sanver (eds.), *Handbook on Approval Voting*, Studies in Choice and Welfare, DOI 10.1007/978-3-642-02839-7_20,

weak orders (i.e., complete and transitive binary relations) over A. We refer to this as the *Arrovian model* (Arrow 1950, 1951).

While many voting rules are covered by the Arrovian model,[2] *Approval Voting* (AV) falls apart: It generates the social outcome by aggregating vectors of subsets of A. Formally speaking, it is an aggregation function $v : (2^A)^N \to 2^A \backslash \{\emptyset\}$ where $S_i \in 2^A$ is conceived as the set of alternatives which are "approved" by $i \in N$. Given any $S \in (2^A)^N$, AV picks the alternatives which are approved by the highest number of individuals. So writing $n(z; S) = \#\{i \in N : z \in S_i\}$ for the number of individuals who approve $z \in A$ at S, we have $v(S) = \{x \in A : n(x; S) \geq n(y; S) \ \forall y \in A\}$.

The literature exhibits various attempts to place AV within the Arrovian model. This is typically done by interpreting AV as a game form μ where 2^A is the common message space of individuals and $v : (2^A)^N \to 2^A \backslash \{\emptyset\}$ is the outcome function. The combination of μ with individuals' preferences over A induces a game whose outcomes are considered. This is a basic mechanism design approach where the approval of an individual is a mere strategic action with no intrinsic meaning. As Dellis (2010), Laslier and Maniquet (2010), Laslier and Sanver (2010b), Nunez (2010) in this volume testify, this interpretation is rich in its variants regarding the modelling and solution of the game. Nevertheless, the same chapters would manifest a dilemma that traps the mechanism design approach: Under natural mechanisms and with mild assumptions over individual preferences, the set of equilibrium outcomes explodes and this set can be refined to the expense of fairly strong assumptions.

We propose to express AV in a framework which partitions $U(A)^N$ into information sets which are finer than those of the Arrovian model. We assume the existence of two cardinal qualifications, "good" and "bad", with a common meaning among individuals. This can be interpreted as the existence of a real number, say 0, whose meaning as a utility measure is common to all individuals. Thus, an information set consists of the ordinally equivalent elements of $U(A)^N$ where 0 is common to all individuals.[3] We call this framework the *extended (Arrovian) model.* In the extended model, the problem can be modeled through an aggregation function $f : W(A \cup \{\emptyset\})^N \to 2^A \backslash \{\emptyset\}$ where $W(A \cup \{\emptyset\})$ is the set of weak orders over $A \cup \{\emptyset\}$. Here the empty-set stands for the separation between good and bad: An alternative which is ranked above (resp., below) the empty set is qualified as good (resp., bad). Henceforth, "approval" is not a strategic action but has an intrinsic meaning: It refers to those alternatives which are qualified as good.

Note that every aggregation function expressed in the Arrovian model can also be expressed in the extended one. In fact, aggregation functions of the Arrovian model coincide with those of the extended model which satisfy the following approval independence condition: We say that $f : W(A \cup \{\emptyset\})^N \to 2^A \backslash \{\emptyset\}$ is *approval*

---

[2] see, for example, Brams and Fishburn (2002).

[3] In other words, $u, v \in U(A)^N$ are in the same information set iff $v = \phi \circ u$ for some $\phi = (\phi_i)_{i \in N}$ where each ϕ_i is monotonically increasing and $\phi_i(0) = 0$.

independent iff $f(R) = f(R')$ for every $R, R' \in W(A \cup \{\emptyset\})^N$ with $x\ R_i\ y \Longleftrightarrow x\ R'_i\ y\ \forall x, y \in A, \forall i \in N$.

Although the literature contains studies which imply the extended model,[4] we are not aware of any formal treatment of it. So we explore the extended model, with particular emphasis on Approval Voting. In Sect. 20.2, we introduce the *majoritarian approval* axiom which we use as a benchmark of the extended model. In Sect. 20.3, we consider four social choice rules, including Approval Voting, under this benchmark. In Sect. 20.4, we evaluate these social choice rules according to two criteria, namely monotonicity and independence. In Sect. 20.5, we make some concluding remarks, including the possibility of further extending the extended model.

20.2 A Benchmark: The Majoritarian Approval Axiom

Sertel and Yilmaz (1999) introduce, within the Arrovian model, a "majoritarian approval" axiom which requires from a social choice rule to pick among the alternatives which receive the "approval" of a majority of voters. This requirement explicitly assumes that a voter "approves" an alternative if and only if he ranks it among the first half of his ordering. Such an artificial meaning attributed to the term "approval" is undesirable, but also inevitable within the informational framework of the Arrovian model. On the other hand, thanks to the additional information incorporated by the extended model, majoritarian approval can be naturally redefined. In fact, within the extended model, it is possible to the aggregate the qualifications "good" and "bad" that individuals attribute to alternatives. In other words, based on individual qualifications attributed to an alternative, it is meaningful to speak about that alternative being "socially good" or "socially bad". To express this more formally, let $q(x) \in \{G, B\}^N$ be a *qualification profile* for $x \in A$, where $q_i(x) = G$ (resp., $q_i(x) = B$) means that individual $i \in N$ qualifies x as good (resp., bad). At every $R \in W(A \cup \{\emptyset\})^N$, we write $q(x; R)$ for the qualification profile for x induced by R, i.e., $q_i(x; R) = G \Longleftrightarrow x\ P_i\ \emptyset$ holds for all $i \in N$.[5] The aggregation of qualification profiles into a social qualification means to map the set $\{G, B\}^N$ into the set $\{G, B\}$. While this is a separate matter of interest, we will take majoritarianism as granted. Let $n^G(x; R) = \#\{i \in N : q_i(x; R) = G\}$ be the number of individuals who qualify x as good at R. We write $\gamma(R) = \{x \in A : n^G(x; R) \geq \frac{n}{2}\}$ for the (possibly empty) set of alternatives which are qualified as "socially good" at R. We say that $f : W(A \cup \{\emptyset\})^N \to 2^A \backslash \{\emptyset\}$ satisfies *majoritarian approval* if and only if we have $f(R) \subseteq \gamma(R)$ at every $R \in W(A \cup \{\emptyset\})^N$ where $\gamma(R) \neq \emptyset$. So, based on

[4] Niemi (1984) distinguishes between "approving alternative x" and "voting for alternative x under Approval Voting", For example, Brams and Sanver (2006) mention the possibility of conceiving approval as an intrinsic part of preference. This idea is developed by Brams and Sanver (2009) who, within the general model, propose two new social choice rules. Peters et al. (2009) define Approval Voting as a social choice rule whose domain is preference and approval profiles.

[5] We write P_i for the strict counterpart of R_i.

how individuals qualify alternatives, majoritarian approval determines the socially good and socially bad ones according to the majority rule, while ruling out the possibility of choosing socially bad alternatives when there are socially good ones.

As we show below, majoritarian approval contradicts approval independence, hence aggregation rules of the Arrovian model all fail majoritarian approval.

Theorem 20.2.1. *Majoritarian approval and approval independence are logically incompatible.*

Proof. Take any $f : W(A \cup \{\emptyset\})^N \to 2^A \backslash \{\emptyset\}$ which is approval independent and satisfies majoritarian approval. Consider a society $N = \{1, 2, 3\}$ and let $R \in W(A \cup \{\emptyset\})^N$ be such that $a\ P_1\ \emptyset\ P_1\ b\ P_1\ c, b\ P_2\ a\ P_2\ \emptyset\ P_2\ c, c\ P_3\ \emptyset\ P_3\ b\ P_3\ a$. Majoritarian approval implies $f(R) = \{a\}$. Now let $R' \in W(A \cup \{\emptyset\})^N$ be such that $a\ P'_1\ \emptyset\ P'_1\ b\ P'_1\ c, b\ P'_2\ \emptyset\ P'_2\ a\ P'_2\ c, c\ P'_3\ b\ P'_3\ \emptyset\ P'_3\ a$. Approval independence implies $f(R') = \{a\}$, which contradicts majoritarian approval. □

20.3 Four "New" Social Choice Rules

Under the majoritarian approval axiom, the collective decision making problem boils down to answering the following two questions:

1. How to refine the set of socially good alternatives, when this set contains more than one alternative?
2. Which alternative to choose when none of them is socially good?

Throughout the chapter, we will refer to these questions as Question 1 and Question 2.

A common answer to both questions is to pick the alternatives which are qualified as good by the highest number of individuals. This is *Approval Voting* which is formally expressed by the aggregation function $f : W(A \cup \{\emptyset\})^N \to 2^A \backslash \{\emptyset\}$ defined as $f(R) = \{x \in A : n^G(x; R) \geq n^G(y; R)\ \forall y \in A\}$ at every $R \in W(A \cup \{\emptyset\})^N$. It is straightforward to check that Approval Voting satisfies majoritarian approval.

Approval Voting, while falling out of the Arrovian model, has a very natural fit to the extended one. In fact, the debate on whether Approval Voting is "indeterminate" or "responsive" now vanishes[6]: The fact is that the informational framework of the Arrovian model is not sufficient to express Approval Voting.[7]

Nevertheless, Approval Voting only uses the information about how individuals qualify the alternatives, hence overlooking the information about rankings. Under Approval Voting, we have $f(R) = f(R')$ for any $R, R' \in W(A \cup \{\emptyset\})^N$ with

[6] Laslier and Sanver (2010a) in this volume give an account of the exchange between Saari and Newenhizen (1988a,b) and Brams et al. (1988a,b).

[7] This is as if the Borda rule is expressed in a model which aggregates the top ranked alternatives of voters into a social outcome - hence needing the assume the rest of individual rankings. See Endriss et al. (2009) for an analysis of ballot languages.

$q_i(x; R) = q_i(x; R') \; \forall x \in A, \forall i \in N$. It goes without saying that this overlooked information can be used to define other social choice criteria. We exemplify two of these which satisfy majoritarian approval and which differ in their answers to Question 1 or Question 2.

First, we revisit an aggregation rule of the Arrovian model, namely the Majoritarian Compromise (MC) of Sertel (1986) which, within the Arrovian model, is defined as follows:

1. The highest-ranked candidate of all voters is considered. If a majority of voters agree on one highest-ranked candidate, this candidate is the MC winner. The procedure stops, and we call this candidate a level 1 winner.
2. If there is no level 1 winner, the next-highest ranked candidate of all voters is considered. If a majority of voters agree on one candidate as either their highest or their next-highest ranked candidate, this candidate is the MC winner. If more than one candidate receives a majority support, then the candidate with highest support is the MC winner. The procedure stops, and we call this candidate a level 2 winner.
3. If there is no level 2 winner, the voters descend – one level at a time – to lower and lower ranks, stopping when, for the first time, one or more candidates receive a majority support. If exactly one candidate receives a majority support, then this candidate is the MC winner. If more than one candidate receives a majority support, then the candidate with the highest majority support is the MC winner.

We know from Sertel (1986), Sertel and Yilmaz (1999) and from Brams and Kilgour (2001) that the MC winner always arises at a level which does not exceed $\frac{\#A}{2}$. It is worth noting that MC understands "majority" in a weak sense, so as to refer to a coalition whose cardinality is at least as big as the cardinality of its complement.

As one can see in Hoag and Hallett (1926, pp. 485–491), MC is the reinvention of a voting rule, known as *Bucklin voting*, invented by James W. Bucklin, a lawyer and founder of Grand Junction, Colorado, who proposed his system for Grand Junction in the early twentieth century, where it was used from 1909 to 1922 – as well as in other cities – but it is no longer used today. Interestingly, Bucklin asks voters to rank as many of the alternatives they wish, but not necessarily all of them. Given the available rankings of voters, Bucklin voting operates precisely as MC, with the impossibility of descending further in the rankings of certain voters who did not rank all alternatives. Clearly, under Bucklin voting, one can reach the lowest ranked alternative of each voter and still not get a majority, in which case the alternatives with the highest support are elected. Although Bucklin voting is formally absent in the designation of good and bad candidates, those candidates that a voter ranks can be implicitly assumed to be the good ones and that the voter qualifies as bad those he did not care to rank. Thus Bucklin voting can indeed be seen as an adaptation of MC to the extended model, where the descent in a voter's ranking stops when the empty-set is reached. If the descent reaches the empty-set in all voters' rankings and yet no candidate is qualified as socially good, then the alternatives which are qualified good by the highest number of individuals are chosen. We call this adaptation of MC, Majoritarian Approval Compromise (MAC). To illustrate how MAC operates in the extended model, consider the following preference profile with four alternatives and nine voters:

3 voters $a| \; b \; c \; d$
2 voters $b \; a \; c| \; d$
2 voters $c| \; a \; b \; d$
2 voters $d \; b \; c| \; a$

The orderings go from left to right, i.e., the first 3 voters prefer a to b, b to c and c to d, etc. The symbol "|" represents the empty set, i.e., separating the good alternatives from the bad ones. So the first three voters see alternative a as good and the rest as bad, etc. In this profile, initially a gets an approval of 3 while b, c and d get an approval of two voters. So none of the alternatives receives a majority approval of 5. We can lower the stick for the b and d voters only (as the a and c voters reached the border between what is good and bad). Now a gets five votes, b gets four votes, c and d get two votes. Hence a is the MAC winner.[8]

MAC satisfies majoritarian approval. Moreover, it coincides with AV when there are no socially good alternatives. In other words, MAC and AV agree in their answer to Question 2, by picking the alternatives which are qualified as good by the highest number of individuals. On the other hand, they answer Question 1 differently: Among the alternatives which are socially good, AV chooses those which are qualified as good by the highest number of individuals (e.g., alternative c in the above example) while MAC picks those which are qualified as socially good at the earliest level.

Preference-Approval Voting (PAV)[9] is a social choice rule which also differs in its answer to Question 1: It refines the set of socially good outcomes through the construction of the pairwise majority relation among these. When socially good alternatives are multiple, it constructs the pairwise majority relation among the set of socially good alternatives; picks the Condorcet winner if it exists and otherwise, among the alternatives in the top-cycle picks those which are qualified as good by the highest number of individuals. Clearly, PAV satisfies majoritarian approval.

As a final example, we present *Approval Voting with a runoff* (AVR). Given any $R \in W(A \cup \{\emptyset\})^N$, let $\rho(R) = \{x, y\}$ be the pair of alternatives – called *runoff winners* – which receive the highest approval, i.e, $n^G(x; R) \geq n^G(z; R)$ and

[8] Brams and Sanver (2009) consider the problem of introducing new social choice rules within the general model and what they propose under the name of *Fallback Voting* is what we call MAC in this paper. We also wish to note the similarity between MAC and "fallback bargaining with an impasse" which is a bargaining solution introduced and analyzed by Brams and Doherty (1993) and Brams and Kilgour (2001).

[9] Preference-Approval Voting is proposed by Brams and Sanver (2009) and further studied by Erdelyi et al. (2008).

$n^G(y; R) \geq n^G(z; R)$ hold for any $z \in A \backslash \{x, y\}$.[10] AVR picks the pairwise majority winner among the runoff winners. Remark that AVR is an adaptation of the well-known *plurality with a runoff* defined within the Arrovian model where the runoff winners is the pair of alternatives which are ranked at the top by the highest number of voters. When $\#\gamma(R) > 1$, we have $\rho(R) \subseteq \gamma(R)$, hence the AVR winner is approved by a majority. On the other hand, when $\#\gamma(R) = 1$, AVR may fail to pick the (unique) alternative which is approved by a majority, hence failing majoritarian approval.[11]

We summarize below the behavior of the four social choice rules, as a function of the cardinality of $\gamma(R)$:

	$\gamma(P) = \emptyset$	$\#\gamma(P) = 1$	$\#\gamma(P) > 1$
AV	most approved alternative in A	$\gamma(P)$	most approved alternative in $\gamma(P)$
MAC	most approved alternative in A	$\gamma(P)$	the alternative which gets "earliest" in $\gamma(P)$
PAV	most approved alternative in A	$\gamma(P)$	most approved alternative in the top-cycle of the pairwise majority relation over $\gamma(P)$
AVR	majority winner in $\rho(P)$	majority winner in $\rho(P)$	majority winner in $\rho(P)$

In the next section, we evaluate the four social choice rules vis-à-vis the satisfaction of two properties, namely monotonicity and independence.

20.4 Monotonicity and Independence

Among the variety of monotonicity conditions introduced within the Arrovian model, we consider the weakest one which requires that raising an alternative x in individual preference rankings without changing the preference relation on pairs of alternatives that do not include x, cannot have an effect on the election outcome

[10] Such a pair need not be unique of course. For expositional simplicity, we assume an exogeneous total order of alternatives which breaks the ties between the alternatives that receive the same number of approvals.

[11] For example, at the preference profile

1 voter $a|b$
1 voter $b\ a\ |$
1 voter $|\ b\ a$

with three voters and two alternatives, b is the AVR winner while a is the only alternative which is approved by a majority.

which is detrimental to x. To state this formally, given any $x \in A$ and any $R, R' \in W(A)^N$, we say that R' *is a lifting of* x *with respect to* R if and only if for every $i \in N$ we have $[x\ R_i\ y \Longrightarrow x\ R'_i\ y\ \forall y \in A]$, $[x\ P_i\ y \Longrightarrow x\ P'_i\ y\ \forall y \in A]$ and $[y\ R_i\ z \Longleftrightarrow y\ R'_i\ z\ \forall y, z \in A \backslash \{x\}]$. A social choice rule $f : W(A)^N \to 2^A \backslash \{\emptyset\}$ is *monotonic* if and only if $x \in f(R) \Longrightarrow x \in f(R')$ whenever R' is a lifting of x with respect to R.[12] We adapt monotonicity to the extended framework as follows: Given any $x \in A$ and any $R, R' \in W(A \cup \{\emptyset\})^N$, we say that R' *is a lifting of* x *with respect to* R if and only if for every $i \in N$ we have $[x\ R_i\ y \Longrightarrow x\ R'_i\ y\ \forall y \in A]$, $[x\ P_i\ y \Longrightarrow x\ P'_i\ y\ \forall y \in A]$, $[x\ P_i\ \emptyset \Longrightarrow x\ P'_i\ \emptyset]$ and $[y\ P_i\ z \Longleftrightarrow y\ P'_i\ z\ \forall y, z \in (A \backslash \{x\}) \cup \{\emptyset\}]$. A social choice rule $f : W(A \cup \{\emptyset\})^N \to 2^A \backslash \{\emptyset\}$ is *monotonic* if and only if $x \in f(R) \Longrightarrow x \in f(R')$ whenever R' is a lifting of x with respect to R. Note that whenever R' is a lifting of x with respect to R, we have $n^G(x; R') \geq n^G(x; R)$ and $n^G(y; R') = n^G(y; R)\ \forall y \in A \backslash \{x\}$, which implies the following result, whose proof is left to the reader.

Theorem 20.4.1. *Approval Voting, Majoritarian Approval Compromise, Preference-Approval Voting and Approval Voting with a runoff are all monotonic.*

So monotonicity does not discriminate among the four voting rules we consider. However, in contrast to Approval Voting with a run-off which is monotonic within the extended framework, Plurality with a run-off fails monotonicity within the Arrovian framework (see p. 235 of Moulin 1988).

To define independence, we consider some alternative x^* which is not in A and we write $B = A \cup \{x^*\}$. Writing $W(B \cup \{\emptyset\})$ for the set of weak orders over $B \cup \{\emptyset\}$, from now on, we conceive a social choice rule as a mapping $f : W(A \cup \{\emptyset\})^N \cup W(B \cup \{\emptyset\})^N \to 2^A \backslash \{\emptyset\}$ such that $x^* \in f(R)$ only if $R \in W(B \cup \{\emptyset\})^N$. Note that all four voting rules introduced in Sect. 20.3 are also defined as a social choice rule $f : W(B \cup \{\emptyset\})^N \to 2^A \backslash \{\emptyset\}$. Hence, they are naturally defined as a social choice rule $f : W(A \cup \{\emptyset\})^N \cup W(B \cup \{\emptyset\})^N \to 2^A \backslash \{\emptyset\}$.

We say that $R \in W(A \cup \{\emptyset\})$ and $R' \in W(B \cup \{\emptyset\})$ *agree* if and only if for every $i \in N$ and for every $x, y \in A$, we have $x\ R_i\ y \Longleftrightarrow x\ R'_i\ y$ and $x\ R_i\ \emptyset \Longleftrightarrow x\ R'_i\ \emptyset$. We call x^* a *spoiler* iff $x^* \notin f(R') \neq f(R)$ at some $R \in W(A \cup \{\emptyset\})$ and $R' \in W(B \cup \{\emptyset\})$ which agree. So x^* is called a spoiler if its presence as an alternative can change the social choice without x^* being chosen. A social choice rule $f : W(A \cup \{\emptyset\})^N \cup W(B \cup \{\emptyset\})^N \to 2^A \backslash \{\emptyset\}$ satisfies *independence* iff f does not admit any spoiler x^*.[13]

---

[12] This condition, dating back to Fishburn (1982), is originally defined for social choice rules which pick a single alternative at every preference profile. As Sanver and Zwicker (2009) discuss, there is a variety of ways to adapt it to the set-valued context, such as those proposed by Barberà (1977) and Peleg (1979, 1981, 1984).

[13] Independence is a well-known choice theoretic property called "Postulate 5*" by Chernoff (1954), "Strong Superset Property" by Bordes (1979), "absorbence" by Sertel and van der Bellen (1979), "Outcast" by Aizerman and Aleskerov (1995). This is also the independence condition which Nash (1950) imposes over a bargaining solution.

Theorem 20.4.2. *Approval Voting satisfies independence.*

Proof. Let $f : W(A \cup \{\emptyset\})^N \cup W(B \cup \{\emptyset\})^N \rightarrow 2^A \backslash \{\emptyset\}$ be Approval Voting. Take any $R \in W(A \cup \{\emptyset\})$ and $R' \in W(B \cup \{\emptyset\})$ which agree. So $\{x \in A : n^G(x; R) \geq n^G(y; R)\ \forall y \in A\} = \{x \in A : n^G(x; R') \geq n^G(y; R')\ \forall y \in A\}$. Thus, if $x^* \notin f(R')$, then $f(R) = f(R')$, establishing the independence of Approval Voting. □

Theorem 20.4.3. *Majoritarian Approval Compromise, Preference-Approval Voting and Approval Voting with a runoff fail independence.*

Proof. To show that MAC fails independence, consider the following preference profile R with five voters and two alternatives

3 voters $a\ b|$
2 voters $b|\ a$

where a is the unique MAC winner. Now consider the preference profile R' with

2 voters $a\ b|\ x^*$
1 voter $x^* a\ b|$
2 voters $b\ |x^* a$

where b is the unique MAC winner. Moreover R and R' agree, hence MAC fails independence.

To show that PAV fails independence, consider the following preference profile R with three voters and two alternatives:

2 voters $a\ b|$
1 voter $b|\ a$

Both a and b are socially good and a majority of voters prefer a to b, so PAV picks a. Now consider the preference profile R' with

1 voter $a\ b|\ x^*$
1 voter $b\ x^*|a$
1 voter $x^*\ a\ b|$

R and R' agree. All three alternatives are socially good at R' and there is a majority cycle over them, hence PAV picks the one which receives the highest approval which is b, hence failing independence.

To show that AVR fails independence, consider the following preference profile R with nine voters and three alternatives:

4 voters $a|\ b\ c$
3 voters $b|\ a\ c$
2 voters $c|\ b\ a$

where the runoff winners are $\{a, b\}$ among which the pairwise majority winner b is the unique AVR winner. Now consider the preference profile R' with

4 voters $a\ |b\ c\ x^*$
3 voters $x^* b\ |a\ c$
2 voters $c|\ b\ a\ x^*$

where the runoff winners are $\{a, x^*\}$ among which the pairwise majority winner a is the unique AVR winner. As R and R' agree, AVR fails independence. □

Nevertheless, MAC, PAV and AVR can be evaluated according to the "popularity" of the spoiler they admit. After all, social choice rules that admit spoilers with little public support are more open to manipulation via artificial candidacies than those where the spoiler must have a reasonably high public support. We show that while MAC performs very poor in this regard, under PAV and AVR, a spoiler must have a reasonably high public support.[14]

Theorem 20.4.4. (i) *Under Majoritarian Approval Compromise, for any number of voters, there may be a spoiler who is approved by only one voter.*

(ii) *Under Preference-Approval Voting,* x^* *is a spoiler only if* x^* *is socially qualified as good.*

(iii) *Under Approval Voting with runoff,* x^* *is a spoiler only if* x^* *is a runoff winner.*

Proof. To show (i), consider a preference profile $R \in W(A \cup \{\emptyset\})$ where the society is split in two coalitions K and $N \backslash K$ whose rankings are as follows:

Voters in K: $a\ b|$...............

Voters in $N \backslash K$: $b|\ a$ ....

There are at least two alternatives called a and b. Voters in K qualify a and b as good; voters in $N \backslash K$ qualify b as good. It does not matter whether there are other alternatives and if so, how they are ranked. We also let the cardinality of K and $N \backslash K$ differ by at most one while $\#K \geq \#N \backslash K$. So at R, if $\#K = \#N \backslash K$, then $\{a, b\}$ is the MAC winner and if $\#K > \#N \backslash K$, then $\{a\}$ is the MAC winner. Now take a voter $i \in K$ and consider the preference profile and $R' \in W(B \cup \{\emptyset\})$ with

Voters in $K \backslash \{i\}$: $a\ b|\ x^*$.........

Voter i: $x^* a\ b|$........

Voters in $N \backslash K$: $b\ |x^* a$............

At R', if $\#K = \#N \backslash K$, then $\{b\}$ is the MAC winner and if $\#K > \#N \backslash K$, then $\{a, b\}$ is the MAC winner. Note that R and R' agree while x^* is not chosen at R'. Hence x^* is a spoiler. Moreover, x^* is approved by only one voter.

To show (ii), let $f : W(A \cup \{\emptyset\})^N \cup W(B \cup \{\emptyset\})^N \to 2^A \backslash \{\emptyset\}$ be PAV. Take any $R \in W(A \cup \{\emptyset\})$ and $R' \in W(B \cup \{\emptyset\})$ which agree while $x^* \notin f(R')$. Note that if $\#\gamma(R) \in \{0, 1\}$, then $f(R) = f(R')$, hence x^* is not a spoiler. Now let $\#\gamma(R) \geq 2$. If $x^* \notin \gamma(R')$, then $\gamma(R) = \gamma(R')$, implying $f(R) = f(R')$, hence x^* is not a spoiler.

To show (iii), let $f : W(A \cup \{\emptyset\})^N \cup W(B \cup \{\emptyset\})^N \to 2^A \backslash \{\emptyset\}$ be AVR. Take any $R \in W(A \cup \{\emptyset\})$ and $R' \in W(B \cup \{\emptyset\})$ which agree. Suppose $x^* \notin \rho(R')$. As R and R' agree, we have $\rho(R) = \rho(R')$ and also $f(R) = f(R')$, hence x^* is not a spoiler. □

[14] This is in contrast to Plurality with runoff which, as defined in the Arrovian model, is hurt by the existence of spoilers with very low public support. More precisely, in the Arrovian model, independent of the number of voters, it is possible to construct an example where the spoiler is the best alternative for two voters and the worst alternative for the rest of the voters. So when monotonicity and independence are the salient criteria to evaluate social choice rules, Approval Voting with runoff presents a neat improvement over Plurality with runoff. This justifies a comment in a similar direction made by Rida Laraki at the workshop on "Reforming the French Presidential Electoral System: Experiments on Electoral Reform", held at CEVIPOF, Sciences-Po, Paris, on 15–16 June 2009.

20.5 Concluding Remarks

Approval Voting calls for an extension of the Arrovian model by incorporating elements of cardinality and interpersonal comparability into individual preferences, through assuming the existence of a common zero. This naturally occurs in certain environments, such as matching models (see, for example, Roth and Sotomayor (1990)), where "being self-matched" is the common zero. However, in general, a common zero is implied by the existence of a common meaning attributed to "good" and "bad". This is a minimal divergence from the Arrovian model whose information sets are refined by the use of monotonic transformations which have one fixed point.[15]

The Arrovian model can be further extended through the use of monotonic transformations having multiple fixed points.[16] This incorporates further elements of cardinality and interpersonal comparability. In fact, at the extreme case of requiring every point to be fixed, the identity function becomes the only allowed monotonic transformation, hence getting back to the existence of an absolute scale to measure utilities.

Extending the Arrovian model invites interesting philosophical questions some of which are discussed by Ng (1992). Moreover, as the degree of incorporated cardinality and interpersonal comparability can be measured by the number of fixed points imposed over the monotonic transformations, the extent to which, if any, the Arrovian model can be extended invites interesting experimental questions as well. In any case, we see these extensions as interesting conceptual tools which, as this section suggests, allow to revisit and better understand certain concepts of social choice theory.

Acknowledgements I owe the ideas expressed in this paper to my collaboration with Steven Brams to whom I am very grateful. I am also grateful to Jean-Francois Laslier for his thoughtful comments on the initial version of the paper. The first formal presentation of these ideas occurred at the workshop on "New Approaches to Voting and Social Choice", 25–26 May 2009, Tilburg University. We thank Harrie de Swart and the participants. My research received financial aid from a project (# 107K560) supported by the Scientific and Technological Research Council of Turkey (TUBITAK) as well as from Turkish Academy of Sciences Distinguished Young Scientist Award Program (TUBA-GEBIP), which I gratefully acknowledge. Of course, the author is responsible from all possible errors.

[15] Instead of using any monotonic transformation. See Footnotes 1 and 3.

[16] In fact, the literature contains studies and proposals of social choice rules that call for further extensions. (see, for example, Hillinger (2005), Aleskerov et al. (2007), Balinski and Laraki (2007)).

References

Aizerman, M., & Aleskerov, F. (1995). *Theory of choice*. Amsterdam: North-Holland.
Aleskerov, F., Yakuba, V., & Yuzbashev D. (2007). A 'threshold aggregation' of three-graded rankings. *Mathematical Social Sciences, 53*(1), 106–110.
Arrow, K. J. (1950). A difficulty in the concept of social welfare. *Journal of Political Economy, 58*, 328–346.
Arrow, K. J. (1951). *Social choice and individual values*. New York: Wiley.
Balinski, M., & Laraki, R. (2007). A theory of measuring, electing and ranking. *Proceedings of the National Academy of Sciences USA, 104*(21), 8720–8725.
Barberà, S. (1977). The manipulation of social choice mechanisms that do not leave "too much" to chance. *Econometrica, 45*, 1573–1588.
Bordes, G. (1979). Some more results on consistency, rationality and collective choice. In J. J. Laffont (ed.), *Aggregation and revelation of preferences*. Amsterdam: North-Holland.
Bossert, W., & Weymark, J. A. (2004). Utility in social choice. In S. Barberà, P. J. Hammond, & C. Seidl (Eds.), *Handbook of utility theory* (Vol. 2). Dordrecht: Kluwer.
Brams, S. J. & Doherty, A. E. (1993). Intransigence in negotiations: The dynamics of disagreement. *Journal of Conflict Resolution, 37*(4), 692–708.
Brams, S. J. & Fishburn, P. C. (2002). Voting procedures. In K. J. Arrow, A. K. Sen, & K. Suzumura (Eds.), *Handbook of social choice and welfare* (Vol. 1). Amsterdam: North-Holland.
Brams, S. J., Fishburn, P. C., & Merrill, S. (1988a). The responsiveness of Approval Voting: Comments on Saari and van Newenhizen. *Public Choice, 59*, 121–131.
Brams, S. J., Fishburn, P. C., & Merrill, S. (1988b). Rejoinder to Saari and Van Newenhizen. *Public Choice, 59*, 149.
Brams, S. J., & Kilgour, D. M. (2001). Fallback bargaining. *Group Decision and Negotiation, 10*(4), 287–316.
Brams, S. J. & Sanver, M. R. (2006). Critical strategies under Approval Voting: Who gets ruled in and who gets ruled out. *Electoral Studies, 25*(2), 287–305.
Brams, S. J. & Sanver, M. R. (2009). Voting systems that combine approval and preference. In S. Brams, W. V. Gehrlein, & F. S. Roberts (Eds.), *The mathematics of preference, choice and order*. Berlin: Springer.
Chernoff, H. (1954). Rational selection of decision functions. *Econometrica, 22*, 422–443.
Dellis, A. (2010). Policy moderation and endogeneous candidacy in Approval Voting elections. In J. F. Laslier & M. R. Sanver (Eds.), *Handbook of Approval Voting*. Berlin: Springer.
Endriss, U., Pini, M. S., Rossi, F., & Venable, K. B. (2009). *Preference aggregation over restricted ballot languages: Sincerity and strategy-proofness* (mimeo).
Erdelyi, G., Nowak, M., & Rothe, J. (2008). Sincere-strategy preference-based Approval Voting broadly resists control. In E. EOchmanski & J. Tyszkiewicz (Eds.), *Mathematical foundations of computer science 2008*. Berlin: Springer.
Fishburn, P. C. (1982). Monotonicity paradoxes in the theory of elections. *Discrete Applied Mathematics, 4*(2), 119–134.
Hillinger, C. (2005). The case for utilitarian voting. *Homo Oeconomicus, 23*, 295–321.
Hoag, C., & Hallett, G. (1926). *Proportional representation*. New York: The Macmillan Company.
Laslier, J. F., & Maniquet, F. (2010). Classical electoral competition under Approval Voting. In J. F. Laslier & M. R. Sanver (eds.), *Handbook of Approval Voting*. Berlin: Springer.
Laslier, J. F., & Sanver, M. R. (2010a). Introduction. In J. F. Laslier, & M. R. Sanver (Eds.), *Handbook of Approval Voting*. Berlin: Springer.
Laslier, J. F., & Sanver, M. R. (2010b). The basic Approval Voting game. In J. F. Laslier & M. R. Sanver (Eds.), *Handbook of Approval Voting*. Berlin: Springer.
Moulin, H. (1988). Axioms of cooperative decision making. Cambridge: Cambridge University Press.
Nash, J. F. (1950). The bargaining problem. *Econometrica, 18*(2), 155–162.

Ng, Y. K. (1992). Utilitarianism and interpersonal comparison: Some implications of a materialist solution to the world knot. *Social Choice and Welfare, 9*(1), 1–15.

Niemi, R. G. (1984). The problem of strategic behavior under Approval Voting. *American Political Science Review, 78*, 952–958.

Nunez, M. (2010). Approval Voting in large electorates. In J. F. Laslier, & M. R. Sanver (Eds.), *Handbook of Approval Voting*. Berlin: Springer.

Peleg, B. (1979). Game theoretic analysis of voting schemes. In O. Moeschlin & D. Pallaschke (Eds.), *Game theory and related topics*. Amsterdam: North-Holland.

Peleg, B. (1981). Monotonicity properties of social choice correspondences. In O. Moeschlin & D. Pallaschke (Eds.), *Game theory and mathematical economics*. Amsterdam: North Holland.

Peleg, B. (1984). *Game theoretic analysis of voting in committees*. Cambridge: Cambridge University Press.

Peters, H., Roy, S., & Storcken, T. (2009). On the manipulability of Approval Voting and related scoring rules. unpublished manuscript.

Roth, A., & Sotomayor, M. A. O. (1990). *Two-sided matching: A study in game theoretic modelling and analysis*. Cambridge: Cambridge University Press.

Saari, D. G. (1989). A dictionary for voting paradoxes. *Journal of Economic Theory, 48*, 443–475.

Saari, D. G., & Newenhizen, J. V. (1988a). The problem of indeterminacy in approval, multiple, and truncated voting systems. *Public Choice, 59*, 101–120.

Saari, D. G. & Newenhizen, J. V. (1988b). Is Approval Voting an unmitigated evil: A response to Brams, Fishburn and Merrill. *Public Choice, 59*, 133–147.

Sanver, M. R., & Zwicker, W. S. (2009). *Monotonicity properties and their adaptation to irresolute social choice rules* (mimeo).

Sen, A. (1986). Social choice theory. In K. J. Arrow & M. D. Intriligator (Eds.), *Handbook of mathematical economics* (Vol. 3). Amsterdam: North-Holland.

Sertel, M. R. (1986). unpublished lecture notes. Bogaziçi University.

Sertel, M. R., & van der Bellen, A. (1979). Synopsis in the theory of choice. *Econometrica, 47*(6), 1367–1389.

Sertel, M. R. & Yilmaz, B. (1999). The Majoritarian Compromise is majoritarian-optimal and subgame-perfect implementable. *Social Choice and Welfare 16*(4), 615–627.

LaVergne, TN USA
28 July 2010
191167LV00004BA/10/P

9 783642 028380